International Association of Geodesy Symposia

Klaus-Peter Schwarz, Series Editor

Springer
Berlin
Heidelberg
New York
Barcelona
Hong Kong
London
Milan
Paris
Singapore
Tokyo

International Association of Geodesy Symposia

Klaus-Peter Schwarz, Series Editor

Symposium 101: Global and Regional Geodynamics
Symposium 102: Global Positioning System: An Overview
Symposium 103: Gravity, Gradiometry, and Gravimetry
Symposium 104: Sea Surface Topography and the Geoid
Symposium 105: Earth Rotation and Coordinate Reference Frames
Symposium 106: Determination of the Geoid: Present and Future
Symposium 107: Kinematic Systems in Geodesy, Surveying, and Remote Sensing
Symposium 108: Application of Geodesy to Engineering
Symposium 109: Permanent Satellite Tracking Networks for Geodesy and Geodynamics
Symposium 110: From Mars to Greenland: Charting Gravity with Space and Airborne Instruments
Symposium 111: Recent Geodetic and Gravimetric Research in Latin America
Symposium 112: Geodesy and Physics of the Earth: Geodetic Contributions to Geodynamics
Symposium 113: Gravity and Geoid
Symposium 114: Geodetic Theory Today
Symposium 115: GPS Trends in Precise Terrestrial, Airborne, and Spaceborne Applications
Symposium 116: Global Gravity Field and Its Temporal Variations
Symposium 117: Gravity, Geoid and Marine Geodesy
Symposium 118: Advances in Positioning and Reference Frames
Symposium 119: Geodesy on the Move
Symposium 120: Towards an Integrated Global Geodetic Observation System (IGGOS)
Symposium 121: Geodesy Beyond 2000: The Challenges of the First Decade

Towards an Integrated Global Geodetic Observing System (IGGOS)

IAG Section II Symposium
Munich, October 5–9, 1998

Edited by

Reinhard Rummel
Hermann Drewes
Wolfgang Bosch
Helmut Hornik

Springer

Volume Editors

Prof. Dr.-Ing. Reinhard Rummel
Institut für Astronomische
und Physikalische Geodäsie
Technische Universität München
Arcisstraße 21
D-80290 München
Germany
(E-mail: rummel@step.iapg.verm.tu-muenchen.de)

Prof. Dr.-Ing. Hermann Drewes
Deutsches Geodätisches Forschungsinstitut
Marstallplatz 8
D-80539 München
Germany
(E-mail: drewes@dgfi.badw.de)

Dr.-Ing. Wolfgang Bosch
Deutsches Geodätisches Forschungsinstitut
Marstallplatz 8
D-80539 München
Germany
(E-mail: bosch@dgfi.badw.de)

Dipl.-Ing. Helmut Hornik
Deutsche Geodätische Kommission
Marstallplatz 8
D-80539 München
Germany
(E-mail: hornik@dgfi.badw.de)

Series Editor

Prof. Dr. Klaus-Peter Schwarz
University of Calgary
Department of Geomatics Engineering
2500 University Drive N.W.
Calgary, Alberta T2N 1N4
Canada

Library of Congress Cataloging-in-Publication Data
Towards an integrated global geodetic observing system (IGGOS): Munich, Germany, 5-9 October 1998 / ed. by Reinhard Rummel ... - Berlin; Heidelberg; New York; Barcelona; Hong Kong; London; Milan; Paris; Singapore; Tokyo: Springer, 2000
(International Association of Geodesy symposia; Symposium no. 120)

ISBN-13: 978-3-642-64107-7 e-ISBN-13: 978-3-642-59745-9
DOI: 10.1007/978-3-642-59745-9

Springer-Verlag Berlin Heidelberg New York
a member of BertelsmannSpringer Science+Business Media GmbH

Softcover reprint of the hardcover 1st edition 2000

Typesetting: Camera ready by editors
Cover-layout: design & production GmbH, Heidelberg

SPIN 10757926 32/3138xz - 5 4 3 2 1 0 - Printed on acid-free paper

Preface

The international symposium *Towards an Integrated Global Geodetic Observing System* was an initiative of section II *Advanced Space Technology* of the *International Association of Geodesy (IAG)*. It took place in the building of the Bavarian Academy of Sciences in Munich from October 5 – 9, 1998. About 130 scientists from 24 countries participated in the symposium. It was organized jointly by the Deutsches Geodätisches Forschungsinstitut and the Institut für Astronomische und Physikalische Geodäsie/Technische Universität München.

The objective of the symposium was an analysis of the state-of-art of geodetic space techniques and an outlook into the possibility of the establishment of a global integrated observing system. In view of the fact that most geodetic space techniques have reached or approach the one part per billion relative precision level the question arises what their joint and coordinated operation could offer to the study of system Earth. For this purpose the symposium was divided into four themes, i.e. reference system and datum integration, strength and weakness of space techniques, upcoming gravity field satellite missions, and implications for Earth sciences. Each of these themes was addressed by invited lectures of prominent scientists with sufficient time left for discussion. The oral sessions were complemented by poster sessions. A panel discussion at the closing session completed the program.

The symposium triggered a very intensive discussion inside section II as well as the whole of IAG about the future strategy and structure of IAG. Only recently at the general assembly of the International Union of Geodesy and Geophysics in Birmingham in July 1999 this topic received serious attention and will be investigated further.

The organizers of the symposium are very thankful for the support and help they received. They acknowledge the help of the staff members and the support from the Bavarian Ministry of Sciences, Research and Arts, the Bavarian Academy of Sciences and its president, Professor Nöth, the association of friends of the academy, the city of Munich, the International Association of Geodesy, and the International Union of Geodesy and Geophysics. It is hoped that with the publication of this volume in the official IAG Symposia Series the important thoughts and discussions at this symposium are made available to many interested readers. We hope that in the near future geodesy will start to further strengthen its impact on Earth sciences through the successful operation of a *Global Integrated Geodetic and Geodynamic Observing System*.

Munich, April 2000 *Reinhard Rummel, Hermann Drewes, Wolfgang Bosch and Helmut Hornik*

Contents

POSTERS - Session B: Strengths and Weaknesses of Space Techniques 143

Symposium Programme

Monday, Oct. 5

Opening Session

Conference Hall 10:30 - 12:30

time/ poster	*presentation: author(s) (underlined: lecturer), title*	*page*
10:40	H. Drewes, *Secretary of CSTG*: Opening	
10:55	H. Nöth, *President of the Bayerische Akademie der Wissenschaften*: Welcome address	
11:10	T. Strobl, *Dean of the faculty for Bauingenieur- und Vermessungswesen, Technische Universität München*: Welcome address	
11:30	K.P. Schwarz, *President of IAG*: Geodesy Beyond 2000	
12:00	R. Rummel, *President of Section II of IAG*: Towards an Integrated Global Geodetic Observing System (IGGOS)	

Scientific Sessions

Session A.1: Reference System and Datums

Convenors: M. Feissel and J.R. Ray

Conference Hall 14:30 - 16:00

14:30	M. Feissel, R. Biancale, A.-M. Gontier: Integrated Reference Frames for Geodynamics: an Introduction	
15:00	G. Petit: Importance of a Common Framework for the Realization of Space-Time Reference Systems	3
15:30	C. Boucher: Relating Global and Regional Terrestrial Reference Frames	

Poster Session A.P

Reference System and Datum Integration

Convenors: M. Feissel and J.R. Ray

Posters on display: Monday, Oct. 5, and Tuesday, Oct. 6

Poster Hall 16:30 - 18:00

A-01	Z. Altamimi, C. Boucher, P. Sillard: Quality Analysis of some IGS Weekly Combined Solutions with Respect to ITRF	93
A-02	A. El-Mowafy: Problems in realization of a uniform reference system for integrating terrestrial and GPS observations in Egypt	
A-03	M. Bursa, J. Kouba, A. Müller, K. Radej, S. True, V. Vatr, M. Vojtiakova: Determination of Geopotential Differences between local vertical datums and realization of a World Height System	
A-04	W. Bosch, J. Klokočnik, C. Wagner, J. Kostelecký: Geosat and ERS-1 Datum Offsets Relative to Topex/Poseidon Estimated Simultaneously with Geopotential Corrections from Multi-Satellite Crossover Altimetry	96
A-05	S.M.A. Costa, D.S. Beattie, K. D. Pereira: The Integration of Brazilian Horizontal Network to SIRGAS – Preliminary Results	99
A-06	S. Ogi, M. Murakami: Construction of the New Japan Datum Using Space Geodetic Technologies	103
A-07	Y. Kuroishi: Deviation of Japanese Vertical Datum and Reference Surface with Respect to the Global Geoid	106
A-08	U. Roßbach: Reference Systems Used in Global Navigation Satellite Systems	110
A-09	R. Ferland, P. Tetreault, C. Huot, D. Hutchison, J. Kouba: Recent Contribution to the ITRF and its Realization in Canada	114
A-10	M. Craymer, R. Ferland, R. Snay: Realization and Unification of NAD83 in Canada and the U.S. via the ITRF	118
A-11	W. Pachelski: Modelling the adjustment and maintenance of the Polish Levelling System	
A-12	D. Angermann, J. Klotz, Ch. Reigber: Geodetic Datum Definition of the SAGA Network	122
A-13	V.E. Zharov, M.V. Sazhin, T.A. Kalinina: Weak Gravitational Lensing and Instability of the Celestial Reference Frame	126
A-14	R. Devoti, M. Fermi, C. Ferraro, R. Lanotte, V. Luceri, A. Nardi, R. Pacione, P. Rutigliano, C. Sciarretta, G. Bianco, F. Vespe: Height Time Series of the Italian GPS Fiducial Network compared with VLBI and SLR Estimations	130
A-15	G. Boedecker: The IAGBN as part of a Global Observing System	
A-16	R. Kolenkiewicz, D.E. Smith, M.H. Torrence, P.J. Dunn: Reference System Scale defined by using SLR Data	
A-17	E.C. Pavlis, S. Mertikas, F. Kouroumbali, P.G. Drakopoulos: Tectonics, Sea-Level Monitoring and Altimeter Calibration with a Regional GPS Array	
A-18	P. Sillard: No-Net Rotation Tisserand Condition Implementation	

A-19	H. Drewes, K. Kaniuth, W. Seemüller, K. Stuber, H. Tremel, N. Hernandez, M. Hoyer, E. Wildermann, L. P. Fortes, K. D. Pereira: Monitoring the Continental Reference Frame in South America	134
A-20	M. Kumar, K.J. Burke: A practical Approach to realize Global Vertical Datum with the use of Geoid	
A-21	E.W. Grafarend, A.A. Ardalan: World Geodetic Datum 2000	138

Tuesday, Oct. 6

Session A.2 – Reference System and Datums

Convenors: M. Feissel and J.R. Ray

Conference Hall 09:00 – 10:30

09:00	R.S. Nerem, R.J. Eanes, J.C. Ries, G.T. Mitchum: The use of a precise reference frame in sea level change studies	8
09:30	R. Scharroo, E. Schrama, R. Haagmans: Combination of Space Techniques into one Integrated Processing Model	13
10:00	J.R. Ray: Towards an Integrated Global Reference System: Geodesy as a Utility	19

Session B.1 – Strengths and Weaknesses of Space Techniques

Convenors: G. Beutler and Ch. Reigber

Conference Hall 11:00 – 12:30

11:00	G. Beutler, H. Drewes, Ch. Reigber, R. Rummel: Space Techniques and their Coordination within IAG at Present and in Future	22
11:30	R. Eanes: Contributions of SLR to Space Geodesy	
12:30	H. Schuh: Contributions of VLBI to Space Geodesy	33

Session B.2 – Strengths and Weaknesses of Space Techniques

Convenors: G. Beutler and Ch. Reigber

Conference Hall 14:30 – 16:00

14:30	D. Geudtner, M. Zink: Applications of Spaceborne SAR Interferometry	
15:00	C.K. Shum, M. Anzenhofer, J.C. Ries: Contribution of Altimetry Missions to Space Geodesy	
15:30	M. Rothacher: Towards an Integrated Global Geodetic Observing System	41

Poster Session B.P – Strengths and Weaknesses of Space Techniques

Convenors: G. Beutler and Ch. Reigber

Posters on display: Monday and Tuesday

Poster Hall 16:30 – 18:00

B-01	G. Liebsch, R. Dietrich: Validation of TOPEX/POSEIDON Measurements in the Southern Baltic Sea	145
B-02	J. Campbell, B. Görres, A. Nothnagel: On the Inherent Vertical Accuracy in the European VLBI and GPS network	
B-03	M. Costes, P. Vincent: DORIS – Contribution of an Orbit Determination System to Accurate and Long Term Altimetry from Space	150
B-04	Y. Mireault, J. Kouba: IGS Combinations of Polar Motion, Length of Day and Universial Time	154
B-05	W.R. Dick, M. Mareyen, M. Meinig, K. Oppitz, B. Richter: Stability of SLR Solutions	
B-06	C. Ma, J.M. Gipson: VLBI Applied to Reference Frames	
B-07	Y. Koyama, R. Ichikawa, T. Kondo, N. Kurihara, Y. Takahashi, T. Yoshino, K. Sebata, M. Furuya: Comparison of Site Velocities Measured by VLBI and GPS in the Key Stone Project Network	158
B-08	J. Müller, D. Egger, B. Reichhoff, M. Schneider, U. Schreiber: Large Lunar Laser Ranging at its Best	161
B-09	U. Schreiber, M. Schneider, C.H. Rowe, G.E. Stedman, S.J. Cooper, B.T. King, D.N. Wright, W. Schlüter, H. Seeger: Large Perimeter Ring Lasers for Geoscience	
B-11	F. Bouille, J.-F. Cretaux, A. Cazenave, L. Soudarin: Geocenter Variations derived from five years of data of the DORIS Space System. Comparision with Surface loading data.	
B-12	J.-F. Cretaux, L. Soudarin, A. Cazenave: Vertical Rates of the DORIS Stations	
B-13	C. Fosu, B. Eissfeller, G.W. Hein: Combining GPS and CCD to Modernise Astronomical Levelling	165
B-14	E. Fragner, R. Weber, M. Ploner: Orbit Modeling Based on Laser and CCD-Observations	169
B-15	C.D. Chadwell, J.A. Hildebrand, F.N. Spiess: Oceanic Stations for a Global Geodetic Observing System	
B-16	A. Shaker, D. Alnaggar, A.A. Saad: Unification of GPS Work in Egypt	173

B-17	T. Yoshino, J. Amagai, K. Sebata, M. Furuya, H. Kunimori, T. Otsubo, T. Kondo, Y. Takahashi, H. Kiuchi, Y. Koyama, R. Ichikawa, T. Gotoh, H. Hase, R. Kilger, W. Schlüter: Intercomparison of Terrestrial Coordinates Obtained by Space Geodetic Techniques at the Colocated Stations in the Key Stone Project and Wettzell	
B-18	R. Devoti, M. Fermi, C. Ferraro, R. Lanotte, V. Luceri, A. Nardi, R. Pacione, P. Rutigliano, C. Sciarretta, G. Bianco, F. Vespe: Earth Orientation Parameters Measured by Space Geodesy Techniques	177
B-19	J. Klokocnik, J. Kostelecky, C.A. Wagner: Residual Errors in Altimetry Data Detected by Combinations of Single- and Dual- Satellite Crossovers	
B-20	R.E. Neilan, G. Beutler, J. Kouba: A contribution to the global geodetic observing system, resources of the International GPS Service (IGS)	
B-21	T.A. Springer, G. Beutler, M. Rothacher: Combination of GPS and SLR Measurements	
B-22	W. Schlüter, A. Boer, R. Dassing, H. Hase, R. Kilger: TIGO-Transportable Integrated Geodetic Observatory, a fundamental station supporting the global reference system	
B-23	H. Hase: New Method for the Selection of Additional Sites for the Homogenisation of an Inhomogeneous Cospherical Point Distribution	180
B-24	A. E. NIELL: Westford VLBI to GPS Vertical Tie and Implications for the TRF	184

Wednesday, Oct. 7

Session C.1 – Upcoming Gravity Field Satellite Missions

Convenors: K.H. Ilk and B. Tapley

Conference Hall 09:00 - 10:30

09:00	K.H. Ilk: Envisaging a new era of gravity field research	53
09:30	Ch. Reigber, H. Lühr, P. Schwintzer: Status of the CHAMP mission	63
10:00	B. Tapley: Status of the GRACE mission	

Poster Session C.P – Upcoming Gravity Field Satellite Missions

Convenors: K.H. Ilk and B. Tapley

Posters on display: Wednesday and Thursday

Poster Hall 11:00 - 12:30

C-01	J. Kusche, K.H. Ilk, S. Rudolph: Impact of terrestrial data on future satellite gravity field solutions	189
C-02	V.N. Strakhov, U. Schäfer, A.V. Strahkov, G. Harnisch: High-resolution regional gravitational models with N>1000 using irregular distributed real data	
C-03	F. Sanso, A. Albertella, G. Bianco, A. Della Torre, M. Fermi, V. Iafolla, A. Lenti, F. Migliaccio, A. Milani, A. Rossi: SAGE – An Italian Project of Satellite Accelerometry	193
C-04	P. Schwintzer, Z. Kang, F. Perosanz: Accelerometry Aboard CHAMP	197
C-05	H. Oberndorfer, J. Müller, R. Dorobantu, C. Gerlach, R. Rummel, N. Sneeuw, R. Koop, P. Visser, P. Hoyng, A. Selig, M. Smit: Simulation of the GOCE Gravity Field Mission	201
C-06	N. Sneeuw, C. Gerlach, J. Müller, H. Oberndorfer, R. Rummel: Fundamentals and Applications of the Gravity Field Mission GOCE	205
C-07	R. Pail, W.-D. Schuh: Effects of Inhomogeneous Data Coverage on Spectral Analysis	209
C-08	E.C. Pavlis, B.F. Chao, C.W. Hwang, C.C. Liu, C.K. Shum, C.L. Tseng, M. Yang: Geodetic Applications of the ROCSAT-3/COSMIC Mission	214
C-09	A.W. Moore, D.A. Stowers, M.R. Marcin, R. Khachikyan, I.L. Harris, J.F. Zumberge: JPL's Approch in Utilizing the Global GPS Ground Network to Support Low-Earth Orbiting Missions	
C-10	J.F. Zumberge, I.L. Harris, R. Khachikyan, E.R. Kursinski, M.R. Marcin, A.W. Moore, Stowers, D.A., M.M. Watkins: Requirements of a Global GPS Ground Network to Support Low-Earth Orbiting Missions	
C-11	Th. Gruber, Ch. Reigber, J. Wünsch: Estimation of Ocean Mass Redistribution by Means of Altimetry and Circulation Models and its Impact on the Gravity Field	218
C-12	Ch. Reigber, R. Galas, R. König, N. Jakowski, J. Wickert, A. Wehrenpfennig: GFZ and DLR contribution to a GPS ground network to support the CHAMP mission	222
C-13	H. Rebhan, J. Johannessen, M. Aguirre, A. Tobias: The ESA Gravity Field and Steady-State Ocean Circulation Explorer Mission – Impact on Solid Earth Physics	225

Session C.2 – Upcoming Gravity Field Satellite Missions

Convenors: K.H. Ilk and B. Tapley

Conference Hall 14:30 - 16:00

14:30	R. Rummel, J. Müller, H. Oberndorfer, N. Sneeuw: Satellite Gravity Gradiometry with GOCE	66

15:00	B. Chao: Gravity Field Satellite Missions for Earth Sciences Research	
15:30	B. Melbourne: Summary	

Session D.1 – Implication for Earth Sciences

Convenor: S. Zerbini

Conference Hall 16:30 – 18:00

16:30	S. Zerbini: The Contribution of an IGGOS to Earth Sciences	
17:00	G. DiDonato, A.M. Negredo, R. Sabadini, L.L.A. Vermeersen: Deformation Patterns due to Tectonic Processes	
17:30	L. Vermeersen, R. Sabadini, R. Devoti, V. Luceri, P. Rutigliano, C. Sciarretta, G. Bianco: Joint mantle viscosity inversion of glacial rebound-induced variations in low-degree geoid harmonics and polar wander	

Thursday, Oct. 8

Session D.2 – Implication for Earth Sciences

Convenor: S. Zerbini

Conference Hall 09:00 – 10:30

09:00	A. Cazenave, F. Mercier: Combining space geodesy-derived global Earth's parameters and oceanographic and climatologic data to study the spatio-temporal mass redistribution in surface reservoirs	
09:30	H.-P. Plag: Integration of geodetic techniques into a global Earth monitoring system and its implication for Earth system sciences	73
10:00	K. Lambeck: The Integration of Geodetic Measurements into an Earth Science Observing Program: The Example of Glacial Rebound	84

Poster Session D.P – Implication for Earth Sciences

Convenor: S. Zerbini

Posters on display: Wednesday and Thursday

Poster Hall 11:00 – 12:30

D-01	Yu.V. Barkin: Regular character of the plate motion: implication for earth sciences	231
D-02	Yu.V. Barkin: About global rotation of the lithosphere	234
D-03	O. Titov: Least Squares Collocation method for space geodetic data analysis	238
D-04	R. S. Nerem, R. J. Eanes, J. L. Chen, P. F. Thompson: Seasonal Variations of the Gravity Field determined from Satellite Laser Ranging	
D-06	A. Tealeb: Programmes for Monitoring Crustal Deformations and Seismic Hazard Assessment in Egypt using Geodetic Techniques	
D-07	K. Sakr, H. Khalil, S.M. Mahmoud, A.S. Mohamed: Analysis of GPS measurements around Nasser Lake, Aswan, Egypt	
D-08	S.M. Mahmoud, Y. Fuji, H. Khalil, K. Sakr, A. Rayan: Strain Analysis deduced from the results of GPS Mesurements at the area around Cairo and Southern Part of Delta, Egypt	
D-09	H.-G. Scherneck, J. Johansson, R. Haas: BIFROST Project: Studies of Variations of Absolute Sea-Level in Conjunction with the Post-glacial Rebound of Fennoscandia	241
D-10	M. Becker, G. Weber: Atmospheric modelling in the German GPS permanent network GREF and its use for Geodesy and meteorology	
D-11	A.K. Kasenda, A.M. Komara, S. Sutisna: The Indonesian gravity field and the geoid model	245
D-12	M. Negusini, B. Richter, C. Romagnoli, W. Schwahn, S. Zerbini: An example for an Integrated Local Geodetic Observing System: the Medicina-Porto Corsini experiment	
D-13	K. Elango, P. Soma, K. Rameshchandra, S. Rangarajan: ISRO initiative for space geodesy & geodynamics in India	248

Session E.1 – Towards IGGOS

Convenors: R. Rummel and H. Drewes

Conference Hall 14:30 – 16:00

14:30	I. Mueller, Convenors of session A – D Plenary/podium discussion, summaries	

Session E.2 – Towards IGGOS

Convenors: R. Rummel and H. Drewes

Conference Hall 16:30 – 18:00

16:30	Convenors of session A – D Plenary/podium discussion, conclusions	

Friday, Oct. 9

whole day

09:00	Excursion to Fundamental Space Station Wettzell	

Minutes of the Plenary Discussion

prepared by R. RUMMEL

The plenary discussion is moderated by Professor **IVAN MUELLER** (IIM). He proposes to divide the session into the three themes

(1) scientific issues related to the four sessions of the symposium,
(2) need for an Integrated Geodetic/Geodynamic Observing System (IGGOS), and
(3) consequences of an IGGOS for IAG.

1. Scientific Issues

Reference frame

FEISSEL: Any combination of space techniques into one system requires the availability of reliable **local ties**. The responsibility for this is with CSTG. What is the state-of-art in this matter?

This question reappeared in one way or the other several times. NEILAN emphasizes the need for the very accurate and reproducible determination of the eccentricities between the observing systems. She also points out that while the necessity increases funds for this purpose and expertise decreases. KOUBA adds that in some cases the exact determination of reference points of systems is highly complicated (e.g. the determination of the electric center of a GPS antenna).

BEUTLER acknowledges the importance of the problem and agrees that it still needs more attention. He points out that local ties should be the responsibility of the (newly founded) services and in particular of the ISGN.

(During the excursion to Wettzell after the symposium there was opportunity to discuss this issue at depth. At Burg Eck – a medieval castle – the following recommendation was formulated: CSTG sets up a commission of experts that defines a set of standards and procedures on how ties between measurement systems are to be carried out and established. An international group of experts is established that carries out local ties according to these guidelines and gives official certificates.)

RUMMEL: Are the current **cross-checks between the various space techniques** sufficient? FEISSEL says that cross-check is one of the central objectives of IERS. BEUTLER confirms that today's level of consistency is due to the work of IERS. What is needed, however, is a real combination of the complementary techniques in one consistent estimation model. ALTAMINI distinguishes the cross-check in horizontal from that in vertical direction. He emphasizes that the real problem is in vertical direction and much needs to be done to improve it.

RUMMEL: How can one establish for satellite **altimetry** global uniformity and **stability over decades** in one reference system for the purpose of sea level monitoring?

SHUM says that this is an important research item. To maintain one system over decades (and extending it over several consecutive missions) is extremely demanding. Only through the definition of comprehensive procedures of cross-calibration and with the support of IGS and GLOSS stability at the 1mm/y level seems feasible. KOUBA adds that in his view none of the geodetic space techniques can be assumed to guarantee a 1mm/y time stability (due to biases, drifts, leakages, data gaps etc.)

BOSCH: IAG has initiated services for GPS, SLR/LLR and VLBI. The services turn out to be highly successful. It almost looks as if altimetry and INSAR are regarded second class, despite their great successes and their importance. Should one not consider the establishment of an **altimetry service**? IIM adds the question of whether data products are standardized and provided in a user friendly form?

SHUM explains that the altimetry data distribution is currently organized on a mission basis. There exist e.g. data centers for TOPEX/POSEIDON and independently for ERS-1 and ERS-2. The basic characteristics of the data records remained the same for all altimetry missions (so-called Geophysical Data Records). There is no close cooperation between these mission related data centers nor do they regularly meet. However, all data centers provide their data in well defined and user friendly formats.

ROTHACHER shows a slide addressing the problem of **consistency**. He distinguishes network level, modeling level and parameter/observation level. Consistency is required on all three levels and the global reference system is ultimately to be defined by all available techniques. RAY thinks that consistency over decades is not yet feasible. The reference system is still affected by relocation of individual stations and geographical station distribution. DORIS and GPS could help in this respect. BEUTLER points out the important role of SLR in this respect.

SCHWARZ: The statement of a 10^{-9} relative precision may be useful for positioning techniques. Such a measure seems not appropriate for the **gravity field** where the relative precision at individual points is to be distinguished from that of the field quantity (continuum). In his opinion a field precision of 10^{-6} (relative e.g. to a geoid value of 100m) seems adequate.

Geodesy and Geodynamics

MUELLER says, geodesy knows how to measure, to combine different techniques and to estimate parameters. But does it know **how to serve geophysics**? How should an integrated modeling approach look like, considering the more *artistic* approach to modeling in geophysics (see the presentations by SABADINI and by LAMBECK). LAMBECK responds in saying that geophysicists usually take from other fields - including geodesy - what serves their purpose. They are often not well prepared to use geodetic equipment or information. Thus it is important to present geodetic information in a way that is well understood. Because of dealing with geological time scales questions concerning reference frames, highest measurement precision etc. are not of utmost concern. In RUMMEL's opinion this depends on the kind of geophysical problem under investigation. For earth rotation research the question of reference system is the very core, for sea level monitoring and its interpretation the whole range of critical items discussed during this symposium (datum, sensor systems, ocean circulation, climate, ice rebound etc.) is of highest relevance. KOUBA advocates for maximum precision in geodesy, for *geodesy's noise may very well be another discipline's signal*. PLAG says that modelers need their data sets all in one reference system and consistent in space and time. BEUTLER concludes that spatial and temporal consistency of geodetic data as well as easy access in a comprehensive form needs more focus in geodesy.

MUELLER shows an older viewgraph that attributes most of the available geodetic measurement techniques to geophysical problems to which they can potentially contribute. Metaphor: *Look at the earth as a fish, then this division into parts makes a fish soup and the question is, how to reconstruct the fish from the soup.*

LAMBECK: At this point solid earth geophysics looks mainly into phenomena, in most cases comprehensive modeling e.g. of the lithosphere would be too complex and the uncertainties too high. However, if one considers the many regional kinematic studies the questions one is faced with are very similar and it could very well be worthwhile to integrate them into one global system. PLAG goes a step further in saying that many areas of geophysics are in a period of transition from purely phenomenological studies to comprehensive modeling. Sooner or later a modular approach to earth system modeling will be necessary. ILK demonstrates with a viewgraph that already now a rather complex system is required to come from gravity and magnetic field satellite measurements to the use of these measurements in solid earth, ice, ocean and climate modeling.

2. Need of an Integrated Geodetic/Geodynamic Observing System

MUELLER: Does geodesy need an IGGOS? BEUTLER, taking a bottom up approach, is convinced that IGGOS represents a logical and consequent next challenge. It so-to-say extrapolates what has happened during the past years with the individual space techniques and their combination. If one takes for example earth rotation, angular momentum exchange and the high correlation with meteorological data the prospect of what such an integrated system would be able to deliver becomes obvious. Most of its components are in operation already today. The right approach will be to formulate a proposal (mission statement) and investigate how such a proposal can be implemented partly as service and partly as research task. KOUBA supports these arguments. He mentions several working items of IERS such as geophysical fluids and center of mass motion, which already have an integrated nature.

3. IGGOS and IAG structure

MUELLER says that currently the whole arsenal of geodetic measurement, modeling and interpretation activities is divided into five sections. Fundamental areas of IAG involvement are the development of techniques, gravity/geoid, reference system problems, and those parts of geodynamics that are closely related to geodetic parameters. One might add in the future topography. There is a general agreement that it is time to bring these things together in a more efficient way.

FEISSEL responds that she is happy with geodesy but much less with IAG. She proposes a new structure divided into three sections that coordinate

- research,
- services that develop and distribute geodetic products, and
- projects that initiate new activities.

The number of bodies should be reduced significantly. IGGOS would be a project that could serve as focus of a large variety of geodetic activities. DREWES wonders what the task of the sections would be and whether the commissions would still exist.

SCHWARZ could imagine to center the IAG structure around something like IGGOS (under the condition that the observing system is comprehensive and can serve as focus of all current (geodetic/geodynamical) activities inside of IAG. However, one would also have to define the adequate structure for the engineering part of geodesy.

BEUTLER wants first the fundamental question to be answered whether in view of IGGOS a reconstruction of IAG is appropriate or not (zero question). If yes, he would expect a number of proposals (apart from the one by FEISSEL) on the optimal structure of IAG.

Conclusion: A proposal is to be formulated for an integrated geodetic/geodynamic observing system by section II of IAG. At the next executive meeting of IAG this proposal has to be discussed and hopefully adopted. The executive committee looks then into the question of a reorganization of IAG.

Participants

The participants' list comprises: name, postal address, phone, fax and e-mail number

Albertella, Alberta
Politecnico di Milano, Dip. di Ingegneria Idraulica Ambientale e de Rivelamento, Piazza Leonardo da Vinci 32, I-20133 Milano, Italy
+39-2-23-996509, +39-2-23996530, alberta@ipmtf4.topo.polimi.it

Alnaggar, Dalal
National Research Center, Survey Research Institute, 308 Pyramides st., 12111, Giza, Egypt
+202-5867174, +202 5849283, sriidnet@brainy1.ie-eg.com

Altamimi, Zuheir
Institut Géographique National ENSG/LAREG, 6-8 Avenue Blaise Pascal - Cité Descartes, F-77455 Champs-sur-Marne, France
+33 1 64153255, +33 1 64153253, altamimi@ensg.ign.fr

Alves Costa, Sonia Maria
Instituto Brasileiro de Geografía e Estatistica IBGE, Av. Brasil 15671 - Parada de Lucas, Rio de Janeiro, Brazil
+55-21-3918217, +55-21-4812747, soniamaria@ibge.gov.br

Andersen, Per Helge
FFI & ITA / UIO, Division of Electronics, P.O.Box 25, N-2007 Kjeller, Norway
+47-6380-7407, +47-63807212, per-helge.andersen@ffi.no

Angermann, Detlef
GeoForschungsZentrum Potsdam, Telegrafenberg C2, D-14473 Potsdam, Germany
+49-331-2881113, +49-331-2881169, dang@gfz-potsdam.de

Ardalan, Alireza Azmoudeh
Universität Stuttgart, Geodätisches Institut, Geschwister-Scholl-Str. 24 D, D-70174 Stuttgart, Germany
+49-711-121-3390, +49-711-121-3285

Baker, Trevor F.
Proudman Oceanographic Laboratory, Bidston Observatory, Birkenhead Merseyside L43 7RA, Great Britain
+44-151-653-8633, +44-151-653-6269, tfb@pol.ac.uk

Barkin, Yury
Sternberg Astronomical Institute, Universitetskij prospekt 13, RUS-119899 Moscow, Russia
+7-095-939-50-24, +7-095-932-8841, barkin@sai.msu.su

Bauch, Alfred
TU München,
Institut für Astronomische und Physikalische Geodäsie, Arcisstr. 21, D-80333 München, Germany
+49-89-2892-3192, +49-89-2892-3178

Beck, Norman
NRCan - Geodetic Survey, 615 Booth St., rm 428, Ottawa, Ontario K1A OE9, Canada
+1-613-995-2923, +1-613-995-3215, beck@nrcan.gc.ca

Beutler, Gerhard
Universität Bern, Astronomisches Institut, Sidlerstr. 5, CH-3012 Bern, Switzerland
+41-31-631 8591, +41-31-631 3869, beutler@aiub.unibe.ch

Boedecker, Gerd
Bayerische Kommission für die Internationale Erdmessung, Marstallplatz 8, D-80539 München, Germany
+49-89-23031-212, +49-89-23031-100, boe@bek.badw-muenchen.de

Bosch, Wolfgang
Deutsches Geodätisches Forschungsinstitut, Marstallplatz 8, D-80539 München, Germany
+49 89 23031-115, +49 89 23031-240, bosch@dgfi.badw-muenchen.de

Boucher, Claude
Institut Géographique National IGN/ENSG/LAREG, 6 et 8, Av. Blaise Pascal, Cité Descartes, Champs sur Marne, F-77455 Marne-La-Vallee Cedex 2, France
+33 1 64153250, +33 1 64153253, boucher@ensg.ign.fr

Campbell, James
Universität Bonn, Geodätisches Institut, Nussallee 17, D-53115 Bonn, Germany
+49-228-733565, +49-228-732988, campbell@kuestner.geod.uni-bonn.de

Cazenave, Anny
Centre National d'Etudes Spatiales, GRGS, 18, av. Edouard Belin, F-31400 Toulouse-Cedex, France
+33-5-61-332922, +33-5-61-253205, cazenave@cnes.fr

Chao, Benjamin F.
NASA Goddard Space Flight Center, Code 926, Greenbelt MD 20771, USA
+1-301-286-6120, +1-301-286-1760, chao@denali.gsfc.nasa.gov

Costes, Murielle
Centre National d'Etudes Spatiales, 18, Av. Edouard Belin, F-31401 Toulouse, France
+33-5-61282260, +33-5-61 282595, murielle.costes@cnes.fr

Craymer, Michael
Geomatics Canada, Geodetic Survey, 615 Booth St., Ottawa, Ontario K1A OE9, Canada
+1-613-947-1829, +1-613-992-6628, craymer@nrcan.gc.ca

Cretaux, Jean-Francois
Centre National d'Etudes Spatiales / GRGS, 18 av. Edouard Belin, F-31401 Toulouse Cedex 4, France
+33-5-61253205, jean-francois-cretaux@cnes.fr

Dichtl, Günter
TU München,
Institut für Astronomische und Physikalische Geodäsie (IAPG), Arcisstr. 21, D-80333 München, Germany
+49-89-28923182, +49-89-28923178

Dietrich, Reinhard
TU Dresden, Institut für Planetare Geodäsie,
D-01062 Dresden, Germany
+49-351-4634652, +49-351-4637063,
dietrich@ipg.geo.tu-dresden.de

Dow, John M.
ESA / European Space Operation Centre (ESOC),
Robert-Bosch-Str. 5, D-64293 Darmstadt, Germany
+49-6151-902272, +49-6151-902271, jdow@esoc.esa.de

Drewes, Hermann
Deutsches Geodätisches Forschungsinstitut, Marstallplatz 8,
D-80539 München, Germany
+49-89-23031-106, +49-89-23031-240,
drewes@dgfi.badw-muenchen.de

Driel, Peter van
Shell International B & P, Volmerlaan 8, P.O. Box 60,
NL-2280AB Rijswijk, The Netherlands
+31-70-3112105, +31-70-3113366, p.vandriel@lr.tudelft.nl

Eanes, Richard
University of Texas at Austin, Center for Space Research,
3925 W. Braker Lane, Austin TX 78759-5321, USA
+1-512-471-7560, +1-512-471-3570, eanes@csr.utexas.edu

Egger, Dieter
TU München, Forschungseinrichtung Satellitengeodäsie,
Arcisstr. 21, D-80333 München, Germany
+49-89-28923183, +49-89-28923178,
Dieter.Egger@tu-muenchen.de

Ehrnsperger, Walter
Bayerische Kommission für die Internationale Erdmessung,
Marstallplatz 8, D-80539 München, Germany
+49-89-23031-111, +49-89-23031-100,
a2101aa@bek.badw-muenchen.de

Elango, K.
PRARE, SLR & GPS Systems, ISTRAC / ISRO,
1st Cross, Peenya Industrial Estate, Bangalore 560 058, India
+91-80-8395129, +91-80-839-4249, elango@istrac.gov.in

Engen, Bjørn
Norwegian Mapping Authority, Geodetic Institute,
Kartverksveien, N-3500 Hønefoss, Norway
+47-32118100, +47-32118101, bjorn.engen@gdiv.statkart.no

Engsager, Karsten
National Survey and Cadastre Denmark, Rentemestervej 8,
DK-2400 Copenhagen, Denmark
+45-35-875269, +45-35-875052, ke@kms.dk

Feissel, Martine
Observatoire de Paris, 61, Av. de l'Observatoire,
F-75014 Paris, France
+33-1-40512015, +33-1-40512291, feissel@obspm.fr

Ferland, Remi
Natural Resources of Canada, 456-615 Booth Street,
Ottawa K1 AOE9, Canada
+1-613-995-4002, +1-613-995-3215,
ferland@geod.nrcan.ge.ca

Fisher, Steven
Jet Propulsion Laboratory, UNAVCO, 3340 Mitchell Lane,
Boulder CO. 80301, USA
+1-303-497-8046, +1-303-497-8028, sfisher@unavco.edu

Flechtner, Frank
GeoForschungsZentrum Potsdam,
Aussenstelle Oberpfaffenhofen, Münchener Str. 20,
D-82230 Oberpfaffenhofen, Germany
+49-8153-1355, +49-8153-28-1840, prs@dfd.dlr.de

Fragner, Elisabeth
Technische Universität Wien, Theoretische Geodäsie,
Gusshausstr. 27-29, A-1040 Wien, Austria
+43-1-58801-3697, +43-1-5042155,
efragner@luna.tuwien.ac.at

Galas, Roman
GeoForschungsZentrum Potsdam, Telegrafenberg 17A,
D-14473 Potsdam, Germany
+49-331-288-1179, +49-331-288-1111, galas@gfz-potsdam.de

Gambis, Daniel
Paris Observatory France, 61, Av. de l'Observatoire,
F-75014 Paris, France
+33-1-40512229, gambis@obspm.fr

Giles, Andre
Military Survey, Geodetic Branch, Elmwood Avenue,
Feltham MIDDX TW13 7AH, Great Britain
+44 181 818 2225, +44 181 818 2574

Gruber, Thomas
GeoForschungsZentrum Potsdam,
Aussenstelle Oberpfaffenhofen, c/o DLR Oberpfaffenhofen,
82234 Wessling, Germany
+49-8153-281208, +49-8153-1207, thomas.gruber@dlr.de

Hase, Hayo
Bundesamt für Kartographie und Geodäsie,
Fundamentalstation Wettzell, D-93444 Kötzting, Germany
+49-9941-603104, +49-9941-603-222, hase@wettzell.ifag.de

Hellwich, Olaf
Technische Universität München,
Institut für Photogrammetrie und Fernerkundung,
Theresienstrasse, D-80290 München, Germany
+49 089 289 22677, +49-89-280 9573,
olaf@photo.verm.tu-muenchen.de

Holota, Petr
Research Institute of Geodesy, Topography and Cartography,
Zdiby 98, CZ-25066 Praha-vychod, Czech Republic
+420-2-685 7907, +420-2-685 7056, gope@asu.cas.cz

Hoyer, Melvin
La Universidad del Zulia, Escuela de Ingeniería Geodésica,
Apartado P. 10311, Maracaibo, Venezuela
+58-61-512214, +58-61-512197, mhoyer@europa.ica.luz.ve

Ilk, Karl Heinz
Universität Bonn, Institut für Theoretische Geodäsie,
Nussallee 17, D-53115 Bonn, Germany
+49-228-732629, +49-228-733029,
ilk@theor.geod.uni-bonn.de

Kabelác, Josef
Academy of Sciences of the Czech Rep.,
Lehrstuhl für Höhere Geodäsie, Ruská; 572, CZ-50601 Jicin,
Czech-Republic
+42-433-524301, kabelac@inet.tpc.cz

Kasenda, Adolfientje
Geomatics R & D Center, BAKOSURTANAL,
KM-46, Jl. Raya Jakarta-Bogor, Indonesia
+62-21-8752064, sobar@cbn.net.id

Klokočnik, Jaroslav
Czech Academy of Sciences, Astronomical Institute,
CZ-25165 Ondřejov, Czech Republic
+42-204-85201, +42-2-881611, jklokocn@asu.cas.cz

Koch, Karl-Rudolf
Universität Bonn, Institut für Theoretische Geodäsie,
Nussallee 17, D-53115 Bonn, Germany
+49-228-732626, +49-228-733029,
koch@theor-geod.uni-bonn.de

Kolenkiewicz, Ronald
NASA Goddard Space Flight Center, Code 926,
Greenbelt MD 20771, USA
+1-301-286-5373, +1-301-286-1760,
ronk@ltpmail.gsfc.nasa.gov

Kouba, Jan
University of Berne, Astronomical Institute, Sidlerstr. 5,
CH-3012 Bern, Switzerland
+41-31-631-8592, +41-31-631-3869, koub@aiub.unibe.ch

Koyama, Yasuhiro
Kashima Space Research Center/CRL, 893-1 Hirai, Kashima,
J-314-0012 Ibaraki, Japan
+81-298-64-5968, +81-299-84-7159, koyama@crl.go.jp

Kristiansen, Oddgeir
Statens Kartverk, Geodetic Institute, Kartverksveien 21,
N-3500 Hønefoss, Norway
+47-321-18299, +47-32118101, oddgeir@gdiv.statkart.no

Kuroishi, Yuki
Space Geodesy Laboratory, Geographical Survey Institute,
1 Kitasato, J-305-0811 Tsukuba, Ibaraki, Japan
+81-298-64-5968, +81-298-64-2655, yuki@geos.gsi-mc.go.jp

Kusche, Jürgen
Universität Bonn, Institut für Theoretische Geodäsie,
Nussallee 17, D-53115 Bonn, Germany
+49-228 733578, +49-228 733029, kusche@uni-bonn.de

Lambeck, Kurt
The Australian National University,
Research School of Earth Sciences, Canberra, ACT 0200,
Australia
+61-6-249 5161, +61-6-249 5443, kurt.lambeck@anu.edu.au

Lanotte, Roberto
Centro di Geodesia Spaziale, C.A. Terlecchia,
I-75100 Matera, Italy
+39-835-377238, lanotte@asi.it

Lichtenegger, Herbert
Technical University Graz, Institute of Applied Geodesy,
Steyrergasse 30, A-8010 Graz, Austria
+43 316 8736833, +43 316 8738888,
lichteneg@ftug01.tu-graz.ac.at

Liebsch, Gunter
TU Dresden, Inst. für Planetare Geodäsie, D-01062 Dresden,
Germany
+49-351-4633045, +49-351-4637063,
liebsch@ipg.geo.tu-dresden.de

Ma, Chopo
NASA Goddard Space Flight Center, Code 926,
Greenbelt MD 20008, USA
+1-301-286-3992, +1-301-286-1760, cma@virgo.gsfc.nasa.gov

Mahmoud, Salah Mohamed
National Research Institute of Astronomy and Geophysics,
Helwan - Cairo, Egypt
+202 554 9780, +202 554 8020, salahmm@frcu.eun.eg

Manning, John
Australian Surveying and Land Information Group AUSLIG,
6 Howie Coast Emu Ridge, Canberra 2617, Australia
+61-26-2014352, +61-26-2014366,
johnmanning@auslig.gov.au

Mareyen, Maria
Bundesamt für Kartographie und Geodäsie,
Michendorfer Chaussee 23, D-014473 Potsdam, Germany
+49-331-316629, +49-331-316602, mamy@potsdam.ifag.de

Matsuzaka, Shigeru
Geografical Survey Institute, Kitasato-1,
J-305-0811 Tsukuba, Ibaraki, Japan
+81-298-64 4832, +81-298-64 2655, shigeru@gaos.gsi.go.jp

Mireault, Yves
Geodetic Survey of Canada, Natural Ressources Canada,
615 Booth Street, Room 468, Ottawa K1A 0E9, Canada
+1-613-947-7937, +1-613-995-3215, mireault@geod.emr.ca

Moore, Angelyn W.
Jet Propulsion Laboratory, Caltech, 4800 Oak Grove Drive.
MS 238-600, Pasadena CA 91024, USA
+1-818-354 5434, +1-818-393 4965,
angelyn.w.moore@jpl.nasa.gov

Mueller, Ivan I.
The Ohio States University,
Dept. of Civil & Environmental Engineer. & Geodetic
Science, 4631 Shire Creek Court, Hilliard,
Ohio, OH 43026-2764, USA
+1-614-876-7830, +1-614-292-2957,
mueller@mps.ohio-state.edu

Müller, Jürgen
TU München,
Institut für Astronomische und Physikalische Geodäsie,
Arcisstr. 21, D-80290 München, Germany
+49-89-28923198, +49-89-28923178,
jxmx@alpha.fesg.tu-muenchen.de

Neilan, Ruth
Jet Propulsion Laboratory, 4800 Oak Grove Drive,
Pasadena CA 91109, USA
+1-818-354 8330, +1-818-393 6686, ruth.neilan@jpl.nasa.gov

Nerem, R. Steven
The University of Texas in Austin, Center for Space Research,
R 1000, Austin TX 78712, USA
+1-512-471-5696, +1-512-471-3570, nerem@csr.utexas.edu

Niell, Arthur
Haystack Observatory, Westford MA 01886, USA
+1-781-981-5416, +1-781-981-0590, aniell@haystack.mit.edu

Nothnagel, Axel
Universität Bonn, Geodätisches Institut, Nussallee 17,
D-53115 Bonn, Germany
+49-228-733574, +49-228-732988, nothnagel@uni-bonn.de

Oberndorfer, Helmut
TU München,
Institut für Astronomische und Physikalische Geodäsie,
Arcisstr. 21, D-80333 München, Germany
+49-89-2892-0, helmut@alpha.fesg.tu-muenchen.de

Ogi, Shoichi M.A.
Geographical Survey Institute, Kitasato 1,
J-305-0811 Tsukuba, Ibaraki, Japan
+81-298-64-4832, +81-298-64 2655

Pachelski, Wojciech
Space Research Centre, Planetary Geodesy, Bartycka 18a,
PL-00 716 Warsaw, Poland
+48 22 403766, +48-22403131, wp@cbk.waw.pl

Pagiatakis, Spiros
Natural Resources of Canada, 615 Booth Street,
Ottawa K1A OE9, Canada
+1-613-995-8720, +1-613-992-1468,
pakiatakis@geod.nrcan.gc.ca

Pavlis, Erricos C.
NASA Goddard Space Flight Center,
Joint Center for Earth System Technology, Code 926,
Greenbelt MD 20771-0001, USA
+1-301-286-4880, +1-301-286-1760,
epavlis@lighthouse.JCET.umbc.edu

Penna, Nigel
University of Nottingham, Institut of Engineering,
Surveying and Space Geodesy (IESSG),
Nottingham NG7 2RD, U.K.
+44-115-951-3921, +44-115-951-3881,
nigel.penna@nottingham.ac.uk

Petit, Gerhard
Bureau International des Poids et Mesures (BIPM),
Pavillon de Breteuil, F-92312 Sevres, France
+33-1-45077067, +33-1-45077059, gpetit@bipm.fr

Plag, Hans-Peter
Statens Kartverk, Norwegian Mapping Authority,
Kartverksveien, N-3500 Hønefoss, Norway
+47-32 118100, +47-32118101, plag@gdiv.statkart.no

Ray, James R.
U.S. Naval Observatory, Earth Orientation Dept.,
3450 Massachusetts Av. NW,
Washington DC 20392-5420, USA
+1-202-762-1444, +1-202-762-1563, jimr@maia.usno.navy.mil

Rebhan, Helge
ESA - ESTEC, Postbox 299, NL-2200 AG Noordwijk,
The Netherlands
+31-715655846, +31-715655675, hrebhan@gw.estec.esa.nl

Reigber, Christoph
GeoForschungsZentrum Potsdam, Bereich 1,
Telegrafenberg A17, D-14473 Potsdam, Germany
+49-331-2881100, +49-331-2881111,
reigber@gfz-potsdam.de

Richter, Bernd
Bundesamt für Kartographie und Geodäsie,
Richard-Strauss-Allee 11, D-60598 Frankfurt a. M., Germany
+49-69-6333273, +49-69-6333425, richter@ifag.de

Rossbach, Udo
Universität der Bundeswehr München,
Institute of Geodesy and Navigation, D-85577 Neubiberg,
Germany
+49-89-6004 3425, +49-89-6004 3019,
udo.rossbach@unibw-muenchen.de

Rothacher, Markus
University of Berne, Astronomical Institute, Sidlerstrasse 5,
Ch-3012 Bern, Switzerland
+41-31-631 8591, +41-31-631-3869, rothacher@aiub.unibe.ch

Rudolph, Stefan
Universität Bonn, Institut für Theoretische Geodäsie,
Nussallee 17, D-53115 Bonn, Germany
+49-228-733578, +49-228-733029,
stephan@theor.geod.uni-bonn.de

Rummel, Reiner
Technische Universität München,
Institut für Astronomische und Physikalische Geodäsie,
Arcisstr. 21, D-80333 München, Germany
+49-89-2892-3190, +49-89-2892-3178,
rummel@step.iapg.verm.tu-muenchen.de

Sabadini, Roberto
Università di Milano, Dipt. di Geofisica,
Via L. Cicognara 2, I-20129 Milano, Italy
+39-2-23691502, bob@sabadini.geofisica.unimi.it

Sakr, Kamal
National Research Institute of Astronomy and Geophysics,
Helwan-Cairo, Egypt
+202-554 3111, +202-554 8020, kamal.mos@frcu.eun.eg

Salbach, Holger
TU Braunschweig, Institut für Geodäsie und Photogrammetrie,
Gauss-Str. 22, D-38106 Braunschweig, Germany
+49-531-3917491, +49-531-3917499, h.salbach@tu-bs.de

Sánchez Rodriguez, Laura Marlene
Instituto Geográfico Agustin Codazzi, Calle 66 A # 17-06,
0217, Santafé de Bogotá, Colombia
+57-1-368-3666, +57-1-368-0991, lsanchez@igac. gov.co

Sansò, Fernando
Politechnico di Milano,
Dipt. di Ingegneria Idraulica Ambientale e de Rilevamento,
Piazza Leonardo da Vinci 32, I-20133 Milano, Italy
+39-2-23-996506, +39-2-23996530,
iges@ipmtf4.topo.polimi.it

Sarti, Pierguido
University of Bologna, DISTART, V. le Risorgimento N. 2,
I-40136 Bologna, Italy
+39-51-6443102, +39-51-6448073, psarti@astbo1.bo.cnr.it

Schäfer, Uwe
Bundesamt für Kartographie und Geodäsie,
Aussenstelle Potsdam, Michendorfer Chaussee 23,
D-14473 Potsdam, Germany
+49-331-316607, +49-331-316602, ufer@potsdam.ifag.de

Scharroo, Remko
Delft University of Technology, Kluyverweg 1,
NL-2629 HS Delft, The Netherlands
+31-15-278 1483, +31-15-278 5322,
remko.scharroo@lr.tudelft.nl

Scheinert, Mirko
TU Dresden, Institut für Planetare Geodäsie,
D-01062 Dresden, Germany
+49-351-4633683, +49-351-4637063,
mirko@ipg.geo.tu-dresden.de

Schlüter, Wolfgang
Bundesamt für Kartographie u. Geodäsie,
Fundamentalstation Wettzell, D-93444 Kötzting, Germany
+49-9941-603107, +49-9941-603-222,
schlueter@wettzell.ifag.de

Schreiber, Ulrich
TU München, Fundamentalstation Wettzell,
D-93444 Kötzting, Germany
+49-9941-603 113, +49-9941-603-222,
schreiber@wettzell.ifag.de

Schuh, Harald
Deutsches Geodätisches Forschungsinstitut, Marstallplatz 8,
D-80539 München, Germany
+49-89-23031-214, +49-89-23031-240,
schuh@dgfi.badw-muenchen.de

Schuh, Wolf-Dieter
TU Graz, Theoretische Geodäsie, Steyrergasse 30,
A-8010 Graz, Austria
+43- 316- 873-6347, +43-316-8736845,
schuh@geomatics.tu-graz.ac.at

Schutz, Bob
University of Texas, Center for Space Research,
3925 West Braker Lane, Suite 200, Austin TX 78759-5321,
USA
+1-512- 471 4267, +1-512-471 3570, schutz@csr.utexas.edu

Schwarz, Klaus Peter
The University of Calgary, Dept. of Geomatics Engineering,
2500 University Drive N.W., Calgary, Alberta T2N 1N4,
Canada
+1-403-220-7377, +1-403-284-1980,
schwarz@ensu.ucalgary.ca

Schwintzer, Peter
GeoForschungsZentrum Potsdam, Bereich 13, Haus 17,
Telegrafenberg A17, D-14473 Potsdam, Germany
+49-331-288-1131, +49-331-288-1111, psch@gfz-potsdam.de

Shum, C.K.
Ohio State University,
Civil & Environmental Engineering &Geodetic Science,
470 Hitchcock Hall, 2070 Mil. Ave, Columbus OH-43210,
USA
+1-614-292-7118, +1-614-292-5196, ckshum@osu.edu

Sillard, Patrick
Institut Géographique National,
Ecole Nationale des Sciences Géographiques, 6 et 8 avenue
Blaise Pascal, F-77455 Marne la Vallée, Cedex 2 - France
altamimi@ensg.ign.fr

Šimek, Jaroslav
Research Institute of Geodesy, Topography and Cartography,
CZ-25066 Zdiby, Czech-Republic
+420-204-649-235, +420-204-649236, gope@asu.cas.cz

Sneeuw, Nico
TU München,
Institut für Astronomische und Physikalische Geodäsie,
Arcisstr. 21, D-80290 München, Germany
+49-89-28923193, +49-89-28923178,
sneeuw@step.iapg.verm.tu-muenchen.de

Soijer, Marco Willem
Delft University of Technology, Kluyverweg 1,
NL-2629 HS Delft, The Netherlands
+31-15-278 2594, +31-15-278 6480, m.w.soijer@ir.tudelft.nl

Springer, Tim
University of Berne, Astronomical Institute, Sidlerstr. 5,
CH-3012 Bern, Switzerland
+41-31-631-8592, +41-31-631-3869,
tim.springer@aiub.unibe.ch

Tapley, Byron D.
University of Texas at Austin, Center for Space Research,
Austin TX 78712, USA
+1-512-471-5573, +1-512-471-3570, tapley@csr.utexas.edu

Thalhammer, Markus
TU München,
Institut für Astronomische und Physikalische Geodäsie,
Arcisstr. 21, D-80290 München, Germany
+49-89-28923180, +49-89-28923178,
thama@step.iapg.verm.tu-muenchen.de

Titov, Oleg
Saint-Peterburg State University, Astrometry,
Bibliotechnaya pl.,2,Peterhof, RUS-198904 Sankt Peterburg,
Russia
+7-812-1598125, +7-812-428 7129,
serene@pmfd.ru, surkis@ipa.rssi.ru

Titz, Helmut
Technische Universität Wien, Theoretische Geodäsie,
Gusshausstr. 27-29, A-1040 Wien, Austria
+43-1-58801-3697, +43-1-5042155

Vandenberg, Nancy R.
NASA / Goddard Space Flight Center, NVI, Inc.,
640 Golden Triangle Av., Greenbelt MD 20110, USA
+1-301-286-9019, +1-301-286-0213,
nrv@gemini.gsfc.nasa.gov

Vespe, Francesco
Agenzia Spaziale Italiana, Centro Geodesia Spaziale,
I-75100 Matera, Italy
+39-835-377211, +39-835-339005, vespe@asi.it

Völksen, Christof
Universität Hannover, Institut für Erdmessung,
Schneiderberg 50, D-30167 Hannover, Germany
+49-511-7623892, +49-511-7624006,
voelksen@mbox.ife.uni-hannover.de

Weber, Robert
Technische Universität Wien, Theoretische Geodäsie,
Gusshausstr. 27-29, A-1040 Wien, Austria
+43-1-58801 3795, +43-1-504 21 55,
rweber@luna.tuwien.ac.at

Wende, Werner
Bayerische Kommission für die Internationale Erdmessung,
Marstallplatz 8, D-80539 München, Germany
+49-89-23031-111, +49-89-23031-100,
wende@bek.badw-muenchen.de

Wilson, Clark
NASA Headquarters, Code YO, Washington DC 20546-0001,
USA
+1-202-358-1373, +1-202-358-2770, cwilson@hq.nasa.gov

Yoshino, Taizoh
Communication Research Laboratory, Keystone Project Team,
4-2-1 Nukui-kitamachi, Koganei-shi, J-184-8795 Tokyo,
Japan
+81-42-327-7560, +81-42 327 6077, yosh@crl,go.jp

Zebhauser, Benedikt E.
TU München,
Institut für Astronomische und Physikalische Geodäsie,
Arcisstr. 21, D-80290 München, Germany
+49-89-28923181, +49-89-28923178,
bene@step.iapg.verm.tu-muenchen.de

Zerbini, Susanna
University of Bologna, Dept. of Physics,
Viale Berti Pichat 8, I-40127 Bologna, Italy
+39-51-6305019, +39-51-6305058, zerbini@astbo1.bo.cnr.it

Zharov, Vladimir
Sternberg State Astronomical Institute,
Laboratory of Gravimetry, Universitetskij prospect, 13,
RUS-119899 Moscow, Russia
+7-095-9395024, +7-095-9328841, zharov@sai.msu.ru

Zielinski, Janusz
Space Research Centre, Polish Academy of Sciences,
Planetary Geodesy, ul.Bartycka 18A,
PL-00 716 Warsaw, Poland
+48-22-6511808, +48-22-403131, jbz@cbk.waw.pl

Zink, Manfred
DLR, Oberpfaffenhofen, D-82234 Wessling, Germany
+49-8153-282386, +49-8153-281449, manfred.zink@dlr.de

Zumberge, James F.
Jet Propulsion Laboratory, 4800 Oak Grove Drive,
Pasadena CA 91109, USA
+1-818-354-6734, +1-818-393-4965,
James.F.Zumberge@jpl.nasa.gov

Position Papers

Importance of a Common Framework for the Realization of Space-Time Reference Systems

G. PETIT[1]

Abstract

General relativity establishes a distinction between proper quantities which are directly measurable and coordinate quantities which depend on some conventions. While basic mesurements are proper quantities (in nearly all cases either time or frequency is measured), coordinates are the necessary tool with which to study, by modeling, the interactions of physical phenomenon with the measurements. They are also the basic instrument for exchanging and summarizing the results of the measurements, mostly through the coordinates of objects (geodetic station, radio-source ...) realizing a space reference.

The paper addresses the issue of how the realization of space-time references may be affected by the choice of coordinate conventions. Some techniques used in space geodesy and astrometry will be studied, stressing how the model chosen, or other assumptions made explicitly or implicitly, influence the results (coordinate quantities). Particular emphasis will be on the issues on time. Indeed the prospects in time metrology are such that new definitions should be drawn to make the best use of the next generation of clocks and be able to compare them. In all fields, however, it is of ever increasing importance for all users to deal with properly defined quantities and common conventions, since the measurement uncertainty in any technique is bound to decrease. For this reason, the Bureau International des Poids et Mesures and the International Astronomical Union have created in 1997 the Joint Committee on relativity for space-time reference systems and metrology, which has among its tasks "to establish definitions and conventions to provide a coherent relativistic frame for all activities in space-time references and metrology at a sufficient level of uncertainty".

1. Introduction

In its 1991 General Assembly the International Astronomical Union (IAU) explicitly adopted the general theory of relativity as the theoretical framework for the definition and realization of space-time reference frames (IAU 1991). Barycentric and geocentric coordinate time scales and the relativistic transformations between them were defined, together with procedures for their realization.

Coordinate time scales in the vicinity of the Earth can be realized by a weighted average of atomic clocks operating on the Earth and on terrestrial satellites. For this purpose the clocks have to be synchronised and syntonised using a relativistic theory at the level of accuracy required. Due to the present and foreseen technical progresses, this is one of the first cases where an extension of the IAU(1991) framework may be needed.

We examine in section 2 the present situation (IAU1991) for the barycentric and geocentric system, some problems that arise and some possible solutions, and in section 3 how the framework (metric and corresponding transformations) could be extended to account for the present limitations in the barycentric system. The same issue for the geocentric system is addressed in section 4. Finally section 5 deals with the future of clocks, making a link with the theoretical discussions of the preceding sections. This paper is based on work done in the Joint Committee on relativity for space-time reference systems and metrology, which has been created by the Bureau International des Poids et Mesures and the International Astronomical Union and which is presently chaired by the author. However it does not represent the conclusions of the work of the Joint Committee which should be reported in the future to its founding organizations.

2. The IAU(1991) framework for the barycentric and geocentric system

2.1. The IAU(1991) framework

The IAU Resolution A4 (1991) provides the framework to define reference systems centered on some ensembles of bodies, notably the solar system (barycentric reference system BRS) and the Earth (geocentric reference system GRS). In doing so, it defined in its Recommendation III, new time scales TCB and TCG as coordinate times of the barycentric and geocentric systems, and differing from the previously defined TDB and TDT.

From the IAU(1991) metric we obtain the relation between proper and coordinate quantities. For example, using the geocentric non-rotating frame with TCG as coordinate time we obtain

$$d\tau/dTCG = 1 - [U(\mathbf{w}) + u^2/2]/c^2 + O(c^{-4}) \quad (1.1)$$

where $(cTCG, w^k)$ are coordinates in the geocentric frame with $\mathbf{w}$ representing the triplet w^k and $(w^k w^k)^{1/2}=r$. The potential at the position of the clock $U(\mathbf{w})$ is the sum of the Earth's potential and tidal potentials of external bodies, and $u = ((dw^i/dTCG)(dw^i/dTCG))^{1/2}$ is the coordinate speed of the clock in the geocentric, non-rotating frame. Similarly using the barycentric frame with TCB as coordinate time we obtain

[1] Gérard Petit, Bureau International des Poids et Mesures, 92312 Sèvres Cedex France; e-mail: gpetit@bipm.fr

$$d\tau/dTCB = 1 - [U'(\mathbf{x}) + v^2/2]/c^2 + O(c^{-4}) \quad (1.2)$$

where $(cTCB, x^k)$ are coordinates in the barycentric frame. The potential at the position of the clock $U'(\mathbf{x})$ is the sum of the potentials of all bodies of the solar system, and v is the coordinate speed of the clock in the barycentric system.

For the transformation between coordinate times, it is specified in the notes to the Recommendation that

$$\text{TCB - TDB} = L_B \times (\text{JD} - 2443144.5) \times 86400 \quad (2.1)$$

which defines L_B in the IAU(1991) formalism and that, as an approximation,

$$\text{TCB - TCG} = L_C \times (\text{JD} - 2443144.5) \times 86400 + v_E \cdot (x - x_E) / c^2 + P \quad (2.2)$$

where x is the barycentric position of the point of interest, x_E and v_E are the barycentric position and velocity of the geocenter and P represents periodic terms. This formula defines L_C.

In Recommendation IV, TT is defined as differing from TCG by a constant rate. It is specified that

$$\text{TCG - TT} = L_G \times (\text{JD} - 2443144.5) \times 86400 \quad (2.3)$$

which defines L_G, and that $L_B = L_C + L_G$.

2.2. Some limitations of IAU(1991)

This Resolution was a great step towards a consistent framework in which reference systems may be defined and realized. However it has suffered from two facts that may be easy to point out after the fact but were not so obvious at the time of its inception. The first one is that changing from existing conventions, even not so good ones, to new conventions requires a major effort. As a consequence, in some cases using different conventions produces different results (coordinates). In the most simple case, the difference is only a scale factor which may be traced and corrected, e.g. in VLBI (see section 2.2.). In other cases (e.g. combination of geometric and dynamical techniques) the differences are more difficult to estimate.

The second fact is that the Resolution did not specify sufficiently its conditions of application and/or did not consider conditions of application that were sufficiently ahead of the technical capabilities, although provisions were made to allow their extension. We shall briefly envision in section 2.3. more precise definitions without extending the scope of the Resolution and possible extensions in section 3.

2.3. Scale factor problems

All space geodesy techniques that make use of time measurements (i.e. nearly all techniques) must use formulas of the type (1.i) to transform the measured proper durations into modelled coordinate quantities and formulas of the type (2.i) to transform between different coordinate quantities. In doing so, it is necessary to use a complete and coherent model and set of formulas; otherwise it may become difficult to interpret the coordinate results.

A typical example is the case of VLBI: The VLBI model allows to compute the calculated VLBI delay from the coordinates of the stations, sources, etc... By comparing the calculated with the observed delays one determines better coordinates etc... This procedure requires consistency between the model and the reference system in which the coordinates are expressed. The VLBI model has been discussed thoroughly in 1991 when a "consensus model" was elaborated (EUBANKS 1991).

When using the IAU(1991) metric, the null equation for propagation of photons yields the original (also found in (IERS 1992)) formulation of the consensus model, relating a TCG coordinate delay d_{TCG} to station coordinates w_{TCG}. Most VLBI software packages use this model. However the observed delays from the correlator are not "TCG like" but "TT like" ($d_{TT}/d_{TCG} = 1 - U_E/c^2$). Therefore the station coordinates one obtains are also TT like and present a scale factor of order 7×10^{-10} with respect to TCG. Since the IUGG recommended in 1991 that spatial coordinates be consistent with TCG and the IAU(1991) metric, coordinates provided by a VLBI analysis should be scaled by $(1+U_E/c^2)$, if the above assumptions apply to this particular analysis.

To complicate the matter, a modified version of the consensus model was specified in (IERS 1996), in the aim of obtaining, from TT observed delays, coordinates which provide distances "consistent with physical distances on the geoid" (IERS 1996 p. 87). This is an unfortunate choice because, in addition to being inconsistent with IAU/IUGG conventions, it does not make sense to realize this aim since VLBI baselines are not on the geoid.

In summary, much care must be taken in using VLBI coordinates otherwise there is room for a scale error of 1.4×10^{-9} or 7×10^{-10}, an amount which is not negligible.

2.4. More precise definitions within the IAU(1991) framework

The Recommendations were formulated in the first PN approximation of general relativity, but keeping only terms in c^{-2} in the metric. In this context it is clear that defining L_B and L_C as average rates by formulas (2.1) and (2.2) is unambiguous down to an uncertainty level of order 2×10^{-16}. Similarly L_G can be defined by (2.3) down to the level of 1×10^{-18}. However, since the symbol JD does not mean any particular time scale, inconsistencies appear when trying to extend to the order c^{-4} the definition and realization of these quantities which were originally defined to the order c^{-2} (2.1 to 2.3).

Since those constants exist and have some use it is better to provide clear definitions for them (e.g. 3.i below) but L_B and L_C should not be used when the utmost accuracy is needed in their determination. In this respect, when transformations are needed with utmost accuracy it is necessary to consider the full space-time transformations (see section 3.) rather than use the average rates as in formulas (2.1) and (2.2). Following (FUKUSHIMA 1995) we propose to use as rigorous definitions, in order to overcome the above problems:

$DTCG / DTCB = 1 - L_C$ (3.1)

$DTT / DTCG = 1 - L_G$ (3.2)

$DTT / DTCB = 1 - L_B$ (3.3)

where, for L_B and L_C, D means a long term average taken at the geocenter.

One of the problems before IAU(1991) is that TDB does not have a good definition, and IAU(1991) does not allow to interpret L_B as a defining constant for TDB. But we could change this and turn a specific value of L_B into a defining constant thus providing in retrospect a good definition of TDB.

3. The case of the barycentric system: extension of the metric

Although the IAU(1991) metric contains only terms in c^{-2}, it has been specified (in the notes of Recommendation 1) that higher order terms may be added to the metric as deemed necessary by users without changing the rest of the Recommendation. In this aim we follow the presentation of (DAMOUR, SOFFEL and XU 1991), in which the general form of the post-Newtonian metric in the coordinate system (t,x) is

$$g_{00} = -1 + 2w/c^2 - 2w^2/c^4 + O(c^{-6})$$

$$g_{0i} = -4w_i/c^3 + O(c^{-5})$$

$$g_{ij} = d_{ij}(1+2w/c^2) + O(c^{-4}) \qquad (4)$$

where w is the scalar potential and w_i the vector potential. There exists a gauge freedom for the time coordinate i.e. if the time coordinate t is transformed as $t' = t + l(t,x)/c^{-4}$ there exists a solution (w', w'_i) of the field equations related to (w, w_i) through the gauge function λ. Extending the IAU(1991) metric therefore implies, in principle, some assumptions on the gauge function. The most common choices are the harmonic and standard PN gauges. Therefore we consider the magnitude of the coordinate difference implied by these two different choices.

Let us consider the coordinate transformation between TCG and TCB at the geocenter. From the metric, we can write a differential formula under the general form

$$dTCG/dTCB = 1 - (U_E+v_E^2/2)/c^2 - f(x_E,v_E,S)/c^4 \qquad (5)$$

where U_E is the gravitational potential of all bodies except the Earth. The function f regroups the terms in c^{-4} and depends on the gauge function λ. S is the set of solar system parameters necessary to compute the function f to the required accuracy. It can be shown (WOLF and PETIT 1998, KLIONER 1998) that, in a nearly all situations in the solar system and for a class of gauges including the standard PN and harmonic ones, formula (5) may be integrated and written in the following form:

$$TCB - TCG = 1/c^2 \int_{(Teph)} (U_E+v_E^2/2)\, dt / (1-1.55\times10^{-8}) + 1.10\times10^{-16}\times(T-T_0) \qquad (6.1)$$

where the integral is to be provided by a solar system ephemeris: This "time ephemeris" is described for example in (FUKUSHIMA 1995). For events not at the geocenter, location dependent terms should be added to the transformation:

$$TCB - TCG = 1/c^2 \int_{(Teph)} (U_E+v_E^2/2)\, dt / (1-1.55\times10^{-8}) + 1.10\times10^{-16}\times(T-T_0)+(x-x_E).v_E/c^2 \qquad (6.2)$$

Differences between the two gauges considered above amount to less than 0.3 ps on time and less than 3×10^{-18} on rate for practical situations in the vicinity of the Earth. They may be larger by more than one order of magnitude in special cases, such as a close flyby of Jupiter. Therefore, provided that the conditions of application are clearly stated, it should be possible to specify an extended metric compatible with the general form (4) with some gauge freedom left (no definitive choice mandatory).

In practice, limitations in realizing the transformations of the form (6.i) are presently at the level of a rate uncertainty of order 5×10^{-18} to 1×10^{-17} due to uncertainty in the masses of the sun, planets and asteroids (FUKUSHIMA 1995) when the Earth's potential is not predominant (transformation TCG to TCB, or proper time to TCB not in the immediate vicinity of the Earth). Limitations are of order 1×10^{-17} in the immediate vicinity of the Earth due to uncertainty in the Earth's potential. In view of future improvements, it might be necessary to further discuss the issue of gauges above, and to consider their effect at the level of 1×10^{-18} and below.

4. The case of the geocentric system

4.1. Realization of TCG and TT

In the geocentric system, terms in c^{-4} in the metric are smaller than 1×10^{-18} so the IAU(1991) metric does not need extension. The rate of a clock with respect to coordinate time (TCG) in the vicinity of the Earth, including all terms larger than one part in 10^{18} is given by:

$$dt/dTCG = 1 - [U(w) + u^2/2 + U_E(x_E+w) - U_E(x_E) - U_{E,k}(x_E)w^k]/c^2 \qquad (7.1)$$

where U is the Earth gravitational potential. Another possibility (specially for clocks on Earth) is to transform directly the proper time to TT, using the gravity potential W and its value at the geoid W_0. The formula is

$$dt/dTT = 1 - [W - W_0 + U_E(x_E+w) - U_E(x_E) - U_{E,k}(x_E)w^k]/c^2 \qquad (7.2)$$

Orders of magnitude of the individual terms in (7.1) and (7.2) and their calculation at the required accuracy are considered in detail in (WOLF and PETIT 1995, PETIT and WOLF 1997, 1998). To summarize, the most practical way of realizing TT or TCG from clocks in the vicinity of the Earth is:

(i) For clocks on satellites: transform the proper time of the clock to TCG using the IAU(1991) metric cf. equation (7.1). Then apply the conventional TT/TCG rate to obtain TT.

(ii) For clocks on the Earth's surface: Determine the gravity potential difference between the position of the clock and the geoid. Transform to TT according to equation (7.2). TCG can then be obtained by application of the conventional TT/TCG rate.

When applying these methods one should take care that the same conventional tidal model (tide free, zero frequency or mean-tide) is used coherently. In (i) this means that the coordinates used for the satellite ephemerides should be coherent with the geopotential model, in (ii) that the geodetic positions used are coherent with the geoid model.

4.2. Definition of TT

The gravity potential on the geoid W_0 is determined with an uncertainty of order 1 m^2s^{-2} (BURSA 1995) so that the corresponding rate uncertainty in the definition of TT is $1x10^{-17}$. If very good clocks are available in space it may be desirable to define a new scale TT^+ that would ensure continuity with TT but would not be limited by such an uncertainty in its definition.

One solution is to define TT^+ as differing from TCG by a fixed constant rate and take for this rate difference a conventional value. To ensure the best continuity, this conventional value should be chosen as the best estimation of W_0/c at the time of the change of definition. At the same time it could be convenient to turn this W_0 value into a defining constant for GRS2000 (GROTEN 1998) under discussion as the successor of GRS80. This would in effect amount to a relativistic definition of the geoid as suggested by (BJERHAMMAR 1985, SOFFEL et al. 1988, KOPEJKIN 1991).

5. Future of clocks and realization of coordinate time in the vicinity of the Earth

Coordinate time scales are established by transformation of the proper time of the most stable and accurate clocks, using the procedures defined in the preceding sections. Presently the most stable clocks reach a fractional frequency stability of a few parts in 10^{16} for averaging durations of a few hours (see below). These clocks are based on trapped ions or cold atoms and do not have yet a long history of continuous measurements. For this reason, on the longer term (one month and above), the stability is either estimated from clocks of less recent design such as commercial Cs clocks, hydrogen masers or continuously operating primary Cs clocks. Another estimate is from the reported accuracy of the recently designed primary Cs standards: the accuracy of the Cs fountain at the LPTF in Paris has been evaluated to be $2x10^{-15}$ (SIMON et al. 1997). The long term stability is passed to the atomic time scales such as TAI and TT(BIPM), with some improvement due to averaging: It is estimated to be about $1x10^{-15}$ for an averaging duration of one month.

Further improvements in the LPTF's fountain and the development of a second device capable of operating either as a fountain on Earth or as a low velocity beam in zero gravity should make attainable an accuracy of $1x10^{-16}$ (CLAIRON et al. 1995, MANN et al. 1998). Linear Hg^+ ion traps are being developed at the JPL in Pasadena (California, USA). Measurements of their stability give $\sigma_y(\tau) = 7x10^{-14}\ \tau^{1/2}$ for τ up to 10000 s and $\sigma_y(\tau)$ remains below 10^{-15} for τ up to 10 days (TJOELKER et al. 1995). Linear Yb^+ ion traps are being developed at the CSIRO in Sidney (Australia) (FISK et al. 1996). Recent measurements give $s_y(t) = 5x10^{-14}\ t^{-1/2}$ for τ up to 20000 s. Such developments promise that the medium term stability of atomic time scales will approach 10^{-16} in coming years. Similar long term stability is expected from the new frequency standards, but it will take some time for this to be acknowledged.

It is therefore expected that an accuracy (and long term stability) of $1x10^{-16}$ and a short term stability of parts in 10^{17} should be reached soon. When this happens, one of the main problems will be to compare such clocks in a sufficiently short measurement time. These questions are being considered within the ACES project (Atomic Clock Ensemble in Space) of the European Space Agency (ACES 1998). Established time transfer methods such as the one way GPS phase technique or the two way satellite time transfer technique should allow to reach $1x10^{-16}$ at intercontinental distances in a few days of measurement. Other methods such as T2L2 (Time Transfer by Laser Link) or a three-way radio link could be candidates for an improvement.

The existence of such clocks in space and on the ground, and time transfer techniques will have many consequences (ACES 1998). In the field of space reference systems, in addition to providing stable references that may help to improve space geodesy measurements, they will provide for the first time an independent and absolute measurement of the gravitational potential and of quantities linked to it (geoid, heights).

6. Conclusions

We can express some guidelines that might help to solve some of the existing problems of mutual comprehension in the field of space geodesy and astronomy. Those relating specifically to the barycentric system are the following:

- Provide, in order to remove ambiguities in the IAU (1991) formalism, clear definitions for L_B and L_C based on differential formulas. In addition provide in retrospect a good definition for TDB by choosing a specific value for L_B. Recommend NOT to use these constants for coordinate transformations when utmost accuracy is needed.
- Specify an extended conventional metric for the BRS in the formalism of General Relativity without parametrization providing an uncertainty of 0.3 ps for time and $3x10^{-18}$ for rate. Formulas should be valid in the vicinity of the Earth at least up to the geostationary orbit. The conventional metric could have no explicit gauge-dependent terms neither in g_{00} nor in g_{0i}. Precise conditions of applications should state in which cases this conventional choice should be distinguished from metrics based on specific gauge conditions.
- Specify that, in the vicinity of the Earth, the recommended realization of TCB is via the formula (6.2) above.

For the geocentric system:

- Turn a specific value of L_G into a defining constant for TT, instead of using the geoid. Choosing for $L_G = W_0/c^2$ the same value of W_0 as a defining constant for a new Geodetic Reference System, provide a homogeneous and consistent set of conventions, and a relativistic definition of the geoid.

These new conventions should, for a few years, allow to lead the technical needs by some amount rather than trail behind. However one should always be conscious that some primary hypothesis may be challenged. In the first place, there are some cases where taking the barycentric system as the primary "inertial" system may be not adequate. Some examples where such a modification may be necessary are: the sensitivity of VLBI observations to the galactic acceleration of the solar system (EUBANKS 1993); the sensitivity of future astrometric observations to gravitational lensing in the galaxy (GONTCHAROV 1998); the sensitivity of binary pulsar time (PETIT and TAVELLA 1996) to the motion of the pulsar or the solar system in the galactic potential. In addition, there is always the possibility that General Relativity may not remain the fundamental description of nature on the macroscopic scale.

Acknowledgements

I thank the members of the BIPM/IAU Joint Committee on relativity for space-time reference systems and metrology who contributed to the elaboration of the ideas presented in the paper. Special thanks are due to my colleague Peter Wolf for his continuous collaboration to this work.

References

ACES (1992): Proc. ACES Utilisation Workshop, Busca G. (ed), ESA.

BJERHAMMAR A. (1985): Bull. Géod. 59, 207.

BURSA M. (1995): *Report of the IAG SC3 Fundamental Constants*, XXI IAG General Assembly.

CLAIRON A. et al. (1995): *Preliminary accuracy evaluation of a cesium fountain frequency standard.* Proc. Fifth Symp. on Frequency Standards and Metrology, J.C. Bergquist (ed.), World Scientific, 49-59.

DAMOUR T., SOFFEL M., XU C. (1991): *General relativistic celectial echanics. 1. Method and definition of reference systems.* Phys. Rev. D 43, 3273.

EUBANKS T.M.E. (ed.) (1991: Proc. USNO Workshop on relativistic models for use in space geodesy.

FISK P.T.H., SELLARS M.J., LAWN M.A., COLES C. (1995): *A microwave frequency standard based on trapped, buffer gas-cooled $^{171}Yb^{+}$ ions.* Proc. Fifth Symp. on Frequency Standards and Metrology, J.C. Bergquist (ed.), World Scientific, 27-32.

FUKUSHIMA T. (1995): *Time ephemeris.* Astron. Astrophys. 294, 895.

GONTCHAROV G. (1998): *Future galactic dynamical reference frame.* Journées1998 Systèmes de Références spatio-temporels, Obs. de Paris.

GROTEN, E. (1998): Circular letter 98/1, IAG SC3 Fundamental Constants.

KOPEJKIN, S.M. (1991): manuscripta geodetica 16, 301.

IAU (1991): IAU Transactions Vol. XXIB, Kluwer pub.

IERS (1992): *IERS Standards (1992).* IERS TN13, McCarthy D.D. (ed.), Obs. de Paris.

IERS (1996): *IERS Conventions (1996).* IERS TN21, McCarthy D.D. (ed.), Obs. de Paris.

KLIONER S. (1998): Working document JCR03.c.

MANN A.G. et al. (1998): Proc. 1998 IEEE Frequency Control Symposium, p. 13.

PETIT G., TAVELLA P. (1996): *Pulsars and time scales.* Astron. Astrophys. 308, 290.

PETIT G., WOLF P. (1997): *Computation of the relativistic rate shift of a frequency standard.* IEEE Trans. IM 46, 2 201.

SIMON E. et al. (1997): Proc. 11th European Frequency and Time Forum, p. 43.

SOFFEL et al. (1988): manuscripta geodetica 13, 143.

TJOELKER R.J., PRESTAGE J.D., MALEKI L., (1995): *Record frequency stability with mercury in a linear ion trap.* Proc. Fifth Symp. on Frequency Standards and Metrology, J.C. Bergquist (ed.), World Scientific, 33-38.

WOLF P., PETIT G. (1995): *Relativistic theory for clock syntonization and the realization of geocentric coordinate times.* Astron. Astrophys. 304, 653.

WOLF P., PETIT G. (1998): Working document JCR03.a.

The Use of a Precise Reference Frame in Sea Level Change Studies

R. S. NEREM[1], R. J. EANES[1], J. C. RIES[1], G. T. MITCHUM[2]

Abstract

Long-term sea level variations are an important indicator of global climate change, and their measurement can provide critical information for determining the socio-economic impact of sea level change on coastal land use. Sea level change over the past century is studied almost exclusively using tide gauge records, but they suffer from the unknown effects of land motion and poor spatial distribution. The advent of satellite altimetry, and especially the launch of TOPEX/POSEIDON in 1992, overcomes many of the limitations of tide gauge measurements in that the measurements are global and are tied to the Earth's center-of-mass in a precise reference frame. However, uncertainties still arise with regard to the long-term performance of the instruments, and the maintenance of the reference frame from one altimeter mission to the next. Here, we discuss the importance of the reference frame for altimeter measurements of sea level. Our results show that care must be taken to maintain the reference frame across multiple geodetic techniques and multiple decades if climate change signals are to be detected. The movement of the tide gauges and the geodetic observatories tracking the altimeter satellites (SLR, DORIS, GPS, etc.) must be precisely monitored through a well-organized international effort.

Introduction

Since its launch in August 1992, TOPEX/POSEIDON (T/P) has been making near global (±66¡ latitude) measurement of sea level on a groundtrack that repeats every 10 days and has a 300 km spacing at the equator (FU et al., 1994). While the point-to-point precision of the measurements is roughly 3 cm, when averaged over the globe, mean sea level may be computed with a precision of 4 mm or better (MINSTER et al., 1995; NEREM, 1995a; NEREM, 1995b).

For this study, we employed TOPEX and POSEIDON data (Merged Geophysical Data Records Revision B) from Cycles 10-221, which covers the time period from the end of 1992 to mid-1998. All of the usual corrections have been applied to the data (NEREM, 1995b), including the oscillator and internal calibration estimates (HAYNE et al., 1994; NEREM, 1997), with the exception that no correction was applied for the inverted barometer effect (NEREM, 1995b). In addition, we applied a correction (KEIHM et al., 1998) for the known drift in the TMR measurements, which causes an increase in mean sea level rise of roughly 1.2 mm/year (HAINES and BAR-SEVER, 1998) after the correction is applied.

Global mean sea level was computed from this corrected data set using a simple equi-area weighted average (NEREM, 1995b). The resulting 10-day mean sea level estimates are shown in Figure 1. Errors in the ocean tide correction, the ionosphere correction, and other diurnally-varying quantities tend to alias to near a 60-day period in mean sea level, consequently we have reduced these errors by smoothing the 10-day estimates using a 60-day boxcar filter. Figure 1 shows a 20 mm rise and subsequent fall of mean sea level over 1997, which roughly coincides with the 1997-98 ENSO event. NEREM (1998) shows that this rise coincides with a 0.4¡C rise in global mean sea surface temperature, and that this occurs whenever there is an ENSO event. An EOF decomposition of the sea level maps (HENDRICKS et al., 1996; NEREM et al., 1998) shows that the rise is mostly represented by the two leading "ENSO" modes, suggesting a physical connection with this phenomena.

Calibrating Satellite Altimeter Measurements

Once the satellite is on-orbit, the monitoring of the performance of the instruments, and the changes in this performance over time, is largely accomplished through comparisons with in situ measurements. Long-term errors in the raw altimeter measurements (oscillator errors, instrument drift, etc.), the environmental corrections (ionosphere, troposphere, inverted barometer, tides, sea state, etc.), and the orbit all have the potential to masquerade as long-term sea level change signals. While a variety of different techniques have been attempted, comparisons with the global network of tide gauges have been shown to produce the most useful assessment of the long-term measurement performance (CHAMBERS et al., 1998; MITCHUM, 1994; MITCHUM, 1998).

In this technique, the altimeter sea level measurements in the vicinity of each tide gauge are used to compute a sea level difference (altimeter minus tide gauge sea level). Over each 10-day T/P cycle, all of these sea level differences are averaged together to compute a single mean estimate of the TOPEX-tide gauge sea level difference during that 10-day period. This process accounts for correlation of the differences (such as occurs when one tide gauge is close to another) and the relative performance of each tide gauge (coastal gauges are generally worse than island gauges for

[1] Robert Steven Nerem, Richard J. Eanes, John C. Ries, University of Texas, Center for Space Research/C0605, PO Box 1085, Austin Texas 78712, USA; Tel: +1-512-471-5573, Fax: +1-512-471-3788;e-mail:nerem@csr.utexas.edu / eanes@csr.utexas.edu / ries@csr.utexas.edu

[2] G. T. Mitchum, Department of Marine Sciences, The University of South Florida

example, due to coastal oceanographic effects not observed by the altimeter).

In order for this technique to have its highest accuracy, vertical land movement at each tide gauge must be properly accounted for. In the future, we hope that this land movement can be monitored using space geodetic techniques (GPS, DORIS, etc.), but this is presently only possible at a limited number of tide gauges. Therefore, the land motion at each tide gauge is estimated by computing the difference of the long term change in sea level at the each tide gauge with the current best estimate of long term sea level change (1.8 mm/year) (DOUGLAS, 1991). This may seem a bit ad hoc, but is the best we can do at the present.

The resulting cycle-by-cycle TOPEX-tide gauge calibration estimates (MITCHUM, 1998) are shown in Figure 2. We applied these calibration estimates directly to the raw altimeter data. No correction was applied to the POSEIDON data, but there are so few POSEIDON cycles, that this data had little impact on the results of this study. Figure 3 shows the calibrated and uncalibrated estimates of global mean sea level from T/P, along with global mean SST anomalies (REYNOLDS and SMITH, 1994). Note that the calibration estimates tend to improve the comparison with the SST variations, especially in the middle part of the time series.

The presence of large ENSO variations in global mean sea level indicates that a long time series of measurements will be required in order to detect the much small variations caused by climate change. NEREM et al. (1998) have simulated the effect of the observed ENSO variability on global mean sea level over the past 100 years, and they estimate that at least a decade of T/P class altimeter data will be required to determine the rate of sea level rise to an accuracy of 0.5 mm/year. In addition, 2-3 decades of such data will be needed to detect an acceleration of mean sea level change.

The TOPEX/POSEIDON Reference Frame

Clearly the results from T/P are approaching the precision where serious consideration of the reference frame, and its long-term maintenance needs to be addressed. The reference frame must be maintained over multiple decades, and thus multiple satellite altimeter missions, which will depend on multiple tracking techniques (SLR, DORIS, GPS, etc.). This will be a challenging task, since there are well-known differences between each of the major space geodetic techniques and the reference frame they realize individually. In addition, the motion of the tide gauges used for altimeter calibration needs to be determined in this same reference frame.

Currently, the T/P reference frame is not formally based on the International Terrestrial Reference Frame (ITRF), as ITRF92 was considered inadequate at the beginning of the mission (due principally to a lack of DORIS station positions). Currently, the SLR and DORIS station coordinates used for the precision orbit computations (TAPLEY et al., 1994) are based on the CSR95L01/95D02 solution, which includes 3-D velocities for most sites. Geocenter motion, which is the motion of the center-of-mass of the Earth in this crust-fixed reference frame, is not currently modeled. CSR polar motion estimates derived from SLR tracking of Lageos are also employed. The station coordinates have been updated twice during the T/P mission, with a negligible impact on the orbit computations. The latest ITRF solution (ITRF96) is very close to the T/P adopted station coordinates, and future ITRF solutions can be considered for adoption by T/P and future missions.

For studies of long-term sea level change using satellite altimetry, errors in the station velocities are the main concern. We decided to test the sensitivity of the computed T/P orbits to changes in the station velocities that define the evolution of the reference frame over time. This was accomplished by reprocessing a sample of the T/P orbits spanning 5 years using different realizations of the reference frame, and then comparing the recomputed orbit to the nominal orbit. Note that a uniform change in the station velocities is equivalent to a secular drift error in the geocenter location. Table 1 summarizes these simulations for different amounts of error in the station positions. Random station errors have only a very small effect on a dynamically computed orbit (orbits computed using GPS in a kinematic mode will be much more susceptible). X/Y geocenter motion does not have a strong effect on the computed satellite orbit because the satellite dynamics are insensitive to small diurnal (with respect to the orbit) station position variations. Z geocenter motion does affect the computed orbit nearly one-for-one, since it masquerades as long-period orbit error, and can be accommodated by the adjustment of initial conditions in the orbit determination process.

Because the Earth is not completed covered by ocean, error due to Z geocenter motion maps into global mean sea level at the 10% level, while the X/Y motions average nearly to zero. This occurs because as the Earth rotates underneath the orbit, the X/Y motions average out, while the Z motions do not. Z geocenter motion maps into northern hemisphere mean sea level at the 40% level, and into southern hemisphere mean sea level at the 50% level (since the magnitude of the Z motion is a maximum at the poles, and zero at the equator).

There is currently some debate concerning how the oceans respond to geocenter motion, i.e., does the ocean follow the Earth's center-of-mass, or the motion of the crust? For periods of a month and longer, the oceans probably follow the center-of-mass of the Earth, thus sea level measurements made with satellite altimeters may contain geocenter signals. The distinction is important, since our reference frame is fixed to the crust. The true orbit will follow the Earth's center-of-mass, but as noted above, the computed orbit will tend to follow the reference frame in Z (rather than the center-of-mass) but not in X or Y. Tide gauges move with the crust, and thus only observe geocenter motions through their direct effect on sea level. This difference between how altimeter and tide gauge measurements are affected by geocenter motion can have important consequences when calibrating altimeter measurements using the tide gauges.

Currently, Satellite Laser Ranging (SLR) gives the best estimates of geocenter motion (mainly at seasonal frequencies), although DORIS and GPS are also being used. Annual variations in geocenter motion have been observed using SLR, and are relatively well understood (CHEN et al., 1998), although they are not currently modeled in the definition of the ITRF. X/Y geocenter motion is more easily determined from satellite tracking data because of the diurnal variations it causes in the data. Z geocenter motion is more difficult to determine because its signature in the tracking data is similar to long period orbit error. Secular geocenter motion cannot be separated from station velocities, and is normally accounted for in the latter.

Conclusions

T/P is providing extremely precise estimates of sea level change, and these estimates will be continued by follow-on missions such as Jason-1, Envisat, etc. Tide gauges have a critical role in monitoring the performance of satellite altimeters on-orbit, in addition to providing a link to historical estimates of sea level change. Random station location errors are not of serious concern for satellite altimetry; however systematic drifts, such as geocenter motion and regional deformation of the reference frame, are potentially quite harmful in Z. In addition, how the oceans respond directly to the geocenter motion is still uncertain, and this determines how the error maps into altimeter measurements of sea level, and how tide gauges sense this error.

Results from the analysis of T/P altimeter data indicate that multiple decades of high accuracy altimeter data will be required to detect background climate change signals in the presence of much larger ENSO interannual variability. This implies that sea level measurements from a series of different altimeter missions will be required, and it can be expected that these missions will employ different tracking techniques as we see a shift from more traditional techniques (SLR, DORIS) towards GPS. Therefore, we must ensure the consistency of the reference frame across multiple decades using different geodetic techniques in order for satellite altimetry to eventually detect sea level variations caused by climate change.

Acknowledgments

The investigation was supported by NASA through a TOPEX/POSEIDON Extended Mission Science Investigation.

References

CHAMBERS D.P., RIES J.C., SHUM C.K., TAPLEY B.D.: *On the use of tide gauges to determine altimeter drift*, J. Geophys. Res., 103 (CC6), 12,885-12,890, 1998.

CHEN J.L., WILSON C.R., EANES R.J., NEREM R.S.: *Geophysical Interpretation of Observed Geocenter Motions*, J. Geophy. Res., in press, 1998.

DOUGLAS B.C.: *Global sea level rise*, J. Geophys. Res., 96 (C4), 6981-6992, 1991.

FU L.-L., CHRISTENSEN E.J., YAMARONE C.A. JR., LEFEBVRE M., MENARD Y.M.E.N., DORRER M., ESCUDIER P.: *TOPEX/POSEIDON mission overview*, J. Geophys. Res., 99 (C12), 24,369-24,382, 1994.

HAINES B.J., BAR-SEVER Y.E.: *Monitoring the TOPEX microwave radiometer with GPS: Stability of columnar water vapor measurements*, Geophys. Res. Lett., 25 (19), 3563-3566, 1998.

HAYNE G.S., HANCOCK D.W., PURDY C.L.: *TOPEX altimeter range stability estimates from calibration mode data*, TOPEX/POSEIDON Research News, 3, 18-22, 1994.

HENDRICKS J.R., LEBEN R.R., BORN G.H., KOBLINSKY C.J.: *Empirical orthogonal function analysis of global TOPEX/POSEIDON altimeter data and implications for detection of global sea level rise*, J. Geophys. Res., 101 (C6), 14,131-14,146, 1996.

KEIHM S., ZLOTNICKI V., RUF C., HAINES B.: *TMR Drift and Scale Error Assessment*, Draft Report, Jet Propulsion Laboratory, 1-11, 1998.

MINSTER J.-F., BROSSIER C., ROGEL P.: *Variation of the mean sea level from TOPEX/POSEIDON data*, J. Geophys. Res., 100 (C12), 25,153-25,162, 1995.

MITCHUM G.T.: *Comparison of TOPEX sea surface heights and tide gauge sea levels*, J. Geophys. Res., 99 (C12), 24,541-24,554, 1994.

Monitoring the Stability of Satellite Altimeters with Tide Gauges, J. Atmos. and Oceanic Tech., 15 (June), 721-730, 1998.

NEREM R.S.: *Global Mean Sea Level Variations from TOPEX/POSEIDON Altimeter Data*, Science, 268, 708-710, 1995a.

Measuring global mean sea level variations using TOPEX/POSEIDON altimeter data, J. Geophys. Res., 100 (C12), 25,135-25,152, 1995b.

Global Mean Sea Level Change: Correction, Science, 275 (February 21), 1053, 1997.

NEREM R.S., CHAMBERS D.P., LEULIETTE E.W., MITCHUM G.T., GIESE B.S.: *Variations in Global Mean Sea Level Associated with the 1997-98 ENSO Event*, Geophys. Res. Lett., in review, 1998.

REYNOLDS R.W., SMITH T.S.: *Improved global sea surface temperature analysis*, J. Climate, 7, 929-948, 1994.

TAPLEY B.D., RIES J.C., DAVIS G.W., EANES R.J., SCHUTZ B.E., SHUM C.K., WATKINS M.M., MARSHALL J.A., NEREM R.S., PUTNEY B.H., KLOSKO S.M., LUTHCKE S.B., PAVLIS D., WILLIAMSON R.G., ZELENSKY N.P.: *Precision orbit determination for TOPEX/POSEIDON*, J. Geophys. Res., 99 (C12), 24,383-24,404, 1994.

Table 1. TOPEX/POSEIDON Reference Frame Tests

	Simulation Case	1	2	3	4
T/P Orbit Error	X (mm/year)	0.07	0.3	0.2	0.06
	Y (mm/year)	0.07	0.1	0.03	0.02
	Z (mm/year)	2.0	9.0	12.0	1.0
	Radial RMS (mm)	9	28	52	5

Case 1: 3 cm random errors in every station coordinate (SLR and DORIS).
Case 2: Systematic 1 cm/year error in each station coordinate velocity.
Case 3: Reduce all station velocities to zero.
Case 4. Switch to ITRF96 from nominal SLR/DORIS coordinates.

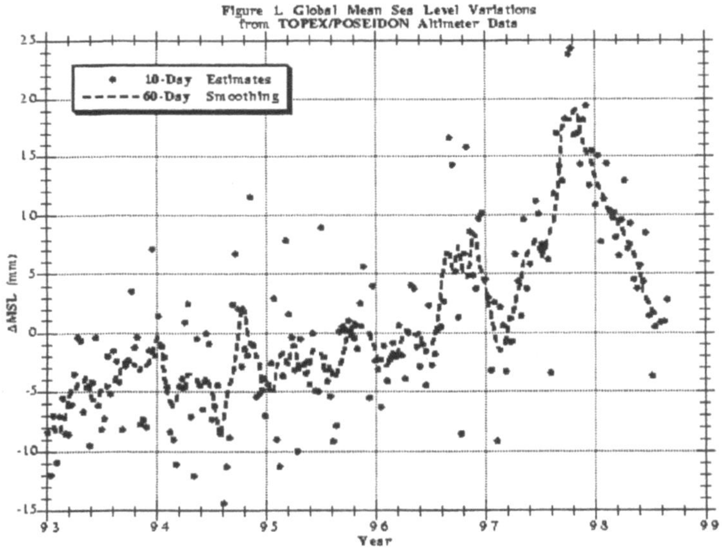

Figure 1. Global Mean Sea Level Variations from TOPEX/POSEIDON Altimeter Data

Figure 2. TOPEX - Tide Gauge Sea Level Differences Averaged Over 10-Day Cycles

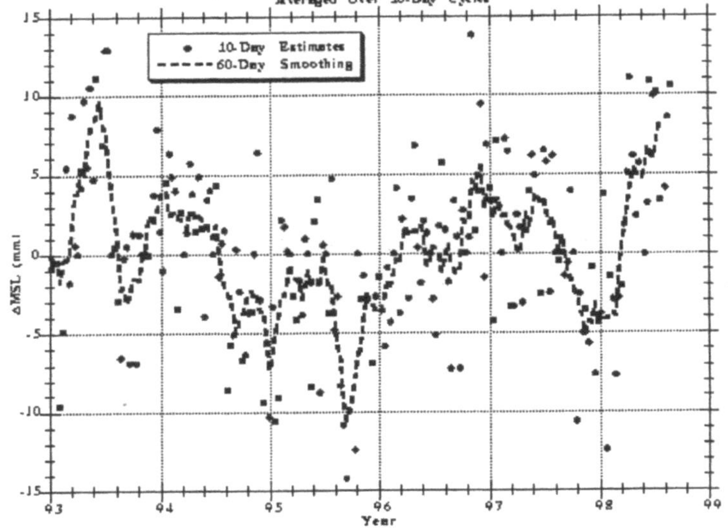

Figure 3. Comparison of Global Mean Anomalies of Sea Level and Sea Surface Temperature

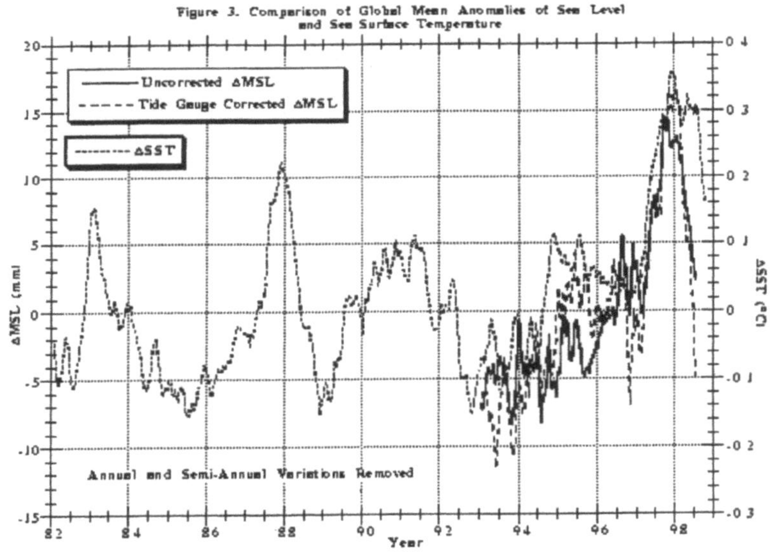

Combination of Space Techniques into one Integrated Processing Model

R. SCHARROO*, E. J. O. SCHRAMA, R. H. N. HAAGMANS†

Abstract

In the determination of the reference system satellite orbit determination plays an important role. With purely geometric techniques it is not possible to determine e.g. the position of the Earth's centre of mass. However, the introduction of dynamic techniques, such as satellite tracking, is accompanied by a range of parameter definitions and assumptions. These make the determination of the reference system and related parameters dependent on the assumed parameter values, but also on the type of tracking data that is used.

In conclusion it becomes obvious that the accurate establishment of reference frames and datums, precise orbit determination of several satellites, satellite tracking and gravity field determination are heavily interlinked subjects that require an integrated processing.

Introduction

In the modern age of geodesy we can distinguish between *conventional geodetic techniques*, like triangulation and leveling, and *space geodetic techniques*, like satellite laser ranging (SLR) and Very Long Baseline Interferometry (VLBI). The distinction lies in the use of an outside source (outside the earth) to establish a reference for determining the main geodetic parameters:

position: marker and station coordinates and their movements

reference frame: the Earth's centre-of-mass and its movement

Earth orientation: pole position, polar wander and length of day

gravity and geoid: and the associated deflection of the vertical and height systems

Conventional geodetic techniques

Conventional geodetic techniques are all *geometric*. This means that they are independent of the choice of other geodetic parameters. For example, by triangulation one can accurately determine the distance between various points in a network, irrespective of ones notion of polar motion or gravity. In principle this holds also for a conventional technique like relative gravimetry. This strong character of conventional geodetic techniques is also their weakness. The independence of other geodetic parameters can also be interpreted as a lack of external reference, which leads to three major drawbacks in conventional geodetic techniques. First, these geometric techniques are mostly *relative*: distances between markers or stations are measured, not their *absolute* positions in a global reference frame. Second, parameters related to the global reference, like earth rotation, can not be resolved. And third, without an intermittent external reference, measurements require line-off-sight connections, which make them most useful for *regional* networks and essential to hook local networks to global networks, but unsuitable to cover continental-wide or intercontinental distances.

Space geodetic techniques

To cover distances at a *global* scale space geodetic techniques employ an celestial target as a reference, this may be a natural target (quasar) or an artificial target (satellite). Currently, popular space geodetic techniques are

SLR: Satellite Laser Ranging

VLBI: Very-Long Baseline Interferometry

GPS: Global Positioning System (and its counterpart Glonass)

DORIS: Doppler Orbitography and Radiolocation Integrated by Satellite

PRARE: Precise Range And Range-rate Equipment

Altimetry: Altitude measurements from satellites

SST: Satellite-to-Satellite Tracking

Gravimetry: Gravity measurements from space

Each of these techniques has its advantages and disadvantages (Table 1). Some of them are more suitable for positioning, others for determining the Earth orientation (polar motion) or for satellite orbit determination; some perform better at smaller scales, others at global scales. The techniques are discussed in short in the next sections. It will become clear that not a single technique is able to determine all four groups of geodetic parameters (position, reference frame, earth orientation and gravity) at all required spatial scales. Hence, combination of techniques is of high importance.

*Remko Scharroo; Delft University of Technology, Delft Institute for Earth-Oriented Space Research (DEOS), Kluyverweg 1, 2629 HS Delft, The Netherlands, Email: remko@deos.tudelft.nl, Tel: +31 15 278 1483, Fax: +31 15 278 5322

†Ejo Schrama and Roger Haagmans; also DEOS, Thijsseweg 11, 2629 JA Delft, The Netherlands

	Positioning relative/absolute	Reference Frame	Earth Orientation	Gravity and Geoid
	Conventional Techniques			
leveling	(rel.) orthometric height	– –	– –	small-scale
triangulation	(rel.) regional	– –	– –	– –
	Space Techniques			
VLBI	(rel.) global	–	+ +	– –
SLR	(abs.) global	+ +	+/–	large-scale
DORIS	(abs.) global	+/–	+/–	large-scale
GPS	(rel.) regional/(abs.) global	+	+/–	– –
altimetry	(abs.) ellipsoidal height	+	– –	all scales
gravimetry	– –	–	–	intermediate scales
SST	– –	–	–	intermediate scales

Table 1. Comparison of various geodetic techniques and their contribution to the recovery of geodetic parameters (position, reference frame, Earth orientation, gravity) at regional and global scales, relative and absolute. '++' is very good, '+' is good, '+/–' is average, '–' is poor, '– –' is bad.

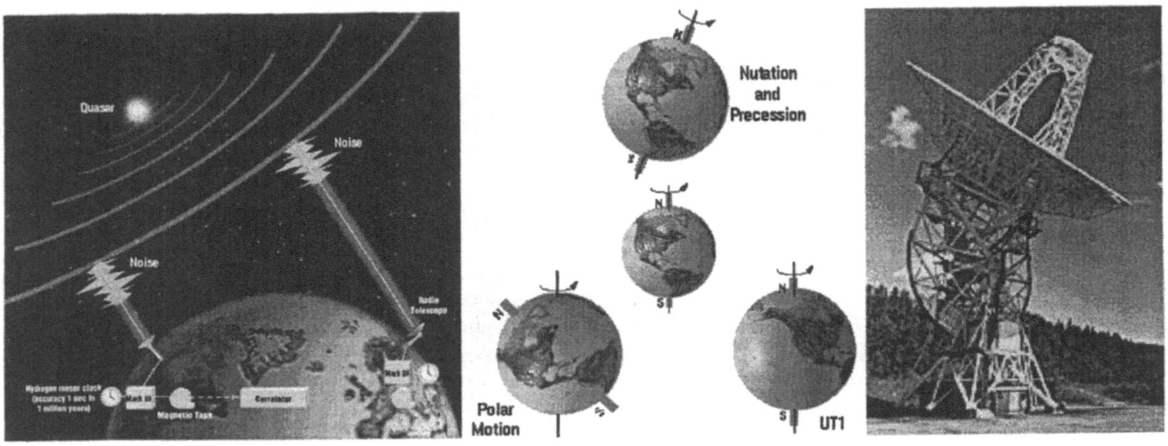

Figure 1. The concept of VLBI (left); earth orientation (middle); a radio telescope at Fairbanks, Alaska (right).

Very-Long Baseline Interferometry (VLBI)

VLBI (Figure 1) is the strongest *geometric* measurement technique for global positioning and polar motion. It uses celestial radio sources (quasars) as an external reference. The signal (mixed with noise) is received by two radio telescopes simultaneously. After correlating the signals received by the two stations, a difference in the reception times can be computed. This value determines the distance between the two telescopes in the direction of the line-of-sight of the quasar. In combination with other lines-of-sight, by selecting other quasars or simply due to the earth rotation, the three-dimensional distance vector (baseline) between the stations can be computed, even at very long distances. In fact, the accuracy of the technique is virtually distance independent.

Because the line-of-sight changes while the Earth rotates it is possible to build up a three-dimensional baseline. Inversely, if the baseline is known the location of the Earth's spin axis in inertial space (precession and nutation), as well as relative to Earth's crust (polar motion) and the rotation rate (the length of day), can be determined with high accuracy.

But obviously, the station positions remain to be relative. This means that a priori reference positions are needed to tie the VLBI-constructed baselines into a global reference frame. Co-located SLR stations may perform this function.

Satellite Laser Ranging (SLR)

SLR (Figure 2) is a strong tool for global positioning, crustal dynamics, gravity modelling and orbit determination. In principle, all can be solved simultaneously. The external reference this time are not fixed points in the sky, like the quasars are for VLBI, but moving targets, satellites equipped with corner-cube laser retro-reflectors. The SLR instrumentation is much less expensive than radio telescopes. Also, there is a wide range of satellites available as targets. However, the variety of satellites, ranging from high-flying small passive canon-balls studded with reflectors to low-flying school-bus sized complex satellites with only a few reflectors, also complicates things. Each satellite almost requires a "life-time" of experience to get measurement (correction) errors down to below the

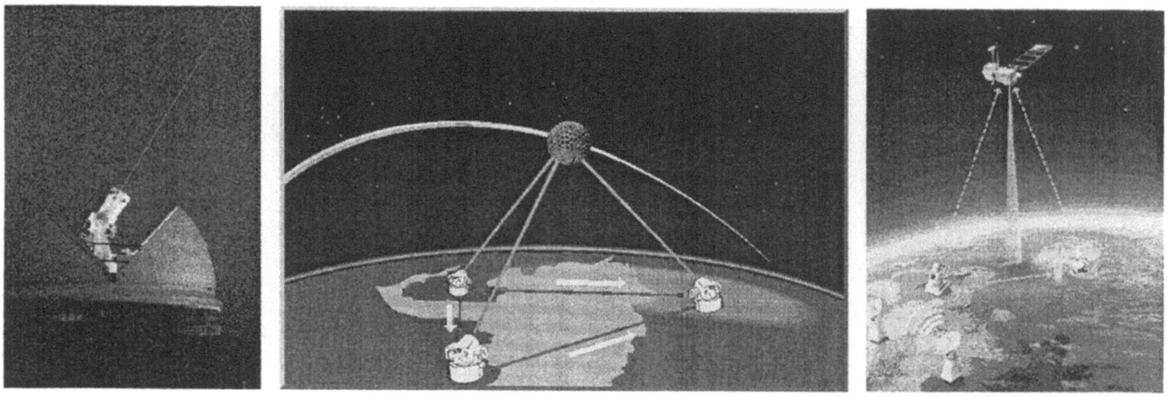

Figure 2. The laser ranging system at Herstmonceux, England (left); the concept of geometric positioning with SLR (middle); the use of SLR for absolute calibration of altimeters (right).

centimetre level.

The positioning of laser stations allows the monitoring of tectonic motion, post-glacial rebound and Earth rotation at any scale. Accuracy again is nearly independent of the baseline length. But tectonic motions can only be determined up to a certain point. Assumptions about the net rotation of the tectonic plates are needed to distinguish it from changes in the Earth rotation. VLBI-derived polar motions are very useful as reference or a priori knowledge. Finally, the global reference frame is determined by the combination of positions of the global SLR network. Unfortunately, the distribution of the stations is heavily biased towards the Northern Hemisphere, allowing significant errors on the southern hemisphere.

The global reference frame is tied to the combination of satellite orbits. Each of them is in principle fixed in space, with alterations due to the Earth's gravity field. So a priori knowledge of the *dynamic* behaviour of the satellite and the gravity field is needed; hence SLR is not a geometric technique. When the effect of gravity is known, the orbits provide a stable reference. Inversely, when the station positions are know, the knowledge of the gravity field can be enhanced by laser ranging. Yet, as shown in the Appendix it is not possible to determine both the reference frame and the gravity field from one single orbit, or even from various orbits at similar altitude. Hence, the need for the processing of various orbits simultaneously is eminent.

Although the interaction with (the knowledge of) the gravity field makes SLR chiefly a non-geometric technique, it still can be considered a geometric technique for very short baselines. When three stations are located sufficiently close, they will be able to track the same satellite simultaneously. The three ranges between the stations and the satellite fully determine the three baselines. Even when the stations do not hit the target at exactly the same time, the dynamic behaviour of the satellite in a short time interval can be modelled very precisely by just a few parameters. Given sufficient near-simultaneous measurements an accurate *geometric* network of *relative* positions can be determined. Again, this approach can be inverted and provide the geometrically determined satellite positions, without the need for accurate knowledge of the dynamics. This approach was used effectively for the determination of the position of the altimeter of ERS–1 on its overflight of an oceanographic tower near Venice. While the knowledge of the gravity field at that time (1991) still resulted in global orbits with a radial accuracy of about 1 metre, the geometric approach, using the laser ranging of several nearby SLR stations, provided orbits at the few centimetre level.

Global Positioning System (GPS)

GPS overcomes some of the disadvantages of SLR: since it works at radio frequencies it can be used also when it is cloudy and the receivers are relatively cheap and very mobile. This means that a much denser network can be established and nearly any location can be used to install a GPS station. However, in contrast to laser the radio signals are strongly delayed in the atmosphere, both by neutral (dry and wet) particles in the troposphere and by charged particles in the ionosphere. Second, the orbits of the GPS satellites that serve as a reference do not always have sufficient long-term precision. And on top of that the antennas of the ground equipment as well as on the satellite experience complicated biases, varying from antenna type to antenna type and from satellite generation to generation. This means that long-term monitoring of global tectonics as well as reference frame determination with GPS is still coping with significant problems.

At short baselines, where the effect of the satellite orbits and atmosphere is less severe, very accurate baselines can be determined. With the aid of the permanent IGS network of GPS stations it becomes possible always to construct relatively short baselines with those stations which are then tied, through GPS and other co-located space geodetic systems into a global reference frame. Here, again, a network of VLBI and SLR stations can provide a global reference within with GPS serves to densify the network.

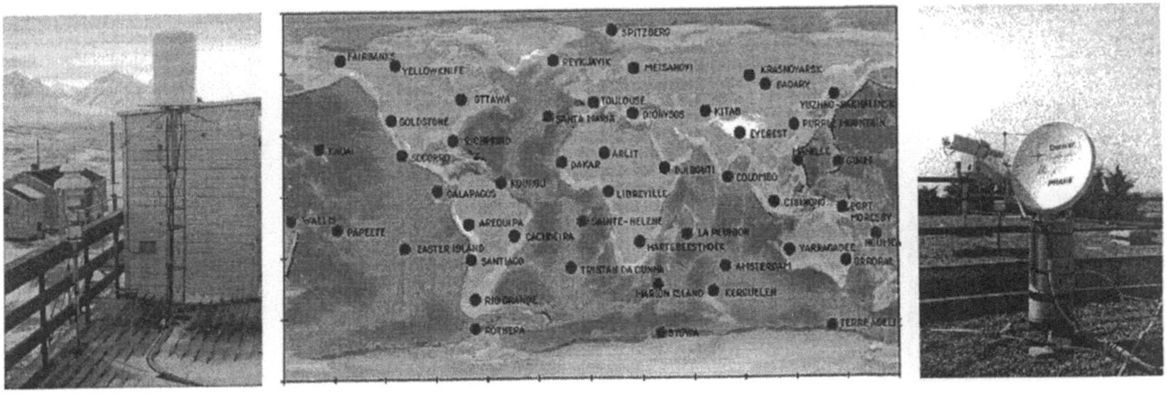

Figure 3. A DORIS beacon at Ny Ålesund, Spitsbergen (left); the global distribution of DORIS beacons (middle); a PRARE transponder at Matera, Italy (right).

Doppler tracking systems

The Doppler tracking systems DORIS and PRARE (Figure 3 are tracking devices designed mainly for the orbit determination of satellites. But they also allow station positioning. Both of them operated at the radio-frequency and thus have all-weather capabilities.

DORIS consists of low-cost ground beacons and a sophisticated space-borne receiver. Because of this design, a dense and homogeneous network of 50 beacons could be installed in a relatively short period. Also, consecutive improvements to the accuracy and capabilities could be incorporated in new generations of the on-board hardware and software and flown on later satellites. This without alterations to the beacons.

The sheer abundance of DORIS tracking data to a single satellite (due to the dense network and all-weather capability) compensates for the lesser intrinsic precision compared to laser measurements. This make DORIS a valiant tool for orbit determination. But as a positioning system Doppler measurements are less suitable. The Doppler measurement is one of (relative) velocity, not range, and thus has very poor geometric capabilities, even on short baselines. So only with very accurate knowledge of the orbit dynamics any positioning information can be extracted. On top of that system biases are poorly controlled, hampering the positioning accuracy.

For PRARE the Doppler measurement is inverted. This time the beacon is on board of the satellite and the receiver on ground. The receiver also functions as a transponder, sending the signal back to the satellite, allowing an integrated Doppler or range measurement. Unfortunately, this design makes both the space-borne and the ground components expensive (the transponder has to have an antenna pointing at and following the satellite). Add to this the failure of the first PRARE instrument on ERS–1 and the future for PRARE is marked.

Satellite altimetry

Although satellite altimetry (Figure 4) is not often considered as either a positioning system or a satellite tracking device, it actually is both, using the ocean as a natural transponder. From a satellite the altimeter monitors many facets of the "global" sea level: ocean tides, sea level rise, ocean dynamics at various spatial scales, gravity and geoid. At the same time it provides a link between coastal tide gauges and the open ocean.

But for *absolute* positioning of the sea level altimetry needs a precise orbital altitude as reference, based on SLR, DORIS, PRARE and/or GPS tracking. In turn the altimeter measurements themselves can aid the orbit determination: the sea level difference measured at two crossing tracks are for a large part determined by errors in the computed orbital altitudes. Minimizing these crossover height difference leads to more precise orbits. However, this mainly reduces orbit errors induced by poor modelling or knowledge of non-conservative forces (drag and solar radiation pressure are important effects on the sizeable and low-flying altimeter satellites) and extends only to the limit of the precision of the gravity field.

This means that in general gravity model errors will slip through the orbit into sea level, and, because they are time-invariant, into the mean sea surface and geoid. These errors will be invisible to repeat track analyses since they are repetitive from repeat cycle to repeat cycle. But on crossing tracks the effect of gravity model errors are well visible, except for the effect of the zonal parameters. After isolating the gravity field induced radial errors in this fashion, a subset of gravity field parameters can be adjusted. It is demonstrated on ERS–1 and ERS–2 that the resulting tuned gravity field provides more precise orbits for these satellites, decreases also the SLR range residuals, creates smaller DORIS residuals in orbits for the SPOT satellites and even performs slightly better for another altimeter satellite: TOPEX/POSEIDON.

Still better, however, is to the combine the tuning with SLR and PRARE residuals. Again a proof of the necessity to combine various geodetic techniques for gravity field improvement, even for a single satellite.

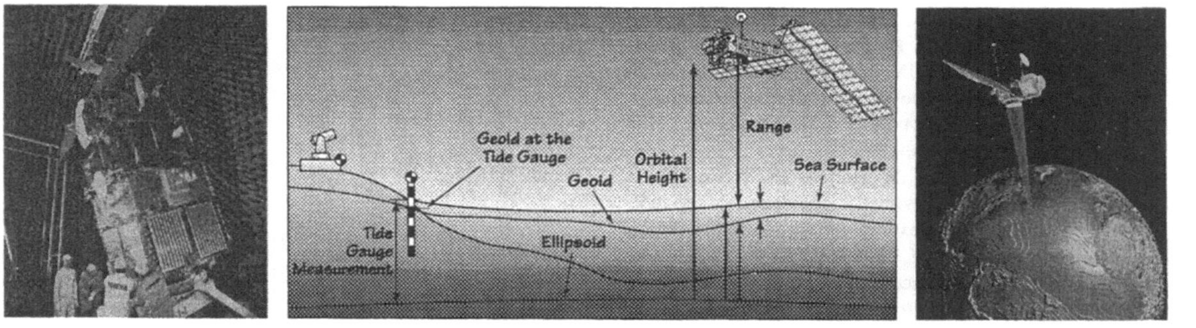

Figure 4. The ERS–1 spacecraft carries a radar altimeter (left); the concept of satellite altimetry (middle); the TOPEX/POSEIDON satellite measuring the global sea level (right).

SST, gravimetry and gravity gradiometry

SST, gravimetry and gravity gradiometry are proposed for three satellites to be launched in the next decade: CHAMP, GRACE, and GOCE. Especially the later two are expected to contribute significantly to the improvement of our knowledge of the gravity field at spatial scales from about 100-40,000 km, and the perception of gravity field variations at the larger scales. This will be a unique opportunity since no satellite before has been launched aimed at doing precisely that.

Towards an integrated geodetic reference system

The discussion of the space geodetic and conventional geodetic techniques shows the weaknesses and strengths of the various measurement types and measurement systems. None of the systems is able to determine station positions, the global reference frame, earth rotation, and gravity or geoid with sufficient precision. Only the combined use of the data, harmonisation of reference frames, and sharing of products and resources, and intercomparison of complementary results will lead to the cross-fertilisation needed to achieve a precise integrated geodetic reference system, one of which all geodetic communities can benefit.

Appendix: Gravity and the scale of the reference frame

Assume you were attempting to determine station positions with SLR from LAGEOS-type satellites at an altitude of 5800 km (r=12171 km) and define by this your global reference frame. The current value for the gravitational parameter is μ=398600.4415$\pm$0.002 km^3/s^2.

Any small increase of the orbital radius Δr results in an lengthening of the orbital period ΔT, which can be determined by differentiating the third law of Kepler:

$$T = 2\pi\sqrt{\frac{r^3}{\mu}} \Rightarrow \frac{\Delta T}{T} = \frac{3}{2}\frac{\Delta r}{r}$$

After one revolution this already means that the satellite lags a distance

$$2\pi r\frac{\Delta T}{T} = 3\pi\Delta r$$

on the original orbit, a distance much bigger than the original change of the radius. Hence timing of the orbital period is the most important determinator of the orbital radius.

When an error $\Delta\mu$ is made in the gravitational parameter, then the result on the orbital period is found again through differentiation of the third law of Kepler:

$$T = 2\pi\sqrt{\frac{r^3}{\mu}} \Rightarrow \frac{\Delta T}{T} = -\frac{1}{2}\frac{\Delta\mu}{\mu}$$

In consequence a 6 ppb uncertainty in μ (0.002 km^3/s^2) leads to a 3 ppb error in the orbital period (80 μs) and a 9 ppb error in the orbital radius (7 mm).

This means that an error in the gravitational parameter of 6 ppb is equivalent to a scale error of the orbit of 9 ppb. The 7 mm radial orbit error most likely results in an elevation of all laser stations by the same amount. Only when satellites are tracked at very different altitudes, the error in μ could be separated from the error in the orbital radius and hence the scale of the reference frame.

References and related publication

Boucher, C., and Z. Altamini, International Terrestrial Reference Frame, *GPS World*, *7*(9), 71–74, 1996.

Cazenave, A., J. J. Valette, and C. Boucher, Positioning results with DORIS data on SPOT-2 after first year of mission, *J. Geophys. Res.*, *97*, 7109–7119, 1992.

Cazenave, A., P. Gegout, L. Soudarin, K. Dominh, F. Barlier, P. Exertier, and Y. Boudon, Geodetic results from Lageos and DORIS satellite data, contribution of space geodetic to geodynamics; crustal dynamics, geodynamics, *AGU Series*, *23*, 81–98, 1993.

Cohen, S. C., and D. E. Smith, LAGEOS scientific results: Introduction, *J. Geophys. Res.*, *90*(B11), 9217–9220, 1985.

Crétaux, J. F., L.Soudarin, A. Cazenave, and F. Bouillé, Present-day plate motions derived from the DORIS space system, *J. Geophys. Res.*, *103*, 30,167–30,181, 1998.

Francis, C. R., A. Caporali, L. Cavaleri, A. Cenci, P. Ciotto, L. Ciraolo, W. Gurtner, F. H. Massmann, D. del Rosso, R. Scharroo, P. Spalla, and E. Vermaat, The Calibration of the ERS–1 Radar Altimeter – The Venice Calibration Campaign, *ESA Report ER–RP–ESA–RA–0257 issue 2.0*, ESA/ESTEC, Noordwijk, The Netherlands, 1993.

Haagmans, R. H. N., A. J. T. de Bruijne, and E. de Min, A procedure for combining gravimetric geoid models and independent geoid data, with an example in the North Sea region, *DEOS Progress Letters, 98.1*, 89–99, 1998.

IERS, Earth orientation, reference frames and atmospheric excitation functions, *IERS Technical Note 19*, 1994 IERS annual report, 1995.

Scharroo, R., and P. N. A. M. Visser, ERS Tandem Mission orbits: is 5 cm still a challenge?, in *Proceedings of the Third ERS Symposium, Eur. Space Agency Spec. Publ., ESA SP-414*, vol. 3, edited by T.-D. Guyenne and D. Danesy, pp. 1643–1648, 1997.

Scharroo, R., and P. N. A. M. Visser, Precise orbit determination and gravity field improvement for the ERS satellites, *J. Geophys. Res.*, *103*(C4), 8113–8127, 1998.

Towards an Integrated Global Reference System: Geodesy as a Utility

J. R. RAY[1]

Abstract

The global reference system provides a framework that allows users to determine geocentric coordinates with a precision of 1 cm or better. The overall accuracy of the system is probably ~5 mm in long-term position and about 1 to 2 mm/yr in rates. This is similar to the 1-day precision of the best measurement systems, an indication that systematic errors dominate over random observational noise in the long term. Nonetheless, the current system satisfies the requirements of all but the most demanding users and it continues to improve. This achievement, combined with the relatively easy and cost-effective access provided by GPS, is transforming the global reference system from a research tool into a geodetic utility. The trend is likely to continue and perhaps accelerate. The IAG is uniquely suited to promote and guide this development in such a way as to ensure the highest standards of user service are met while protecting the interests of scientific researchers. However, doing so will require new organizational structures to link the observing techniques and to provide improved user interfaces and services.

1. Global Reference System

A working definition for a "global reference system" adopted here is the complete process which provides geocentric coordinates {X(t), Y(t), Z(t), t} for properly instrumented points near the Earth's surface. Note that these coordinates should be expressible in an *inertial frame* as well as an Earth crust-fixed (rotating) frame. Inertial coordinates are needed for geophysical studies as well as for accurate satellite navigation. Note also that the present definition is explicitly 4-dimensional and not merely spatial. This is a recognition that time-transfer technology is emerging which permits accuracies similar to positioning accuracies, primarily via GPS. Indeed, the technologies are merging.

A global reference system consists of the following components:

- celestial reference frame (ICRF, GPS orbits)
- terrestrial reference frame (ITRF)
- Earth orientation parameters (EOPs)
- standards, conventions, constants
- catalogs of reference frame information
- tools for user access

Several practical realizations of some of these components exist, but only the International Earth Rotation Service (IERS) maintains a self-consistent, rigorously well defined, and complete system. Notable, in particular, are the International Celestial Reference Frame (ICRF), recognized by the International Astronomical Union (IAU), and International Terrestrial Reference Frame (ITRF), sanctioned by the International Association of Geodesy (IAG). GPS orbits, and potentially those of other satellites, provide an alternative intermediate celestial framework that is especially accessible for many users. In terms of the actual quality of realization of the global reference system, the current state-of-the-art can be summarized, admittedly subjectively, as having:

- a *precision* at 1-day resolution (where good data exist) of
  - ~ 3 mm horizontally
  - ~ 1 cm vertically
- and a long-term integrated *accuracy* of
  - ~ 0.5 to ~1 ppb globally
  - ~ 3 mm (0.1 mas) for horizontal coordinates and EOPs
  - ~ 1 to 2 mm/yr for rates

This precision estimate reflects approximately the demonstrated repeatability observed using global GPS networks, but is also generally representative of results obtainable from other space geodetic techniques under favorable circumstances. The accuracy estimate above is more debatable and is based on the best level of agreement seen when comparing global results from independent techniques.

If these precision and accuracy estimates are reasonable, even roughly, then it is clear that the potential accuracy gain that might be expected by long-term integration is not being realized in practice. In other words, we can conclude that the overall accuracy of the current global reference system is limited by systematic errors at roughly the 3-mm level.

On the other hand, the coverage of the reference system is truly global and its densification is progressing rapidly. Progress has been steady and impressive, as seen, for example, in successive realizations of the ITRF (see the *Annual Reports* and *Technical Notes* of the IERS). Associated with the progress in realization has been the development of cost-effective, widespread, and easier user access, principally through the use of GPS data together with the products of the International GPS Service (IGS). Generally, however, the very highest precision available to researchers is not readily available yet to most users due to the interven-

[1] James R. Ray, U.S. Naval Observatory, Earth Orientation Dept., 3450 Massachusetts Av. NW, Washington DC 20392-5420, USA, +1-202-762-1444, +1-202-762-1563, jimr@maia.usno.navy.mil

tion of the systematic errors discussed above and because of technique subtleties which are not well documented.

2. Global Reference System Applications

Historically, the demands of geophysics researchers have driven the accuracy requirements for the global reference system. Their interests have included:

- plate tectonics and surface deformations
- Earth's interior (core, mantle, rheology, ...)
- fluid flows and angular momentum transport
- gravity field and its variations

To a large extent, the geophysical needs have been substantially satisfied. In doing so, a framework has been built suitable for other applications. Prime among these are the various disciplines involved in studying the Earth's environment, including:

- global change, generally
- changes in sea level, continental water storage, ice mass...
- global warming
- monitoring natural hazards
- atmospheric temperature and water vapor profiles
- oceanography (topography, circulation, tides, ...)

A global reference system has natural relevance to classical geodesy and positioning, such as mapping and the maintenance of national or regional datums. However, because of overriding concerns related to local legal requirements or other institutional restrictions, global systems have traditionally had little impact. This is now changing rapidly, though, with the increasing demands for seamless, reliable, and safe navigation across national frontiers. Likewise, integrating and relating various geographical information system (GIS) databases is also facilitated by a unified global reference system. For the most part, these needs are being addressed using satellite-based systems, usually GPS, an inherently global methodology.

Another emerging application, likely to become much more significant in the future, is for time and frequency transfer. The positioning requirements needed to support time transfer have generally been rather modest. The International Atomic Time scale (TAI) uses techniques with estimated accuracies of a few ns (~1 m equivalent). However, new satellite-based approaches are under development which will probably improve time transfer by at least an order of magnitude. Indeed, GPS and similar systems rely equally upon both spatial and temporal elements. Were it not for the intentional degradation of GPS timing by S/A effects (which are scheduled to be discontinued by 2006), time transfer capabilities would already be much more advanced. Soon it will be possible for routine time dissemination at a level better than that available between timing laboratories today. Potential uses of such capabilities are only just being considered.

It is reasonable to expect that other applications of a high-accuracy global reference system will continue to emerge, particularly as its access and availability to general users expands. The significant point for the geodetic community is to anticipate these developments and to provide simplified tools and interfaces for non-experts.

3. Geodesy as Global Utility

The title of this symposium envisions a global geodetic *observing* system. However, it is clear from the many diverse applications being developed, which depend on an accurate and consistent global reference system, that the system is not merely used for observing. It should be much more – essentially, a global utility. In this view, the global reference system is a geodetic infrastructure, akin to the Internet's role as a global information infrastructure.

It seems safe to conclude that most current accuracy requirements are now being satisfied. Only the most demanding needs are still unmet, such as for monitoring sea level change. Technique development must there-fore continue. Moreover, new applications will most certainly continue to emerge, particularly involving time transfer and high-accuracy navigation.

The modern expansion of the realm of geodesy is due to the combination of ultra-high accuracy with far easier and cost-effective global access (typically, via GPS). There is little reason to expect any change of this trend in the near future.

4. Ingredients of Global Reference System

The practice used by the IERS in formulating its global reference system is to combine the products of all the available space geodetic techniques, striving to maximize the strengths of each while minimizing systematic errors. Besides GPS, the techniques currently used are very long baseline interferometry (VLBI), satellite laser ranging (SLR), and Doppler orbit determination and radiopositioning integrated on satellite (DORIS). Their strengths as contributors to the global reference system can be summarized as:

- VLBI ⇒ ICRF, ITRF/ICRF connection via UT1 and celestial pole, ITRF scale, long-term EOP rates
- SLR ⇒ geocenter, ITRF scale, long-term EOP rates
- GPS ⇒ ITRF densification, polar motion, "easy" global access
- DORIS ⇒ ITRF densification, nearly uniform global coverage
- all ⇒ ITRF coordinates and velocities

In principle, interferometric synthetic aperture radar (INSAR) is capable of providing very high-resolution information on ITRF deformations. Gravimetry and satellite altimetry also complement the other techniques with additional rich information. However, these are not generally being integrated in a globally systematic way yet. Eventually, GPS and DORIS should contribute also to determination of the geocenter, ITRF scale, and long-term EOP rates. In this sense, the mix of techniques is dynamic and the optimal combination will continue to evolve.

Colocations between techniques are *critical.* These are sites deploying multiple techniques sufficiently close to one another that the inter-system vector separations can be very accurately measured. They are absolutely essential for inter-technique combinations, which depend upon accurate intra-site tie vectors. Colocation also ensures that external systematic errors are common to the various independent systems. In this way, such sites are invaluable for studying sources of systematic error and assessing their magnitudes.

Among the important tasks of the geodesy community are the development of tools to perform the global reference system combinations optimally and self-consistently and to distribute the results to users via convenient interfaces. Both of these are areas of very active efforts.

5. Weaknesses of Current Reference System

Generally, the existing global reference system is well suited for a broad range of applications, as discussed above. Some problems exist, however, which should be confronted.

Technique-related systematic errors almost certainly set the current accuracy limits in most cases. That is, random observational noise is usually not the controlling error source. While the current accuracies are probably adequate for most uses, improvements are needed for certain very demanding applications. Systematic errors are often poorly characterized or quantified, which can lead to precision estimates being mistakenly used for highly optimistic accuracy expectations. Error sources which may be well known to technique experts are sometimes not well documented for the general user, which can lead to degraded or inconsistent results. Such incidents can, among other things, undermine confidence in the global reference system and its components.

Interfaces among the space geodetic techniques should be improved. In part, this is needed to focus attention and resources on studies of systematic errors. Rivalries between techniques have probably hurt more than helped in the past. In addition, improved coordination would be useful in developing better inter-technique combination methods and in promoting more and better fundamental sites with multi-technique colocations. As described already, colocation sites are critical to a robust global reference system and the current set is deficient. Only six ITRF sites exist with colocations of all four of the IERS techniques and none of these is fully satisfactory. In many cases, the necessary intra-site vectors are not well determined.

Interfaces to the user community are inadequate. The mechanisms for delivery of the global reference system components are the accumulated result of a variety of disjoint efforts by a number of different groups. They are intended for use by geodetic "experts," not general users. The interfaces need to be unified and made more user-friendly, comprehensive, and simpler. Ideally, an internationally recognized central authority should be responsible for this important activity.

The CSTG (IAG Commission VIII, Commission on International Coordination of Space Techniques for Geodesy and Geodynamics) is certainly well suited to promote investigations of systematic errors and improvements within techniques, and the IERS does provide some useful interfaces between techniques and to users. However, neither organization seems adequate to serve the full range of needs posed by a global geodetic utility, at least not as they are currently structured. The mission of each group is focused heavily on research interests and both depend basically on the goodwill and the voluntary collaborative support of scientific institutions. For the global reference system to be acceptable as a geodetic utility, part of a reliable international infrastructure, a higher profile organizational stature, with more dependable delivery mechanisms, seems essential.

6. Recommendations

Given the confluence of events and technology which have shaped the development of geodesy over the last couple of decades, the IAG is in a unique position to promote the global reference system, which it has helped to develop, as a new utility of international scope. To do this, the IAG should consider creating a new service organization spanning the space geodetic technique services (IGS, ILRS, IVS, etc). Such an umbrella service would provide the following functions:

- provide strong ties among IGS, ILRS, IVS, ...
- perform central coordination
- form combined products and provide quality assurance
- promote colocations and inter-technique links
- investigate technique-dependent systematic errors
- add a time dissemination component to provide a fully 4-dimensional reference system
- replace IERS and CSTG (or combine and reorganize them)
- provide unified user interfaces and services
- garner long-term inter-governmental support
- perhaps establish a link to the U.N. (*e.g.*, via UNESCO)

The biggest challenge among these tasks will probably involve the design and implementation of a suitable suite of user interfaces appropriate for a global utility. Depending on the extent and growth of the user base, it may be necessary to enlist commercial partners to do this job satisfactorily. In this case, other problems must be confronted. Alternatively, governmental and inter-governmental bodies might be solicited for long-term support.

The appropriate role for the IAG is to actively promote geodesy and the global reference system as a public utility while striving to ensure that interests of scientific researchers continue to be met. The IAG is uniquely situated in having the stature, scope, and expertise necessary to achieve these somewhat conflicting goals.

Space Techniques and their Coordination within IAG at Present and in Future

G. BEUTLER[1], H. DREWES[2], CH. REIGBER[3], R. RUMMEL[4]

Abstract

In this article we first analyze the performance of international institutions playing a major role in *Space Geodesy*, where we will primarily focus on those contained in *it Section II* of the *International Association of Geodesy (IAG)*. The time period considered is that of the years 1995 to 1999, between the XXIst and the XXIInd General Assemblies of the *International Association of Geodesy and Geophysics (IUGG)*.

Based on this analysis we will come up with some recommendations concerning the future of organizations in Space Geodesy. These recommendations are meant to improve the structure of the mentioned organizations and strengthen the collaboration between different branches with the goal to eliminate weaknesses of individual Space Techniques. Also they should give an indication where, according to the authors' opinion the emphasis in research should lie in the near future.

Introduction

General Considerations

The time period between the two *IUGG* General Assemblies (1995 in Boulder, USA and 1999 in Birmingham, UK) is a fascinating period in Space Geodesy. In this article we review what was achieved so far and we try to find out where the emphasis should lie in the near future, in particular in the next *IAG/IUGG* "legislature period" between 1999 and 2003.

On the organizational level we have seen the remarkable development of the *IGS*, successful reorganizations of the *SLR/LLR* and of the *VLBI* communities which were accompanied by various attempts to connect different space geodetic techniques – on the observation and on the analysis level.

The "combination task" proves to be a more difficult one. The attempt to create a *FRCN* (Fundamental Reference and Calibration Network) was abandoned. The creation of the *ISGN* (International Space Geodetic Network) which is based on the existing technique-specific networks, is underway. Attempts to combine space geodetic techniques led to the creation of the SINEX format (Software INdependent EXchange format (NEILAN et al., 1996)) which theoretically allows combination of results from different techniques – actual applications are still related to one technique, however.

The *GPS* (Global Positioning System) is well known, and for high accuracy applications the *IGS* plays a key role. Other satellite microwave techniques exist and give excellent results *(e.g.*, for LEO orbit determination). The French *DORIS* and the German *PRARE* system are good examples. Still another example is the Russian *GLONASS* (Global Navigation Satellite System) which may be very well combined with the *GPS*. This aspect is particularly attractive for the *GPS* community and it is in the focus of *IGEX-98*, the International GLONASS Experiment, sponsored by all major organizations of Space Geodesy and Navigation.

In the time period 1995 – 1999 it became clear that space missions focussing on geodetic and neighbouring disciplines are becoming increasingly important. TOPEX/Poseidon ((FU et al., 1994) and (BERTINGER et al., 1994)) and GPS/MET (YUNCK et al., 1996)(to give only two examples) gave impressive demonstrations of the potential of such missions for Earth Sciences. The upcoming series of gravity (and atmosphere related) missions like CHAMP, GRACE, and GOCE, which include spaceborne GPS will be of greatest importance for Space Geodesy in the future.

Synthetic Aperture Radar *(SAR)* has become a highly successful remote sensing technique capable of monitoring geophysical parameters related to the structural properties of the Earth's surface. The interferometric *SAR* from two neighbouring space positions *(INSAR)* allows to extract precise height information. Using successive interferograms in a differential mode *(D-INSAR)*, the detection of small surface variations in the *cm*-level *(e.g.*, caused by geodynamic deformation) has been demonstrated. Because of its independence on solar illumination and penetration of clouds it is a promising technique for the future, both in large scale topographic mapping and for monitoring crustal deformations.

[1] Gerhard Beutler: Astronomisches Institut, Universität Bern, Sidlerstrasse 5, CH - 3012 Bern, Switzerland; Tel. +41 - 31 - 631 85 91, Fax +41 - 31 - 631 38 69, E-mail: beutler@aiub.unibe.ch

[2] Hermann Drewes: Deutsches Geodätisches Forschungsinstitut, Marstallplatz 8, D - 80539 München, Germany; Tel.: +49 - 89 - 23 031 106, Fax: +49 - 89 - 23 031 240, E-mail: drewes@dgfi.badw-muenchen.de

[3] Christoph Reigber: GeoForschungsZentrum Potsdam, Kinematik u. Dynamik der Erde, Telegraphenberg A 17, D - 14473 Potsdam, Germany; Fax: +49-331-288-11 11, Tel.:+49-331-288 - 1100, - 1101; E-Mail: reigber@gfz-potsdam.de

[4] Reinhard Rummel: Institut für Astronomische und Physikalische Geodäsie, Technische Universität München, Arcisstraße 21, D-80290 München, Germany; Tel.:+49 - 89 - 289 - 2 3189/ Sekr. 2 31 90, Fax: +49 - 89 - 289 - 2 31 78, e-mail: rummel@step.iapg.verm.tu-muenchen.de

Whereas the authors of this article are rather optimistic that "classical Space Geodesy" using ground-based observatories will be adequately dealt with in future by the existing *IAG* organizations related to Space Geodesy *(CSTG, IERS, IGS)*, they are convinced that a considerable effort is required to make the spaceborne applications accessible to and accepted by the broader space geodetic community. The article tries to give first indications for activities related to Space Geodesy for the next *IAG/IUGG* time period 1999 - 2003.

Specific Goals of this Analysis

The 1998 Symposium of *IAG* Section II (October 1998 in Munich) is entitled *Towards an Integrated Global Geodetic Observing System*. Four topics, namely *Reference System and Datum Integration, Strength and Weaknesses of Space Techniques, Upcoming Gravity Field Satellite Missions*, and *Implication of an Integrated Observing System in Earth Sciences* are addressed.

The present article is meant to introduce topic *No. 2*, "Strengths and Weaknesses of Space Techniques". The session title is inspired by the observation that (a) many, if not most, papers, reports, *etc.*, dealing with Space Geodesy are technique-related *resp.* technique-specific, and that (b) the principal goal of such, in general very valuable, contributions (naturally) is to demonstrate the outstanding performance of a particular technique. When reading such contributions an "outsider" may easily get the impression that other techniques are not as important - although it is usually not the intention of the authors of such contributions to draw this conclusion.

There are exceptions to this observation: the *IERS* (International Earth Rotation Service), *e.g.*, understood itself since its creation as a multi-technique service; and the currently responsible team of *CSTG* (Commission VIII of *IAG* and Commission B.2 of COSPAR on the "International Coordination of Space Techniques for Geodesy and Geodynamics") tries to put the emphasis on the combination of space geodetic observations and analysis of all techniques. The topic is a difficult one and will be addressed in the next section of this paper (subsection on *FRCN and ISGN*).

In geodesy, in particular in Space Geodesy, service oriented work seems to be rather successful. *IERS* and *IGS* may be viewed as good examples. A special session at the *IAG* Scientific Assembly in 1997 in Rio de Janeiro revealed that there are, as a matter of fact, many more than just the two examples mentioned within *IAG*. We refer to the booklet (MUELLER, 1998a) for more information. In view of the success of the services in geodesy there is the temptation to organize more and more *IAG* activities in a service-like manner.

When speaking of a service, one has to ask the questions (a) service for whom?, and (b) what are the products of the service? The services considered here are scientific services, services performed by scientists for scientists - and a broader user community. In most cases the community of *geodesy* and *geodynamics* is "the principal customer" (remarkable interdisciplinary developments will be mentioned in the next section). Serving the scientific community is an ambitious goal: Scientists usually have clear views of their needs and they have the tendency to complain when these needs are not properly addressed.

Be this as it may, the International Association of Geodesy *(IAG)* is considering to reorganize its general structure right now, and, not surprisingly, service-like structures are seriously considered in this process, as well. It is not by accident that the Intersection Symposium G6 *Geodesy beyond 2000 - the Challenges of the First Decade* announced for *IUGG99*, the XXII General Assembly of the International Union of Geodesy and Geodynamics *(IUGG)* is among other topics considering "a structure for *IAG* that can meet the future scientific challenges". Our contribution obviously addresses similar problems on the lower level of Section II of *IAG*. Maybe that the experiences gained in the reorganization of major branches in Space Geodesy will be of interest for *IAG* and for the symposium mentioned.

Three quarters of the time period 1995 - 1999 have elapsed and the moment seems appropriate to review *IAG*-related Space Geodetic organizations, to summarize what was (not) achieved, and to propose what should or might be done in the near future.

Making plans for the future is fashionable at all times, in particular (?) in times of reduced funding. If this observation is correct, it is certainly appropriate to come up with plans for Space Geodesy right now.

Our contribution to *IGGOS* (Integrated Global Geodetic Observing System) is "only" considering structures and organizations set up for the benefit of Space Geodesy. We believe that this "narrow view" is justified because all other contributors to this symposium will strictly address science. Our approach is driven by the wish to stimulate the discussion for a firm program for important areas in Space Geodesy for the next *IAG/IUGG* period 1999 - 2003. Our considerations and conclusions are based on previous plans and progress reports, namely on (BEUTLER and DREWES, 1996, 1997), (REIGBER and FEISSEL, 1997), and (RUMMEL, 1998).

In the next section we briefly review space geodetic activities within *IAG* Section II, in particular within *CSTG* and *IGS*. We also include the *IERS* in our overview, although this service is attached to Section V (Geodynamics) of *IAG*. We wanted to include all major *IAG*-related organizations in Space Geodesy to make an attempt of coming up with reasonable proposals for a meaningful development (the first decade in the next millennium ?). Such plans and the rules proposed for their realization will be presented in the last section of our article.

Developments in Space Geodesy in the Period 1995 - 1999

Space techniques and space geodetic applications are "well distributed" over the *IAG* structure. This distribution should not be viewed as result of *IAG* or even higher wisdom, but rather as a result of a historical development. Let us try to summarize those (major) parts of *IAG* which are clearly related to Space Geodesy today:

- **Commission X** on Global and Regional Geodetic Networks of **Section I** is nowadays completely based on space geodetic networks and methods. There are also obvious links to the realization of regional and global reference frames (let us mention the *EUREF* subcommission and the *SIRGAS Project* where this aspect is of primary interest. Reference frames, by the way, are declared as a primary topic of Section V of *IAG*,
- **Section II** on Advanced Space Technology is (of course) heavily involved in Space Geodesy. Theoretically this involvement is technique-related. This is true for Commission VIII *(CSTG)*, dealing with space technologies *per se*. Special Commission 6 *(Wegener)* addresses questions related to regional geodynamics from a scientific point of view, Special Commission 7 addresses Satellite Gravity Gradiometry; Special Study groups deal with *SAR*, spaceborne *GPS*, and orbit determination. The *IGS* is formally attached to Section II although it is a cornerstone for the actual realization of the terrestrial reference frame(s), for Earth rotation, *etc.* – topics in the center of interest of Section V.
- **Section V** (Geodynamics) heavily relies on Space Geodetic Techniques, today. The topics addressed include of course many items of theoretical nature. Of particular interest to *Space Geodesy* is the *IERS* which uniquely relies on Space Geodetic Techniques.

Key components for Space Geodesy within *IAG* are *CSTG* and the two services *IERS* and *IGS*. This section first gives a report on the important developments within these three *IAG* units in the time period 1995 – 1998. The section concludes with the discussion of a proposal developed by the President of *IAG* Section II (RUMMEL, 1998).

CSTG, IGS, and IERS

Let us start by briefly characterizing the three mentioned organizations and by adding a few general comments concerning their mission and structure:

- **CSTG:** The Commission on International Coordination of Space Techniques for Geodesy and Geodynamics is responsible to develop links between groups engaged in Space Geodesy and Geodynamics, coordinate their work, elaborate projects implying international cooperation, ... (from (RUMMEL, 1996)).
- **IERS:** The International Earth Rotation Service is responsible for defining and maintaining the conventional terrestrial and celestial reference frames, and for determining the transformation parameters between them (from (REIGBER and FEISSEL, 1997, p. 3 and p. 11)).
- **IGS:** The International GPS Service for Geodynamics is responsible for making available GPS data from its global network, producing and disseminating high accuracy GPS orbits, Earth rotation parameters, station coordinates, atmospheric information, *etc.* (from IGS Terms of Reference, (IGS Central Bureau, 1998)).

The missions of the two services are crystal-clear, whereas the definition of *CSTG* allows – theoretically – for many different interpretations. It is therefore clear that the actual work performed within the Commission may change considerably from one *IAG* four-year period to the next (this could very well be observed in the past). From the *IAG* point of view *CSTG* is neither service-oriented nor should it be *research-oriented* (in the sense of a Special Study Group), it should stimulate research-oriented projects requiring international coordination and collaboration, however.

Let us mention that *reorganization* and *structure changes* were addressed by all organizations mentioned (but not in a coordinated way). This may be coincidence, but it may also indicate that there is a need for reorganization in Space Geodesy. Such phases may be stimulating (comparable to policy re-definitions after a political election) provided that they are not too frequent. We hope that the new structures will prove their usefulness for the next one or two *IAG periods*. The reorganization and restructuring process will have to be brought on the next higher level now, where, *e.g.*, the *IAG* section structure is considered.

CSTG. The commission was formally established in 1979. The history of its creation and activities till 1995 is documented in (MUELLER, 1996). The general structure of *CSTG* is, *e.g.*, given in (RUMMEL, 1996) or in (BEUTLER and DREWES, 1996). *CSTG* is associated with *IAG* and *COSPAR* as Commission VIII of *IAG* and Subcommission B.2 of *COSPAR*. Whereas *CSTG* is well established within *IAG* it is at present not well established within *COSPAR*. A first promising attempt to improve the situation was made by asking *Dr. John Dow* to join the *CSTG* Executive Committee as *CSTG Liaison Officer*. He is chairman of the *COSPAR* Panel on Satellite Dynamics *(PSD)*.

The second attempt to improve relationships consisted of "familiarizing" *COSPAR* (again) with geodesy and geodynamics (a) in general by a one-hour interdisciplinary lecture held at the 32nd Scientific Assembly of *COSPAR* in Nagoya Japan (July, 1998) (BEUTLER et al., 1998) and (b) in particular by organizing a common *PSD/CSTG* symposium at the same assembly in Japan.

Claude Boucher as the chairman of the *IERS Terrestrial Frame Section* was invited to join the Executive Committee in 1997 on request of the *IERS* Directing Board. This should strengthen the relations between *IERS* and *CSTG*. The step is to be viewed as an *ad hoc* measure. The *CSTG* Executive Committee should decide whether or not to make this an official practice in future (after 1999).

ILRS: International Laser Ranging Service: The *SLR* and *LLR* subcommissions, "old established" subcommissions of *CSTG*, were combined into one *CSTG* subcommission on Satellite and Lunar Laser Ranging at the XXIst General Assembly of *IUGG* in Boulder (July 1995) with Dr. John Degnan from Goddard Space Flight Center as subcommission chair. Ideas to streamline Laser Ranging and to move towards a modified subcommission structure comparable to the *IGS* were first discussed at the 1996 Porto Meeting of the *SLR/LLR* subcommission. A concrete proposal by the *CSTG* President was then brought forward to the

SLR/LLR steering committee at its meeting attached to the 10th International Workshop on Laser Ranging Instrumentation in Shanghai (November 1996). The proposal was to investigate the feasibility of an *ILRS*, modeled in part after the *IGS*. Terms of Reference for the new service and the charters for four working groups were drafted by members of the steering committee, presented to and accepted by the *CSTG* Executive Committee in May 1997, then presented, discussed and modified by the *SLR/LLR* community in Maratea, Italy (June 1997). After a "final" steering committee meeting in London, UK (September 1997) a joint *CSTG/IERS* Call for Participation could be sent out through the *SLR* Electronic Mail and through the *IERS* on Friday, February 13, 1998 (!).

The purpose of the new service is very well characterized in the Call for Participation (DEGNAN, 1998). The response to this joint *CSTG/IERS* call for was most encouraging. The proposals were reviewed and (with few exceptions) accepted at an *SLR/LLR* steering committee meeting on Saturday, 18 April, 1998 in Nice. On Sunday, April 19 the *CSTG* Executive Committee and afterwards the *IERS* Directing Board formally approved the establishment of the new service. This procedure of establishing the *ILRS* is a nice example for a fruitful collaboration of *CSTG* and *IERS*. Dr. John Degnan, chairman of the *SLR/LLR* subcommission in the crucial time period, was to a great extent responsible for the efficiency of the process. The space geodetic community owes him many thanks. For more information concerning the early phase of the *ILRS* development we refer to (DEGNAN, 1997)

Epilogue: On September 25, 1998 the ILRS Governing Board held its first meeting in Deggendorf (attached to the 11th Workshop on Laser Ranging) and elected ... as its first chairman. We congratulate the Laser community for the efficient creation of this new service and wish the *ILRS* and its chairman all the best for the future.

IVS: International VLBI Service for Geodesy and Astrometry. The activities eventually leading to the *IVS* were probably initiated at the 1997 *CSTG* Executive Committee Meeting on May 29 in Baltimore. Tom Clark, chairman of the VLBI subcommission at that time, presented the *CORE/IVN* project "Geodetic VLBI for the next Millennium" at that meeting, where he could tell the community that the program, aiming at daily *VLBI* observations of a subset of the *VLBI* network (with a clever schedule to include the entire network), got funded by *NASA*. (CLARK et al., 1997). The discussion then focussed on the question whether the *CSTG VLBI* subcommission should strive for a more service oriented structure. There were strong pro and contra arguments concerning such a development. No decisions were taken at that meeting.

Later on, however, the *VLBI* subcommission decided to move in a similar direction as the *SLR/LLR* subcommission and the *IGS*. A first draft for the Terms of Reference was released in September 1997. In early 1998 an electronic workshop discussed the draft by e-mail and came up with a definite proposal for the Terms of Reference for the *IVS*. They address the objectives, data products, components, and coordination. The principal components of the new organization are the Network of Stations, Operation Centers, Correlators, Analysis Centers, Data Centers, Technology Development Centers, Central Bureau, and Directing Board. The coordination shall be done by Technology, Network, and Analysis Coordinators.

The *VLBI* proposals for the development of the *IVS* were presented at the *CSTG* Executive Committee Meeting in Nice (April 19, 1998) and they were fully endorsed. The developments were subsequently approved by the *IERS* Directing Board (actually on the same day in the afternoon). Meanwhile the joint CSTG/IERS *Call for Participation* for the International *VLBI* Service for Geodesy and Astrometry (*IVS*) went out on June 17, 1998.

Epilogue: the *IVS* Call for Participation has 20 September 1998 as a deadline. It is expected that the proposals will be evaluated at Munich *IAG* Section II Symposium (October 4-9, 1998 in Munich) and that the *IVS* will become fully operational before the 1999 IUGG General Assembly. Again we observe an excellent cooperation between *CSTG* and *IERS*. Let us not forget, however, that the principal work had to be done in the *VLBI* subcommission of *CSTG*. We are in particular indebted to Dr. Tom A. Clark (chairman of the subcommission till 18 April, 1998), Prof. James Campbell, the new chairman, and to Dr. Nancy Vandenberg who has many responsibilities in the development of this new scientific entity.

From FRCN (Fundamental Reference and Calibration Network) to ISGN (International Space Geodetic Network). A lot of work was invested in the establishment of a global space geodetic network combining (in one way or another) many (if possible all) modern observation methods at all its sites (MUELLER et al., 1996).

At the *IAG* Executive Committee meeting of November 10, 1995 the decision was taken to set up an IAG/CSTG working group with the goals to

- come up with a scientific and logistic rationale for an *FRCN*,
- present a report to the *IAG* Executive Committee at the 1997 Spring Meeting.

The procedures how to set up the working group are summarized in (BEUTLER, MUELLER, and SCHWARZ, 1996). The working group consisted of fourteen members and two permanent guests. The FRCN had several meetings and a workshop in 1996 and early 1997; it eventually came up with its final report which was presented to the *IAG* Executive Committee on 28 April, 1997. The full report is reproduced in *CSTG* Bulletin No. 13 (DREWES and BEUTLER, 1997).

The final conclusion and recommendation of the working group was *that there are no compelling scientific reasons to identify stations as FRCN stations. It was argued that an International Space Geodetic Network (ISGN) as a combination of all space geodetic networks already exists and that this network is well suited to take over most of the tasks that were considered for the FRCN.*

Based on this conclusion the following recommendations were made:

- The *ISGN* shall be formally established.
- The *ISGN* shall be an open network. [...]
- Criteria have to be defined to include a station into the *ISGN*. [...]
- *CSTG* shall take care of the final definition of the *ISGN*.
- *IERS* will make routine use of the *ISGN* data and will promote the network maintenance and optimization in cooperation with and in support of *CSTG*

The report, in particular its conclusions, were accepted by the *IAG* Executive Committee at its meeting on April 28, 1997 in Potsdam.

The next step of establishing *ISGN* was made at the joint meeting of *CSTG, IERS* and *IGS* on September 5, 1997 in Rio de Janeiro (attached to 1997 *IAG* Scientific Meeting). The decision was taken to set up a small working group with Hermann Drewes as chairman, Claude Boucher (*IERS/ITRF* specialist), Ivan Mueller (*FRCN* initiator), John Bosworth (*CSTG* sites specialist), and Tom Herring (*CSTG* analysis specialist) as additional members. The following goals were specified:

- Define the criteria for including a station into the ISGN) (sites criteria, analysis criteria, IERS requirements).
- Install a better cooperation between individual techniques.

Draft versions for these criteria were defined and presented to the *IERS* Directing Board at its meeting on April 19, 1998. The next step will consist of compiling the *ISGN* station list and to ask the stations for cooperation.

International GLONASS Experiment (IGEX-98): The *CSTG* Subcommission on Precise Satellite Microwave Systems always was very much interested in the Russian *GLONASS* (Global Navigation Satellite System), the Russian counterpart of the *GPS*. At the joint meeting of *CSTG, IERS*, and *IGS* on September 5, 1997 in Rio de Janeiro it was decided to plan a GLONASS test campaign to prove the usefulness of the system as an addition to GPS. The applicability for navigation, time and frequency transfer, and for precise geodetic positioning should be tested. It was agreed that organizing such a campaign would be a typical project for *CSTG*. It was also clear, on the other hand, that *IGS* participation and infrastructure would be crucial for the success of the experiment.

A small steering committee consisting of Pascal Willis (chair), Gerhard Beutler, Werner Gurtner, Günter Hein, Ruth Neilan, and Jim Slater was formed and started immediately with the preparations of the first global *GLONASS* tracking campaign - tentatively scheduled for the last three months in 1998. It is a lucky circumstance that *ION*, the Institute of Navigation, has a strong interest in the project. The draft plan to organize a *GLONASS* test campaign was presented at the 1997 *ION GPS-97*, the 10th International Technical Meeting of the Satellite Division of the Institute of Navigation in Kansas City (September 1997).

The next step of the steering committee consisted of writing a Call for Participation for *IGEX-98*. After a review performed separately by *CSTG, IGS*, and *ION* the official Call for Participation was sent out in February 1998. The proposals were received, reviewed and the participants notified of the acceptance at the *resp.* after a steering committee meeting on June 29, 1998 in Paris. For more information we refer to (WILLIS et al., 1998)

IGS. The *IGS* became an official *IAG* service on January 1, 1994. It was recognized as an official *FAGS* service (Federation of Astronomical and Geophysical Data Analysis Services) in 1996. The *IGS* is based on a network of about 200 stations, which are not evenly distributed over the globe, but only very few real problem areas remain today (Africa probably being the most important example).

According to the Terms of Reference *the primary objective of the IGS is to provide a service to support, through GPS data products, geodetic and geophysical research activities. [...] IGS collects, archives, and distributes GPS observation data sets of sufficient accuracy to satisfy the objectives of a wide range of applications and experimentation. These datasets are used by the IGS to generate the following data products:*

- high accuracy GPS satellite ephemeries,
- Earth rotation parameters,
- GPS satellite and tracking station clock information,
- ionospheric information,
- tropospheric information.

We should distinguish three sets of products:

- rapid products (and predictions) available 23 hours after the observations
- final products (available after about 10 days)
- products that are made available less frequently (*e.g.*, sets of coordinates and velocities).

The rapid products, in particular the daily values for polar motion and length of day are cornerstones for the *IERS* rapid products and predictions. For more information concerning the "basic characteristics" and the development of the *IGS* we refer, *e.g.*, to (MUELLER, 1992), (BEUTLER et al. 1996) and (NEILAN et al., 1997). Based on this characterization we may state that the *IGS* is

- based on a single technique,
- using satellite geodetic techniques (like *SLR*),
- the basis for all high accuracy work using the *GPS*,
- the ideal, low cost collocating tool for all global (and regional) (Space) geodetic and geophysical networks,
- exploiting its network in every respect (not only in the geodetic domain).

Let us address a few of these general characteristics: The *IGS* is based on one satellite system, the Global Positioning System *GPS*. We should keep in mind that the *GPS* was primarily designed for navigation and that it has, as every navigation system, a limited lifetime. The *GPS* is expected to be available for at least the next thirty years - which still is a short time period compared to the longer periods we are interested in global geodynamics. *VLBI* and *SLR* have no such limitations. This latter statement indicates the importance of these techniques to maintain long term stability and also for calibration. We conclude that state of the art receivers must be deployed at all *SLR, VLBI, etc.*, sites and included into routine *IGS* processing.

Every technique - even one that seems perfect - is exposed to systematical error sources. The problems of the *GPS* lie exactly in those fields where so called "absolute information" is involved (although much progress was made in this field). A good example is the estimation of the geocenter. One often notices considerable changes in "accurate" absolute results when processing strategies or orbit models are modified. This demonstrates that *GPS* gives (for such absolute information) precision and consistency, but not accuracy.

On the positive side we point out that *IGS*, thanks to the accuracy of its products, provides the ideal tools for all regional and local geodetic work, and that it is very well suited to connect other space geodetic networks. This is probably why no compelling reasons for the creation of an *FRCN* could be found: *GPS* is capable of collocating (geometrically) the *SLR, VLBI, DORIS, PRARE, gravimeter, etc.* networks with an accuracy sufficient for most problems addressed today.

It is remarkable that recently the *IGS* development went into the direction of a truly interdisciplinary service. This is, *e.g.*, reflected by the change of the Terms of Reference in December 1995, where (among other changes) atmosphere issues were more explicitly addressed - as a result of work performed by people "at home" in the areas of geodesy and meteorology (see, *e.g.*, (BEVIS et al., 1992)). We also have seen the formal creation of the *Troposphere Combination Center* at *GFZ* (GeoForschungsZentrum) in 1997 and the *IGS* Ionosphere Working Group in 1998. These developments will lead to a full exploitation of the *IGS* network for atmosphere research.

Another interdisciplinary development which is more closely related to the original *IGS* charter is the *IGS/BIPM* pilot project aiming at the exploitation of the *IGS* network for time and frequency transfer. Still another unexpected development was the formal proof that the *IGS* network can be used to determine short-period nutation terms (ROTHACHER et al., 1998).

It is amazing that this service, starting as a satellite geodetic service for geodesy and geodynamics in 1992, is an interdisciplinary service today.

Let us also mention that the *IGS* archives of the *GPS* observations of its entire global network since (about) 1992 contains material of greatest value for Earth sciences. When/if this material will be reprocessed one day using the "latest and greatest" software tools there can be no doubt that invaluable high accuracy, high resolution time series of station coordinates, Earth rotation parameters, *etc.* will result - certainly best high resolution data for the years 1993 - 2000.

Our summary shows that the *IGS* still is in full development. Many breath-taking developments were only briefly mentioned, others, like all the spaceborne applications of the *GPS*, will impact the service in the near future. The upcoming gravity missions *CHAMP, GRACE, GOCE* are, *e.g.*, all relying in one way or another on spaceborne *GPS* receivers. For more information we refer to (BEUTLER et al., 1998).

The *IGS* still is a young service. Nevertheless, and despite its impact on Earth Sciences, there are problem areas and questions related to the future direction which were discussed again and again. This was the main reason to organize the first IGS Retreat in December 1998. The main result of this retreat are action items and recommendations (MUELLER, 1998b) which in turn will result in a major revision of the terms of reference by the end of 1998. The workshop revealed that the *IGS* wants to move into the direction of interdisciplinarity. This is, *e.g.*, reflected by the facts that in future *IGS* will stand for "International GPS Service" and no longer for "International GPS Service for Geodynamics" and that spaceborne applications of *GPS* will be an important issue within the *IGS*.

IERS. Let us remind ourselves that the *IERS* is a service in Section V of *IAG* and that it was created by *IUGG* and *IAU*. It plays, however, a very important role in Space Geodesy. It is therefore necessary and appropriate to briefly review its recent achievements here. As opposed to the *IGS* the *IERS* is based on many techniques, as a matter of fact on all techniques having demonstrated their capability to contribute to the terrestrial and/or the celestial reference frame with state of the art accuracy. The *IGS* is thus one of the services *resp.* techniques contributing to the *IERS*. The *IERS* plans to rely on coordinating centers to produce technique-specific results (*e.g.*, sets of coordinates, velocities, Earth rotation parameters, *etc.*). One key activity of the *IERS* then consists of comparing and combining technique specific time series. It is exactly on this level that one expects to detect biases and/or discrepancies between different series.

We quote parts of the foreword of (REIGBER and FEISSEL, 1997) to further characterize the development and mission of the *IERS*:

The IERS was created in 1987 by the IAU and the IUGG, replacing the Bureau International de l'Heure (BIH) and the Polar Motion Service (IPMS), as a result of the ten-year international MERIT Campaign. The IERS started its activities in January 1988.

IERS was given the missions to define and maintain a terrestrial reference system based on the most precise space geodesy techniques at that time, to define and maintain a celestial reference system based on the directions of extra galactic radio sources and its tie to other celestial reference frames, and to monitor the Earth's rotation, i.e., the parameters that connect the terrestrial and the celestial reference systems as a function of time.

When operations began, the observational techniques available for these tasks were VLBI, LLR, and SLR. The

Directing Board included representatives of the Coordinating Centers for these observing techniques, of the Central Bureau, and of IUGG, IAU and FAGS.

[...] The main decisions of the Directing Board in the first eight years of activity were the setting up of permanent monitoring of global atmospheric dynamical parameters in 1989, the adoption of the GPS as an additional observation technique in 1989-93 and of DORIS (Doppler Orbitography by Radio positioning Integrated on Satellite) in 1994-96, and the redefinition of VLBI contributions in 1995-96.

In 1995, the IERS had achieved its primary missions, by providing globally distributed terrestrial and celestial references and the Earth's orientation with an accuracy equivalent to about one centimeter on the surface or the Earth.

This latter proud statement underlines that the initial phase of the *IERS* was extremely successful. The Directing Board found this moment appropriate to think about the future direction of the service. This was why the 1996 *IERS* Workshop was dedicated to a thorough review of the *IERS* mission and to possible extensions. This workshop actually produced a number of recommendations and action items, which are documented in (REIGBER and FEISSEL, 1997). Let us again quote from the foreword of this report to summarize the key activities emerging from the 1996 *IERS* Workshop (we are only reproducing a selection of the activities of general interest in abbreviated form. The full list is given in (REIGBER and FEISSEL, 1997)):

- Works concerning the International Celestial Reference Frame *ICRF* and its tie to the HIPPARCOS catalogue,
- monitoring global geophysical fluids,
- monitoring the motion of the Earth's center of gravity,
- enhancing the role of coordinators and extending their field of responsibility to general themes of importance to *IERS*,
- developing or improving technical abilities for the global combination of products from analyses of the various observing techniques, and fostering internal competition within *IERS* on this aspect
- ...

The implementation of many of these recommendations may be followed in the *IERS* Gazette, an *IERS* e-mail publication series which was started on April 5, 1996 and which allows it easily to track the essential *IERS* events since that time.

Gazette *No. 10* (2 Jan 1997) issued a Call for Participation (*CfP*) for the *IERS* Geocenter Motion Campaign. This *CfP* has to be seen as the concrete reaction to the third of the above recommended activities. The *CfP* was sent out by Jim Ray. The planned activities should cover essentially the entire year 1997. The results were reviewed at the 1997 *AGU* (American Geophysical Union) Fall meeting in San Francisco. Gazette *No. 26 (12 Jan 1998)* summarized the results: *The IERS Working Group on the ITRF datum met during the Fall AGU Meeting last month in San Francisco and considered results presented during the special geodesy sessions on "Investigations of Geocenter Motions" as well as other results from the Geocenter Campaign.*

The overall impression of the work presented thus far is that the net motion of the terrestrial reference frame relative to the Earth's center of mass is detectable but small, probably no more than about 1 cm in any component. The diurnal and semidiurnal tidal variations appear to be well determined and in good agreement with modern ocean tidal models. There seems to be some general agreement among the techniques in detecting seasonal variations, although more work remains to be done in this area. Geophysical computations of the expected motions based on global fluid motions are only roughly consistent with the observations.

[...]. It is expected that appropriate changes will be made in the next edition of the "IERS Conventions". [...].

In Gazette *No. 14* (21 Feb 1997) the *IERS* Directing Board launched the *CfP* for the Coordinating Center "Monitoring Global Geophysical Fluids". This *CfP* was successful, the result is reflected in Gazette *No. 22*: *GSFC*, Goddard Space Flight Center, was given the responsibility as Coordinating Center, Special Bureaus were created for the Atmosphere, the Ocean, Ocean tides, Continental Hydrology, Mantle Fluid Variations, the Core, and Geocenter plus Gravitational field. These special Bureaus are located mostly in the US, one (for the Core) in Europe (Belgium and France).

An interesting, typical *IERS* activity was launched in Gazette *No. 21* (30 Sep 1997) by David Salstein. He asked for investigations of interactions between El Niño and Southern Oscillation on one hand and Earth rotation variations on the other hand. Such topics can only be addressed thanks to the long *IERS* time series of Atmospheric Angular Momentum *AAM* and high resolution Earth Rotation Parameters (*ERP*) data. The correlations between such series indeed are striking and it can be stated that if events like El Niño may be detected in one of the series they must be visible in the other one, as well.

In Gazette *No. 25* (23 Dec 1997) the *IERS* Central Bureau and the *VLBI* Coordinator could announce that at its 23rd General Assembly in August 1997 in Kyoto, Japan the *IAU* (International Astronomical Union decided that as of January 1, 1998, the *IAU* celestial reference system shall be the International Celestial Reference System (*ICRS*) [...] and that the tasks resulting from the maintenance of the corresponding reference frame and its tie to other frames are shared between the *IERS VLBI* Coordinating Center and the *IERS* Central Bureau.

Gazettes *Nos. 28, 30* and *33* contain the *CFPs* for the *ILRS*, the *IGEX-98*, and the *IVS* and underline the good agreement between *IERS* and *CSTG*. These *CfPs* for *ILRS* and *IVS* certainly "enhance the role of the *IERS* coordinators". We are thus probably left with *developing or improving technical abilities for the global combination of products from analyses of the various observing techniques, and fostering internal competition within IERS on this aspect* as "pièce de résistance" for the near future.

Epilogue: EVENTS OF THE IERS RETREAT ... (to be addressed AFTER the retreat).

Global Integrated Geodetic and Geodynamic Observing System (GIGGOS)

So far, we tried to give an overview of what was achieved within *IAG* in Space Geodesy in recent years. We reported on activities of major components like *CSTG*, *IGS*, and *IERS*. These components periodically review their plans, discuss future directions, try to get coordinated (where necessary), and "somehow" implicitly or explicitly find their way into the future. When going through the above reports it becomes clear that an amazing amount of work was performed in this "bottom up" design of Space Geodesy.

There is of course a different approach, which might be called "top down", where one tries to come up with a grand design, in our case this might be a grand design of Space Geodesy, or even of geodesy "per se" and to confront such a design with the actual development which went on in a particular field.

Both approaches, "top down" and "bottom up", are necessary to develop sound plans for the future. In a certain sense, the *retreats* as performed, *e.g.*, by *IERS* and *IGS* in the time period considered may be seen as analyses of the first kind - admittedly after a long phase of "bottom up" activities.

(RUMMEL, 1998) adopts the "top down" approach to analyze the state of Space Geodesy (or even geodesy in general) and to make recommendations for the future. The recommendations aim at a very high level, the *IAG* or even the *IUGG* level – but there are of course interactions between this highest level and the work which is performed within the *IAG* sections, commissions, and study groups.

Let us briefly summarize the essential elements in the analysis (RUMMEL, 1998) and conclude with a few remarks. The complete report is included in these proceedings.

The *Global Integrated Geodetic and Geodynamic Observing System (GIGGOS)* is considered to be geodesy's *resp.* *IAG*'s contribution to large international science programs like, *e.g.*, the *WCRP* (World Climate Research Program), the *IGBP* (International Geosphere Biosphere Program). The *IAG* contribution should provide the geodetic component to Earth system research, very much like what *GOOS* (Global Ocean Observing System) does for the oceanographic part. In these programs the emphasis obviously lies on climate change and environment. (RUMMEL, 1998) states that programs leave to a large extent open how solid Earth and ice processes as well as their interaction with ocean and atmosphere are to be quantified and modeled, and that geodesy is in a position to fill this gap by offering a very concrete and central element, namely the *GIGGOS*, the central objectives of which are:

- To provide in a well defined and reproducible *global terrestrial frame*,
- the integral effect on *Earth rotation* of all angular momentum exchange inside the Earth, between land, ice, hydrosphere and atmosphere, and between Sun, Moon, and planets,
- the *geometrical shape* of the Earth's surface (solid Earth, ice, and oceans), globally or regionally, *and* its *temporal variations*, whether they are horizontal or vertical, secular, periodical or sudden, and
- by adding the *Earth's gravity field* – stationary and time variable – mass balance, fluxes, and circulation.

The above four elements of *GIGGOS* are briefly referred to as *(1) frame*, *(2) geokinematics*, *(3) Earth rotation*, and *(4) gravity field*. The aimed at precision of the observing system must be of the order of about 10^{-9}, provide a high spatial and temporal resolution, consistency and stability over decades.

(RUMMEL, 1998) wants the *GIGGOS* to play a strategic role for the *IAG* in two respects:

- *GIGGOS* should be seen as the contribution of geodesy to the large scale international programs mentioned above. This contribution should be recognized by all associations within *IUGG* or *ICSU*.
- *GIGGOS* could serve as a common and very challenging focal point for "practically all"(?) current research activities inside IAG.

For more information the reader is referred to (RUMMEL, 1998).

Many, if not most, of the elements of the *GIGGOS* already exist and are dealt with in a very satisfactory way, by the *IAG* entities, namely *IERS*, *IGS*, *CSTG*. We mention the celestial and terrestrial reference systems, the time series of Earth rotation parameters connecting the two systems, we also mention the development of the *ISGN* which should be able by design to take over most of the observational part of *GIGGOS*. Regional activities are very well addressed as well within *IAG* – let us mention the *WEGENER Special Commission* or the *EUREF Subcommission* and the *SIRGAS Project*.

What is really new, is the vision to see all these activities as one entity.

If we would only consider the first three of the four components of *GIGGOS*, we could safely state that *GIGGOS* is almost in place. The gravity field is an exception: our knowledge (as seen from the point of view of satellite geodesy) goes back essentially to optical and *SLR* tracking of geodetic satellites (like *Lageos 1,2, Starlette, etc.*). The aimed at accuracy (of 10^{-9}), with a very high time and space resolution, has not yet been achieved.

The new gravity-oriented satellite missions will dramatically improve our knowledge of the gravity field, however. This aspect will have to be seriously addressed in the development of Space Geodesy.

Plans for the Near Future (1999 - 2003)

In the preceding sections we saw that much work was (is being) performed within the time period 1995 - 1999 within the *IAG* sections. It is our task to make sure that this work will be carried on in the next *IAG* legislature period 1999 - 2003 with a similar dedication. We propose to observe the following principles for the future work:

- Take into account work performed in the past and avoid duplications, repetitions, *etc.* to the extent possible.

- For new activities make the attempt to assign them to the existing structure (services, commissions, *etc.*) before creating new entities.
- Define new activities and new directions now, and then try to foster project chairs, *etc.* to fill the required positions. Send out Calls for Proposals for Section II activities for the next "legislature period".

The first two principles are meant to minimize friction and frustrations. The third principle is an attempt to improve the efficiency of research-oriented activities within *IAG*. We imagine that the proposals presented below (and which cannot be attributed (according to principle No. 2) to a service-like structure) will be published, *e.g.*, in the *Journal of Geodesy* as a Call for (concrete) proposals. Such proposals should be coordinated between the Section II Commissions and Services, and the *IERS*.

One has to take into account, of course, that a new "crew", with different opinions and priorities, will take over the work in *IAG* Sections and the Commissions of *IAG* in 1999. We have seen in the past that a lot depends on the amount of time the steering committees, section and commission presidents, and secretaries, are willing and/or able to invest into *IAG* work. The transition period from one *IAG* period to the next is crucial. One of the primary goals of this section is the attempt to make this transition in 1999 as smooth as possible for Section II of *IAG*. It is important to note that the situation is different for service-like structures. *IUGG* General Assemblies do not affect services very much, which is due to their higher degree of independence. Continuity, but also changes in policy, are therefore much easier to be achieved within service-like structures. Increased flexibility was one of the key reason for restructuring the *SLR/LLR* and the *VLBI* subcommissions of *CSTG*. It is thus clear that we have to focus on Commissions and on Sections in our recommendations.

Separate recommendations are presented for *CSTG* and Section II. A few remarks are made concerning the *IAG* as a whole.

Recommendations concerning CSTG

From the preceding years we have seen that formalities, like the publication of the *CSTG* Bulletins, an electronical mail service, announcement of regular meetings, *etc.* are of vital importance for a loose organization like *CSTG* . That things ran relatively smoothly in the period 1995 – 1999 was thanks to the creation of the position of the *CSTG* secretary who "happened to have" a suitable infrastructure to perform this work. It will be of greatest importance to have access to a similar infrastructure in future, as well. In general, it cannot be assumed that a newly elected *CSTG* president has such facilities at hand. We therefore propose:

Recommendation 1: An official *CSTG Secretariat* offering Bureau functions to *CSTG* shall be created through a Call for Participation. The secretariat shall be a long-term commitments. The secretariat will be represented in the *CSTG* Executive Committee. Secretary and Commission president may be identical. Duties would include editing and publishing the *CSTG Bulletins*, organization of *CSTG* meetings (executive committee, public meetings at *IUGG, COSPAR, etc.* meetings), maintenance of electronic mail and WWW-facilities, *etc.*. A detailed list of duties would have to be published in the *CfP*.

CSTG work by definition is closely linked to the *IERS* and the *IGS*. If *CSTG* wants to successfully play its coordinating role in Space Geodesy, a certain degree of coordination with the two important units in space geodesy is mandatory. Also, it was realized, that relationships with *COSPAR* are crucial. The subsequent two recommendation addresses these relationships.

Recommendation 2: *CSTG* shall organize one symposium and one open Executive Committee Meeting at *COSPAR* Scientific Assembles. The first *COSPAR* related *CSTG* events after an *IUGG* Assembly shall be organized by the "previous" *CSTG* Executive Committee. Close cooperation with the *COSPAR* panel on Satellite Dynamics *(PSD)* is strongly recommended. In the time period 1995 – 1999 the *ILRS* (International Laser Ranging Service) and the *IVS* (International VLBI Service) were created as *services within CSTG*. The reason for that was pragmatic: It would have been impossible to change the *IAG* structure to give these structures a similar status like, *e.g.*, the *IGS*, which is an *IAG* service, formally associated with Section II, but otherwise directly under *IAG*.

Recommendation 3: The *CSTG* Executive Committee for 1999 to 2003 shall develop a proposal for the future status of the *ILRS* and *IVS* within *IAG* and their relationships to other scientific organizations. A concrete proposal shall be presented to the *IAG* Executive Committee at its regular meeting in the year 2001.

Organizing *IGEX-98*, the International GLONASS Experiment, was a key activity of the satellite microwave subcommission of *CSTG*. We propose that this activity shall continue in the next period.

Recommendation 4: The *CSTG* Subcommission on Satellite Microwave Systems shall conclude the experiment *IGEX-98*, evaluate the experiment at a workshop, and come up with a concrete proposal for the future *CSTG, IGS, ION* and *IERS* involvement in *GLONASS*. Such a proposal should include a discussion of continuous global *GLONASS* observations and analysis. This proposal shall be developed in close collaboration with *IGS*, *ION*, and *IERS*.

When looking at the structure of Section II (see (RUMMEL, 1996)) one recognizes, that there are four commissions and special commissions and "only" three special study groups *(SSGs)*. Study groups tend to be isolated from "the rest of the section". We therefore propose:

Recommendation 5: There will be no special study groups in Section II in future. These shall be replaced by *Projects* which are associated to the Commissions and Special Commissions of Section II.

Remarks and Wishes concerning IAG, its Services and its Section II

Changing the structure of Section II affects the structures of other sections, thus of *IAG*. We believe that a change of structure is inevitable, and this is reflected in our considerations below. We are of course aware of the fact that the input of all sections (not only that of Section II) is required for such an analysis. Obviously, at this point, such a general review cannot be given. We thus focus on the most urgent wishes related to Section II. We include aspects related to other sections only, if (in our opinion) Section II "interests" are involved.

It is also problematic to present recommendations here which affect the *IAG* Services (like, *e.g.*, the *IGS* or the *IERS*). In view of the fact that these *IAG* elements are independent and responsible for their own activities, we hesitate to present such recommendations. We have wishes, however: It is easy to predict, and it was pointed out several times in our article, that *space borne applications for the GPS* will be of vital importance in the near future. Therefore we present the following:

Wish addressed to the IGS: In view of many upcoming space missions involving space borne applications of the *GPS*, the *IGS* is encouraged to get seriously involved in these activities and to strive for making this type of research "easily" accessible to the broader geodetic community.

Another wish is related to the *IERS*, the *ILRS*, the *IGS*, and the *IVS*. We have pointed out that the combination aspect in the analysis of space geodetic data is of central importance in future. We have stated that not too much progress was made in the time period 1995 to 1999. It is clear from the theoretical point of view, however, that results derived from different space geodetic techniques containing common parameters (*e.g.*, Earth Rotation Parameters (ERPs), troposphere parameters, *etc.*) should be correctly combined on the level of the normal equation matrices (and not by some averaging methods). In order to exploit the full power of all techniques involved we have the following:

Wish addressed to the IERS, IGS, ILRS, and IVS: In view of the strength and weaknesses of individual space geodetic techniques *one* common working group (or other entity) should be created to allow for a mathematically correct combination of results stemming from different space geodetic techniques. In view of its multi-technique background the *IERS* should take the lead in this effort.

Participation of Japanese groups active in the "keystone project" should be ensured. The present status concerning the creation of the *ISGN* was presented in the previous section. This work must be continued and concluded by *CSTG* in collaboration with the *IERS*:

The developments in the time period 1995 - 1999 within Section II have shown that the "boundaries" given to Section II are no longer opportune or that they are even problematic: The creation of the *ILRS* and the *IVS* show a clear trend away from pure technique-oriented *resp.* pure science- *resp.* geodynamics-oriented activities. The examples of the *IGS* and the *IERS* have shown that a combination of the two aspects is required for services. These examples were followed by other space geodetic techniques through the creation of the *ILRS, IVS*.

From the technique point of view it is quite clear that *permanent observations* using global, technique specific networks and/or networks of the *ISGN*-type are intimately linked to the questions of global geodynamics, in particular to the area related to the *realization* of celestial and terrestrial reference frames, to Earth rotation, *etc.* We briefly reviewed space geodetic activities within *IAG* at the beginning of the previous section and we find the situation rather unsatisfactory.

IAG is divided into five sections. The idea behind this structure has been conceived by our predecessors. It reflects the logical line from the development of techniques and experiments *(Sections I, II, III)*, via methodology *(Section IV)* to the interpretation of the results *(Section V)*. Thereby measurement techniques were divided into the three areas *classical, space*, and *gravimetric*. Meanwhile - without underestimating the need of terrestrial and airborne techniques - space methods have revolutionized geodesy and have penetrated almost every corner of all five *IAG* sections. The overwhelming success of space methods in Earth sciences implies that today geodesy is much closer to geodynamics, glaciology, and physical oceanography. In addition, we are in a period where the goals of geodesy increasingly form one unity; let us, *e.g.*, think in terms of a Global Integrated Geodetic and Geodynamic Observing System. Thus, we may ask whether we arrived at a situation where the current section structure is no longer optimal or even counterproductive. Should the development of techniques, the required methodology, and the interpretation of results not form a new entity? What role do we expect from governmental research institutes and what from universities? How should they collaborate? How can we close geographical gaps in global observing networks?

Such considerations let us believe that the *IAG* section structure should be reviewed. *Space Geodesy* at present is "scattered" more or less over all sections, where it is clear that the *IAG* Commissions and Services, in particular *CSTG, IERS, IGS, IVS, ILRS, and Commission X* play an important role. It would be nice if this fact could be *also* reflected by the section structure. It is not the place here to come up with a concrete proposal, but from the point of view of Section II it would be highly desirable to make an attempt for a restructuring process within IAG sections.

Acknowledgments

The authors are indebted to Prof. Ivan I. Mueller and to Drs. Jan Kouba and Markus Rothacher for critically reading and commenting this manuscript.

References

BERTINGER et al. (1994): *GPS Precise Tracking of Topex/ Poseidon results and implications* in Journal of Geophysical Research, Vol. 99, No. C12, pp. 24449-24464

BEUTLER G., DREWES H. (1996): *The Role of CSTG Today and in the Near Future,* in *CSTG Bulletin No. 12*, pp. 21-30, Deutsches Geodätisches Forschungsinstitut, Munich.

BEUTLER G., DREWES H. (1997): *CSTG Activities between 1995 and 1997 – Review and Position Paper -,* in *CSTG Bulletin No. 13*, pp. 7-14, Deutsches Geodätisches Forschungsinstitut, Munich.

BEUTLER G., MUELLER I.I., SCHWARZ K.P. (1996): *Memorandum for the Record: Establishment of the IAG/CSTG Working GROUP "IUGG Fundamental Reference and Calibration Network",* in *CSTG Bulletin No. 13*, pp. 95-96, Deutsches Geodätisches Forschungsinstitut, Munich.

BEUTLER G., MUELLER I.I., NEILAN R.E. (1996): *The International GPS Service for Geodynamics (IGS): The Story,* in *International Association of Geodesy Symposia, No. 115, GPS Trends in Precise Terrestrial, Airborne, and Spaceborne Applications.* pp. 3-13, Springer- Verlag, ISBN 3-540-60872-6.

BEUTLER G., KOUBA J., NEILAN R.E. (1998): *The International GPS Service: An Interdisciplinary Service in Support of Earth Sciences.* submitted for publication in Advances in Space Research.

BEVIS M., BUSINGER S., HERRING T.A., ROCKEN C., ANTHES R.A., WARE R.H (1992): *GPS Meteorology: Remote Sensing of Atmospheric Water Vapor using the Global Positioning System,* in *Journal of Geophysical Research*, Vol. 97, pp. 75-94.

CLARK T.A., MA C., VANDENBERG N.R., GIPSON J.M., NIELL A.E. (1997): *CORE/IVN Geodetic VLBI for the New Millennium – Continuous Observations of the Rotation of the Earth using the International VLBI Network -,* in *CSTG Bulletin*, No. 13, pp. 37-41, Deutsches Geodätisches Forschungsinstitut, Munich.

DEGNAN J. (1997): *Pro 1997 CSTG Satellite and Lunar Laser Ranging – Subcommission Report -.* in CSTG Bulletin, No. 13, pp. 47-59, Deutsches Geodätisches Forschungsinstitut, Munich.

DEGNAN J. (1998): *Joint CSTG/IERS Call for Participation. in SLR Electronic Mail,* No. 0102.

FU L.-L., CHRISTENSEN E.J., YAMARONE CH. A., LEFEBVRE M., MÉNARD Y., DORRER M., ESCUDIER PH. (1994): *TOPEX/POSEIDON Mission Overview,* in *Journal of Geophysical Research*, Vol. 99, No. C12, pp. 24369-24381.

IGS CENTRAL BUREAU (1998): *IGS Directory,* January 1998. Available through IGS Central Bureau at JPL, Pasaedna, USA.

MUELLER I.I. (1992): *Planning an International Service Using the Global Positioning System (GPS) for Geodynamic Applications,* in IAG Symposium 109: Permanent Satellite Tracking Networks for Geodesy and Geodynamics, Mader G. ed., pp. 1-22, Springer-Verlag.

MUELLER I.I. (1996): *The Commission: a Historical Note,* in *CSTG Bulletin*, No. 12, pp. 7-20, Deutsches Geodätisches Forschungsinstitut, Munich.

MUELLER I.I., MONTAG H., REIGBER CH., WILSON P.(1996): *An IUGG Network of Fundamental Geodynamic Reference and Calibration Stations: Rationale and Recommendations,* in CSTG Bulletin, No. 12, pp. 75-94, Deutsches Geodätisches Forschungsinstitut, Munich.

MUELLER I.I. (1998a): *Science Services: International Association of Geodesy (IAG) / Federation of Astronomical & Geophysical Services (FAGS) in Special Volume,* IAG Scientific Assembly, Rio de Janeiro, Brazil, Sep 3-9, 1997, 62 pages, available through IAG Central Bureau c/o Department of Geophysics, Copenhagen, Denmark.

MUELLER I.I. (1998b): *Recommendations and Action Items of the IGS Governing Board Retreat,* Napa Valley, December 12-14 1997. in Proceedings of the 1998 Analysis Center Workshop, ESA/ESOC Darmstadt, Germany.

NEILAN R.E., ZUMBERGE J.F., BEUTLER G., KOUBA J. (1997): *The International GPS Service: A Global Resource for GPS Applications and Research,* in Institute of Navigation, ION GPS-97, Proceedings, pp. 883-889.

NEILAN R.E., SCOY P.A. VAN, ZUMBERGE J. F. (eds.) (1996):, *SINEX-Solution (Software / Technique) INDEPENDENT EXCHANGE Format Version 1.00 (June 30, 1996)* in Proceedings, 1996 Analysis Center Workshop, pp. 233-276. IGS Central Bureau; also available at Central Bureau Information System ftp://igscb.jpl.nasa.gov/igscb/data/format/sinex.txt.

REIGBER C., FEISSEL M. (1997): *IERS mission, present and future – Report on the 1996 IERS Workshop,* IERS Technical Note, No. 22, Observatoire de Paris.

ROTHACHER, M., BEUTLER G., HERRING T.A., WEBER R. (1998): *Estimation of Nutation using the Global Positioning System,* in Journal of Geophysical Research (accepted for publication, 1998).

RUMMEL R. (1996): *Section II: Advanced Space Technology,* in the Geodesist's Handbook, Journal of Geodesy, Vol. 70, No. 12.

RUMMEL R. (1998): *Global Integrated Geodetic and Geodynamic Observing System (GIGGOS).* P. 253, this volume.

WILLIS P., G. BEUTLER, W. GURTNER, G. HEIN, R.E. NEILAN, J. SLATER (1998): *IGEX: International GLONASS Experiment – Scientific Objectives and Preparation,* in Advances of Space Research (1998).

YUNCK T.P., MELBOURNE W.G. (1996): *Spaceborne GPS for Earth Science,* in International Association of Geodesy Symposia, No. 115, GPS Trends in Precise Terrestrial, Airborne, and Spaceborne Applications. pp. 113-122, Springer- Verlag, ISBN 3-540-60872-6.

Contributions of VLBI to Space Geodesy

H. SCHUH[1]

Abstract

The eminent importance of Very Long Baseline Interferometry (VLBI) in geodesy and geophysics is due to its connection with a quasi-inertial reference frame defined by compact extragalactic radio sources. VLBI is unique in its ability to monitor without hypotheses *all* components of Earth orientation. Thus, the technique is essential for the definition and maintenance of the celestial reference frame, for the determination of Universal Time (UT1), length of day (lod) and polar motion and for monitoring the motion of the celestial pole (precession and nutation). Newest VLBI results are presented on these topics, e.g. determination of the Earth rotation parameters with highest temporal resolution and observation of the Free Core Nutation (FCN). By wavelet analyses of UT1 and nutation time series quasi-periodic effects could be exactly localized. An overview about the international VLBI collaboration is also given.

Keywords: Very Long Baseline Interferometry (VLBI); Earth rotation; length of day; polar motion; nutation; quasi-inertial reference frame; wavelet analysis

1. Introduction

For almost 20 years the technique of geodetic Very Long Baseline Interferometry (VLBI) has been applied to determine regional and global crustal motions and the rotation of the Earth. The geodetic and geophysical interest in VLBI is based on the use of a quasi-inertial reference frame formed by a set of compact extragalactic radio sources. VLBI measures very accurately the angles between the Earth-fixed baseline vectors and the directions to the space-fixed radio sources. Thus, even the most subtle changes of the baseline lengths and of the baseline directions with respect to the inertial system can be detected. Geodynamical phenomena such as polar motion, Universal Time (UT1) variations, precession and nutation, Earth tides, ocean tidal response, and tectonic plate motions can be monitored with highest accuracy. The fundamentals of the VLBI technique and the main elements of a VLBI system were described in many publications, e.g. by MA (1978), CARTER et al. (1985), CLARK et al. (1985), SOVERS (1991), SCHUH and CAMPBELL (1994).

2. The VLBI geometry

Networks of radio telescopes separated by several thousands of kilometers regularly observe compact extragalactic radio sources, such as quasars, which form an ideal reference frame for studies of the Earth's rotation. The difference in arrival times of the signals from the radio sources at the stations yields a time delay τ, which is the principal observable in geodetic VLBI. The geometric time delay of signals arriving at the two radio telescopes is to first order a function of the baseline vector $\boldsymbol{b}$ between the two telescopes and of the unit vector $\boldsymbol{k}$ in the direction of the radio source:

$$\tau = -\frac{1}{c}\,\boldsymbol{b}\,\boldsymbol{k} \qquad (1)$$

The negative sign reflects the conventions used in defining τ and $\boldsymbol{b}$. c is the velocity of light. For practical reasons the baseline vector $\boldsymbol{b}$ is normally expressed by Cartesian coordinates in a three-dimensional terrestrial reference system defined by a number of VLBI radio telescopes while the radio source positions are given in a quasi-inertial reference system in space. In order to express $\boldsymbol{b}$ and $\boldsymbol{k}$ in the same coordinate system, several transformations are necessary (e.g. MA, 1978). These can be applied to either vector. Equation (1) can be written as:

$$\tau = -\frac{1}{c}\,\boldsymbol{b}\,\boldsymbol{WSNP}\,\boldsymbol{k} \qquad (2)$$

where $\boldsymbol{W}$ is the rotation matrix for polar motion (wobble), $\boldsymbol{S}$ is the diurnal spin matrix, $\boldsymbol{N}$ is the nutation matrix and $\boldsymbol{P}$ is the precession matrix. The parameters in eq. (2) or combinations of parameters can be determined from VLBI experiments depending on the configuration of the network and the objectives of the observing session. Present VLBI networks observe up to 25 radio sources several times in a predetermined sequence during a session of 24 hours providing 1000 to 2000 individual delay observables τ.

3. Networks, observing programs and new organisational structures

The capabilities of VLBI for the precise determination of geodetic parameters have been demonstrated by thousands of geodetic VLBI experiments on baselines connecting all major tectonical plates. The global distribution of most of the VLBI stations which are at present used for geodetic applications is shown in figure 1.

In order to combine the efforts in different countries around the world, several programs of international cooperation have been launched in the last two decades, among which the following are the most important ones: In the past, the U.S.A. NASA Crustal Dynamics Project, now succeeded by the Space Geodesy Project, included several international long-term monitoring projects, most of them dedicated to measuring crustal deformations and present-day plate motions. Observations managed by U.S.A. Jet Propulsion Laboratory using the Deep Space Network telescopes were made primarily for spacecraft navigation and could be used

[1] Harald Schuh: Deutsches Geodätisches Forschungsinstitut (DGFI), Marstallplatz 8, D-80539 München, Germany; Email: schuh@dgfi.badw.de

to obtain information about the orientation of the Earth. At present, the U.S.A. National Earth Orientation Service (NEOS) operated by the U.S. Naval Observatory (USNO) manages 24-hour observing sessions in international cooperation at weekly intervals for establishing and maintaining the celestial reference system as well as for monitoring the Earth rotation parameters. The NEOS-A network is formed by the U.S.A. stations in Green Bank and Hawaii, and by Fortaleza (Brazil) and Wettzell (Germany). Either Fairbanks (U.S.A., Alaska) or Ny Ålesund (Norway, Spitsbergen) complement the NEOS-A network whereas Algonquin Park (Canada) occasionally takes part. Daily intensive UT1 sessions (about one hour per day) started in 1984 on the Wettzell (Germany) to Westford (U.S.A.) baseline and were continued with the Wettzell - Green Bank interferometer. The Geodetic Institute of the University of Bonn (Germany), in cooperation with the German Bundesamt für Kartographie und Geodäsie (BKG), manages the IRIS-S observing program. IRIS-S is operated once a month with Wettzell, Westford and the southern hemisphere stations of Hartebeesthoek (South Africa) and Fortaleza (Brazil). Santiago (Chile) took part in IRIS-S from mid 92 to mid 93. The frequency of IRIS-S experiments is low, i.e. one 24-hour session per month, but they provide high precision results owing to extremely large baseline lengths. The IRIS-P network managed by the VLBI group of the National Astronomical Observatory (NAO) in Mizusawa, Japan, carried out regular sessions with the VLBI stations Kokee (U.S.A., Hawaii), Hobart (Australia), Urumqi (China) as well as the 34 m radio telescope at Kashima (Japan). Since 1997, NASA Goddard Space Flight Center (GSFC) has organized global VLBI observations with various radio telescopes as a preparation of the CORE program (Continuous Observations of the Rotation of the Earth). The CORE program will generate a continuously observed Earth rotation data set of unprecedented time resolution and accuracy (CLARK et al., 1997). A possible scenario of overlapping observing networks within the CORE program is shown in figure 2.

From spring 1997 onwards there was an intensive period of communication among members of the VLBI community for setting up an International Geodetic VLBI Service (IVS). According to the *Terms of Reference* of the IVS, its main goals are:

- Fostering VLBI programs as a joint service to support geodetic, geophysical, astrometric and other research and operational activities.
- Promotion of research and development activities in all aspects of the geodetic and astrometric VLBI technique.
- Interaction with the community of users of VLBI products and integration of VLBI into a global Earth observing system.

After the IVS call for proposals had been launched on June 1st, 1998, more than 70 proposals were submitted. The evaluation by CSTG and IERS was done during the IGGOS meeting in Munich in October 1998. Till March 1999 the successful proposers will be notified, elections will be conducted, and the initial Directing Board of IVS will be established. On March 1st, 1999 the IVS will begin its operation. It will be formally introduced at the next IUGG General Assembly in Birmingham (UK) in July, 1999.

4. Results

All of the VLBI programs mentioned above have profited from a broad international cooperation and have produced impressive results. As for plate tectonics the baseline length changes detected by VLBI confirm the plate tectonic models to a surprisingly high degree. A prominent intercontinental baseline with a very long history and several hundreds of measurements is the 6000 km baseline between Wettzell and Westford. The evolution of the baseline length from 1984 till today is plotted in figure 3. The trend of 1.70 ±0.01 cm/year can be determined with high confidence and is in good agreement with the predicted relative tectonic motion of the American and Eurasian plates (NUVEL model rate is 1.89 cm/year). The baseline measurements will soon allow to investigate nonlinear effects in the baseline evolution such as periodic variations or higher order polynomials.

The Earth rotation data observed by VLBI have begun to reveal hitherto unseen phenomena. Examples are the atmospheric excitation of the rotation of the Earth due to the zonal winds in the upper atmosphere and departures from the normal state, e.g. caused by the El Niño events in the southern Pacific. Figure 4 shows the remarkably smooth trace of the instantaneous pole from 1980 to 1999 which consists of the annual and the Chandlerian components of polar motion. The VLBI data set provided the first opportunity to investigate short-period polar motion, even the diurnal and subdiurnal oscillations of the pole can be seen, if the time resolution of the pole coordinates is increased in the VLBI solution. The UT1 variations are dominated by the seasonal terms, an annual and a semiannual oscillation; but there is also a rich spectrum of tidally induced short-period oscillations (YODER et al., 1981), which could be seen by VLBI (ROBERTSON et al., 1985; CAMPBELL and SCHUH, 1986). Even the short-period tidal influence of the oceans (BROSCHE et al., 1989) could be detected in the VLBI measurements by analysing a large number of observing sessions (BROSCHE et al., 1991). Today, the highest temporal resolution of the Earth rotation parameters can be achieved by VLBI. Figure 5 shows UT1 parameters determined every 5-7 minutes during the 5-day VLBI campaign CONT in September 1996. The results were obtained by TITOV (1998) by means of the least-squares collocation method. Such data sets offer the opportunity to search for tiny variations in the rotation of the Earth, e.g. those which are supposed to be triggered by Earthquakes. As mentioned before, also the precession and nutation angles can be estimated from VLBI data (HERRING et al., 1986). The deviations of the nutation in longitude and obliquity from the IAU 1980 Model (Wahr's theory for an elastic Earth) show a prominent annual oscillation with an amplitude of about 6 mas (fig. 6) and further periodicities.

A very important contribution of VLBI to astronomy and geodesy is the realization of the International Celestial Reference Frame (ICRF). The quasi-inertial reference frame obtained by international collaboration of many VLBI groups is based on the radio positions of 212 extragalactic sources distributed over the entire sky (MA et al., 1998). The ICRF was adopted by the IAU on January 1st, 1998. It replaces the previous fundamental celestial reference frame defined by the FK5 optical catalogue.

5. Wavelet analysis of Earth rotation parameters observed by VLBI

There are several remarkable advantages of the wavelet transform, which was introduced by GROSSMANN and MORLET (1984), in comparison with 'classical' methods based on the Fourier transform. In the wavelet transform, the harmonic oscillation is replaced by a wavelet function which is adapted to the signal to be analysed by shifting it along the time axis and scaling it along the frequency axis. It allows the time localization of an unstable quasi-periodic signal within a given data set. Hence, the wavelet transform is an excellent tool for analysing signals with time-varying amplitudes and/or periods.

In the following, two examples will be given for the application of the wavelet transform to Earth orientation data observed by VLBI. In the nutation data obtained by VLBI, the retrograde Free Core Nutation (FCN) can be seen clearly around a period of 430 days. Figure 7 shows the wavelet spectrum of the observed complex-valued nutation parameters ($\Delta\varepsilon$ - i $\Delta\psi$ sinε_0) (SCHMIDT and SCHUH, 1999). Before the wavelet analysis was carried out, the IERS correction model (1996 IERS Annual Report, 1997), which contains 25 astronomical terms including an annual variation, had been subtracted from the data. The wavelet transformation reveals a significant decrease of energy in the range of the FCN since 1989/1990. Possible reasons for this decrease are given in a recent paper by DEHANT et al. (1997). From a closer look at the output of the wavelet analysis of the nutation series, it can be seen that not only the energy related to the FCN has changed, but also the period which corresponds to the highest energy in the wavelet spectrum at a particular time has varied in the span from 1984 till 1998. Figure 8 shows the exact periods in the series of the observed ($\Delta\varepsilon$- i $\Delta\psi$ sinε_0) corresponding to the maximum spectral energy in the range of the FCN period. A modulation period of about 4.8 years can be extracted from figure 8, which was discussed in more detail by SCHUH and HAAS (1998).

The high resolution UT1 and polar motion data obtained from the VLBI CONT-sessions by TITOV (1998) as described in the last section were also analysed by the wavelet transformation. First, the main diurnal and semi-diurnal variations due to the ocean tides were removed using the IERS correction model. Then, in most of the datasets the wavelet analysis reveals irregular fluctuations with extremely short periods, e.g. between 5 and 7 hours and even down to 2 to 3 hours. An example of a wavelet spectrum of UT1 is given in figure 9. Possible excitation mechanisms for these variations such as resonances of the Earth due to Earthquakes are under discussion.

6. Conclusions

The eminent importance of VLBI in geodesy and geophysics is due to its connection with an inertial reference frame formed by a set of compact extragalactic radio sources. VLBI is unique in its ability to monitor without hypotheses *all* components of Earth orientation. Thus, the technique is essential for the definition and maintenance of the celestial reference frame, for the determination of Universal Time (UT1), length of day (lod) and polar motion and for monitoring the motion of the celestial pole (precession and nutation). The results of VLBI experiments are needed for many practical applications which require an accurate inertial reference frame, e.g. high-precision navigation and positioning. They are also used for scientific research in many related areas, such as: tidal deformation of the solid Earth and oceanic and atmospheric tidal loading, excitation of UT1 and polar motion down to periods of a few minutes, investigations of the Earth interior and improvements of the terrestrial reference frame.

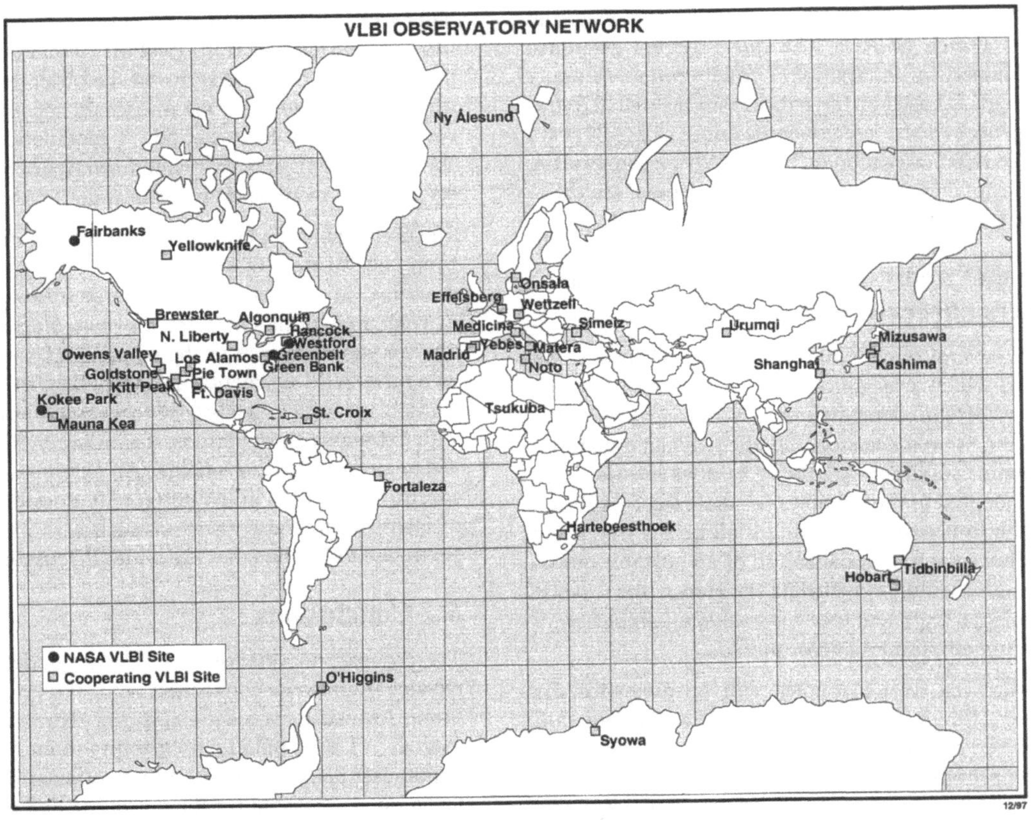

Fig. 1: The geodetic VLBI network (CDDIS Bulletin, Dec. 97)

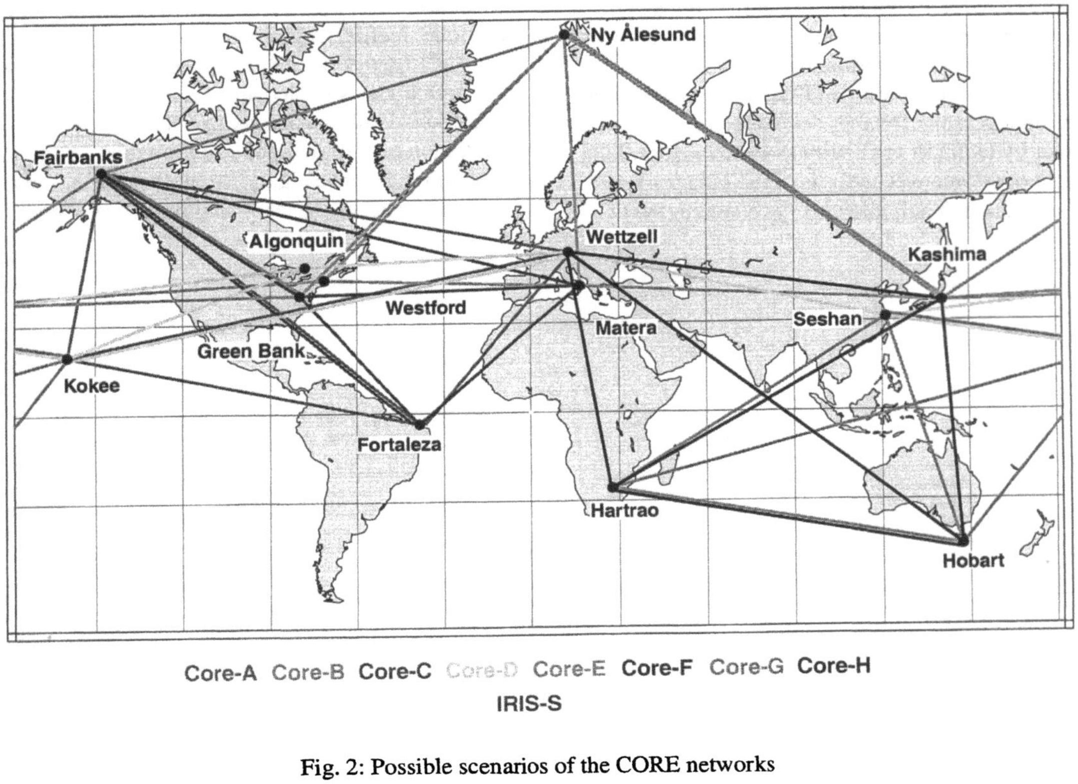

Fig. 2: Possible scenarios of the CORE networks

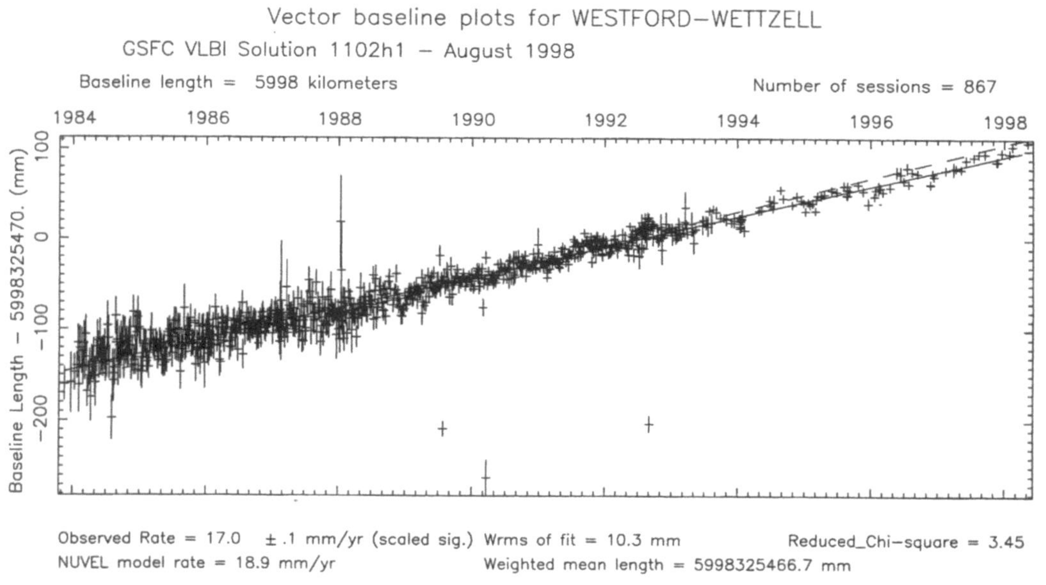

Fig. 3: Baseline evolution Wettzell - Westford

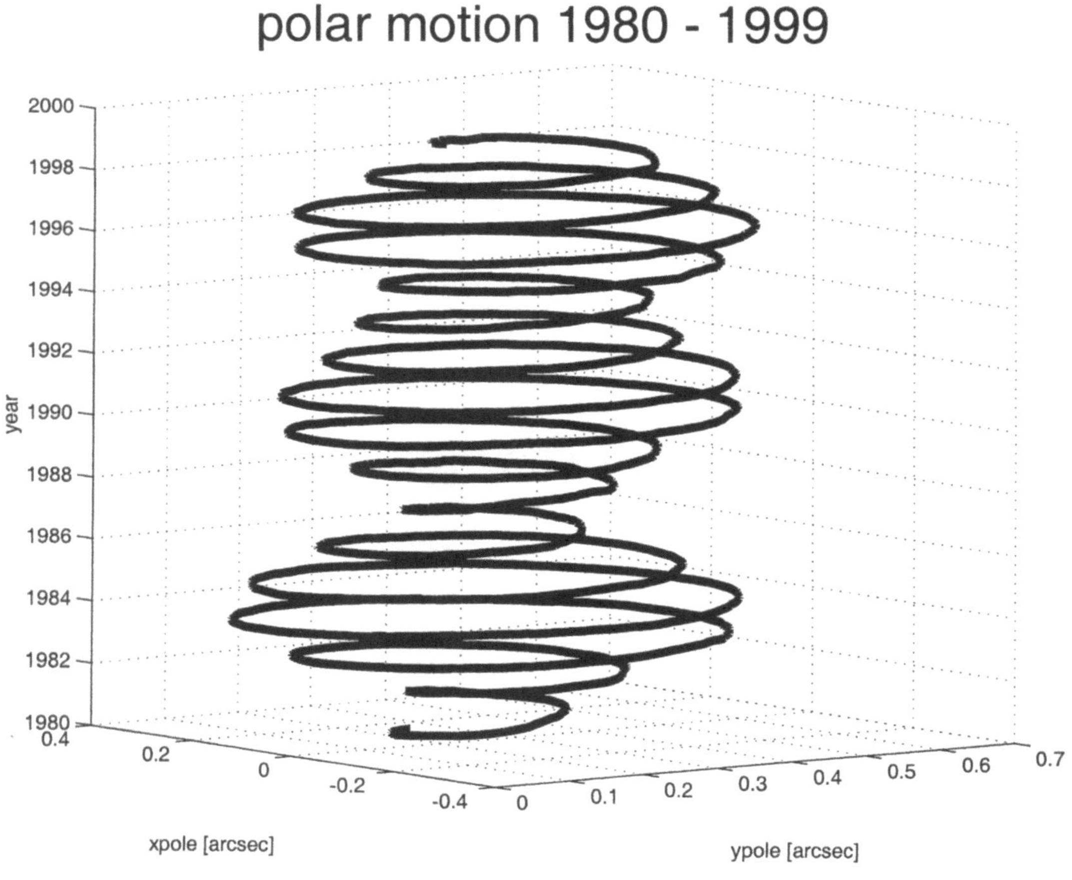

Fig. 4: Polar motion from 1980 till 1999

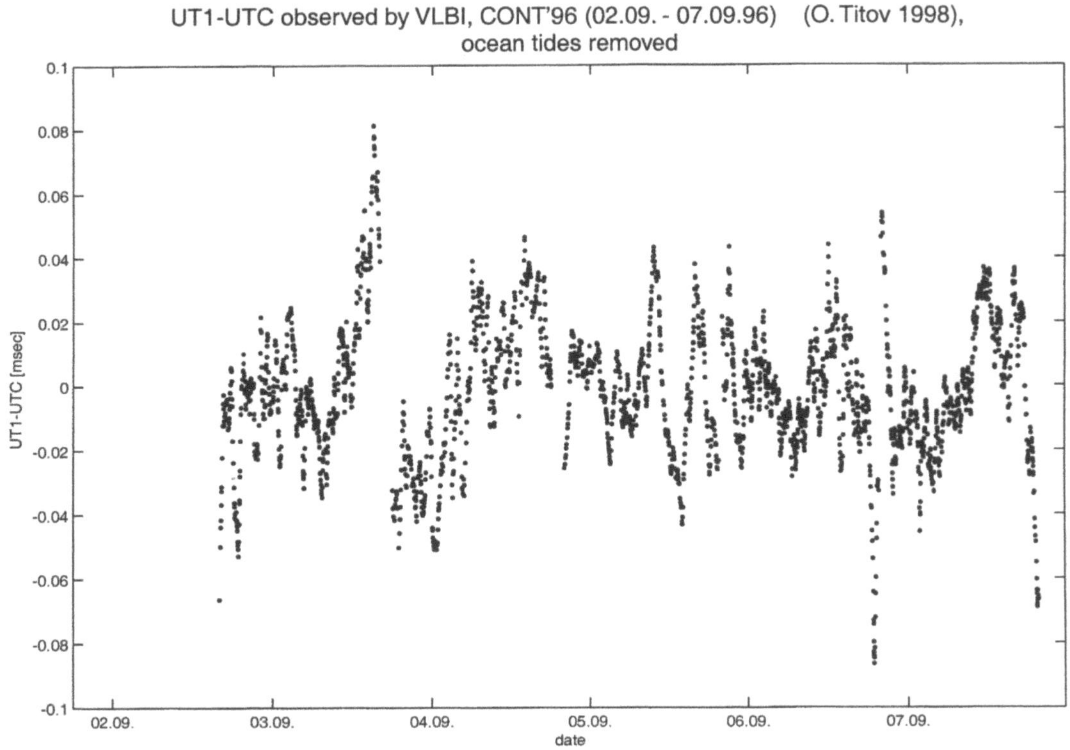

Fig. 5: UT1 parameters with high temporal resolution determined during the 5-day VLBI campaign CONT in September 1996; the results were obtained by TITOV (1998)

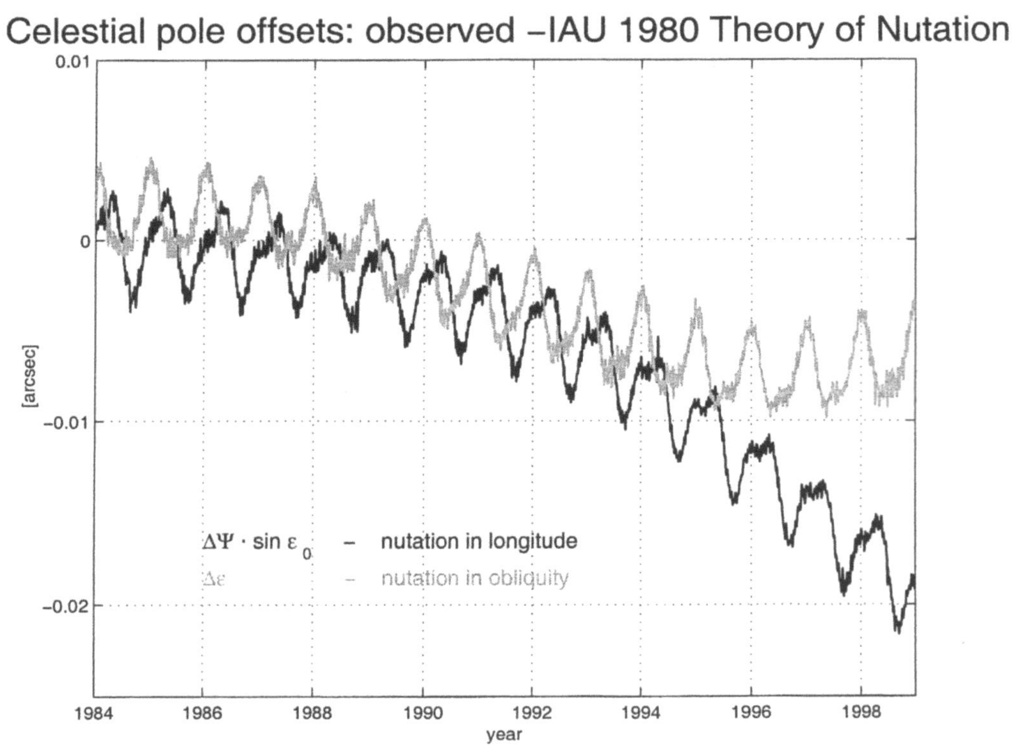

Fig. 6: Nutation time series (mas) observed by VLBI: motion of the celestial pole relative to the IAU 1980 theory of nutation and the IAU 1976 precession (1996 IERS Annual Report)

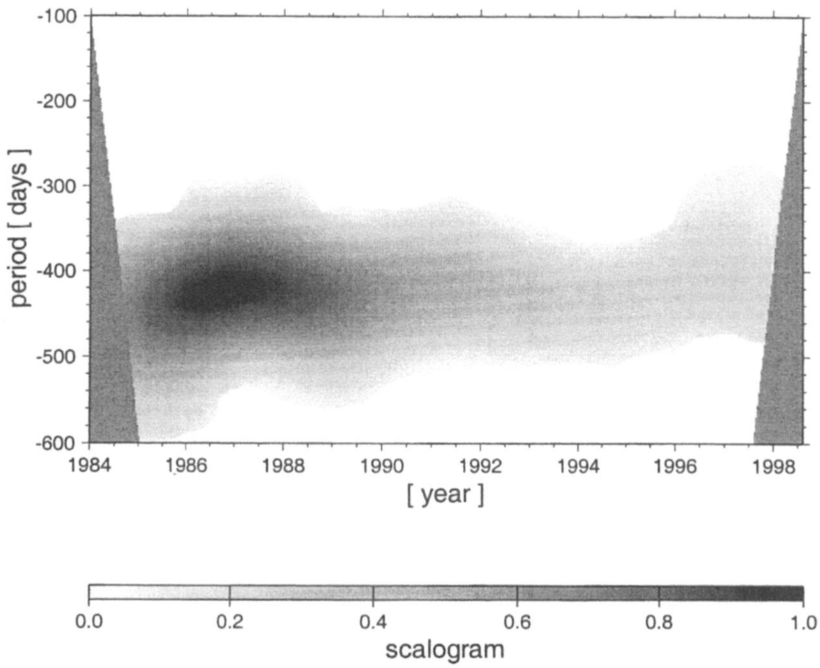

Fig. 7: Retrograde part of the wavelet spectrum of complex-valued nutation data ($\Delta\varepsilon$- i $\Delta\psi\sin\varepsilon_0$) observed by VLBI from 1984 to 1996 (SCHMIDT and SCHUH, 1998)

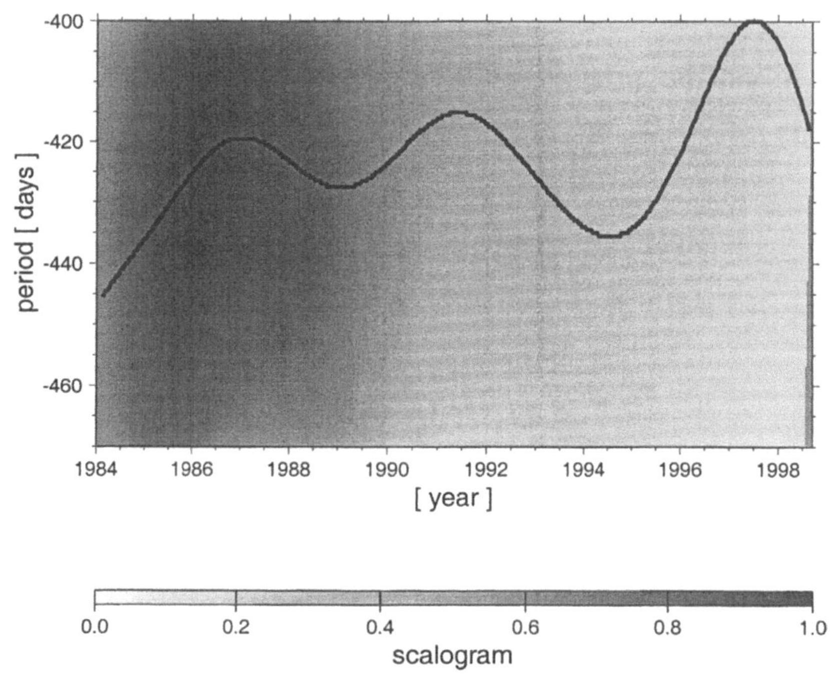

Fig. 8: Period corresponding to the maximum energy of the wavelet spectrum (fig. 7) versus time (ridge function). A modulation of about 4.8 years can be seen clearly

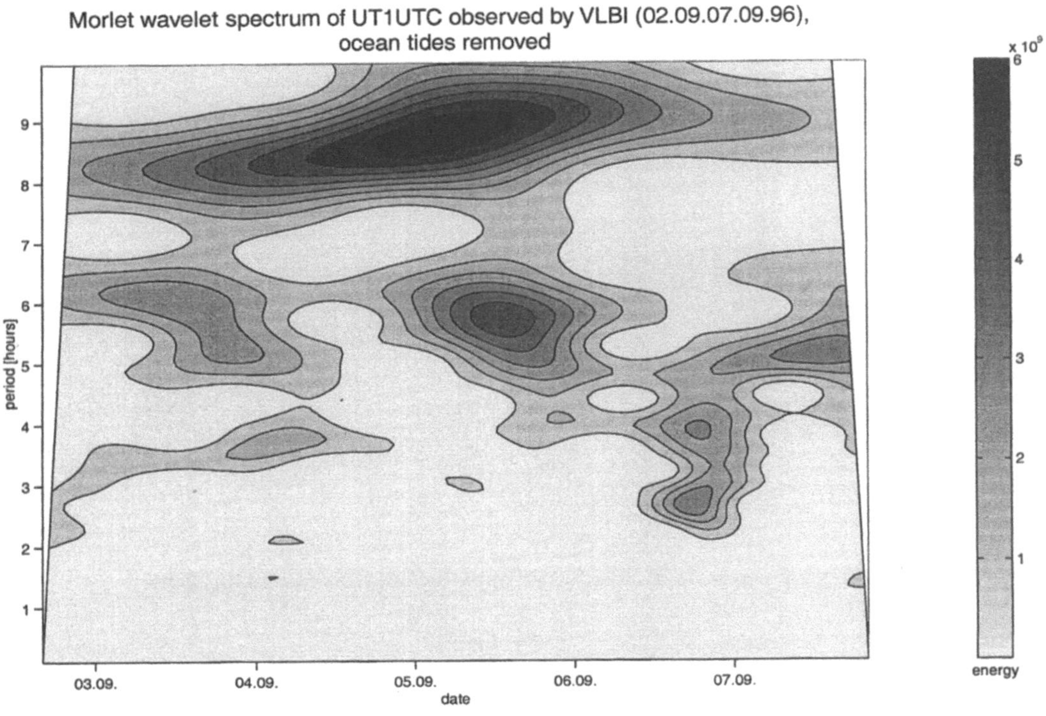

Fig. 9: Wavelet spectrum for periods below 10 hours of the high resolution UT1 series shown in fig. 5

7. References

BROSCHE P., SEILER U., SÜNDERMANN J., WÜNSCH J.: *Periodic changes in Earth's rotation due to oceanic tides,* Astronomy and Astrophysics, 220, 318-320, 1989.

BROSCHE P., WÜNSCH J., CAMPBELL J., SCHUH H.: *Ocean tide effects in Universal Time detected by VLBI,* Astronomy and Astrophysics, 245, 676-682, 1991.

CAMPBELL J., H. SCHUH H.: *Short-period variations of Earth rotation determined by VLBI,* Proc. of the Xth International Symp. on Earth Tides, Madrid, 1985, ed. by R. Vieira, C.S.I.C., 943-951, 1986.

CARTER W.E., ROBERTSON D.S., MACKAY J.R.: *Geodetic Radio Interferometric Surveying: Applications and Results,* J. Geophys. Res., 90, No. B6, 4577-4587, 1985.

CLARK T.A. et al.: *Precision Geodesy using the Mk-III Very-Long-Baseline Interferometer System,* IEEE Transactions on Geoscience and Remote Sensing, Vol. GE-23, No. 4, 438-449, 1985.

CLARK T.A., MA C., VANDENBERG N.R., GIPSON J.M., NIELL A.E.: *CORE/IVN - Geodetic VLBI for the New Millenium,* CSTG Bull. No. 13, Progress Report 1997, ed. by G. Beutler, H. Drewes and H. Hornik, 37-41, 1997.

DEHANT V., FEISSEL M., DEFRAIGNE P., ROOSBEEK F., SOUCHAY J.: *Could the energy near the FCN and the FICN be explained by luni-solar or atmospheric forcing?,* Geophys. J. Int., 130, 324-335, 1997.

GROSSMANN A., MORLET J.: *Decomposition of Hardy functions into square integrable wavelets of constant shape,* SIAM J. Math. Anal., 15(4), 723-736, 1984.

HERRING T.A., GWINN C.R., SHAPIRO I.I.: *Geodesy by Radio Interferometry: Studies of the Forced Nutations of the Earth,* 1. Data Analysis, J. Geophys. Res., 91, B5, 4745-4754, 1986.

MA C.: *Very Long Baseline Interferometry Applied to Polar Motion, Relativity and Geodesy,* Ph.D. Thesis, NASA Technical Memorandum 79582, Greenbelt, USA, 1978.

MA C. et al.: *The International Celestial Reference Frame as Realized by Very Long Baseline Interferometry.* The Astronomical Journal, 116, 516-546, 1998.

ROBERTSON D.S., CARTER W.E., CAMPBELL J., SCHUH H.: *Daily Earth rotation determinations from IRIS very long baseline interferometry,* Nature, 316, 424-427, 1985.

SCHMIDT M., SCHUH H.: *Wavelet-Analyse der mit VLBI beobachteten Nutationsreihen.* Zeitschrift für Vermessungswesen (ZfV), 124, 24-30,1999.

SCHUH H., CAMPBELL J.: *VLBI in geodynamical investigations,* Acta Geod. Geophys. Hung., Vol. 29 (3-4), 397-420, 1994.

SCHUH H., HAAS R.: *Earth Tides in VLBI Observations,* Proc. of the 13th International Symposium on Earth Tides, Brussels, July 1997, ed. by B. Ducarme and P. Pâquet, Série Géophysique Obs. Roy. de Belgique, 101-110, 1998.

SOVERS O.J.: *Observation Model and Parameter Partials for the JPL VLBI Parameter Estimation Software "MODEST"* - 1991, JPL Publication 83-39, Rev. 4, 1991.

TITOV O.: priv. communication, 1998.

YODER C.E., WILLIAMS J.G., PARKE M.E.: *Tidal variations of Earth rotation,* J. Geophys. Res., 86, 881-891, 1981.

... 1996 *IERS Annual Report,* Observatoire de Paris, editor M. Feissel, 1997.

... *CDDIS Bulletin,* Vol. 13, Issue 2, December 1997.

Towards an Integrated Global Geodetic Observing System

M. Rothacher[1]

Abstract

Today several independent techniques (VLBI, SLR, LLR, GPS, GLONASS, DORIS, and PRARE) form the basis of space geodesy, each system depending on its own global network of tracking stations. Before thinking of a thorough combination and integration of all these systems, we compare the results of the individual techniques (site coordinates and velocities, geocenter motion, Earth rotation parameters (ERPs), information about the atmosphere, and gravity) to better understand the strengths and weaknesses of each of the techniques.

In a more general discussion of the steps to be done towards an integrated global geodetic observing system we then identify three different levels that have to be considered in such an integration effort: the integration on the network level, the consistency on the level of the observation models and processing standards, and the level of the combination of common geodetic parameter. The success of the challenging new missions to come like CHAMP, GOCE, GRACE, JASON, COSMIC, etc. will critically depend on the quality and consistency of such an integrated global observing system.

1 Introduction

Today three main observation techniques form the basis of space geodesy:

- Ranging to satellites or the moon using laser pulses; the corresponding techniques are Satellite Laser Ranging (SLR) and Lunar Laser Ranging (LLR).
- Observation of quasars with radio telescopes; the so-called Very Long Baseline Interferometry (VLBI).
- Observations of microwave signals from satellites or satellite systems; the most prominent techniques are the Global Positioning System (GPS), the Russian GLONASS, the French DORIS, and the German PRARE systems.

These basic techniques are complemented by various other observing systems like SAR (Synthetic Aperture Radar), altimetry, surface and air-borne gravity measurements, etc., and new missions to come will add further measurement types to the list of observation techniques.

In the first few sections of this paper we will focus on the present state of comparisons and combinations of the results obtained from these different space geodetic techniques. We will have a look at various areas, where the combination and integration of observing techniques already take place or where they will become very important in the near future. In many of the examples special emphasis will be put on the GPS technique, on the one hand because no paper is specifically dedicated to GPS alone and on the other hand because it is clear that GPS is not only involved in many of the topics of interest here (reference frame, ERPs, orbits, gravity field, atmosphere, time transfer, ...) but will also play a decisive role in the integration of the various observing techniques.

The final section will discuss the question and problems of the combination of the techniques and their integration from a more general point of view and show the steps on the way to an integrated global geodetic observing system (IGGOS).

2 Site Coordinates, Velocities, and Geocenter

2.1 Site Coordinates and Velocities

The establishment and realization of global terrestrial reference frames common to all space geodetic techniques (including optical instruments in former times) has a long history. The first reference frame, the Bureau International de l'Heure (BIH) System, was defined in 1968 by the adoption of the astronomical coordinates for a network of 68 optical instruments [*Guinot and Feissel*, 1969]. Starting in 1984, the BIH System was realized through the coordinates of a set of sites, where space geodesy stations

[1]Markus Rothacher; Astronomical Institute, University of Berne, Sidlerstrasse 5, CH–3012 Berne, Switzerland; Tel. +41-31-631 85 91; Fax +41-31-631 38 69; Email rothacher@aiub.unibe.ch

were operated. This led to the first BIH Terrestrial System 1984 (BTS84)[*BIH*, 1985]. Already at this time a combination was performed including local ties and the site coordinate estimates from four space geodetic techniques, namely VLBI, LLR, SLR, and Doppler tracking of transit satellites. After the establishment of the International Earth Rotation Service (IERS) in 1987, the first IERS Terrestrial Reference Frame (ITRF), called ITRF88, was produced in 1988 using VLBI, SLR, and LLR results.

These historical remarks show that the idea and also the necessity to combine the results all space geodetic techniques to realize a common reference frame are not new at all. Over the years, the number of sites and the accuracy of the results were increasing and thus also the necessity to improve the combination procedures:

- BTS86: For the first time a velocity model (AM0-2) was applied [*Boucher and Altamimi*, 1990].
- ITRF91: The first velocity field was computed (independently from the combination of the site coordinates) using velocity estimates from space geodesy [*Boucher et al.*, 1992].
- ITRF94: The first ITRF realized by taking into account the full variance-covariance matrices of the individual solutions and not just the diagonal elements [*Boucher et al.*, 1996].
- ITRF96: The first *simultaneous* combination of coordinates and velocities using the full variance-covariance matrices and estimating variance components [*Boucher et al.*, 1998].

A comparison of the global solutions from individual space techniques and analysis groups contributing to the ITRF96 is given in Table 1 [*Boucher et al.*, 1998]. We see that the best solutions of each space geodetic technique are performing equally well (with the exception of DORIS) and reach a quality of about 1 cm for the site coordinates (including height) and about 2 mm/year for the site velocities. Because the results of all techniques are at about the same level of precision they all contribute substantially to the combined solution. This combined solution might not be quite as consistent as the solutions of individual techniques or institutions but it is most probably more accurate.

Table 1 also shows that GPS with more than 100 sites contributes very significantly to the ITRF96, especially in view of the fact that many of the VLBI and SLR sites reported in ITRF96 are mobile sites or sites no longer observed. When looking at the resources (finances and manpower) needed to run a space geodetic station equipped with VLBI or SLR, it is clear that GPS is *the* tool for a further densification of the ITRF.

Table 1: RMS difference of site coordinates and velocities estimated by different techniques with respect to the combined ITRF96 solution.

Technique	**Inst.**	**#Sta.**	**Coord. mm**	**Veloc. mm/a**
VLBI	GSFC	120	5.8	1.9
VLBI	GIUB	43	13.6	0.5
VLBI	NOAA	111	14.7	1.9
VLBI	JPL	8	20.7	
SLR	CSR	89	11.1	3.8
SLR	GSFC	38	10.9	1.7
GPS	EMR	36	10.0	3.5
GPS	GFZ	66	16.8	3.3
GPS	CODE	100	7.1	1.9
GPS	EUR	39	2.4	0.3
GPS	EUR	58	2.9	0.3
GPS	MIT	132	8.5	9.2
GPS	NCL	114	5.4	6.3
GPS	JPL	113	9.4	3.8
DORIS	GRGS	48	26.9	8.0
DORIS	CSR	54	26.1	10.6
DORIS	IGN	62	28.3	12.8

Up to now a combination of the site coordinate (and velocity) results of all the major space geodetic techniques has only been done on the level of multi-year solutions. We know, however, that such combined multi-year solutions only contain a tiny, though important fraction of the information available for each site. To obtain all the details contained in the full time series of the site coordinates, it is necessary to combine solutions from as short time intervals as possible (e.g. a week or even a day). For this reason the analysis centers of the IGS started in 1995 to submit weekly coordinate solutions in the SINEX (Solution INdependent EXchange) format to the global data centers. Three global combination centers, so-called Global Network Associate Analysis Centers (GNAACs), combine these weekly SINEX files of all the seven IGS analysis centers into weekly combined IGS solutions. A similar combination effort is also taking place in the framework of EUREF (European Reference Frame), where the weekly GPS solutions of about 10 regional analysis centers are compared and combined into the official EUREF solution [*Bruyninx*, 1997; *Springer et al.*, 1997]. Time series of about 4 years worth of weekly global (and regional) solutions are now available to study variations in the site coordinates.

Analyses of these series by *Herring* [1997], *Davies and Blewitt* [1997] show, that the coordinate series of the individual analysis centers compared to the

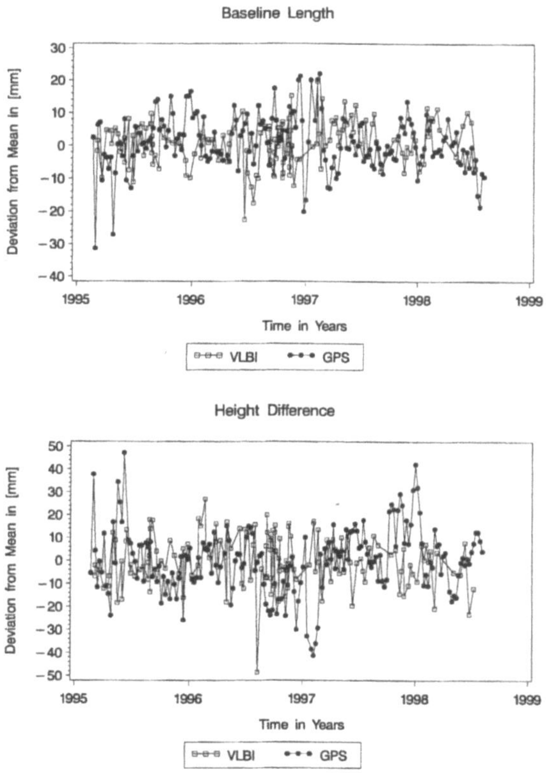

Figure 1: Comparison of time series derived from VLBI and GPS data, respectively, for the baseline Fairbanks–Wettzell (6857 km). The baseline length and the height difference are shown.

combined series are following very much the same systematic variations in time (see [*Herring*, 1997], pages 260, 262, and 264 for examples). It is not clear, however, whether these variations are real. A systematic comparison and combination of weekly solutions from various techniques would be extremely beneficial to understand these effects. To get a first impression of how such a comparison might look like, Figure 1 shows the time series of the length as well as the height difference of the baseline from Fairbanks (Alaska) to Wettzell (Germany) independently computed from VLBI and GPS data. The VLBI baseline results stem from the official solutions of the Goddard Space Flight Center [*Ma*, 1998], the GPS results (weekly solutions) were computed at the Center for Orbit Determination (CODE) in Berne, Switzerland. Neither in the baseline length nor in the height differences any significant correlations may be detected. (The mean has been subtracted in each of the series). This seems to indicate that most of the variations seen in the two series might be "noise" or systematic biases in the estimates of one or both techniques.

It is important to mention here, that only a very few baselines could be found, that have been measured for more than a few months by both techniques, VLBI and GPS. In the case of SLR and GPS the situation is about the same. This leads to the conclusion, that the ties between the networks of the individual techniques is weak and that every VLBI, SLR, (and DORIS) site should therefore be equipped with a GPS receiver to increase the connections between the global networks.

It is to be hoped that with the establishment of the International VLBI Service (IVS) and the International Laser Ranging Service (ILRS) time series of weekly (or monthly) solutions will be generated by VLBI and SLR groups and eventually compared and combined with the corresponding GPS series. From the experience within the IGS it is clear, that a lot of interesting results will emerge from such comparisons and combinations. The results will be beneficial to all techniques involved, and, especially, to future realizations of the ITRF.

2.2 Geocenter

The origin of the global reference frame is defined to be located at the center of mass of the Earth (including oceans and atmosphere), also called the geocenter. Analyses of SLR data have shown, that the ensemble of tracking stations on the crust of the Earth is moving with respect to the center of mass. This motion, viewed from a crust-fixed frame, is called "geocenter motion" and is caused by mass movements in the atmosphere and the oceans.

During the "IERS Analysis Campaign to Investigate Motions of the Geocenter" (call for participation issued in January 1997; coordinated by Jim Ray) a concentrated effort was made to compare and, to a certain extent, combine geocenter estimates from various groups using tracking data from SLR, TOPEX, DORIS, and GPS. In Figure 2 some of the geocenter series (available at `http://maia.unso.navy.mil/geoc.html`) are plotted. The GPS series shown here has been computed by the GNAACs at the Massachusetts Institute of Technology (MIT, Tom Herring) as a combination of the global solutions submitted by the IGS analysis centers on a weekly basis. The SLR series (12-day intervals) was obtained by Richard Eanes using a combination of Lageos-1 and Lageos-2 SLR data. Patrick Sillard at the Institut Géographic National in Paris provided the DORIS geocenter series (monthly values). The TOPEX geocenter results at 10-day intervals, finally, are based on TOPEX SLR data and were made available by Minkang Cheng, University of Texas. Figure 2 reveals the strengths and weaknesses of individual series and/or individual techniques. We see that the two SLR series (from different SLR data sets) are very consistent and have the smallest scatter. In the y-component, variations with an annual period may easily be detected. The quality of the DORIS series from IGN is not yet com-

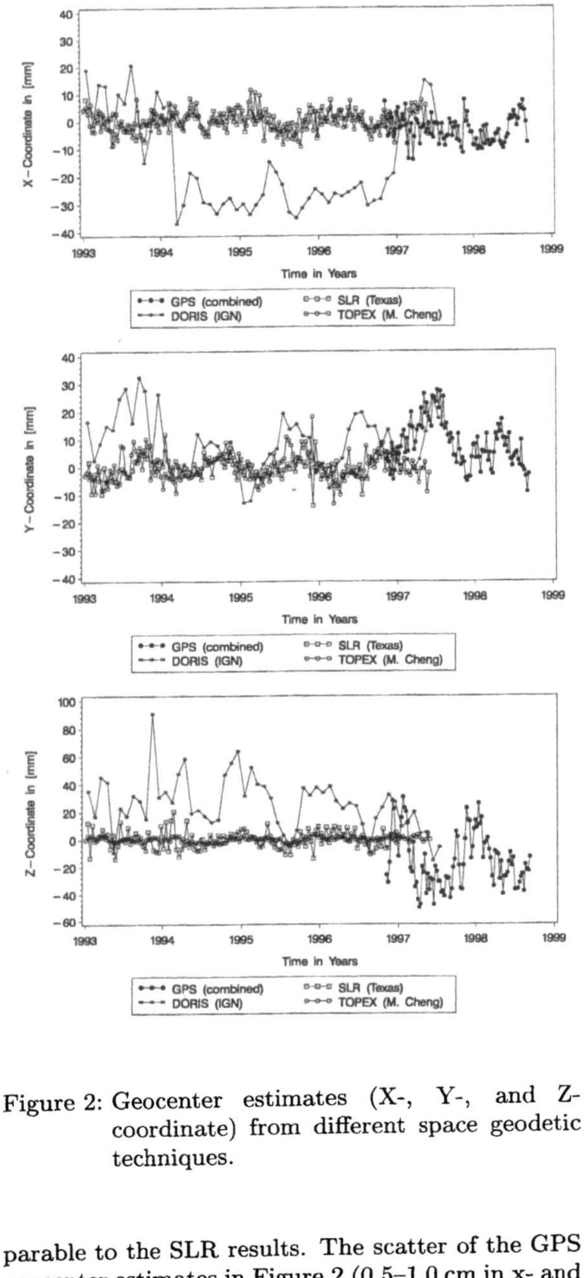

Figure 2: Geocenter estimates (X-, Y-, and Z-coordinate) from different space geodetic techniques.

parable to the SLR results. The scatter of the GPS geocenter estimates in Figure 2 (0.5–1.0 cm in x- and y-component and about 2 cm in z-direction) clearly shows that the GPS satellites, orbiting the Earth at a distance of more than 20'000 km, are too far away from the Earth to allow a precise determination of the location of the geocenter. In addition, Figure 3 reveals that the GPS-based geocenter estimates are also influenced by the processing strategy and the (non-gravitational) orbit model used. Fixing the initial phase ambiguities to integer numbers in the global CODE solutions starting in fall 1994 resulted in a drastic reduction of the scatter in the geocenter z-component. The obvious jump of about 6 cm in the y-component of the geocenter mid 1995 is due to a significant change in the CODE orbit modeling (estimation of pseudo-stochastic pulses for all GPS

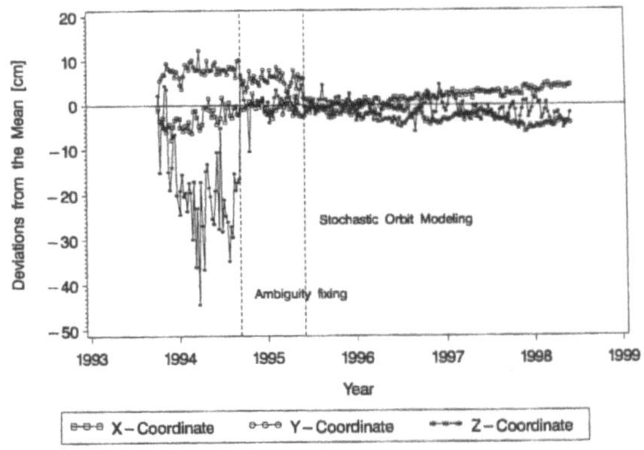

Figure 3: Weekly GPS geocenter estimates of the CODE analysis center over a time interval of 5 years.

satellite orbits once per revolution). It is therefore evident, that only GPS time series based on one and the same processing strategy have the potential, if at all, to contribute to the monitoring of the geocenter.

In the future, most if not all of the Low Earth Orbiting (LEO) satellites to be launched as part of future gravity and atmosphere missions, will be equipped with GPS receivers. Including the data of LEOs into the global GPS solutions will lead to very accurate geocenter estimates, a pre-requisite for, e.g., global sea level monitoring.

3 Earth Rotation Parameters

The first two subsections of this part are focussing on two special Earth Rotation Parameter (ERP) series from GPS and how the results of these series compare and might be combined with the results of other techniques. In the third subsection we will discuss the importance of consistency between ERP series on the one hand and site coordinates and velocities on the other hand.

3.1 Nutation

For a long time it was assumed that satellite geodetic techniques like GPS may not contribute anything to the estimation of nutation angles or nutation amplitudes, because orbital elements have to be estimated simultaneously with the ERPs. It is well known that the angles describing the orientation of the orbital planes of the satellites are fully correlated with UT1 and the nutation angles in longitude and obliquity. The exact relationship between these two sets of parameters (orbit parameters and ERPs) may be found in [*Rothacher et al.*, 1999]. The situation is completely different, however, for the rate of change of UT1 (or Length of Day (LOD)) and the nutation angle rates. There is indeed no fundamental difference

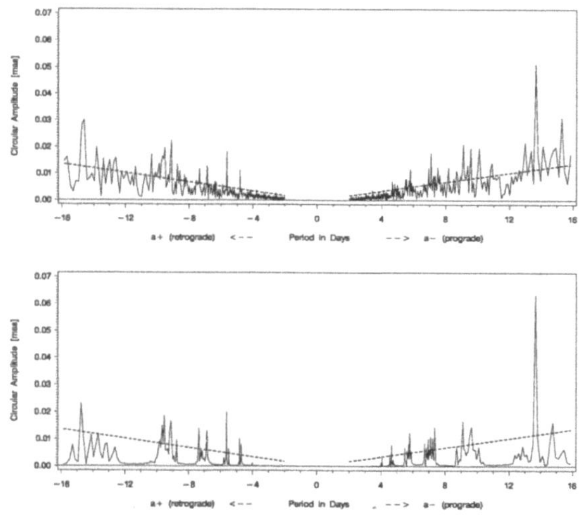

Figure 4: Spectrum of circular nutation amplitudes derived from the GPS nutation rate series of 4 years with respect to the IAU80 model (top) and the differences between the IERS96 and the IAU80 nutation model (bottom).

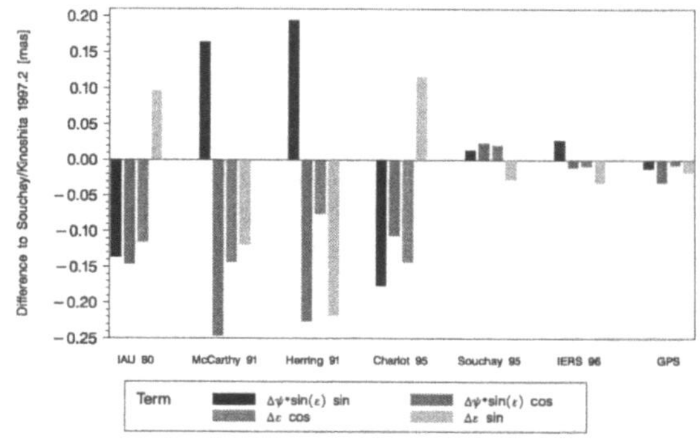

Figure 5: Comparison of the 13.66-day nutation coefficients from different sources with the most recent model by Souchay/Kinoshita 1997.2 .

between the estimation of LOD and the estimation of nutation rates, although LOD values have been determined by satellite techniques for a long time whereas nutation rates were not considered in the analyses.

At the CODE analysis center the routine estimation of nutation rates from global GPS 3-day solutions started in March 1994 and a series of more than 4 years in now available. This series of nutation rates has been analysed in detail by *Rothacher et al.* [1999]. Because "only" nutation *rates* are available from GPS solutions, no information about the long term behavior of nutation may be obtained from the GPS series. Low-frequency variations may only be derived from VLBI and, to a certain extent, from LLR data. GPS may contribute, however, to the nutation terms with periods below about 20 days.

Evidence confirming this statement comes from Figures 4(a) and (b), where a spectrum — converted from longitude and obliquity into prograde and retrograde components, see [*Herring et al.*, 1991] — has been computed (a) from the GPS nutation rate series of 4 years (the nutation rate corrections have been estimated with respect to the IAU80 nutation model) and (b) from the differences between the IERS96 [*McCarthy*, 1996] and the IAU80 model. From these two spectra it is evident, that all the major deficiencies present in the IAU80 model at periods below 16 days, as they were discovered by VLBI analyses already a decade ago, may also be seen by GPS.

We may thus estimate nutation amplitudes for periods below about 16 days from the GPS nutation rate series. That the quality of these amplitude estimates is comparable to that of VLBI analyses is show in Figure 5 for the 13.66-day period, which is very interesting from a geophysical point of view. The amplitude differences of the sine and cosine terms in longitude $\Delta\psi \cdot \sin \epsilon_0$ and obliquity $\Delta\epsilon$ with respect to the recent model by Souchay/Kinoshita 1997.2 [*Herring*, 1997] are given for a variety of different VLBI and LLR determinations contained in the literature (see [*Rothacher et al.*, 1999] for more details). The GPS-derived amplitudes agree as good as those of any of the more recent VLBI analyses (Souchay 95, IERS 96) with the Souchay/Kinoshita 1997.2 model.

Three techniques are therefore contributing to the estimation of nutation amplitudes and thus to the evaluation of today's nutation models: VLBI, LLR, and GPS. The GPS series containing mainly the high-frequency part and the VLBI and LLR series defining the long term behavior ideally complement each other and we may expect that a combination of all these series will be very fruitful and will produce the most accurate nutation amplitude estimates.

3.2 High-Frequency Earth Rotation

Short period variations in the Earth's rotation (periods shorter than 2 days) are mainly caused by changes in the ocean heights and ocean currents due to the gravity tidal potential of the Sun and Moon. These variations have well-known periods around 12 and 24 hours. Apart from these tidal signals also mass movements in the atmosphere are contributing to the high-frequency part of the ERP spectrum, although to a much smaller extent.

When we intend to compare and combine results from different space geodetic techniques concerning high-frequency variations in Earth rotation, we may do this on two very different levels:

- Direct comparison (and combination) of the x- and y-pole coordinates and UT1-UTC values given with a high temporal resolution (e.g. at intervals of 1–2 hours).

- Comparison and combination of the ocean tide amplitudes in polar motion (PM) and UT1 derived directly from long time spans of geodetic data or indirectly from a long series of sub-daily ERPs.

The first type of comparison is more basic and allows the study of variations in Earth rotation seen by the individual techniques independently of the physical origin and periods involved. It has the disadvantage, however, that ERP estimates with a high temporal resolution have to be available from different techniques. At present the estimation of ERPs at 1–2 hour intervals is only possible for VLBI (intensive campaigns, CORE activities of presently two days per week) and for GPS. The comparably sparse amount of data in the case of SLR or DORIS observations prevents the estimation of sub-daily ERPs with a sufficient precision.

That both, VLBI and GPS, are capable of estimating sub-daily ERPs and thus monitoring high-frequency variations in the Earth rotation, has been demonstrated using the data collected during intensive campaigns like CONT'94, CONT'96 etc. (see [*Lichten et al.*, 1992; *Freedman et al.*, 1994; *Weber*, 1996]. The quality of the results obtained by the two independent techniques may be seen in Figure 6 for the x-component of PM and for UT1. For VLBI and GPS 1-hour ERP estimates were computed. Both series follow quite neatly the ocean tide model by *Ray et al.* [1994] (identical with the model C described in [*Chao et al.*, 1996]). In Tables 2 and 3 the RMS differences between the three series are listed as computed over the time interval of the 11 days shown in Figure 6.

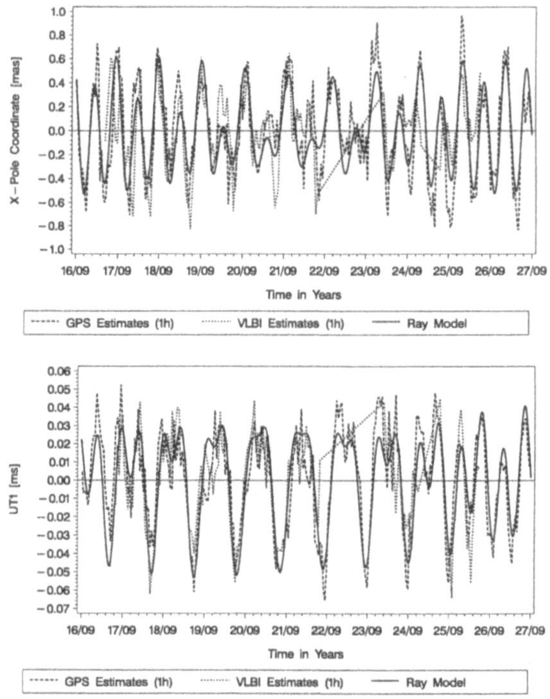

Figure 6: High-frequency variations in Earth rotation during CONT'96 from VLBI and GPS compared to the Ray model derived from altimetry data. The x-component of PM and UT1 are shown.

Table 2: RMS difference in mas between the 1-hour PM series from VLBI and from GPS and the model by Ray derived from altimetry data over a time interval of 11 days.

Values/Model	VLBI	Altimetry
GPS	0.33	0.20
VLBI		0.26

The two tables indicate, that the VLBI and GPS series are still dominated by the noise in the 1-hour estimates, because both series are in better agreement with the model by Ray than with each other. The RMS difference between the VLBI and GPS series is larger than that between one of the series and the Ray model by about a factor of $\sqrt{2}$. This is the factor you might expect in the case when the VLBI and GPS time series are fully independent. When the accuracy of the sub-daily ERP estimates gets better with time, we might start to detect signals common to both series but not present in the altimetry model. We may also expect that the combination of the VLBI and GPS series is more accurate than either of the two individual series.

Let us now have a closer look at the second level of comparison, the comparison of *ocean tide amplitudes* derived from space geodetic techniques. Four different techniques are capable of computing ocean tide "models" for the ERPs from their data sets: Richard Eanes computed ocean tide amplitudes from almost two decades of SLR data (Eanes, private communication), for VLBI such amplitudes are available

Table 3: RMS difference in μs between the 1-hour UT1 series from VLBI and from GPS and the model by Ray derived from altimetry data over a time interval of 11 days.

Values/Model	VLBI	Altimetry
GPS	19	13
VLBI		14

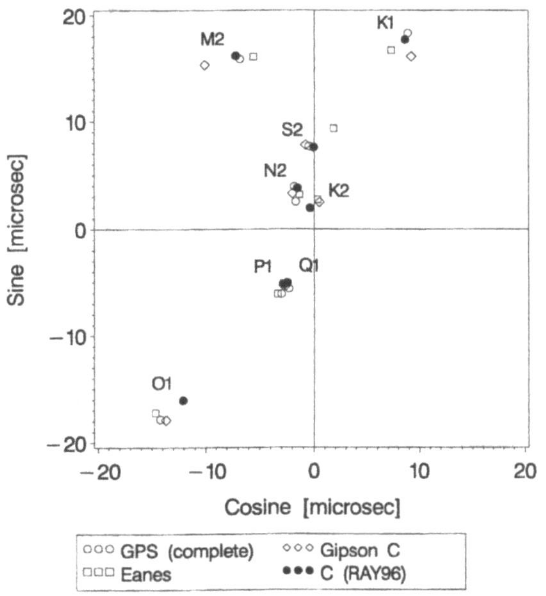

Figure 7: Diurnal and semi-diurnal UT1 amplitudes of the major ocean tide terms derived from four independent space techniques.

from [*Gipson*, 1996], and the altimeter data of the TOPEX/Poseidon mission was analysed in detail by *Chao et al.* [1996] to come up with amplitudes for the major tidal terms present in Earth rotation. In the case of GPS a continuous, uninterrupted series of ERPs with a time resolution of 2 hours is now available for a time span of more than 4 years. This unique series has been produced at CODE and is well documented in [*Rothacher*, 1998] (including detailed comparisons with other techniques).

As an example the UT1 tidal amplitudes derived from these four techniques are shown in Figure 7 in a so-called phasor diagram, where the cosine and the sine amplitudes of each tidal term are plotted against each other. Apart from the fact that in general the amplitudes of the four techniques agree very well, it is interesting to note, that for the large M_2 and K_1 tides the agreement is best between the Ray model and GPS, whereas for the O_1 tide the SLR, VLBI, and GPS results are very close to each other, but differ considerably from the values computed from altimeter data. Such differences between techniques give very important clues on systematic biases in the results of one or the other technique.

The overall comparison of all the estimated tidal amplitudes (more than 50 terms in GPS and VLBI) is summarized in Tables 4 and 5. They contain the RMS difference of the ocean tide amplitudes between the four techniques. We see that the quality of the results of all the four techniques is quite similar and amounts to about 10 μas and 1 μs in PM and UT1 amplitudes, respectively. It is no question, that

Table 4: RMS differences in μas of tidal polar motion amplitudes derived from different space techniques.

Technique	VLBI	SLR	Altim.
GPS	9.4	12.0	9.1
VLBI		9.6	6.5
SLR			9.0

Table 5: RMS differences in μs of tidal UT1 amplitudes derived from different space techniques.

Technique	VLBI	SLR	Altim.
GPS	1.0	1.1	0.9
VLBI		1.2	1.1
SLR			1.2

the best ocean tide amplitudes to be used to correct space geodetic data will be obtained from a rigorous combination of the estimates of the four techniques. More information about these comparisons and the GPS sub-daily ERP series may be found in [*Rothacher*, 1998].

3.3 Consistency of Reference Frames and ERPs

Because the Earth rotation parameters define the transformation between the terrestrial and the celestial reference frame, the consistency of the ERP series with the realization of the terrestrial and celestial frame is of utmost importance. The effect of different ITRF realizations on the determination of ERPs may be seen in Figure 8. The x-component of PM has been computed at CODE using (a) the changing ITRF realizations (e.g. ITRF93 was used from January 1995 to June 1996 in the routine processing) and (b) the ITRF96 (reprocessing starting in June 1993). In Figure 8 the differences between these two series and the IERS Bulletin A values are shown. We see that the different ITRF realizations produce offsets and drifts in the ERP series (net rotations between the coordinate sets defining the reference frames result in ERP offsets; net rotations in the velocity components lead to ERP drifts). Especially ITRF93 deviates significantly from the other ITRF realizations.

It is interesting to note that till 1987, ERP series took part in the combination of the BTS solutions [*Boucher and Altamimi*, 1990]. Due to a change of the combination software, ERPs have no longer been included in the generation of the terrestrial reference

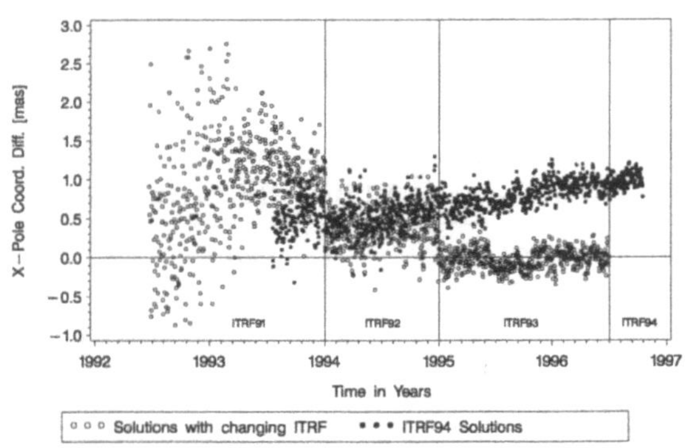

Figure 8: X-component of PM computed using the changing ITRF realizations (routine solution) and based entirely on ITRF96 (reprocessing). The differences of these two series with respect to IERS Bulletin A values are shown.

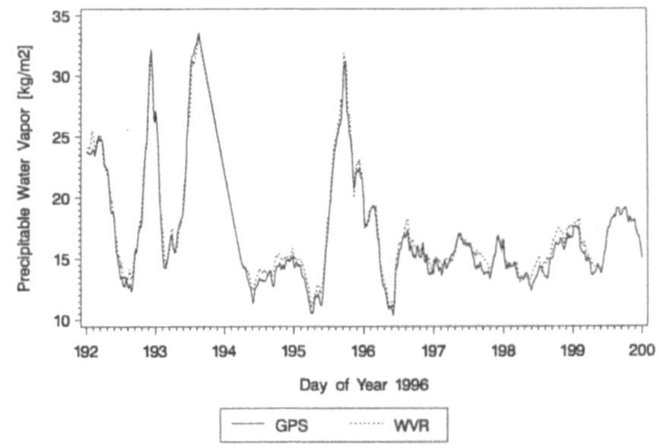

Figure 9: Comparison of the integrated precipitable water vapor above Onsala (Sweden) computed from GPS data and measured by a water vapor radiometer (WVR).

frame solutions after 1987 and the combination of ERPs and site coordinates and velocities has been done (except from an alignment) more or less independently. Only now activities have started (mainly within the IGS) to combine the ITRF solution and the ERPs series as rigorously as possible. Since fall 1998 most of the IGS analysis centers report ERP estimates together with site coordinates in their weekly SINEX files. Some of the GNAACs already perform a correct, simultaneous weekly combination of site coordinates and ERPs, which guarantees the consistency between the site coordinates and the ERPs. A similar procedure should eventually be implemented for the multi-year solutions of IERS. And, as the final goal, the ICRF (the coordinates of the radio sources) should be included in such a combination, too.

4 Atmosphere and Time Transfer

In the last few years the applications of GPS have grown in many directions and the IGS has become an interdisciplinary service. We will just briefly describe two of these new fields: the derivation of information about the atmosphere from GPS data and time transfer with geodetic GPS receivers connected to high-precision masers.

4.1 Atmosphere

The people doing VLBI analyses already knew for a long time, that the delays of the signals caused by the ionosphere and the neutral atmosphere (often simply called troposphere) contain important information about the atmosphere above a site.

Due to the frequency dependence of ionospheric refraction, the difference in the ionospheric delays of two signals emitted at different frequencies can be used to compute the total number of free electron along the path of the signals. Because the global GPS network is quite dense, its data can be used to derive global ionosphere models [*Feltens and Schaer*, 1998; *Schaer et al.*, 1998]. These models are not only of importance for GPS applications (ambiguity resolution, processing of single-frequency receivers) and ionosphere physicists, but also for the calibration of altimeter data, which is affected by ionospheric refraction, too.

The delays exerted on VLBI signals, GPS signals, or signals of other microwave systems, when they pass through the troposphere, may be converted (assuming that accurate pressure and temperature measurements are recorded at the site) into the amount of integrated water vapor above the site. Since water vapor is a critical quantity for weather prediction and global change, a great interest exists today to exploit GPS networks (and in principle also networks of other microwave systems) for meteorological and climatological purposes. As an illustration the water vapor estimates derived from GPS data are compared in Figure 9 to the water vapor measurements recorded by a water vapor radiometer (WVR). The two series of integrated precipitable water vapor above the station Onsala during 8 consecutive days of the year 1996 agree on the level of about 0.6 kg/m^2. The bias between the GPS and WVR series is -0.33 kg/m^2. Water vapor estimates of similar quality have also been reported by VLBI groups. Unfortunately, apart from comparisons, to our knowledge nobody made explicit use yet of the fact that the troposphere zenith delay parameters estimated in VLBI and GPS analyses should be identical for a site equipped with both techniques. This would be another step towards the integration of techniques. To have SLR measurements at the same site should be very beneficial, too, because the SLR observations are almost unaffected by the troposphere (compared, e.g., to GPS) and will thus help to resolve the extremely high correlations between the station heights and the troposphere parameters in VLBI and GPS analyses.

Apart from these ground-based applications, space-borne missions with Low Earth Orbiters (LEOs) carrying GPS receivers on board are opening up an entirely new dimension in atmosphere monitoring. The GPS/MET mission was the first of a series of missions (like CHAMP [*Reigber et al.*, 1998], COSMIC [*COSMIC*, 1999], ...) for atmosphere sounding using the GPS atmospheric occultation technique [*Yunck and Melbourne*, 1996; *Hajj et al.*, 1996]. It is expected that more than 500 atmospheric profiles might be recovered from the observations of a single LEO. To obtain optimum information about the atmosphere (ionosphere as well as troposphere) a combination and integration of the space-borne *and* ground-based techniques should be aimed at. The profiles from LEOs supply the vertical resolution whereas the ground-based segment gives horizontal resolution and calibration information.

4.2 Time Transfer

In December 1997 the IGS/BIPM Project to "Study Accurate Time and Frequency Comparisons" was set up to exploit the full potential of the GPS in the field of time and frequency transfer. It is expected that very accurate time transfer well below the nanosecond level will be achieved in an operational way. The accuracy limiting factor for frequency transfer will only be the measurement noise of the GPS phase observations of 1-2 mm [*Overney et al.*, 1998]. To reach these goals geodetic GPS receivers at timing laboratories have to be directly connected to high-precision clocks and the data should flow into the processing scheme of the IGS. Such a timing network, imbedded into the IGS, is going to revolutionize time and frequency transfer and will add the "fourth dimension" to the global reference frame.

5 Altimetry and Gravity

In this section we would like to give two examples in the field of altimetry and gravity showing the importance of and the benefit resulting from a combination of different observation techniques.

The first example is the TOPEX/Poseidon mission [*Fu et al.*, 1994]. Three independent satellite tracking systems (SLR, DORIS, and GPS) have been utilized to compute the trajectory of the TOPEX/Poseidon spacecraft. The availability of these three different tracking systems for Precise Orbit Determination (POD) allowed and still allows detailed comparisons and analyses of the positions of the spacecraft derived from different observation types and using different processing strategies. The combination of these three tracking systems led to a remarkable improvement in POD strategies for LEOs and the understanding of systematic biases between the systems. This progress is crucial for the goals of the TOPEX/Poseidon mission, especially for the high-precision monitoring of sea level with altimetry. Ocean tide models derived from altimeter data again are essential for the high-frequency variations in Earth rotation (see Section 3.2), showing that all these areas of research are interlaced.

The second example is the computation of the Earth gravity model EGM96 [*Lemoine et al.*, 1998]. To generate this global gravity model a combination and integration of a large variety of data types was necessary: surface gravity data, airborne surveys, altimetry data from TOPEX, ERS-1, and GEOSAT, SLR data from about 20 different satellites, GPS, DORIS, and TDRSS data. It is clear that such a combination of heterogeneous measurements is a tremendous task. The results and techniques developed in this area of research will be the starting point for the missions to come. CHAMP, GOCE, and GRACE will add a new dimension to the space and time resolution of the gravity field. Most of these missions will carry a GPS receiver on board, underlining the integrating role of GPS in this developments.

6 Integration and Combination of the Space Geodetic Techniques

In order to establish an integrated global geodetic observing system, integration efforts have to take place on three different levels:

1. Network level
2. Observation modeling level
3. Parameter estimation level

6.1 Network level

On the network level the various space geodetic techniques should be collocated to the extent possible at so-called fundamental observing sites. This is a "conditio sine qua non", because a common, highly consistent terrestrial reference frame may only be established through stations common to more than one technique. Without accurate local ties between the individual techniques located at one site, the integrating aspect of the fundamental station is lost. Looking at the quality of the site coordinate estimates nowadays, local ties should be known at the 1–2 mm level to really contribute to the integration of techniques, i.e., the local ties should not be the accuracy limiting factor in the realization of a common reference frame. The sites with more than one technique installed, which have a long time history should be considered as absolutely vital for the stability of the terrestrial reference frame. As already mentioned in Section 2.1, there are for instance only a very few baselines common to GPS and VLBI for a time interval of more than a year. To increase the number of collocation sites a GPS receiver should be operated at every SLR and VLBI site, at a reasonable number of DORIS sites, and at the locations of other important observing systems.

6.2 Observation Modeling Level

The integration of the techniques has to take place also on the level of the observation models used. The consistency concerning the models used (e.g. the ocean loading model for sites or the model for sub-daily ERP variations) has to be realized by means of the IERS Conventions. These processing standards, to be followed by all techniques, have to be continuously adapted and refined and are a prerequisite for the final integration level, namely the meaningful comparison and combination of the parameters stemming from different space techniques.

6.3 Parameter Estimation Level

Until now, the results of the various space geodetic techniques were mainly *compared* to each other. To really benefit from the particular advantages of each of the individual techniques, more emphasis has to be put on the *combination* of parameters common to more than one technique. That such comparisons and combinations are very fruitful has been shown, e.g., in TOPEX/Poseidon precise orbit determination using various tracking systems and in the IGS, where the products of seven analysis centers are combined.

Examples of parameter types common to more than one technique are manifold. Table 6 is an attempt to give an overview of the parameter space of interest in the combination of the basic geodetic observation techniques. When looking at Table 6 we see that the number of contributing techniques is largest for the site coordinates and velocities (ITRF) and the ERPs. It is therefore evident that the rigorous combination of these parameters should be the first step in the integration process. Optimum consistency will be reached when all the techniques supply weekly (or monthly) solutions containing all the parameters relevant to the reference frames, namely site coordinates, ERPs, and radio source positions together with the full variance-covariance information. We should keep in mind that the terrestrial reference frame is the basis for all integration aspects and serves as framework for most of the geophysical studies using space geodetic techniques.

Table 6: Parameter space of the basic geodetic observation techniques for combination.

Parameter Type	VLBI	GPS GLO.	DORIS PRARE	SLR	LLR
Radio S. (ICRF)	X				
Nutation	X	(X)			X
Pole X, Y	X	X	X	X	X
UT1	X				
LOD		X	X	X	X
Sub-daily ERPs	X	X			
ERP Ocean Tide Amp.	X	X		X	
Crd.+Vel. (ITRF)	X	X	X	X	X
Geocenter		X	X	X	
Gravity Field		X	X	X	(X)
Orbits		X	X	X	X
POD of LEOs		X	X	X	
Ionosphere	(X)	X	(X)		
Troposphere	X	X	X		
Clocks (Time Tr.)	(X)	X		(X)	

A logical next step is the combination of the troposphere zenith delays from VLBI, GPS, DORIS, and PRARE to improve the consistency and quality of station height estimates.

A much bigger step would then be the integration/combination of orbit parameters and the parameters of the gravity field. This part of the integration effort will become especially challenging and important for the new gravity missions to come.

Let us point out, that only the basic space geodetic observation techniques are considered in Table 6. Many more observing systems as for instance absolute gravimeters, altimeters, water vapor radiometers, SAR, etc. and future space-borne missions not included in Table 6, should be imbedded into an integrated global geodetic observing system.

It is clear, however, that in the near future nobody will be able to rigorously combine *all* the observables and parameter types of an integrated global geodetic observing system. Compromises are necessary in the practical realization. Thus it may not make much sense in general to use the GPS satellites at high altitudes to improve coefficients of the gravity field (except for resonance terms !) or to use the sparse LLR data to determine sub-daily ERPs.

7 Conclusions and Outlook

The comparison and combination/integration of all space geodetic techniques is vital (1) to improve the consistency between the techniques, (2) to realize the best possible ITRF, ICRF, and ERP series, (3) to distinguish between technique-specific biases and geophysical signals, and (4) to gain new insight into the geophysical processes.

Because of the weaknesses and strengths of each individual technique the integration of all techniques into an **Integrated Global Geodetic Observing System** (IGGOS) will be beneficial to all techniques. The resulting reference frames, models, and parameter series will be the basis for a multitude of new missions, missions with the aim to monitor and better understand our changing planet.

IGGOS might remind us of the "Grand Unification", the ambitious goal in elementary particle physics, i.e., there is a long way to go and a lot of work to be done!

References

BIH (1985), Annual Report for 1984, *published for the International Council of Scientific Unions*, Observatoire de Paris & Bureau International des Poids et Mesures, Paris, July 1985.

Boucher, C., and Z. Altamimi (1990), Evolution of the realizations of the Terrestrial Reference System done by the BIH and IERS (1984–1988), *IERS Technical Note 4*, Observatoire de Paris, Paris, December 1990.

Boucher, C., Z. Altamimi, and L. Duhem (1992), ITRF 91 and its associated velocity field, *IERS Technical Note 12*, Observatoire de Paris, Paris, October 1992.

Boucher, C., Z. Altamimi, M. Feissel, and P. Sillard (1996), Results and Analysis of the ITRF94, Central Bureau of IERS, Observatoire de Paris, Paris.

Boucher, C., Z. Altamimi, and P. Sillard (1998), Results and Analysis of the ITRF96, *IERS Technical Note 24*, Observatoire de Paris, Paris, May 1998.

Bruyninx, C. et al. (1997), The Use of the EUREF Permanent GPS Network for the Maintenance of the European Terrestrial Reference Frame, in *Proceedings of the Scientific Assembly of the International Association of Geodesy*, Rio de Janeiro, Brazil, September 3–9, (in preparation).

Chao, B. F., D. R. Ray, J. M. Gipson, G. D. Egbert, and C. Ma (1996), Diurnal/semidiurnal polar motion excited by oceanic tidal angular momentum, *Journal of Geophysical Research, 101*(B9), 20151–20163.

COSMIC (1999), Description of the COSMIC project at `http://www.cosmic.ucar.edu/cosmic/about.htm`.

Davies, P.B.H., and G. Blewitt (1997), Newcastle-Upon-Tyne Global Network Associate Analysis Centre Annual Report, in *IGS 1996 Annual Report*, edited by J.F. Zumberge *et al.*, pp. 237–252, IGS Central Bureau, Jet Propulsion Laboratory, Pasadena, California U.S.A., November 1997.

Feltens, J., and S. Schaer (1998), IGS Products for the Ionosphere, in *Proceedings of the IGS Analysis Center Workshop in Darmstadt, Germany*, February 9–11, 1998.

Freedman, A. P., R. Ibañez-Meier, T. A. Herring, S. M. Lichten, and J. O. Dickey (1994), Subdaily Earth rotation during the Epoch '92 campaign, *Geophysical Research Letters, 21*(9), 769–772.

Fu, L.L., E.J. Christensen, Ch.A. Yamarone, M. Lefebvre, Y. Ménard, M. Dorrer, and Ph. Escudier (1994), TOPEX/Poseidon Mission Overview, *Journal of Geophysical Research, 99*(C12), 24369–24381, December 1994.

Gipson, J. M. (1996), Very long baseline interferometry determination of neglected tidal terms in high-frequency Earth orientation variation, *Journal of Geophysical Research, 101*(B12), 28051–28064.

Guinot, B., and M. Feissel (1969), Annual Report for 1968, *published for the International Council of Scientific Unions*, Bureau International de l'Heure, Paris.

Hajj, G.A., E.R. Kursinski, W.I. Bertiger, S.S. Leroy, T.K. Meehan, L.J. Romans, and J.T. Schofield (1996), Initial Results of GPS-LEO Occultation Measurements of Earth's Atmosphere Obtained with the GPS-MET Experiment, in *International Association of Geodesy Symposia, Symposium No. 115: GPS Trends in Precise Terrestrial, Airborne, and Spaceborne Applications*, edited by G. Beutler *et al.*, pp. 144–153, Springer, ISBN 3-540-60872-6.

Herring, T. A. (1997), Analysis of most recent VLBI nutation offset series available at `ftp://gemini.gsfc.nasa.gov/pub/solutions/1083_aug97/eop_nutation.1083c` (private communication).

Herring, T. A., B. A. Buffett, P. M. Mathews, and I. I. Shapiro (1991), Forced Nutations of the Earth: Influence of Inner Core Dynamics, 3. Very Long Interferometry Data Analysis, *Journal of Geophysical Research, 96*(B5), 8259–8273.

Herring, T.A. (1997), MIT T2 Associate Analysis Center Report, in *IGS 1996 Annual Report*, edited by J.F. Zumberge *et al.*, pp. 255–267, IGS Central Bureau, Jet Propulsion Laboratory, Pasadena, California U.S.A., November 1997.

Lemoine, F.G., S.C. Kenyon, J.K. Factor, R.G. Trimmer, N.K. Pavlis, D.S. Chinn, C.M. Cox, S.M. Klosko, S.B. Luthcke, M.H. Torrence, Y.M. Wang, R.G. Williamson, E.C. Pavlis, R.H. Rapp, and T.R. Olson (1998), The Development of the Joint NASA GSFC and the National Imagery and Mapping Agency (NIMA) Geopotential Model EGM96, *NASA/TP-1998-206861*, NASA, Goddard Space Flight Center, Greenbelt, Maryland 20771, July 1998.

Lichten, S. M., S. L. Marcus, and L. O. Dickey (1992), Subdaily resolution of Earth rotation variations with Global Positioning System measurements, *Geophysical Research Letters, 19*(6), 537–540.

Ma, C. (1998), Recent VLBI analysis available at `ftp://gemini.gsfc.nasa.gov/pub/solutions/1102_aug98/` (private communication).

McCarthy, D.D. (1996), IERS Conventions (1996), *IERS Technical Note 21*, Observatoire de Paris, Paris, July 1996.

Overney, F., L. Prost, G. Dudle, T. Schildknecht, G. Beutler, J. A. Davis, J. M. Furlong, and P. Hetzel (1998), GPS Time Transfer Using Geodetic Receivers (GeTT): Results on European Baselines, in *Proceed-*

ings of the 12th *European Frequency and Time Forum EFTF 98*, Warsaw, Poland, March 10–12 1998.

Ray, R. D., D. J. Steinberg, B. F. Chao, and D. E. Cartwright (1994), Diurnal and semidiurnal variations in the earth's rotation rate induced by oceanic tides, *Science*, *264*, 830–832.

Reigber, Ch., H. Luehr, and P. Schwintzer (1998), Status of the CHAMP Geopotential Mission, in *Proceedings of the IAG Section II Symposium*, Munich (in preparation).

Rothacher, M. (1998), Recent Contributions of GPS to Earth Rotation and Reference Frames, Habilitation (2nd Ph.D.), University Press, University of Berne.

Rothacher, M., G. Beutler, T.A. Herring, and R. Weber (1999), Estimation of Nutation Using the Global Positioning System, *JGR*, *104*(B3), 4835–4859, March 10 1999.

Schaer, S., G. Beutler, and M. Rothacher (1998), Mapping and Predicting the Ionosphere, in *Proceedings of the IGS Analysis Center Workshop in Darmstadt, Germany*, February 9–11, 1998.

Springer, T. A., W. Gurtner, M. Rothacher, and S. Schaer (1997), EUREF Activities at the CODE Analysis Center, in *Proceedings of the 4th International Seminar on GPS in Central Europe*, vol. 4(27), pp. 165–176, Warsaw University of Technology, Warsaw, Poland, May 7–9 1997.

Weber, R. (1996), Monitoring Earth Orientation Variations at the Center for Orbit Determination in Europe (CODE), *Österreichische Zeitschrift für Vermessung&Geoinformation*, *84*(3), 269–275.

Yunck, T.P., and W.G. Melbourne (1996), Spaceborne GPS for Earth Science, in *International Association of Geodesy Symposia, Symposium No. 115: GPS Trends in Precise Terrestrial, Airborne, and Spaceborne Applications*, edited by G. Beutler *et al.*, pp. 113–122, Springer, ISBN 3-540-60872-6.

Envisaging a new era of gravity field research

K.H. ILK[1]

Abstract

The investigation of the Earth's gravity field will enter a new era at the turn of the forthcoming millenium. Despite the fact that a remarkable improvement of our knowledge of the gravity field has been achieved during the past decades the coming years promise another giant step in better understanding the system earth. The surprising success of the Global Positioning System, the breathtaking development of microcomputer technology but also the availability of highly sophisticated sensor components for reasonable prices enable to envisage some promising space borne concepts of gravity field missions. Various gravity sensors with complementary sensitivity characteristics will have the potential of measuring the various spectral parts of the static gravity field with high precision. The snapshots of the gravity field in monthly intervals should allow to detect time dependencies of the gravity field in certain parts of the spectrum.

In a first part the need of an improved gravity field is pointed out and a short overview of the scientific arguments is given. It is shown that the expected results will provide insight into the behavior of important features of the system earth. The improvement of accuracy and resolution can be achieved only by new space-borne gravity measurement techniques. The basic principles are sketched and the development of various mission concepts of three decades are reviewed. All three basic concepts discussed extensively during that time, high-low SST, low-low SST and SGG have a good chance to become realized within the coming years. CHAMP and GRACE will be launched at the turn of the millenium. GOCE will (hopefully) follow some years later. One should be aware of the fact that the coming years will represent an enormous challenge for the geo-sciences, not only with respect to the measurement and processing procedures, but also with respect to the interpretation and inversion of the data.

1. Do we need an improved gravity field?

We are looking back to forty years of satellite geodesy. During that time, millions of satellite observations have been collected. The combination with a huge number of terrestrial gravity data resulted in gravity field models which nobody could expect forty years ago. The latest NASA/DMA model, EGM96, represents the state of the art of such models. It is a spherical harmonics expansion complete up to degree 360. The resolution is about 100km and the accuracy of the geoid varies between 0.5 to 1m worldwide. So, why do we need a new era of gravity field research? Why do we want to go another step ahead? Is it just the human endeavor to approach the limits? Another amazing result of forty years of satellite geodesy: the magic line of one centimeter in defining precise global reference frames seems to be crossed already. Corresponding accuracy can be expected for geodetic positioning.

Indeed, there is a strong and urgent need for a further improvement of our knowledge of the earth's gravity field to come up with an adequate accuracy for the static part of the gravity field. And we need to know the temporal variations of the gravity field, at least for the long wavelength part.

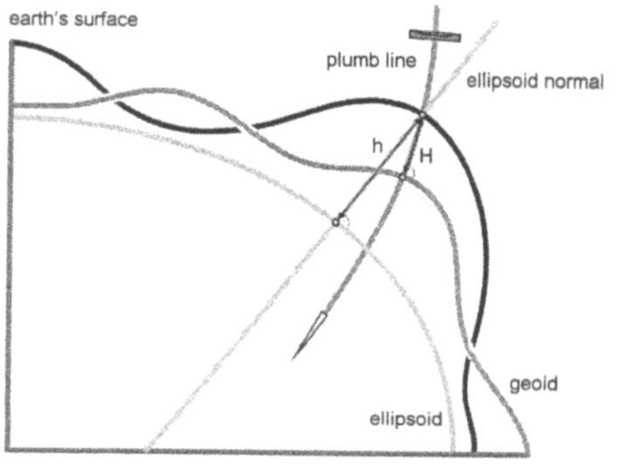

Fig. 1: Ellipsoid height h and orthometric height H.

What do we achieve with such an improved accuracy, e.g., for the geoid? The geoid as an equipotential surface of the gravity field is an important reference surface in geodesy. If we know the geoid with an accuracy of few centimeters for a resolution of 50 to 200km then a variety of geodetic applications can be envisaged. A world wide vertical datum system could be realized and the integration of regional vertical datum systems would be enabled. An improved geoid accuracy is necessary also for investigations of ice-mass balances and scientific tasks which depend on a precise vertical reference. The list of scientific applications could be continued, but the benefit of a centimeter geoid is not a pure scientific one. There are also practical applications which will influence the daily work of the surveyors. It could simplify the labour-intensive geodetic levelling by the so-called levelling by GPS. That is the transformation of GPS-derived ellipsoidal heights h into physical heights H, e.g., normal or orthometric heights (fig.1). In many parts of the earth geometric levelling is simply not possible and GPS-levelling is the only alternative.

[1] K.-H. Ilk: Institut für Theoretische Geodäsie, Universität Bonn, Nußallee 17, 53115 Bonn, Germany; Fax: +49 –228 -733 029, Tel.: +49 –228 -732 628, e-mail: ilk@theor.geod.uni-bonn.de

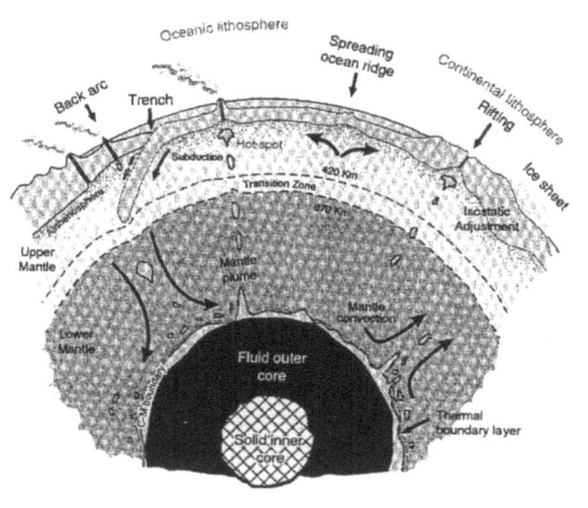

Fig. 2: Schematic representation of the structure of the earth's interior (ESA, 1996)

The earth is a dynamic system which undergoes continuous changes. Not only the topography and the ice regions are affected by various dynamic processes, also inside the earth a continuous redistribution of masses takes place. Examples are mantle convection, plate sub duction, spreading of ocean ridges and isostatic adjustment processes.

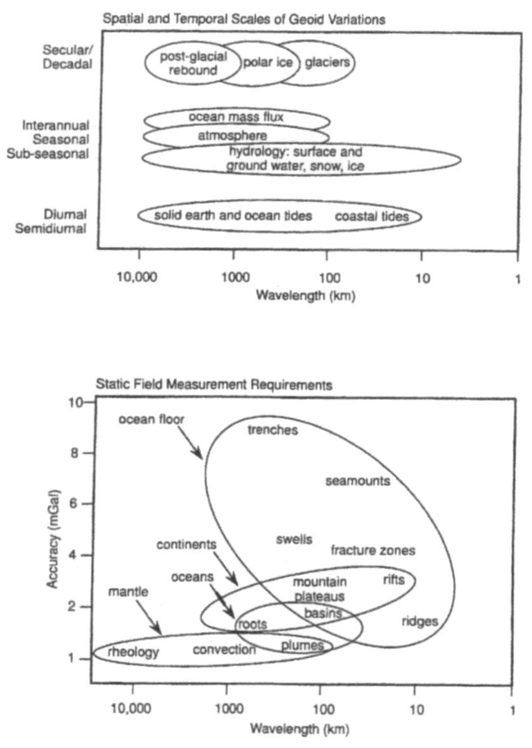

Fig. 3: Geophysical phenomena of the static and temporal constituents of the gravity field in dependency of the accuracy of the resolution in space and time (NRC, 1997).

The modification of mass distribution inside the earth and of the earth's envelope consequently alters the gravity field in the various spectral ranges in space and time. Variations of the outer gravity field reflect the internal density within the earth (fig.3,4, bottom). Therefore, a precise gravity field is an indispensable boundary condition for the internal mass structure of the earth and for geodynamic processes within the earth.

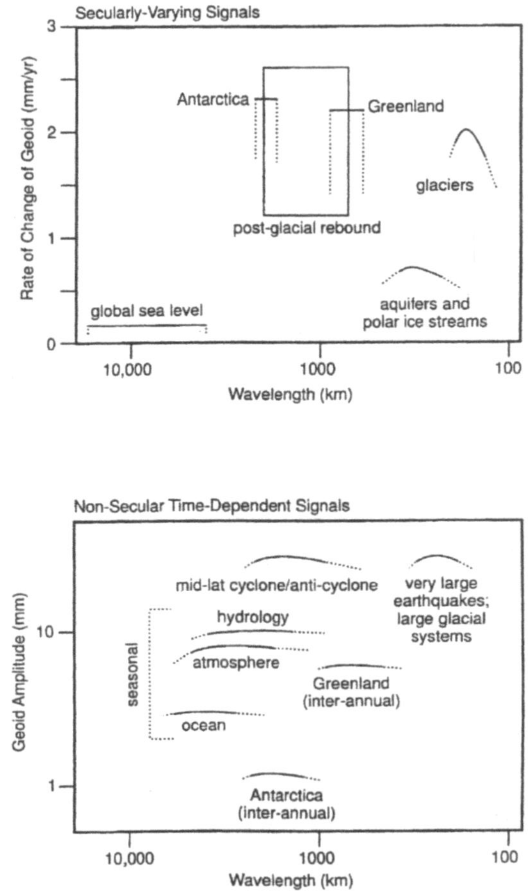

Fig. 4: Sources of variability in the earth's gravity field that would appear to vary linearly in time over a few years of observation (NRC, 1997)

In geological time scales the gravity field of the earth can be considered as a static system. Ninety nine percent of the total gravity field can be considered as being static. Only one percent undergoes a temporal variation of hours to thousands of years (figs.3,4, top). These are tidal phenomena caused by the gravity effects of sun and the moon, interannual, seasonal and subseasonal variations of snow and ice and irregular and secular mass transports in the hydrosphere. Mass variations of the atmosphere cause changes of the surface pressure and contribute to seasonal but also long term variations of the gravity field. Decadal and secular changes of the gravity field are caused by post glacial rebound processes and changes in the cryosphere (polar ice and glaciers).

A substantial improvement in the accuracy of the measurement of the gravity field would help to understand the earth system better than before. If the geodynamic phenomena as for example the thermal structure of the lithosphere, post glacial rebound processes or isostasy and mass balances shall be detected, then we need gravity anomalies with an accuracy of 1 to 2 mGals and with a resolution in space of 50 to 400km (fig.3).

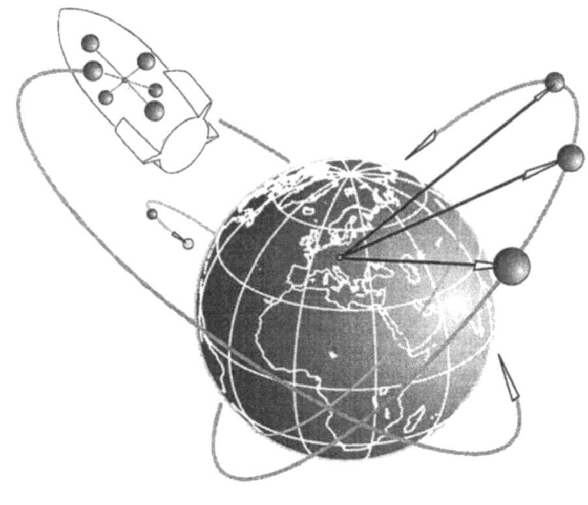

Fig. 5: Free fall techniques of gravity field measurements: principles of SST/SGG, absolute gravity measurement and laser tracking to artificial satellites (from left to right).

There is also a strong and urgent need of a high resolution gravity field in various oceanographic applications. For the detection of mesoscale phenomena with a spatial resolution of 100 to 200km a geoid accuracy of less than two centimeters down to a couple of millimeters is necessary. In case of basin scale phenomena we even need an accuracy of less than a couple of millimeters. This information is necessary to refer the sea surface topography to an equipotential surface, to investigate ocean currents and global ocean circulation and its temporal variations, and finally, to investigate meachanisms of mass and heat transport important for the development of the earth's climate.

2. Gravity Field from Space with Future Satellite Techniques

An improvement of our knowledge of the gravity field of the Earth requires global coverage with consistent gravity functionals. If an eventual time dependency of the gravity field (seasonal variations or secular changes) shall be detected then the observations must be available over a time period of several months either observed continuously or in form of „snapshots" with sufficient „exposure times" in certain time intervals. The classical approach of satellite geodesy consists basically in deriving the spherical harmonics coefficients representing the gravitational potential from an analysis of accumulated orbit perturbations of artificial satellites with different altitudes and orbit inclinations. These observations are collected over a time span of decades. Another important data source, radar altimetry, reflects oceanographic phenomena together with gravity field effects. Finally, terrestrial data to complement the gravity information in the high frequent range are very in homogenous with respect to accuracy, temporal and geographical distribution and datum reference. Therefore, the traditional procedure will not be able to fulfill the requirements necessary for a further improvement of the gravity field.

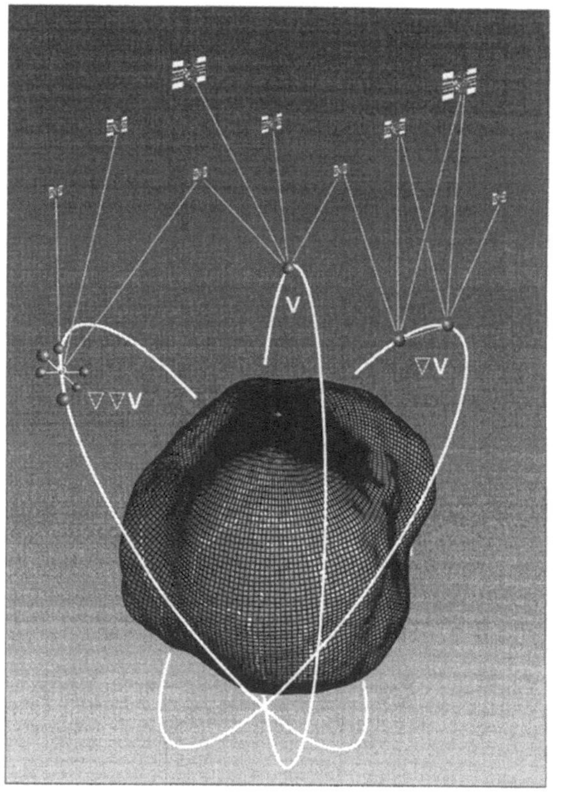

Fig. 6: Various future scenarios of gravity field measurement: SGG, high-low SST and high-low/low-low SST (from left to right).

The new type of observations should detect differential gravity field effects because of its higher sensitivity to irregular high frequent gravity field features. Furthermore, the observations have to be taken by continuous tracking. A basic role in many gravity measurement techniques plays the observation of the free fall motion of a test mass within the gravity field. Whether we perform absolute gravity measurements or laser tracking to a satellite equipped with reflectors the physical model is based on a test mass falling free under gravitational interaction. The alternative idea is not to observe the test mass from an earth fixed reference system but from another system which itself is freely falling. This can be performed e.g. within a spacecraft when a group of test masses is observed with respect to the freely falling center of mass of a spacecraft (fig.5). The basic difference of the various modifications of this concept is the size of the mutual distances of the test masses and their altitude, that means their sensitivity to the earth's gravity field.

There are two concepts under discussion which are based on the same physical principle (fig.6): Satellite-to-satellite Tracking (SST) and Satellite Gravity Gradiometry (SGG).

In both alternatives, the relative motion of test masses are measured, as relative distances and velocities or as relative accelerations. In case of SST the relative motion is measured along the line-of-sight(s) of two (or more) satellites. The concept is possible either in the so-called low-low or in the high-low mode. In the former case the satellites have approximately the same altitude (200 to 400km). In this case both satellites are equally sensitive to gravity field irregularities. In the latter case only one (the gravity field sensitive) satellite is placed into a low orbit while the other (observing) satellite(s) describe orbits with high altitudes. In case of SGG the elements of the gravity gradient or linear combinations thereof are intended to be measured simultaneously, depending on the sensitivity axes realized in the gradiometer instrument. It can be shown that the observations in these three cases can be related to the

- gravitational potential V, in bcase of high-low SST, to the
- gradient of the potential ∇V , in case of low-low SST, and to the
- gradient of the gradient of the gravitational potential $\nabla\nabla V$ (gravitational tensor), in case of SGG.

Of course, the technical realization is extremely difficult and every modification has its own special problem area. Without going into details, the gravity sensitive satellite cannot be modeled anymore by a simple free falling mass point. The new satellite techniques are complicated observation and test systems which interact with the various force fields of the system earth. The physical models have to take these facts into account.

3. Satellite Gravity Mission Preparations in the Past 25 Years

The idea of SST and also SGG goes back to the sixties and various mission scenarios have been investigated since that time. There are hardly any other satellite mission concepts to derive the gravity field of the Earth which are comparably well-founded and seriously investigated prior to its realization as SST and SGG. There were numerous conferences and strategy papers where the benefit of an improved gravity field for various disciplines is pointed out and, consequently, the need of realizing one of these promising space-borne gravity field missions. A short history of gravity field mission preparations (important conferences and reports) in the United States of America as well as in Europe is given in fig.7. In 1969 the first proposal was made to derive the gravity field of the Earth by a dedicated low-low SST mission. As milestones one should mention the Williamstown Report from 1969, the Colorado Springs meeting in 1987, and the Coolfont Report from 1991. Recent documents are (GOCE, 1996) or (NRC, 1997) where updated scientific arguments are given.

1969 First SST proposal (Wolff)
Williamstown Report (Solid Earth and Ocean Physics Conference)
1970
1972 EOPAP (Earth and Ocean Physics Application Programme)
1978 SONG (Space Oceanography, Navigation and Geodynamics)
1979 Application of a Dedicated Gravitational Satellite Mission
1980
1981 Satellite Altimetric Measurements of the Ocean
1983 Geopotential Research Mission – Scientific Rationale
1985 Geodesy – A Look at the Future
1986 SESAME (Solid Earth Science and Application Mission for Europe)
1987 Geophysical and Geodetic Requ. for Global Gravity (Colorado Springs)
ESA/NASA Workshop on a Joint Solid Earth Mission (Matera)
1988 Earth System Science – A Closer View
The Interdisciplinary Role of Space Geodesy (Erice)
1990 Geodesy in the Year 2000
1991 Solid Earth Science in the 1990's (Coolfont Report)
ARISTOTELES Workshop (Anacapri)
Earth Observation User Consultation Meeting
1992 The Future of Spaceborne Altimetry: Ocean and Climate Change
1994 Earth Observation Advisory Committee (EOAC)
Initiation of Earth Explorer and Earth Watch Missions
1995 Airborne Geophysics and Precise Positioning
1996 GOCE: Gravity Field and Steady State Ocean Circulation Mission
CHAMP Phase B-Executive Summary
1997 Satellite Gravity and the Geosphere (US National Research Council)
1998 European Views on Dedicated Gravity Field Missions: GRACE and GOCE

Fig. 7: History of gravity field mission preparations in USA and in Europe.

Nevertheless, a dedicated mission was never realized - only some experiments and first tests of these techniques have been performed (fig.8). One mission scenario for a SST low-low experiment, the Geopotential Research Mission (GRM), was treated in detail in the first half of the eighties. Despite the fact that GRM was not followed up, it initiated a number of activities, also in Europe. Here, the idea of SST and SGG was officially taken up with some delay and led in the mid-eighties to the definition of a Solid Earth Mission. End of the eighties a mission concept, based on the French GRADIO instrument, the so-called ARISTOTELES mission, was discussed in detail. In 1995 it was decided GPS/ ARISTOTELES not to follow up further on. The next realistic chance to place a geodetic gravity field mission emerged with the definition of a proposal to test the equivalence principle, fundamental for Einstein's Gravitational Theory. This so-called STEP mission was proposed as a candidate for the Medium Size Programme M2 as a joint ESA/NASA project. It was proposed again for M3 as a sole ESA project. The STEP geodesy experiment was designed to consist of a one-axis superconducting gravity gradiometer with an expected accuracy of 10^{-4} E $Hz^{-1/2}$ (e.g., PAIK, 1994). The experiment was finally planned to be combined with multi high-low SST. In spring 1996 STEP was not recommended as M3-mission - only a so-called MiniSTEP or GeoSTEP should be realized, most probably as a joint NASA/ESA project.

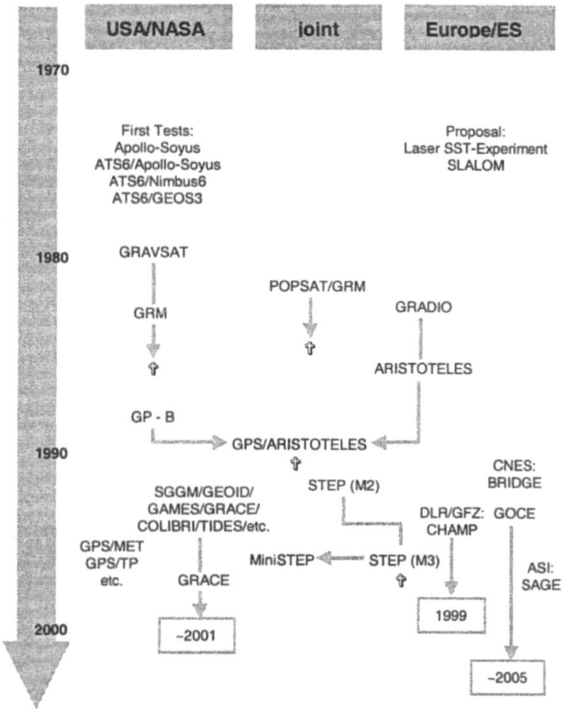

Fig. 8: Mission concepts and mission scenarios in USA and in Europe discussed since 1970.

After the year 1992 a number of „intermediate“ mission concepts were proposed. They should „bridge“ the past and future in gravity field research - and they should be cheaper than the mission proposals before. This is the programmatic idea behind the French mini-satellite concept BRIDGE. It was a CNES project and was considered to be based on SST/GPS or DORIS. Improved technology, in the main parts already available as existing sensors and commercial spacecraft components, and therefore cheaper than expected before gave a new fresh impulse to the idea of a future satellite gravity mission. These are the main criteria of the so-called „small satellite mission“ concepts. Additional cost-reducing actions are the decision to reduce quality standards and test efforts. A further cost reducing factor is the availability of the Global Positioning System (GPS) or any other precise satellite navigation system as GLONASS. By these systems the high altitude component of a high-low SST configuration is provided - and a relatively cheap possibility to determine precise orbits.

The high-low mission concept based on GPS and a LEO (Low Earth Orbiter) were applied in the past for Topex/Poseidon, GPS/Met and the Explorer platform (PAVLIS et al., 1996; RIM et al., 1996). Consequently, the high-low SST links are generally integrated in current mission concepts. Furthermore, it can be shown that SST and SGG are complementary techniques for detecting the gravity field: the long wavelength part can be improved by (high-low but also low-low) SST and the high-frequent part can be detected best by SGG. Therefore, current promising mission concepts consist of a combination of both, high-low SST and SGG or high-low SST and low-low SST. In the former case a low platform carries a gravity gradiometer (preferably a full tensor component gradiometer) and a system of high altitude satellites observes the low earth orbiter. In the latter case the low-low SST configuration is linked to the high altitude satellites of a precise satellite navigation system (GPS and/or GLONASS). Even if the small satellite mission concepts are not able to provide the very high accuracies and resolutions necessary for some applications in the geo-sciences they could decisively contribute to a gravity field improvement.

During the last years, a series of concepts were proposed; some of them were considered as low cost missions, e.g., the American proposals GRACE (DAVIS et al., 1996), COLIBRI or HUMMINGBIRD (Clark et al., 1996; Pavlis et al., 1996), or the German project CHAMP (REIGBER et al., 1996a, 1996b). Typically, the gravity field sensitive spacecraft were planned to be equipped with GPS receivers and accelerometers. A recent proposal is SAGE (Satellite Accelerometry for Gravity field Exploration). It is an Italian mission concept aiming at determining the gravity field of the earth by the high-low SST mode based on very accurate acceleration and position measurements (ASI, 1998). In case of the mission concepts GRACE and TIDES the relative motion of two small free-flying spacecrafts is planned to be measured. TIDES (Tidal Interferometric Detector in Space) is based on laser doppler interferometry using ultra-stable lasers (CHAO et al., 1996). The concept consists of two small free-flying spacecraft in tandem formation about 500km apart and in near-circular orbit. The surface forces are considered to be compensated or measured and taken into account afterwards. Other mission concepts were based on satellite gravity gradiometry for deriving the high frequency part, eventually equipped with GPS receivers to enable satellite-to-satellite tracking to improve also the long wavelength part of the gravity field, as e.g. the American mission concept GEOID or ESA's Earth Explorer Mission candidate GOCE. GEOID (Gravity for Earth, Ocean, and Ice Dynamics) was a spaceflight mission proposed by GSFC and based on the University of Maryland's superconducting gravity gradiometer (BILLS et al., 1996; CANAVAN et al., 1996; SHIRRON et al., 1996). These concepts can be considered as dedicated gravity missions. They have the potential to improve high frequent features of the static part of the gravity field up to a degree of 200 of a spherical harmonics expansion or even more - but they are expected to be more expensive and do not fulfill the criteria of small satellite missions. A comparative study of these various concepts is performed in (SNEEUW and ILK, 1997).

After careful selection procedures two mission concepts were successful and will be realized in the coming years: CHAMP, a German multi-sensor satellite mission with international contributions and GRACE, a combined high-low/low-low SST mission as a joint American-German mission. The European SGG - mission concept GOCE has a realistic chance for realisation and is under intensive preparation.

4. The Coming Missions

In December 1999, the German geoscientific small satellite CHAMP (Challenging Mini-Satellite Payload for Geophysical Research and Application) will be brought into a nearly circular orbit with an inclination of 83° at an altitude of about 400km (REIGBER et al., 1996a, 1996b). In a five years mission the observation system will provide earth system related data with yet unattained accuracy. The following main tasks are envisaged:

- measurement of the stationary and time variable part of the global gravity field in the medium and long wavelength frequency range,
- measurement of the global magnetic field and its temporal variations,
- sounding of the atmosphere and the ionosphere.

The satellite mission will be carried out by DLR (German Space Agency) and GFZ (GeoForschungsZentrum) Potsdam. Of special importance of this mission is the fact that the first time in the history of satellite geodesy important earth related data will be collected at the same time for a time span of five years on a low altitude platform:

- precise intersatellite measurements between the low satellite CHAMP and the high GPS satellites (high-low SST with a range accuracy of $0.1\text{m}/\sqrt{\text{Hz}}$),
- accelerations along three axis to determine the surface and inertial forces of CHAMP (accuracy of acceleration: $1\text{nm}/\text{s}^2\sqrt{\text{Hz}}$),
- orientation measurements by star sensors and by various GPS antennas,
- radio occultation measurements between CHAMP and the GPS satellites,
- measurement of scalar and vector magnetic field strengths,
- measurement of the electrical field by ion drift meter,
- precise earth bound (two-color) laser ranging,
- altimeter measurements between CHAMP and the sea/ice surface by nadir oriented GPS antenna.

The measurement accuracy of the GPS receivers and the altitude of CHAMP between 300 to 500km enables a spatial resolution of the gravity field of approximately 500km. The expected accuracy of the corresponding spectral range is by a factor ten higher than realized in the present gravity field models. That means that the geoid can be determined with an accuracy of one centimeter for wavelengths down to 1000km.

The American-German mission GRACE (Gravity Recovery and Climate Experiment) will open the door to a further improvement of the gravity field with respect to accuracy and resolution. GRACE is a follow-on mission to CHAMP. The mission will consist of two identical

CHAMP satellites without the boom where the magnetometers are placed in case of CHAMP. The intersatellite range-rates in along-track direction between the low satellites (at an altitude of about 400km) will be measured with μm-accuracy. The high-low links to the GPS satellites will have a comparable accuracy as planned for the CHAMP mission. With these measurement accuracies it will be possible to determine the medium and long wavelength part of the geoid down to 500km wavelengths with an accuracy of about 0.1mm. These are two orders better than in case of CHAMP. The launch of GRACE is planned for the year 2001. CHAMP and GRACE will have a parallel mission period of approximately two years. The parallel missions with different orbits will enable additional experiments which will lead to a synergetic effect of both missions. The tasks of GRACE are directed to the:

- determination of the static part of the global gravity field with unattained accuracy in the medium spectral range; geoid accuracies of 0.01mm for a resolution of >5000km and of 0.01 to 0.1mm for a resolution of 500 to 5000km can be expected,
- determination of the time variability of the gravity field in two to four weeks temporal resolution; the accuracy of the change of the geoid in the size of 0.01 to 0.001mm / year can be expected,
- sounding of the atmosphere and the ionosphere.

The GRACE low-low SST mission and the GPS-GRACE high-low SST test system will provide valuable earth related observables over a mission period of five years:

- precise along-track intersatellite measurements between the low satellites of GRACE (low-low SST) in the size of few μm and a mutual distance of both satellites between 200 and 500km,
- precise relative measurements between the GRACE satellites and the GPS satellites,
- accelerations along three axis to determine the surface and inertial forces of CHAMP (accuracy of acceleration: $1\text{nm}/\text{s}^2\sqrt{\text{Hz}}$),
- radio occultation measurements between the low satellites of GRACE and the high GPS satellites at the setting and rising phase,

GOCE (Gravity Field and Steady-State Ocean Circulation Explorer Misson) is one of the candidates to become the first Earth Explorer core mission. It is a high resolution gravity field mission concept and will open a completely new range of spatial scales of the earth's gravitational field spectrum down to 100km wavelength. GOCE is planned to be launched in a nearly circular sun-synchronous orbit with an inclination of $\approx 97°$ at an altitude of around 250 km, carrying a

- dual frequency combined GPS and GLONASS receiver called GRAS (for its utilisation as GNSS receiver for atmospheric sounding), and a
- three-axis gravity gradiometer, either an ambient temperature instrument with a precision of about $4\cdot10^{-3}\text{E}/\sqrt{\text{Hz}}$ for the diagonal components of the gravity tensor, or a cryogenic gradiometer with a performance raughly one order of magnitude better, furthermore a

- star tracker to control the orientation of the gradiometer with an accuracy of about $3 \cdot 10^{-3} \mathrm{rad} / \sqrt{\mathrm{Hz}}$, and an
- equipment to keep the satellite motion drag-free.

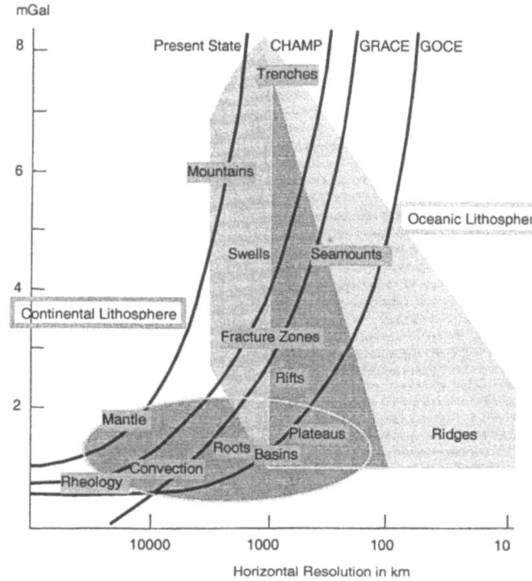

Fig. 9: Required accuracy (estimated at approximately 10 percent of the gravity signal) as a function of horizontal resolution necessary to resolve the quoted geodynamical and tectonic features. The curves represent the potential improvements of knowledge by CHAMP, GRACE and GOCE missions (from ESA, 1998).

The excellent evaluation of this mission concept under nine mission proposals of ESA's Earth Explorer Mission programme justifies the hope that also this mission will be realized within the coming decade.

5. Analysing the System Earth

CHAMP, GRACE and GOCE have the potential to revolutionize the knowledge of the system earth. Not only the static part of the gravity field can be determined with unattained accuracy also an eventual time dependency can be derived. Despite the fact that all three missions have the potential to measure the gravity field by sort of relative measurements between free falling sensors, they are not redundant. Indeed, the characteristics of high-low SST, low-low SST and SGG are rather complementary than competitive (compare, e.g., ESA, 1998, NRC, 1997, Sneeuw and Ilk, 1997).

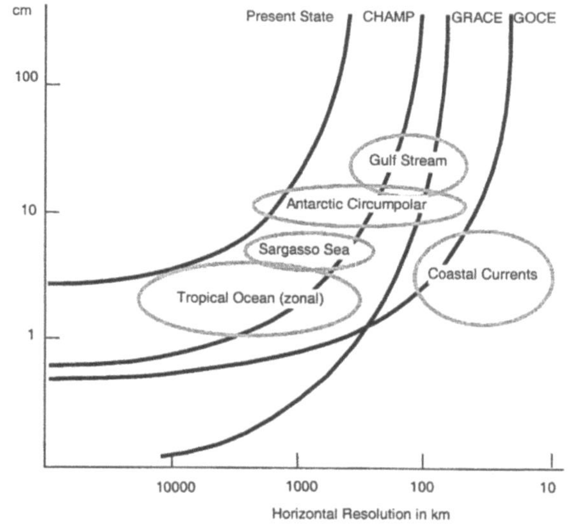

Fig. 10: Required accuracy (estimated at approximately 10 percent of the signal) as a function of horizontal resolution necessary to resolve various ocean circulation features. The curves represent the potential improvements of knowledge by CHAMP, GRACE and GOCE missions (from ESA, 1998).

SST is superior in the lower harmonics below degree and order 50 to 60. A mission like GRACE, therefore, is optimal for studying time-varying gravity effects at moderate wavelengths. The static part of the gravity field up to approximately degree 50 can be expected with high accuracy. A condition to detect temporal effects is a corresponding mission duration of several years. Satellite gradiometry is superior for obtaining high spatial resolution from a moderate mission length. A recent study (ESA, 1998) showed that increase of measurement precision or decrease of altitude results in a clear gain of spatial resolution in case of SGG, while this effect is very moderate in case of SST. A SGG mission like GOCE is superior in the short wavelengths parts of the gravity field up to a spherical harmonics degree of 250. The results of a mission like GOCE start to be better than those of a low-low SST mission from degree 60 to 80 on. A high-low SST mission like CHAMP can provide an improvement in the knowledge of the gravity field of approximately one order of magnitude over present models for wavelengths between 400 to 2000km. Fig. 9 gives an idea of the benefit of all three missions for the study of features with geophysical interest, and fig. 10 shows the progress by these missions in oceanography with respect to ocean circulation features.

One should be aware of the fact that the coming years will represent an enormous challenge for the geo-sciences. First of all with respect to the measurement and the processing procedures. Typical for the future satellite missions is the fact that they are not just point masses moving free within the gravity field as in the past. Instead they are multi-purpose observation and test systems on low flying platforms (fig.11). Various different observables are collected parallel over years providing valuable information of the system earth. The first time in history it is possible to investigate the interactions of the potential fields of the earth. Challenging questions are related to the (figs. 11,12)

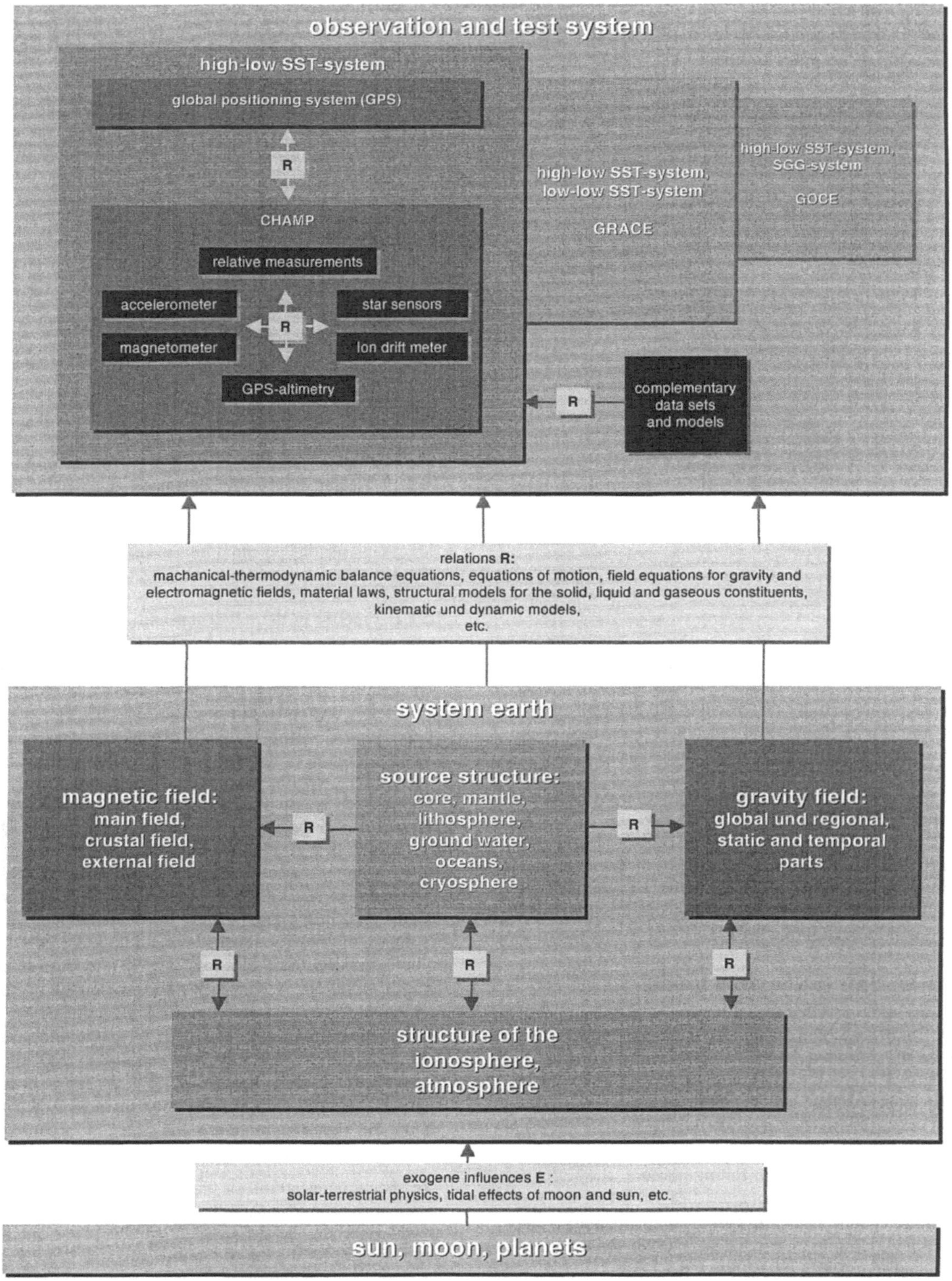

Fig. 11: Testing the system earth.

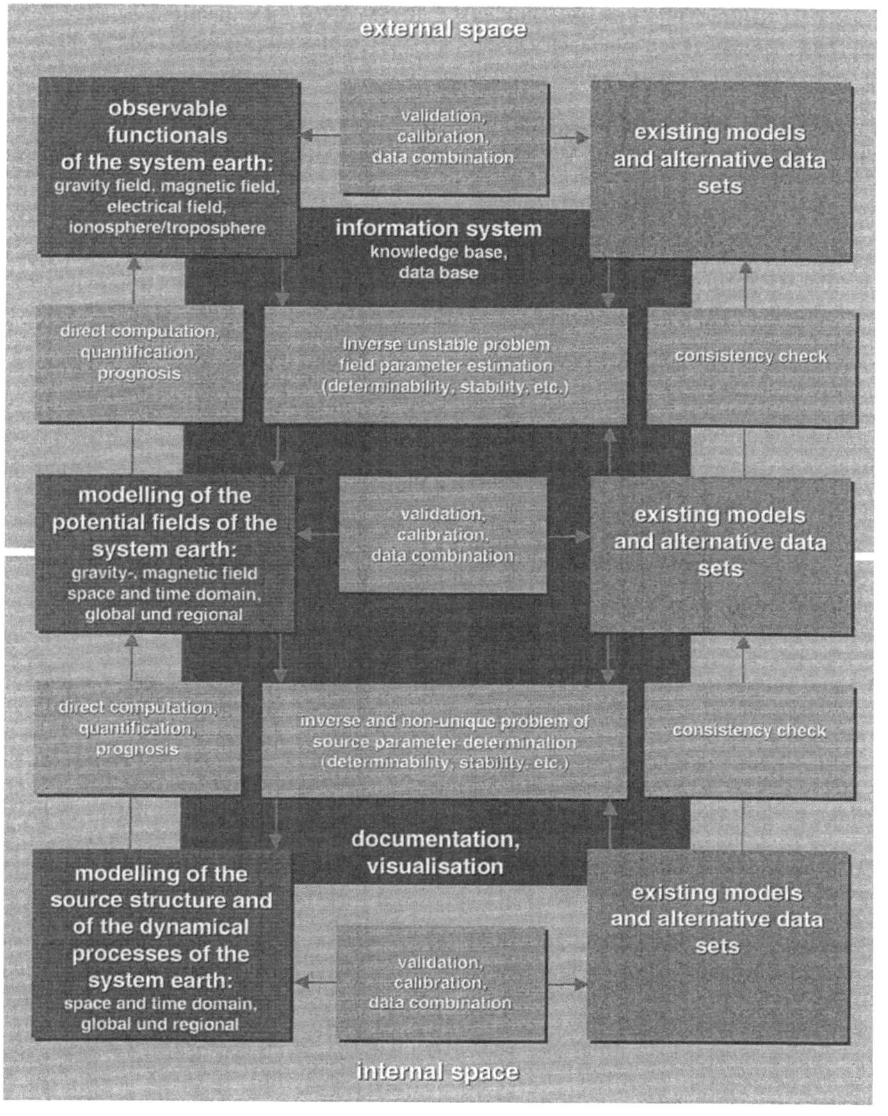

Fig. 12: Data analysis structure

- formulation of balance equations for the various model parameters of the static and temporal parts of the gravity field,
- development of models to take into account the fact that the three-dimensional orbits are measured together with atmospheric data, atitude control data (by star sensors) and accelerations in three directions to eliminate the surface forces,
- data verfication, calibration and combination procedures based on existing terrestrial measurements and available models, and
- tackling the inverse geophysical/geodetic models based on time series of the gravity field and the ionosphere, collected over a time span of severeal years.

Satellite measurements can provide unprecedented views of the earth's gravity field and its changes with time. Together with complementary geophysical data, satellite gravity data represent a "new frontier" in studies of the system earth.

References:

ASI: *SAGE, Phase A Final Report*, International Geoid Service, Milano, Nov. 1998

BILLS B.G., PAIK H.J.: *GEOID mission: Gradiometer Views of Static and Dynamic Gravity Signals*, presented at: AGU Spring Meeting, Baltimore, May 1996

CANAVAN E.R, VOL MOODY M.: *GEOID Mission: Instrument and Error Budget*, presented at: AGU Spring Meeting, Baltimore, May 1996

CHAO B.F., COLOMBO O.L, BENDER P.L.: *Global Gravitational Changes and The Mission Concept of TIDES,* presented at: AGU Spring Meeting, Baltimore, May 1996

CLARK T.A., SKILLMAN D.R., BAUER F.H., O'DONNELL J.R., PAVLIS E.C., LIBERMAN D.I., BISIACCI F.: COLIBRI: *A Low-cost Spacecraft Design for Gravity Change Studies,* presented at: AGU Spring Meeting, Baltimore, May 1996

DAVIS E.S., MELBOURNE W.G., REIGBER CH., TAPLEY B.D., WATKINS M.M.: *GRACE: An SST Mission for Gravity Mapping,* presented at: AGU Spring Meeting, Baltimore, May 1996

ESA: *Geophysical Interpretation of High Resolution Gravity Field Information,* Final report, 1991

ESA: *Gravity Field and Stead-State Ocean Circulation Mission, Reports for Assessment: The Nine Candidate Earth Explorer Missions,* ESA SP-1196(1), ESA Publications Division, ESTEC, Noordwijk, 1996

ESA: *European Views on Dedicated Gravity Field Missions: GRACE and GOCE,* European Space Agency, ESD-MAG-REP-CON-001, May 1998

GOCE: *Gravity Field and Steady-State Ocean Circulation Mission. Reports for Assessment,* The Nine Candidate Earth Explorer Missions, ESA SP-1196(1), 1996

MÜLLER J., SNEEUW N., RUMMEL R.: *Problems and Prospects of the Planned Gravity Missions GOCE and CHAMP,* this volume, Oct. 1996

NASA: *Geophysical and Geodetic Requirements of Global Gravity Field Measurements (1987-2000).* Report of a Gravity Workshop at Colorado Springs, Geodynamics Branch, 1987

NRC: *NRC (National Research Council): Satellite Gravity and the Geosphere.* National Academy Press, Washington D.C., 1997

PAIK H.J.: *Superconducting Gravity Gradiometers on STEP,* in: R. Rummel and P. Schwintzer (eds.), A Major STEP For Geodesy, Report 1994 of the STEP Geodesy Working Group, Munich, Potsdam, Nov. 1994

PAVLIS E.C., CLARK T.A, BILLS B.: *Gravity Modelling With the Hummingbird Constellation,* presented at: AGU Spring Meeting, Baltimore, May 1996

PAVLIS E.C., LEMOINE F.G., OLSON T.R., ROWLANDS D.D., CHAN J.C.: *Earth Gravity Model Improvement From GPS Tracking of Topex/Poseidon,* GPS/Met and Explorer Platform, presented at: AGU Spring Meeting, Baltimore, May 1996

REIGBER CH., KANG Z., KÖNIG R, SCHWINTZER P.: *CHAMP - A Minisatellite Mission for Geopotential and Atmospheric Research,* presented at: AGU Spring Meeting, Baltimore, May 1996a

REIGBER CH., BOCK R., FÖRSTE CH., GRUNWALDT L., JAKOWSKI N., LÜHR H., SCHWINTZER P., TILGNER C.: *CHAMP Phase B,* Executive Summary, Scientific Technical Report STR96/13, GeoForschungsZentrum Potsdam, Potsdam, Nov. 1996b

RIM H.J., SCHUTZ B.E., ABUSALI P.A.M., NAM Y.S., POOLE S.R., RIES J.C., SHUM C.K. TAPLEY B.D: *The Contribution of GPS Data to Modern Gravity Model Improvement,* presented at: AGU Spring Meeting, Baltimore, May 1996

RUMMEL R., SNEEUW N., MÜLLER J.: *Geodetic Requirements and Prospects Study of Gravity Explorer Mission Requirements* (A Simulation Study), Technical Note, prepared for DASA, Dornier Satellitensysteme GmbH, München, Sept. 1995

SHIRRON P., DIPIRRO M., CASTLES S.: *GEOID Mission: Spacecraft and Dewar Design,* presented at: AGU Spring Meeting, Baltimore, May 1996

SNEEUW N.: *Global Gravity Field Error Simulations for STEP-Geodesy,* in: R. Rummel and P. Schwintzer (eds.), A Major STEP For Geodesy, Report 1994 of the STEP Geodesy Working Group, Munich, Potsdam, Nov. 1994

SNEEUW N., RUMMEL R., MÜLLER J.: *The Earth's Gravity Field from the STEP Mission,* Class. Quantum Grav. 13, A113-A117, 1996

SNEEUW N., ILK K. H.: *The Status of Spaceborne gravity field mission consepts: a comparative simulation study,* in J. Segawa, et al. (eds): Gravity, Geoid and Marine Geodesy, IAG Symposia Vol. 117, Springer, Berlin, etc., 1997

Williamstown Report: *The Terrestrial Environment.* Solid-Earth and Ocean Physics, Application of Space and Astronomic Techniques. Report of a Study at Willliamstown, Mass, to the NASA, 1969.

Status of the CHAMP Mission

CH. REIGBER, H. LÜHR, P. SCHWINTZER[1]

Abstract

The German geopotential and atmospheric research mission CHAMP has entered it's development phase C/D in January 1997 under the responsibility of the GFZ Potsdam. The spacecraft is manufactured by an industrial consortium with DASA Jena Optronik as the main contractor. The CHAMP instrumentation is composed of a GPS receiver for continuous high-low GPS/CHAMP tracking, for GPS altimetry and for atmosphere limb sounding (NASA provided), an accelerometer for measuring surface force accelerations (CNES provided), an ion drift meter (USAF provided) and scalar and vector magnetometers. Ground segment components are to a large extent provided by the German Aerospace Center (DLR). Launch of the spacecraft is foreseen for January 2000. The paper gives a short overview on the status of the overall system development.

1. Science Background

he central task of geosciences in the next decades will be studying and trying to understand the Earth as a system: a system composed of solid, fluid and gaseous parts which show large variations in space and time and between of which complex interactions take place on quite different time scales. Characterising such an extensive, complex and heterogeneous system requires very long and synoptic data series of the phenomena taking place within and between the various spheres of the system, beside of course the computer resources required for making use of such enormous data sets in the model development, improvement or validation. Global long term, synoptic data sets can systematically only be acquired by near Earth satellites, which however in general provide a limited resolution because of the reduced sensitivity with altitude. A combination of satellite observations with more regional data sets obtained on ground or in the first few kilometers in the atmosphere is therefore the suitable observation scenario to follow.

On this background the small satellite mission CHAMP with it's multifunctional and complementary payload is aiming for a substantial contribution to the following themes

- Solid Earth: Composition, structure and dynamics of the solid parts and interactions with the ocean and atmosphere
- Ocean and atmosphere: Structure and short term dynamics of the ocean and atmosphere and their interaction with properties of the solid Earth including the global hydrological cycle, weather and climate issues.

Primary task of the CHAMP mission is a substantial improvement in observing the global structure of the Earth's gravitational, magnetic and electrical fields and the space-time variability of these fields. A secondary mission goal is the sounding of the neutral atmosphere and ionosphere by equipment used already for the geopotential recovery.

CHAMP fulfils the criteria of a small satellite mission, i.e., short development time through the usage of existing sensors and commercial subsystem components and reduced cost through protoflight approach, suitable quality standards and streamlined test efforts. GFZ Potsdam (PI: Ch. Reigber) as the primary scientific user of the mission is responsible for the overall project management, system control and instrument engineering.

The mission was studied in its feasibility, designed and specified under lead of GFZ together with an industrial consortium and the German Aerospace Centre (DLR). After completion of the project's definition and specification phase (Phase B), CHAMP entered into its manufacturing and integration phase (Phase C/D) in January 1997. The satellite is built by a consortium of German space firm under the lead of DASA Jena Optronik. The launch is scheduled for January 2000.

2. Mission Characteristics and Ground Facilities

CHAMP will be injected into a 87.3 degrees inclined orbit with an approximate initial altitude of 470 km. By air drag impact and orbital manoeuvres the satellite will decay to about 300 km within it's design lifetime of 5 years. The low altitude orbit supports the spatial resolution of the geopotential field whereas the long mission duration helps to recover temporal field variations. The S/C will be launched on a Russian COSMOS launcher from the launch site Plesetzk near Archangelsk.

The CHAMP ground segment consists of the Mission Control Center (MCC) handling the Mission Operational System (MOS), a Raw Data Center (RDC) receiving the CHAMP science and housekeeping data, a Mission Management and Control Center (MMCC) acting as interface between experimeters, principal investigators and users and the MCC, a high rate GPS, laser station and magnetic field observatories network and a science data system, composed of processing and arching centers (CAPs) at GFZ Potsdam (gravity, magnetics, atmosphere) and DLR Neustrelitz (ionosphere) and the GFZ Earth System Data Index (GFZ-ESDI) system providing data and products to the users. DLR's Space Operations Center (GSOC) is

[1] Christoph Reigber, H. Lühr, Peter Schwintzer: GeoForschungsZentrum Potsdam, Div. 1, Kinematics & Dynamics of the Earth, Telegrafenberg A 6, D-14473 Potsdam, Germany; Phone.: +49 - 331 - 288-1100, Fax: +49 - 331 - 288-1111, E-mail: reigber@gfz-potsdam.de, www: http//www.gfz-potsdam.de/

responsible for the spacecraft control using the ground stations in Weilheim (TT&C) and Neustrelitz (TM). The German remote sensing data center (DFD) will run the raw data archive including the science data preprocessing task. GFZ is responsible for setting up and running the MMCC and the dedicated GPS network. The preliminary and critical design meetings for all of these components took place in 1997/98. Development work is proceeding for the launcher, the launch services and all ground segment components.

3. The Satellite and its Payload

Figure 1 sketches the satellite structure and its payload components. The structure consists of a trapezoidal body of 0.6 m height, 4 m length and 1.6 m / 0.4 m bottom/top width and a 4 m long boom in flight direction. The front panel and rear panel are 70 deg and 20 deg, respectively, inclined towards nadir.

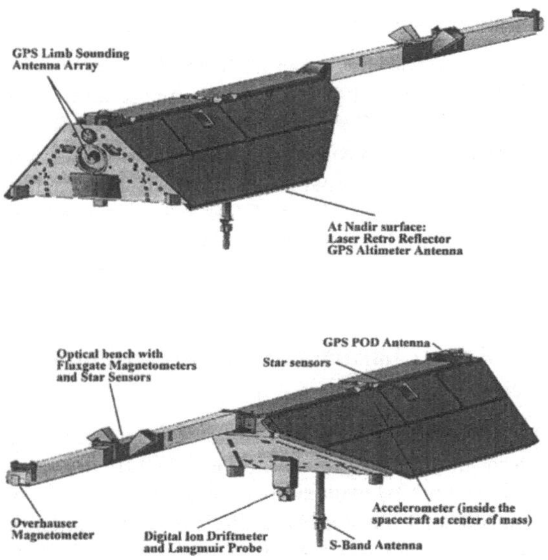

Figure 1. Sketch of the CHAMP satellite and payload accommodation

The mass of the satellite is 500 kg including 32 kg of payload mass. The power consumption will be 150 W of which 45 W is needed to run the instruments.

The CHAMP payload consists of

- a redundant space-borne 16 channel, dual-frequency GPS receiver (JPL, USA) connected to a multi-antenna system for precise satellite-to-satellite tracking between CHAMP and the high-flying GPS satellites (top-antenna: orbit determination, rear-antenna: limb sounding, nadir-antenna: experimental altimetry from surface reflected signals),
- a three-axes accelerometer (ONERA, France) at the spacecraft's center of mass to measure directly the non-gravitational orbit perturbations (air drag, solar and Earth radiation pressure), rigidly connected to two star sensors (DTU, Denmark) for precise inertial orientation information,
- a laser retro-reflector (GFZ Potsdam, HN Berlin, Germany) for additional tracking from ground,
- a magnetometer instrument package consisting of an Overhauser scalar magnetometer (LETI, France) and two Fluxgate vector magnetometers (DTU Denmark), rigidly mounted on an optical bench along the boom together with two star sensors (DTU, Denmark), and
- a Digital Ion Drift Meter (AFRL, USA) to measure the electrical field vector along the orbit.

The nominal operation of these instruments requires a three axes stabilised spacecraft with Earth pointing. Attitude control shall be performed with a pointing accuracy of better than 5° and 0.1°/s for each axis. This is handled by the Attitude and Orbit Control System (AOCS) which also handles the transfer of the S/C in the safe mode in case of reference attitude loss and the Δv manoeuvres for raising or lowering the orbital altitude. The AOCS is composed of a cold gas system with 16 thrusters, three magnetic torquers, 6 coarse Earth and Sun sensors and makes in addition use of measurements coming from the four star images, the Fluxgate Magnetometers and the GPS receiver of the science payload.

The CHAMP satellite and subsystem critical design reviews took place in 1997/98. The structure was manufactured in 1998 and integrated by March 1999. Work on the different payload instruments is in progress and tests on unit level are under way. Flight models were delivered in early 1999 and integrated into the satellite structure in late spring 1999. Integrated system -, radio frequency compatibility - and environmental tests are planned for summer 1999. Delivery of the satellite for launch is foreseen for December 1999.

4. OUTLOOK

CHAMP is the first mission dedicated to simultaneously measure the gravity and the magnetic field, intended to be used in synergetic geophysical studies. By also deriving from radio occultation measurements key parameters on the status of the neutral atmosphere and ionosphere interesting aspects of solar-terrestrial relationships are expected to be derivable. CHAMP will be the forerunner of a new generation of SST- and SGG-gravity missions (GRACE, GOCE) for improving the static satellite gravity field models by one order of magnitude.

CHAMP's five years lifetime in particular will allow to observe large scale secular, seasonal and interannual variations of the gravity field through repeated field recoveries. Examples of such observable signals are post-glacial rebound, ocean water mass redistribution, water/ice mass balance changes and hydrologic and atmospheric mass redistributions. GRACE will especially aim at a further improvement of the spatial resolution and recovery precision of such phenomena.

The CHAMP orbit characteristics (near-polar, low altitude and long duration) constitute ideal conditions for resolving the global magnetic field over a wide range of scales. Besides a remarkable improvement of the main field models (e.g. IGRF as an international standard model), CHAMP will revolutionize the ability of studying crustal magnetic anomalies from space. Advanced instrumentation will allow a resolution to spherical harmonics up to degree and order 65.

The long active mission life time will allow to monitor the broad spectrum of temporal changes. The obtained data will be of interest for modelling the geodynamo, estimating the dynamics at the core/mantle boundary and studying the conductivity distribution in the mantle. Present day magnetic field models have significant deficits in describing the secular variations in the southern hemisphere and in particular over ocean areas. For example, the temporal behaviour of the South Atlantic Anomaly which is of particular interest for space missions is not well known.

The Danish Ørsted mission (launch February 23, 1999) and CHAMP will terminate a long period void of vector magnetic field missions following MAGSAT (1979/80).

The most challenging parts of the CHAMP mission are the variety of payload components, especially the accelerometer and the magnetometers, each one with demanding environmental requirements. They are designed to observe both the gravitational as well as the magnetic geopotential with an unprecedented precision from a single platform in order to get a complementary scientific payback. An international Announcement of Opportunity will be released in summer 1999 for the establishment of an international geoscientific cooperation in addition to the already installed national mission support program. The users' interface and access to the scientific products will be via CHAMP's Information System and Data Centre installed at GFZ Potsdam.

The German Aerospace Center (DLR) is funding the CHAMP project under grant 50EP9087.

Satellite Gravity Gradiometry with GOCE

R. RUMMEL, J. MÜLLER, H. OBERNDORFER, N. SNEEUW[1]

1. Status of GOCE

During the past years the European Space Agency (ESA) has worked on the definition of a comprehensive Earth observation satellite programme for the era after ENVISAT. This "ESA Living Planet" (1998) programme has recently been published. It distinguishes three types of missions: the research oriented medium scale explorer missions, the smaller opportunity missions, and the monitoring type Earth watch missions. In preparation of a pre-selection of the first explorer missions a meeting of the European Earth science community had been organized at Granada / Spain in May 1996. The Gravity Field and Steady-State Ocean Circulation Explorer (GOCE) was one of nine candidates that were presented there. Following a recommendation of the Earth Science Advisory Committee of ESA four of the nine were selected for a phase-A study. They were: the gravity field and steady-state ocean circulation mission, an atmospheric dynamics mission, an Earth radiation experiment, and a land surface processes and interaction mission. Another meeting of the user community is scheduled for October 1999, where two of the four will be selected. They will be prepared in phases B and C for implementation in the time frame between 2004 and 2006.

The phase-A study of GOCE is currently carried out under the leadership of the Italian space company Alenia. It is supported by several studies concerned with a full end-to-end simulation, the preparation of scientific data processing, the impact of GOCE on ocean circulation studies, on geodesy and on sea level and regional tectonic studies.

2. GOCE Objectives and Mission Goals

GOCE is a dedicated gravity field mission. It combines the concept of satellite gradiometry and high-low satellite-to-satellite tracking (SST-hl) between GOCE and the satellites of the U.S. Global Positioning System (GPS) and its Russian counterpart GLONASS. The mission objectives are to determine the Earth's gravity field globally, accurately and with best possible spatial resolution.

More specifically the objectives are

- the determination of gravity anomalies with an accuracy of about 1 to 2 mGal (1mGal $=10^{-5}$ ms^{-2})
- the determination of the geoid accurate to less than 1 cm

and to achieve these accuracies with a global resolution of below 100 km half-wavelength. Results of this quality will serve a series of applications in solid Earth physics, oceanography, ice research, geodesy and sea level studies.

The mission is intended to last 17 months with 12 months experiment time. The orbit will be sun-synchronous with an inclination of 97°, almost circular and as low as 250 km. Thus, the orbit will leave unsurveyed the two polar caps with a spherical radius of 7°. The spacecraft will be kept Earth oriented with its x-axis pointing along-track, the y-axis cross-track and the z-axis radially outwards. The two core instruments are a combined GPS-GLONASS receiver and a three dimensional full tensor gradiometer. The orientation of the satellite will be maintained by active attitude control using proportional thrusters and, as control elements, a star tracker and the gradiometer (see below). The satellite will also be kept drag-free. The mission characteristics are summarized in Table 2-1.

Table 2-1: GOCE Mission Characteristics

ORBIT	
Altitude	260 km (7 months)
	250 km (5 months)
Inclination	97° (sun-synchronous)
Eccentricity	< 0.001 (almost circular)
GRADIOMETER	
Components	V_{xx}, V_{yy}, V_{zz} and V_{xz}
white noise floor	1 - 3 $\mathrm{mE}/\sqrt{\mathrm{Hz}}$ $(1\,\mathrm{E}=10^{-9}\,\mathrm{s}^{-2})$
Corner frequency $1/f$	At 10^{-3} Hz
GPS – GLONASS	
Δx	$3\ \chi\mu/\sqrt{Hz}$
Δy	$2\ \chi\mu/\sqrt{Hz}$
Δz	$1\ \chi\mu/\sqrt{Hz}$
angular control and drag control	

A choice had to be made between an inductive and a capacitive gradiometer design. The former is operating inside a dewar in a superconducting environment, the latter operates at ambient temperatures. The capacitive gradiometer has been selected due to its higher maturity with a performance not much different from the inductive one. The capacitive gradiometer built by ONERA will measure the second order derivatives of the gravitational potential of the Earth. The measurements are based on the principle of differential accelerometry. For

[1] Reinhard Rummel, Jürgen Müller, Helmut Oberndorfer, Nico Sneeuw: Institut für Astronomische und Physikalische Geodäsie, Technische Universität München, Arcisstr. 21, D-80290 München; Tel.: +49 - 89 - 2892-3189, Fax: +49 - 89 - 2892-3178, E-mail: rummel@step.iapg.verm.tu-muenchen.de

this purpose a total of three accelerometer pairs is arranged in a diamond configuration. Each of the accelerometers contains a cubic proof mass which is kept levitated electro-statically by a servo-control mechanism. Each accelerometer is three dimensional with two directions of highest precision and one axis slightly deteriorated. Ideally the accelerometers of each pair should be perfect twins. Then all linear, i.e. common, accelerations acting on the accelerometers would be perfectly removed when taking the difference between the readings of the two along the same axis. In reality a common mode rejection ratio (CMRR) of 10^{-5} is attained here. The gradiometer will reach a precision of about 1 to 3 $mE/Hz^{1/2}$ over the measurment bandwidth ($5 \cdot 10^{-3}$ Hz - 0.1 Hz)

3. Gradiometry Principle

The theoretical foundations of gravity gradiometry are given e.g. in (RUMMEL, 1986). Gravity gradients are the second order derivatives $\partial^2 V/\partial x_i\, \partial x_j = V_{ij}$ of the gravitational potential $\underline{V}$. They form a second-rank tensor with nine components:

$$V_{ij} = \begin{pmatrix} V_{xx} & V_{xy} & V_{xz} \\ V_{yx} & V_{yy} & V_{yz} \\ V_{zx} & V_{zy} & V_{zz} \end{pmatrix}. \qquad (3\text{-}1)$$

This 3 x 3 matrix must be symmetrical (∇V is a curl-free vector field) and in empty space (density $\rho = 0$) its trace must be zero: $V_{xx} + V_{yy} + V_{zz} = 0$ (Laplace equation). Thus, in each point only five of the nine components are independent. The order of magnitude of the components at satellite altitude is $V_{zz} \approx 2800$ E, $V_{xx} \approx V_{yy} \approx 1400$ E, $V_{xz} \approx 10$ E, $V_{yz} \approx V_{xy} \approx 0$.

An accelerometer with three axes $i = 1, 2, 3$ (or x, y, z) senses e.g. the acceleration

$$f_i = -\partial V(0)/\partial x_i - V_{ij}(0)\, dx_j + \Omega_{ik}\, \Omega_{kj}\, dx_j + \dot{\Omega}_{ij}\, dx_j - b_i(0) \qquad (3\text{-}2)$$

with $b_i(0)$ the sum of all non-gravitational accelerations acting on the satellite along the i-axis at the centre of mass 0 (COM).

In the case of GOCE the gravity gradients are derived from the difference of the measured accelerations f_i along e.g. the i-axis of one couple A and B of the accelerometers, separated by a distance Δx_j along the j-axis of the diamond configuration. Then it is (compare again RUMMEL, 1986, p. 336):

One observes that the components V_{ij} are not observable directly but only their combination with two other terms. These two express the effect of centrifugal acceleration ($\Omega\Omega$) and angular acceleration ($\dot{\Omega}$). They are caused by the measurement in a moving frame fixed to the satellite. Closer inspection shows that $\Omega_{ik}\, \Omega_{kj}$ like V_{ij} is symmetric, whereas $\dot{\Omega}_{ij}$ is antisymmetric. It is in detail:

$$\frac{f_i(B) - f_i(A)}{\Delta x_j} = \Gamma_{ij} = -V_{ij} + \Omega_{ik}\, \Omega_{kj} + \dot{\Omega}_{ij}. \qquad (3\text{-}3)$$

$$\Gamma_{ij} = \begin{pmatrix} V_{xx} - (\omega_y^2 + \omega_z^2) & V_{xy} + \omega_x \omega_y + \dot{\omega}_z & V_{xz} + \omega_z \omega_x - \dot{\omega}_y \\ V_{yx} + \omega_x \omega_y - \dot{\omega}_z & V_{yy} - (\omega_z^2 + \omega_x^2) & V_{yz} + \omega_z \omega_y + \dot{\omega}_x \\ V_{zx} + \omega_z \omega_x + \dot{\omega}_y & V_{zy} + \omega_y \omega_z - \dot{\omega}_x & V_{zz} - (\omega_x^2 + \omega_y^2) \end{pmatrix}. \qquad (3\text{-}4)$$

Thus, a separation into symmetric and antisymmetric part yields:

$$\frac{1}{2}(\Gamma_{ij} + \Gamma_{ji}) = V_{ij} + \Omega_{ik}\, \Omega_{kj} \qquad (3\text{-}5)$$

and

$$\frac{1}{2}(\Gamma_{ij} - \Gamma_{ji}) = \dot{\Omega}_{ij} \qquad (3\text{-}6)$$

In addition, one finds:

$$trace\,\Gamma = \Gamma_{xx} + \Gamma_{yy} + \Gamma_{zz} = -2({\omega_x}^2 + {\omega_y}^2 + {\omega_z}^2) = -2\omega^2. \qquad (3\text{-}7)$$

In the above one should keep in mind that ω_y, the angular velocity about a vector perpendicular to the orbital plane, is much greater than ω_x and ω_y. It represents the once-per-revolution rotation of the Earth pointing satellite: $\omega_y \gg \omega_x, \omega_z$.

Now one can draw the following conclusions:

- Ideally, when differencing the accelerations f_i, all non-gravitational accelerations $b_i(0)$ drop out, as well as the gravitational accelerations $\partial V(0)/\partial x_i$, see eq. (3-2).
- The gravitational gradients cannot be sensed in isolation. They are mixed with centrifugal and angular acceleration terms, see eq. (3-4).
- The angular accelerations can be isolated and, after integration and double integration, be employed for the angular control of the space-craft.
- One time-integration of the angular accelerations yields angular velocities which are used to isolate V_{ij} from $\Omega_{ik}\, \Omega_{kj}$.

- For GOCE, with one axis of each accelerometer weaker than the two others, the arrangement of the six 3-D accelerometers is chosen such as to permit precise determination of ω_y, whereas ω_x and ω_z are not determined as well. Consequently, only V_{xx}, V_{yy}, V_{zz} and V_{xz} can be reproduced with highest precision.
- Below $3 \cdot 10^{-2}$ Hz the component V_{zz} is less accurate than the three diagonal components. This is caused by the uncertainty in the orientation angles, which results in a projection of the large
- diagonal components onto the off-diagonals. Nevertheless V_{xz} has a positive effect on the overall results.
- The sum, i.e. the common mode accelerations f_i (A) + f_i (B), are a measure of the non-gravitational accelerations

$$f_i(A)+f_i(B)=-2b_i(0), \qquad (3\text{-}8)$$

if the accelerometers are arranged symmetrically with respect to the centre of mass of the spacecraft, for then $\partial V(0)/\partial x_i \approx 0$, see eq. (3-2). They are then employed for drag-free control (together with the GPS orbits).

The complete logic of the GOCE sensors is shown in Figure 3.1.

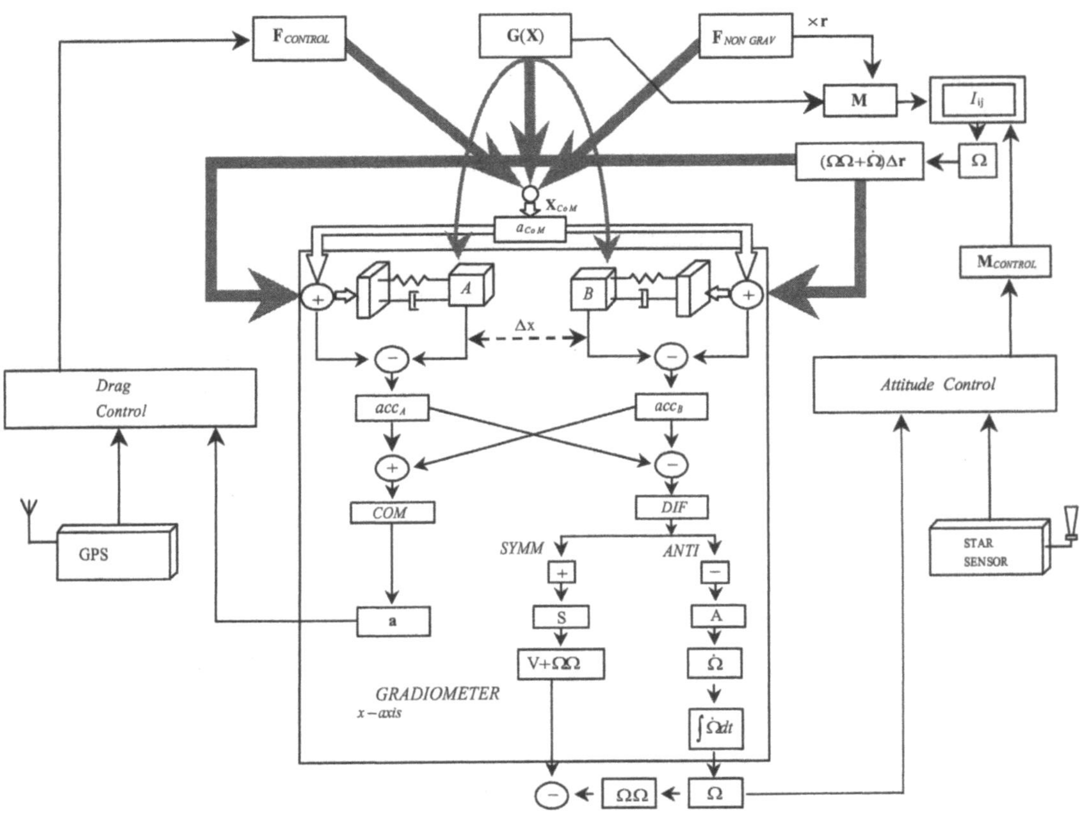

Figure 3-1: The flow chart describes the interaction of forces, sensors and control loops. It is the basis for the closed-loop simulation of GOCE. The central block is the satellite platform with the gradiometer instrument mounted on it. All linear forces acting on the platform/accelerometer frame apply to the center of mass. The left block shows the control of the linear accelerations, the right block the attitude control. The attitude control system is aided by star sensors. The central element is the gradiometer. It consists of 6 accelerometers in a so-called diamond configuration. In the flow chart only two of them are shown. Observables in gradiometry are differences of accelerations over short baselines, which can be converted to gradients of the gravitational field. Since the observations are performed on a rotating platform, inertial forces (centrifugal and Euler) are sensed too. The pure gravity signal V_{ik} is separated from the angular velocity term and the angular acceleration term by using the (anti-)symmetric properties of the tensors involved.

On basis of the above considerations three critical areas emerge that deserve careful attention:

The accelerometers are "no perfect twins". This is caused by imperfect scale factor matching, alignment and positioning along one axis. As a result one needs a comprehensive calibration scheme that includes laboratory, on-board and post-mission tests. Another remedy is to keep linear and angular disturbances (non-gravitational and apparent forces) low by active drag and angular control.

Centrifugal and angular accelerations mix inevitably into the gravitational gradients. Separation of the anti-symmetric from the symmetric tensor part will help to determine ω_y and to a lesser extent ω_x and ω_z. The trace condition will yield the sum of $\omega_x^2 + \omega_y^2 + \omega_z^2$ as control information. Finally star trackers will provide absolute orientation at 10 s intervals. An additional option is the parametrisation of misorientation angles in the context of the data processing. Thus an isolation of the gravitational tensor alone is feasible.

The third error source is the effect of time variable self gravitation of the spacecraft. On the one hand it can be kept below the required threshold by choosing a very stiff instrument and satellite construction (no moving parts). On the other hand thruster fuel consumption is of concern. The choice of ion thrusters for the attitude control and cold gas thrusters for the drag control eliminates this problem.

A simulation of the error behavior of the complete system is given in this volume, see (OBERNDORFER et al., 1999).

4. Gravity Field and Geoid Performance of GOCE

Based on the system simulation realistical performance parameters could be identified for the orbit parameters, the mission duration and the GPS and gradiometry performance, compare Table 2-1. They are used in a second step for a simulation of the gravity field recovery with GOCE data. The gradiometer and GPS / GLONASS performance parameters of Table 2-1 are only approximate. Actually error spectra derived from the system simulation have been used.

More precisely an error variance-covariance propagation simulation is carried out, i.e. based upon prior error estimates for the SST-hl and the gradiometry part the error variance-covariance of a full set of dimensionless potential coefficients $\overline{C}_{\ell m}$ and $\overline{S}_{\ell m}$ is computed. The linear model connecting observables and potential coefficients is that of the so-called timewise approach. Details about the various estimation strategies can be found, e.g. in (RUMMEL et al., 1993, SCHUH, 1996, or SNEEUW et al., 1996).

With the parameters of Table 2-1 a full mission error propagation simulation has been carried out. The maximum spherical degree and order, that has been resolved, is $L = 300$. This corresponds to a spatial half wavelength of $D = 66$ km. The full triangular error spectrum of spherical harmonic coefficients displayed in Fig. 4-1

The degree ℓ is on the vertical axis (left), the order m on the horizontal axis, with the coefficients $\overline{C}_{\ell m}$ and $\overline{S}_{\ell m}$ on the right and left hand side, respectively, of the triangle. The vertical axis can also be associated with spatial scale D (vertical axis at right). The grey scale refers to the number of significant digits of the individual coefficients. It can be seen that all coefficients up to degree and order 300 can be resolved. One also observes the weakness in the determination of the coefficient band with order m close to zero. This effect is caused by the orbit not exactly covering the poles. It will be shown that this effect in the spectrum is perfectly confined to the polar areas, when transforming the errors back from the spectrum to the sphere.

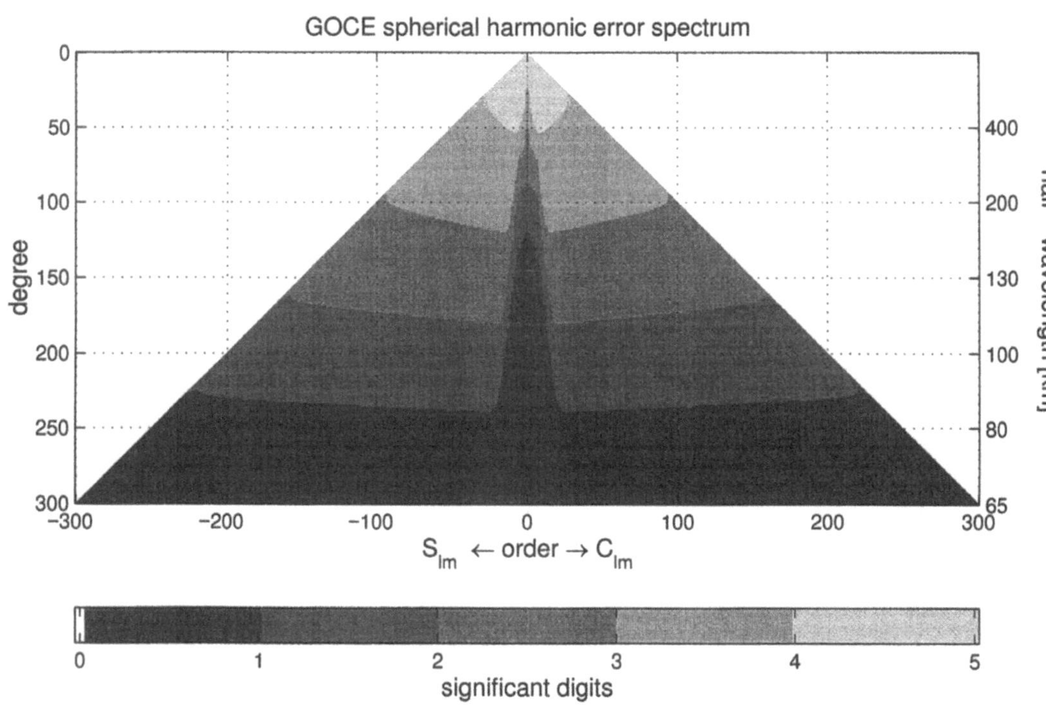

Figure 4-1: Resolvable significant digits of the determined spherical harmonic coefficients: The vertical axis refers to the spherical harmonic degree ℓ (or to the corresponding spatial resolution D), the horizontal axis to the order m of the coefficients.

From this error spectrum Figures 4-2 a-c are derived. They show the average coefficient error per degree (Fig. 4-2 a) and the expected cumulative geoid uncertainty (Fig. 4-2 b) and gravity anomaly uncertainty (Fig. 4-2 c). Figure 4-2 a shows also that the coefficients up to a degree of about 300 can be resolved. The cumulative geoid error reaches the cm-level at $\ell = 240$, the cumulative gravity anomaly error becomes 1 mGal at $\ell = 280$ and 2 mGal at $\ell = 300$. The dotted line in Figures 4-2 b and c show the error contribution per degree (not accumulated).

Another way of looking at the performance is to map the errors as a function of latitude by propagating the covariance error characteristics onto the sphere for the geoid and gravity anomaly. This is shown on Fig. 4-3. The errors are homogeneous except for the polar gaps (~7° radius) where the errors increase by up to about one order of magnitude. Elsewhere, the level of the RMS curves is consistent with the cumulative errors up to the specified maximum degree L (see Fig. 4-3 b, c). For instance, up to degree 200 (100 km resolution) GOCE yields a geoid error below the centimetre level, and a gravity anomaly error just below 0.08 mgal, for the static field.

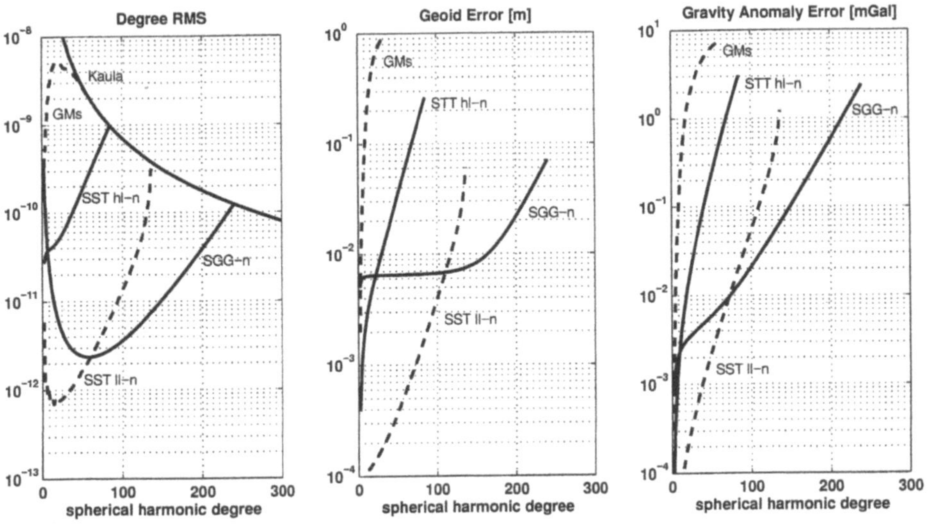

Figure 4-2 a - c: Average coefficient error and expected cumulative geoid and gravity anomaly error per degree. Dotted line: geoid and gravity anomaly error per degree (not accumulated).

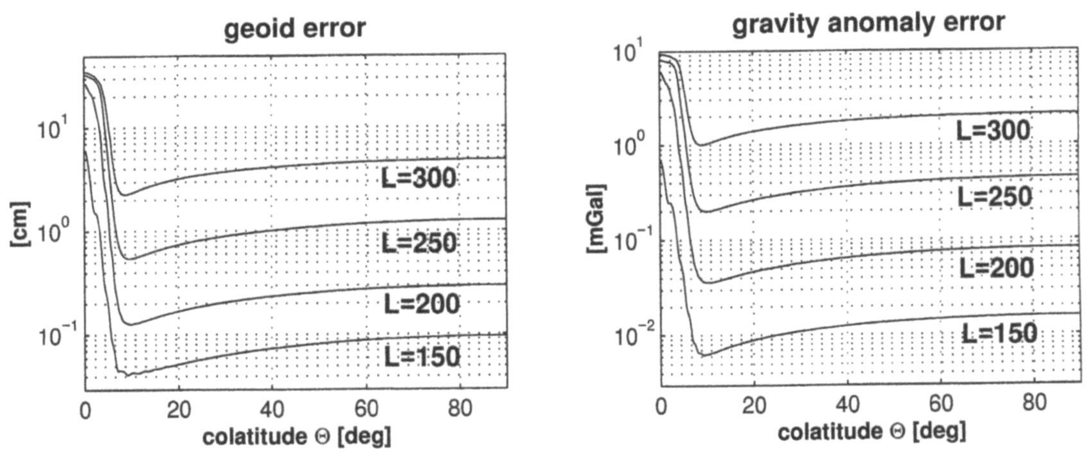

Figure 4-3: Accumulated point error on geoid [a] and gravity anomaly [b], for series expansions up to degree and order L = 150 (D = 135 km), L = 200 (D = 100 km), L = 250 (D = 80 km), L = 300 (D = 65 km), as function of colatitude for the northern hemisphere (results are symmetrical with respect to equator), from the GOCE baseline mission simulation. Note that the errors decrease when going towards the pole due to the accumulation of satellite ground tracks, and then increase due to the polar gap.

Finally, let us remark that the GOCE instrument is three-dimensional, i.e. it simultaneously measures the gravitational field in all three spatial directions. Consequently the errors of the resulting gravity parameters (gravity anomalies or geoid heights) exhibit no preferred direction. Apart from the redundancy the errors are independent and isotropic. This isotropy of the error structure is of great advantage, in particular when the directional structure of the gravitational field is of importance, as it is the case in oceanography where slopes of dynamic topography are to be derived.

5. GOCE and Related Missions

If things work out according to plans a series of three gravity field missions will be flown in the next decade. CHAMP will be the first mission that will combine SST-*hl* with 3-D accelerometry. The GRACE mission will employ the concept of satellite-to-satellite tracking between two low orbiters (SST-*ll*). Thereby the two low orbiters will follow each other in the same orbit at a distance of about 200 km. Their relative motion will be measured very accurately. Both satellites will carry a 3-D accelerometer. Finally, GOCE will apply the principle of satellite gradiometry.

From a theoretical point of view these three experiments can be viewed as measuring

- in the case of CHAMP, the motion of a single test mass in its orbit under the influence of the gravity field of the Earth,
- in the case of GRACE, the relative motion between two test masses, about 200 km apart, and
- the relative motion between the test masses of one gradiometer instrument with typical distances of less than a metre for GOCE.

In mathematical terms this corresponds to the transition from a measured vector quantity along the orbit, to the finite difference between this vector quantity at two orbit points, to its differential. By differentation the small scale part of the gravity field is emphasized while the long wavelength part is attenuated. This way the natural signal attenuation at altitude can by counteracted.

In error propagation studies this rationale translates into typical **slopes** of the error curves for SST-*hl*, SST-*ll* and gradiometry. They are displayed in Figure 5-1 per spherical harmonic degree together with the gravity signal curve. Whereas any change in the system performance would result in a vertical shift of the entire error curve, the slope of the curves is intrinsic to the technique. Therefore, since the system performance of GRACE will be extremely high it will be able resolve long to medium wavelength (below n = 50) with high accuracy. Thus even temporal variations in the Earth's gravity field such as changes in soil moisture, atmospheric pressure, or seasonal ocean changes will become visible at those spatial scales. The error curve of GOCE is much flatter. This allows GOCE to attain highest spatial resolution. For more details it is referend to (BALMINO et al., 1998).

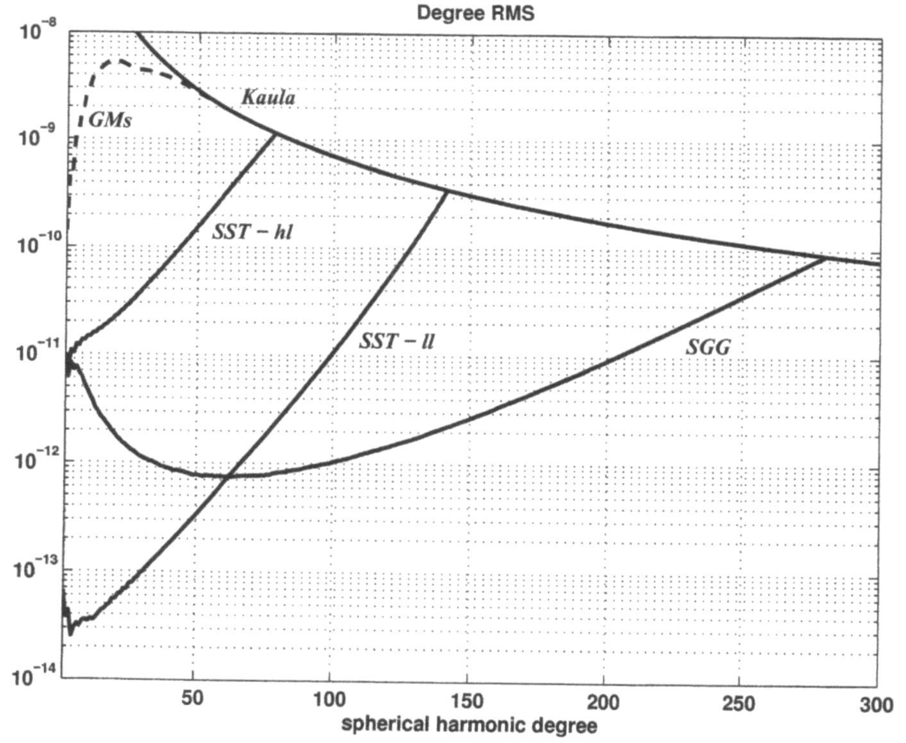

Figure 5-1: Typical error curve per degree for the three techniques SST-*hl*, SST-*ll* and satellite gradiometry, together with the average gravity signal line.

References:

BALMINO G., PEROSANZ F., RUMMEL R., SNEEUW N., SÜNKEL H., WOODWORTH P. (1998): *European Views on Dedicated Gravity Field Missions: GRACE and GOCE*. ESD-MAG-REP-CON-001.

ESA (1998): *The Science and Research Elements of ESA's Living Planet Programme, Earth Science Division*. ESA directorate of scientific programmes, ESA SP – 1227, Noordwijk.

OBERNDORFER H., MÜLLER J., DOROBANTU R., GERLACH C., RUMMEL R., SNEEUW N., KOOP R., VISSER P., HOYNG P., SELIG A., SMIt M. (1999): *Simulation of the GOCE Gravity Field Mission* (this issue).

RUMMEL R. (1986): *Satellite Gradiometry*, in: Lecture notes in Earth Sciences, vol. 7, 318-335, Mathematical and Numerical Techniques in Physical Geodesy, ed. H. Sünkel, Springer, Berlin.

RUMMEL R., SANSÒ F., V. GELDEREN M., BROVELLI M., KOOP R., MIGGLIACCIO F., SCHRAMA E., SACERDOTE F. (1993): *Spherical Harmonic Analysis of Satellite Gradiometry*. Netherlands Geodetic Commission, New Series, 39, Delft.

SCHUH W.-D. (1996): *Tailured Numerical Solution Strategies for the Global Determination of the Earth's Gravity Field*, Geodätische Institute der T.U. Graz, Graz.

SNEEUW N., RUMMEL R., MÜLLER J. (1996): *The Earth's Gravitiy Field from the STEP Mission*, Class. Cantum Grav. (13, A113 – A117.)

SNEEUW N., GERLACH C., MÜLLER J., OBERNDORFER H., RUMMEL R. (1999): *Fundamentals and Applications of the Gravity Field Mission GOCE* (this issue).

Integration of geodetic techniques into a global Earth monitoring system and its implication for Earth system sciences

Hans-Peter Plag*

Abstract

The increasing vulnerability of infrastructure and human lives particularly in areas of the Megacities as well as the broad acceptance of sustainability as guiding principal for societal development is putting new emphasis on global processes and the associated geophysical variability and global change. In Earth system studies, the need for integrated data sets is increasingly obvious, both for observational studies of system processes and the validation of integrated system models. Though rapidly improving, the current scientific knowledge of the Earth system with respect to the main processes, the key indicators and the major forcing factors is still limited. Therefore, a more or less complete monitoring of the system's state and trends is mandatory if a valuable contribution of global monitoring to environmental security and sustainability is to be expected. Thus, integrated global monitoring is a prerequisite for global environmental management of a sustainable Earth. Specialised global observing systems have already been initiated and partly are implemented (e.g. the G3OS, the Global Terrestrial, Climate and Ocean Observing Systems: GTOS, GCOS, GOOS). However, these systems are largely based on space-born remote sensing techniques, while investment in urgently needed in-situ measurements are smaller and often decreasing. Moreover, the integration of ground-based networks, data archiving and distributing facilities to provide integrated data sets of in-situ observations is still at its infancy. Space geodesy is now capable of monitoring variables of potential value for environmental monitoring systems. Moreover, in some of the existing networks, these techniques provide crucial auxiliary observations. It is therefore of fundamental value to fully integrate geodetic monitoring techniques into a global integrated Earth monitoring system, which is a significant step towards an Earth information system for sustainability. This integration should follow the Global Integrated Observing Strategy.

1 Introduction

The Earth appears to be a highly complex system where chemical, physical and biological processes are interacting in nearly all processes determining the system's state. Moreover, over the last few centuries, mankind has evolved to be a dominating factor shaping the Earth surface. Particularly the rapid growth in the number of human beings combined with the presently nearly unlimited availability of easy-to-exploit energy[1] has brought mankind to the position of being powerful enough to affect crucial system processes and eventually force secular change in the mean conditions of the Earth system. Therefore, human economic, social and psychological conditions are likely to have a significant impact on the future development of the system. In such a highly complex situation, it is difficult to predict in any respect, be it directly related to human affairs or more pertaining to the non-human part of the system. However, what can be said is that the big challenge for mankind in the next century is going to be the management of the human activities in a way increasingly approaching sustainable development.

The environmental debate taking place over the last thirty-five years can be characterised by a few milestones, which also illustrate a shift in focus over time. One of the early milestones was the "resource-depletion debate" which is exemplified by the 1972 report of the Club of Rome (Meadows *et al.*, 1972). At that time, the size of the natural resources were seen as the major limitation for the growth of mankind and our economy.

The high social and political risks due to the geographically uneven access to sufficient resources influenced the United Nation's World Commission on Environment and Development in its 1987 report "Our Common Future" to revitalised and globalised the concept of sustainable development[2] (World Commission on Environment and Development, 1987). This so-called Brundtland report marks a turning point to the "sustainability debate", after which the size of available resources were no longer accepted as the only constraint and limitation for economic activities. The normative concept of sustainable development brought ethical considerations into the discussion questioning our rights to deplete natural resources at will. The concept also challenged the basis for our economic theories, which up to then were all based on the assumption of nature being a free gift for humans [3].

The third milestone to be mentioned here is the recognition of another severe limitation to human activity: the capacity

*Norwegian Mapping Authority, Kartverksveien, N-3500 Hønefoss, Norway, phone: +47-32118100, fax: +47-32118101, Email: plag@gdiv.statkart.no

[1]For a discussion of the impact of the availability of low-cost energy on the recent societal development see Pfister (1995).

[2]The reports states that *"Sustainable development is development that meets the needs of the present without compromising the ability of future generations to meet their own needs."*

[3]For a more detailed discussion of the influence of the concept of sustainable development on the current economic paradigms, see e.g. Ahmad *et al.* (1989).

of the Earth system to maintain a homeostasis favourable for human life under heavy anthropogenic interference with system processes. This limitation may be characterised by the "emission-reduction debate". Under the present level of emissions resulting from production, consumption and transportation, the risk of a change in the system's state to conditions unfavourable for human beings is apparent. Probably most prominent in the scientific and public discussion is the risk of climate change.

In this paper, we will first give a brief review of the sustainability and climate change debate with main emphasis on the relevance of Earth monitoring and Earth system science to progress. An analysis of the potential contribution of Earth sciences on the way towards sustainability is given in Section 3. On the basis of this analysis, the need for an integrated sustainable monitoring of the Earth system is underlined in Section 4. The potential contribution of geodesy to an Earth monitoring system is considered in Section 5. Finally, a possible way to overcome the deficiencies of the present monitoring as well as it interaction with research and assessments is outlined in Section 6. In the conclusions, a need is pointed out for an intergovernmental assessment of the emerging monitoring system as a guidance on the way towards an Earth information system for sustainability.

2 Sustainable development and climate change: a brief review

Currently, the global political dialogue is aiming at securing bearable if not favourable conditions for future generations. A major step was the Earth Summit in Rio de Janeiro, Brazil in 1992, which brought 135 countries and heads of states to Rio. There they signed the Framework Convention on Climate Change, the Framework Convention on Biodiversity, and the Agenda 21, which is a commitment to implement the objectives of the Earth Summit. After this Summit the concept of sustainable development has been widely accepted as the guiding principal for societal development, although its interpretation is not unanimous. The world community presently is in the state of operationalisation of the concept, and the ideas still range widely from dramatic lifestyle changes over simple conservation to "business as usual". In this process, the interdependence of the different dimensions of sustainability and the need for an integrated, interdisciplinary approach to the achievement of sustainable development becomes increasingly obvious.

Agenda 21 [4] clearly emphasises the need for information and a complete monitoring of the Earth system. Chapter 40 of Agenda 21 provides information for decision-making. The first paragraph states: "*40.1. In sustainable development, everyone is a user and provider of information considered in the broad sense. That includes data, information, appropriately packaged experience and knowledge. The need for information arises at all levels, from that of senior decision makers at the national and international levels to the grass-roots and individual levels. The following two programme areas need to be implemented to ensure that decisions are based increasingly on sound information:*

(a) Bridging the data gap;

(b) Improving information availability."

In particular, the need for the development of indicators of the system state is pointed out in paragraph 4 of the same chapter: "*Indicators of sustainable development need to be developed to provide solid bases for decision-making at all levels and to contribute to a self-regulating sustainability of integrated environment and development systems.*"

Thus, appropriate tools for the assessment of impacts due to human activities need to be developed as a prerequisite for determining the sustainability of these activities. It is the responsibility of the scientific community to provide such tools and to carry out the necessary assessments. In many national and international research funding programs, this is well reflected. The last decade has seen many activities aiming at the definition of geo-indicators [5]. However, research and monitoring is still not focussing sufficiently on identifying variables of indicative value and providing observations of the full set of these key variables.

A major thread for a sustainable development arises from potential climate changes, which in part are the consequences of a non-sustainable use of fossil energy resources. Recognising the potential risk for an anthropogenic climate change and the huge social and economic consequences of anticipated impacts, the WMO and UN in 1988 institutionalised the Intergovernmental Panel on Climate Change (IPCC), which in its 1990 Assessment of Climate Change and the 1992 Supplement (see references in Houghton *et al.*, 1996) provided an initial basis both for decision-making and research development. In 1996, the Second Assessment of the IPCC (Houghton *et al.*, 1996) summarised the current status with little indication that the risk of anthropogenically induced climate change had been reduced. Several countries initiated national committees to work out means to reduce the impacts of an anticipated global warming. Nevertheless, the scientific community is far from providing the required assessments of measures to the political decision-makers, as was well documented by the recent discussion of the agreement for CO_2-reduction, where no clear scientific statement on the contribution of forests in the Carbon cycle could be given.

[4] See gopher://gopher.un.org:70/00/conf/unced/English/ for the full text in English.

[5] For just one example, see the statement which originated at the International Workshop on Geological Indicators of Rapid Environmental Change, held on July 11–17, 1994, in Corner Brook and Gros Morne National Park, Newfoundland, Canada as published in Episodes, nol & 2, 1994, p.2, where it is stated: *In order to assess the state of any environment, reliable indicators are needed, just as doctors use blood pressure and body temperature as simple, inexpensive guides to human health. Even if causes cannot be determined, we must be able to detect change and warn of dangerous conditions.*

3 Sustainable development: contributions of Earth sciences

As mentioned above, the problem of sustainability has a strong ethical and social component. However, once it is defined and accepted as a basic principle for human activity, sustainability acquires a crucial scientific component. The quest for sustainability puts the scientific community in a position comparable to the physician being ask to design a therapy that will keep a human being in a healthy state or that will restore healthiness in case of illness. Under this principle, any on-going or planned action needs to be assessed in terms of sustainability. This assessment can only be carried out on the basis of a profound knowledge of the Earth system, its processes, and its dominating factors. The contribution of science to sustainable development can be separated into at least two basic topics:

(1) to monitor the Earth system in order to give an indication of the system's state and trend; in this respect it is important to keep in mind that the key variables are still not clearly identified;

(2) to develop higher integrated Earth system models, which are required for the assessment of potential impacts of actions planned to bring the world society on a course towards sustainable development.

A basic prerequisite for the assessment of impacts is a system model designed to simulate the consequences of certain measures. Currently, the most advanced models represent subsystems in terms of the interactions, processes, regions and system components included. Thus, the physical general circulation models (GCM) simulate the circulation in atmosphere and ocean, but do not consider the interaction between chemical constituents and the circulation, i.e. so-called emission scenarios are used to force the circulation model. On the other hand, chemical transport model assume the circulation as given and simulate the distribution of the chemical constituents due to sources and sinks. Most of these models do not account for the biosphere and its interaction with the non-biological system components. Models of the biosphere are basically modelling the carbon transport in the biosphere or the population dynamics as a consequence of environmental conditions. In most of the models used in the study of climate change and the potential impact of the anticipated warming, the solid Earth is only represented as a passive boundary condition not taking part in the dynamics.

The need for higher integrated models is widely recognised. However, the presently available data sets of relevant observations are not sufficient for validation of any higher integrated model.

The recognition of the scientific dimension of climate change and sustainable development has stimulated an ongoing discussion of whether the scientific approach still widely used in studies of global change can ever contribute valuably to the problem of sustainability. This approach, which is best termed as the reductionist's approach, breaks down research into specialised disciplines and studies small, strongly reduced, and highly simplified subsytems of the Earth. An increasing number of scientists are criticising this approach as inappropriate and requesting a more wholistic approach embedded in a system viewpoint.

However, helpful methodological developments to overcome the deficiencies of the reductionist's approach were not to be seen until recently, when first a group at the University of Bern (see Pfister, 1995) used a syndrome description to describe phenomena associated with our energy use. The concept of syndromes is known from medicine and is further elaborated in a recent article by Schellnhuber *et al.* (1997). These authors lay the methodolocial and mathematical basis for an analysis of global change taking into account our limited knowledge of the system and the large uncertainties associated with observations of key variables. This approach, which assigns a high weight to the spatial characteristics of global change, may well be apt to overcome the problems of the current scientific approach in global change research.

It is important to note here that in the context of most classical discussions of global climate change and activities to mitigate the impact of future changes, emphasis is on the effect of Greenhouse gas emissions on the global temperature while the effect of many other anthropogenic interactions are widely neglected. Consequently, most measures also concentrate on the reduction of these emissions. This seems to reflect a rather simplicistic view of the Earth system, which assumes that a few switches (e.g. CO_2 emission) are sufficient to gain control over major processes and, ultimately, the system's state.

However, looking at the complexity of the Earth system, it must be questioned whether this strongly reduced "control panel" is sufficient for a sustainable management of the Earth. It may be worthwhile to compare this Earth system control panel to similar panels required to control technical processes. As an example, we may imagine the control panel of a large modern airplane. To control the relatively simple and well-understood process of flying an airplane, a large number of variables have to be monitored continuously at many locations within the system to ensure that changes in the system's state are detected early enough to allow taking measures counteracting undesired changes. In addition to the numerous displays, the control panel has a number of red lights, which flash whenever critical situations occur. Moreover, there are switches to initiate well-defined and controllable emergency actions if the displays and red lights warrant them. Nevertheless, a small risk remains that specific combinations of conditions might result in a serious accident.

Considering the control panel of the Earth system as sketched in Figure 1, we have to state that only a few variables of the system are monitored, most of them, however, spatially incomplete and many with a rather recent start. Despite the well established fact that the Earth is a bio-geochemical system, only a few chemical variables are monitored, most of them with insufficient geographical distribution. Some physical variables related to the thermodynamic state of the system are collected in near-global networks. Geophysical and geodetic variables are increasingly monitored. Considering the power acquired by mankind, astonishingly few variables related to human activities are monitored and available at the "control panel". Political decisions to control economic activities most often are still reduced to one variable, the gross national product, which is taken to be indicative for the health of a national economy[6]. Nevertheless, the key variables and the "red lights" of the system remain largely unknown. Moreover, only a few switches to affect the system's state are available, and their long-term effects are poorly understood.

From this comparison, it is clear that the "flight" of the Earth system carries some potentially much higher risks than in a modern airplane. In particular, one might be justified in asking whether current scientific input into the process of political decision-making is based on firm ground. Assessments of the current situation and potential risks like the IPCC assessments do not clearly and openly state the limitation of our monitoring system, modelling capabilities and impact assessment tools. In most assessments, quantitative knowledge is preferred over qualitative knowledge.

An important question in the current global change debate is whether science, with its limited ability to model the Earth system, is able to provide significant quantitative results as input for the decision process. Examples of such an input are the projection of global warming and sea level rise due to predefined emission scenarios (see, e.g. chapters 6 and 7 in Houghton *et al.*, 1996). In public, these projections are interpreted to indicate that it is sufficient to reduce industrial CO_2 emission by a certain amount to ensure that global warming will not result in economically devastating consequences (see Houghton *et al.*, 1996, for a detailed discussion). However, it must be questioned whether these projections provide a firm and scientifically justified basis for such a conclusion which is then used as the basis for international decisions and action plans.

In such a situation, where detailed quantitative constraints for impacts and consequences of proposed actions are difficult to attain, it may be more sensible to consider qualitative knowledge with a high priority. Here, we refer to results obtained from observational and modelling studies not directly convertible into numbers applicable in the course of decisions. A rather simplified example is a patient with considerable risk of a heart attack. It is impossible to provide quantitative information to this patient stating how many stress-free or stress-reduced hours, or how large a reduction in smoking is required, to avoid the event of an heart-attack. It is, however, possible to provide the qualitative input that a stress-reduction, sensible food, and less smoking reduces the risk of an heart-attack. This qualitative result is deduced on the basis of observational studies as well as a broad understanding of the nature of the problem and the physiology of a human being. Similarly, physiological studies of the Earth system based on models as well as integrated data sets may provide qualitative results more important for designing the best therapy to achieve sustainability than quantitative projections based on the current levels of understanding.

Many published examples (see e.g. Schellnhuber and von Bloh, 1993; Rahmstorf, 1995) demonstrate the likely effect of perturbations on systems that are in a state of a dynamical equilibrium (homeostasis). This effect is not a gradual nearly linear response to external forcing or perturbations but more likely a rapid transition to a different state whenever thresholds are exceeded.

For the Earth system, there may be many thresholds for numerous different external and internal perturbations. Exceeding such thresholds may be responsible for the accretion and disintegration of the large ice sheets, and they may be the cause of rapid climate fluctuations believed to be documented in ice cores (Greenland Ice-core Project (GRIP) Members, 1993) and also found in observational data of the last 100 years (Ellsaesser *et al.*, 1986). We do not know any of these thresholds of the complex Earth system. However, we know that we have been changing the Earth's surface to a great extend (for an impressive documentation of the changes, see Turner II *et al.*, 1990). The qualitative knowledge of the Earth system combined with the abundant documentation of our effective interference suggest the conclusion that we may well be close to thresholds separating us from severe risks. The danger of hidden risks awaiting us has been raised in many publications (e.g. Broeker, 1987), however, the scientific community has not succeeded in popularising this point in the social and political process of decision-making.

Moreover, present monitoring activities, including data archiving and information production routines are not adequate to allow for the timely detection of any severe rapid trends in the Earth system. The basic uncertainty in our knowledge concerning the interannual to decadal variability of the system as pointed out in the recent IPCC assessment (Houghton *et al.*, 1996), is a major obstacle in providing such information.

4 The need for sustainable integrated monitoring of the Earth system

As outlined in the previous section, sufficient monitoring of the Earth system is one of the cornerstones required to approach and secure the sustainability of mankind's existence and to control interference with the system's processes. The last decade has seen the emergence of many global or regional programmes and activities directed towards monitoring of the environment (Figure 2). Monitoring the Earth system presently is strongly subdivided and organised according to disciplines and subsystems. A major disadvantage of this lack of integration is the nearly complete absence of the integrated data sets required for the study of system processes. Consequently, science programmes or

[6] There is, however, considerable research going on to define a more diverse set of variables to measure "well-being" in the context of the sustainable development debate.

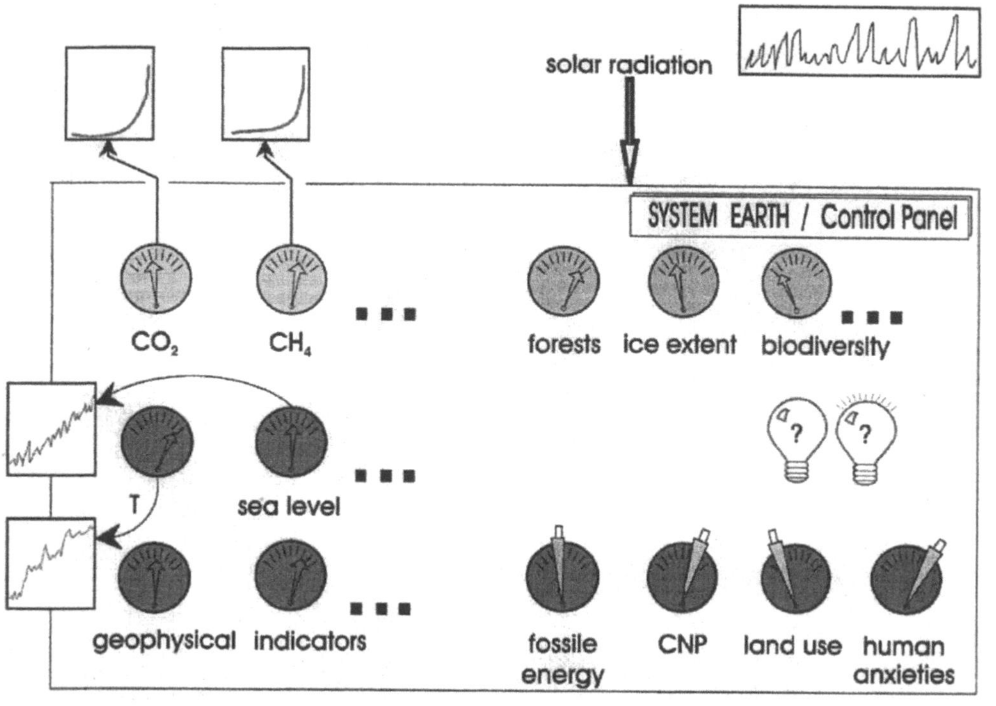

Figure 1: Sketch of the Earth system's control panel.
The "control panel" currently used in science for the Earth system is astonishingly simple, with only a few chemical, biological and physical monitoring parameters, far less parameters pertaining to the state within the antroposphere, and still less control switches. The red lights still remain unknown.

projects aiming at a better understanding of system processes are forced to build up such integrated databases first. An Integrated Global Observing Strategy (IGOS) is drafted only in the remote sensing community, and the three Global Observing Systems (G3OS) are developed according to this strategy. In research, several large programmes like the International Geosphere-Biosphere Programme (IGBP) and the World Climate Research Programme (WCRP) are looking at specific aspects of the global system, while other programmes (such as the International Lithosphere Programme, ILP) still concentrate on subsystems. Assessment is partly organised on a global level, but smaller national and even non-governmental institutes contribute significantly to the assessment of the state of the Earth system.0 Policy-makers crucially depend on the results of these assessments. Thus, these assessments are the interface between both monitoring and research and those determining the course of global development. The direction of funding is largely determined by decision makers and depends on the priorities resulting from the assessment. It should be noted that the assessment presently still draws mainly on the results of the research programmes and not on operational monitoring.

Currently, the monitoring system is characterised by a number of sub-networks with spatial and temporal heterogeneities and with little coordination and cooperation across disciplinary boundaries (see Figure 2). The ground-based component consists of meteorological, hydrological, oceanographical, geophysical, geodetic and chemical networks, with the number of operational stations varying in time. Additionally, a significant amount of data is collected in campaign-type measurements at varying time intervals and locations.

All these sub-networks produce data sets which are inhomogeneous due to spatial and temporal heterogeneities in the station distribution, and due to variations in the observation procedures including the sensors and recording equipment. Problems due to these inhomogeneities are exemplified in Ellsaesser *et al.* (1986) using the station temperature observations at land and sea sites. For a sustainable monitoring, the problem of long-term homogeneity is a crucial one.

Over the last two decades, a strong space-borne component has been introduced into the monitoring. The nearly complete coverage of most of the remote sensing satellites has greatly improved monitoring and opened new doors to understanding system processes. However, in terms of sustainable monitoring, the limited life time of the satellites and sensors, and the high costs of most of the missions, are severe limitations likely to introduce temporal heterogeneities into the data sets. In many cases, only sin-

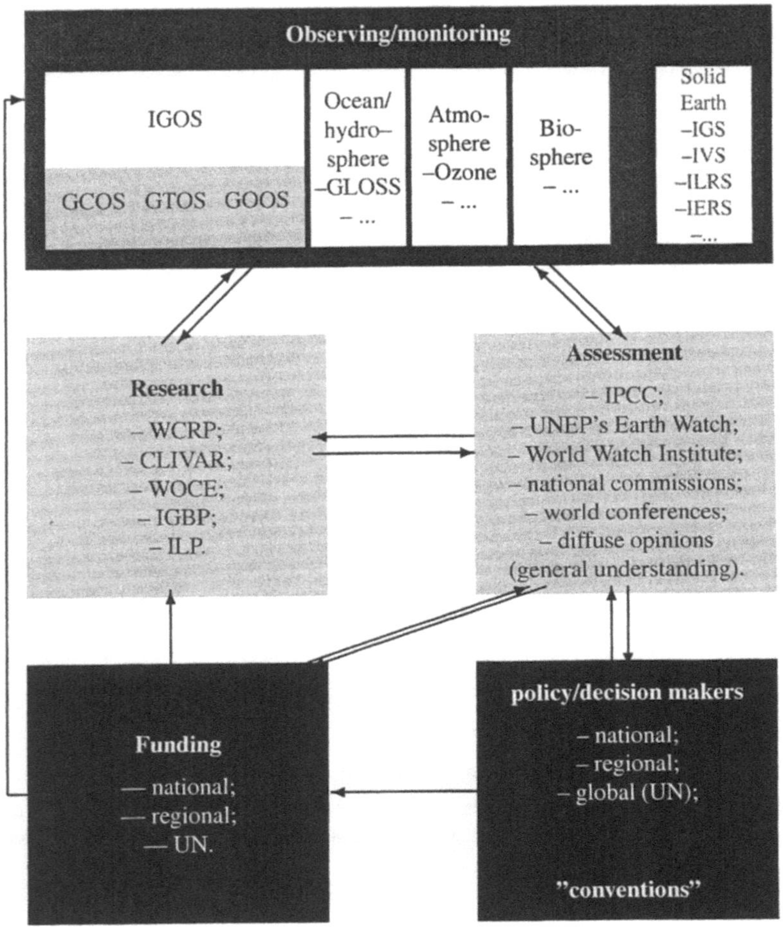

Figure 2: Present monitoring of the Earth in relation to research programmes and system assessment. For a discussion of the figure, see text in Section 4.

gle sensors exist, and the danger of processing errors and miss-interpretation is high (as demonstrated by the error in TOPEX/Poseidon processing, see Nerem *et al.*, 1997).

Particularly within Earth monitoring based on remote sensing, progress has been made recently towards an IGOS (see, e.g., Williams and Townshend, 1998). Within the framework of the emerging global observing systems (see, e.g., Dahl, 1998), much has been done over the last two years to develop the IGOS. A document has been drafted by the key players in global observation, and this document forms a valuable basis for the discussion in the present paper. The most recent draft is available on http://www.unep.ch/earthw/igosstr.htm. The drivers for IGOS are the scale of the issues (global climate change, sustainable development) to be addressed; the cost of space components for remote sensing of the Earth environment; the logistics especially for in-situ data; and the need for data integration from multiple sources for products of use to decision makers, science, and society at large. For key variables of the Earth system, IGOS attempts to provide long-term continuity, adequate data archives and accessibility, consistency of data records, and the ancillary data required for data quality assessment.

These principles of IGOS provide the framework for a coherent response of the monitoring system to the integrated user requirements. Under IGOS, an operational system guaranteeing the long-term continuity of observations to support scientific research can be achieved. IGOS intends to build upon existing strategies for international observation programs, focusing on the identification of areas where the existing systems can be improved, where duplication of observations can be reduced and gaps in observations and data sets can be identified. Moreover, IGOS facilitates improved high-level product developments, and capacity building in developing countries. Thus, if effectively implemented, IGOS appears to be the strategy for providing the observational basis for a future Earth information system. A key issue identified in IGOS is the need to transform many observational activities from their research states into operational monitoring.

In summary, comprehensive monitoring of the Earth system is a crucial prerequisite which needs to be established within the research community and transformed into operational activities. Taking into account the nature of the task at hand, the necessary properties of a sustainable monitoring include
— long-term stability,
— operational mode,
— homogeneity in time,
— multi-parameter sites,
— global coverage and participation, and
— integrated observation and data sets.

5 Geodesy's potential contribution to Earth system monitoring

The present capabilities of space-geodetic techniques are described in detail in several of the publications in this issue. Based on these analyses, the basic contribution of geodesy, and particularly of space geodesy, to Earth system monitoring, can be identified as

- the maintenance of stable reference frames for all position-related information (both terrestrial and celestial);
- the monitoring of the kinematics of the Earth's surface, including deformations;
- the monitoring of the Earth's rotation and its variation in time;
- the observation of variations in the Earth's gravity as a measure of mass movements as well as a quantity required to observe other parameters such as sea surface topography;
- the monitoring of other environmental parameters (e.g. GPS-MET);
- the provision of observing techniques, particularly satellite-borne ones.

Presently, geodesy is realising these contributions with

- global geodetic networks;
- services built upon these networks;
- dedicated satellite missions.

Within space-geodetic observation techniques, over the last decade, a very rapid development has occurred leading to unprecedented possibilities to measure positions (see, e.g., Plag *et al.*, 1998). Several space-geodetic methods including Satellite Laser Ranging (SLR), Very Long Baseline Interferometry (VLBI) and the Global Positioning System (GPS), have reached maturity where the transition from research to operational states appears to be feasible. These techniques have the potential to provide for reliable monitoring of key parameters describing the Earth's system, such as surface displacements and deformations, earth orientation, integrated precipitable water vapour content of the atmosphere (IPWC), and electron density in the ionosphere.

Currently, permanent infrastructure potentially valuable for space-geodetic measurements has been and is being installed increasingly for several purposes. Most prominent are networks of permanent GPS sites continuously recording data of several or all visible satellites (denoted here as CGPS sites). Most of these networks originally were established for dedicated purposes, such as differential GPS networks for navigation systems, high precision networks for global reference purposes (i.e. the IGS[7]) and as regional densifications of the global network (e.g. the EUREF[8]), or for geodynamical investigations, pilot-project type networks to derive meteorological information from the GPS data (e.g. the SuomiNet in USA[9]), as well as small station nets for high-precision multi-parameter local studies. Recently, a trend is surfacing towards multi-purpose use of some of these networks.

Some of these nets are operational and often supported or operated by governmental institutions, while others are more associated with research and, as such, operated by university groups or research institutes. Some of the installations are complying to standards for space-geodetic sites and only data storage and availability restricts the possible general scientific or broader use of the recorded observations. Other sites meanwhile do not meet these standards and would require upgrading in order to be of value in geodetic monitoring. The absence of overall coordination in the development and use of these networks represents an unnecessary economic waste. Under the present circumstances there is certainly a loss of potential primary and secondary benefits both in terms of science and economy. Developing an integrated observing strategy would lay the basis for the implementation of a multi-purpose coordinated network of observing stations including those established for commercial purposes. For space-geodetic networks, this strategy currently is not available or being developed. However, as described in the previous section, in other areas of Earth monitoring progress has been made recently towards an IGOS. To develop an acceptable way to bring the space-geodetic networks in accord with IGOS could ensure the quality and availability of the collected space-geodetic data required for monitoring and scientific exploitation.

Examples where operational infrastructure provides data that is of high relevance for science can be found within meteorology, where sites operated particularly at airports (where local conditions are required for safety purposes)

[7] see http://http://igscb.jpl.nasa.gov/index.html
[8] see http://www.oma.be/KSB-ORB/EUREF/eurefhome.html
[9] see http://www.unidata.ucar.edu/SuomiNet/SuomiNet.html

not only provide the basis for regional weather forecasts but also result in long meteorological time series which are now the observational basis for climate studies at time scales up to several decades.

Another example is tide gauges, which are often operated to ensure the safety of ship traffic. This operational monitoring resulted in a database central in studying climate variability over the last 100 years. Analyses of this database provided crucial constraints for models of the cryosphere-ocean interaction and the assessment of the impact of future global change.

To enable maximum scientific exploitation of the data being collected at space-geodetic sites established to serve both scientific and commercial purposes, the installation has to comply to specified standards. Moreover, data collection, processing, and storage has to provide for data integrity, homogeneity in time, and accessibility. Providing for multipurpose use of the data will help to avoid overlaps in installation. The current development within CGPS is clearly in danger of having established a number of different, partially overlapping, networks. Instead of having an increasing number of CGPS sites for dedicated purposes, a future scenario could (and should) be a multi-purpose CGPS network as an integral part of the general infrastructure of society. Such a network would collect and provide real-time data for navigation purposes, near real-time environmental parameters such as IPWV, electron density, and surface displacements used e.g. as input for forecasts of relevant environmental conditions and possibly hazard warnings, and post-time data for scientific studies. Operated over a longer time period, such an infrastructure would create an environmental database as a legacy for future generations of scientists in the same way as we today partly live on the databases from infrastructure established by far-sighted people more than a hundred years ago.

A multi-purpose CGPS network preferably would be established on a global basis. To secure operation over a longer time period, a clear international dedication and national support would be required. However, for a step-by-step implementation, regional approaches are very valuable. Therefore, the strategy for such a network needs to allow for a modular structure.

In the on-going discussion within IAG to step towards an Integrated Global Geodetic Observing System (IGGOS, see e.g. Beutler *et al.*, 1999), a multi-purpose CGPS network could be a path-finder in the same way as IGS was a path-finder for services based upon single space-geodetic techniques. The strategy developed for a multi-purpose CGPS network naturally would also have to account for multi-sensor sites, and thus foster the integration of techniques.

6 Towards sustainable integrated global Earth monitoring

Any monitoring system can be considered as composed of three different networks which can be visualised as a triangle with each corner of equal importance for the performance and success of the system (Fig. 3). This triangle should be centred around the needs of the users of the system, which may be scientists or society at large. On one corner of the triangle, the design of a *virtual network*, including its objectives, needs to be such that the products correspond to scientific or societal needs. Thus, the first step in developing the virtual network should be the identification of the user's requirements, that is, the design process should be user-driven. The *physical network*, including the single stations, the communication tools, the data processing and the analysis, has to meet the specifications resulting from the virtual network and, particularly, the objectives of the network. Quality control of performance and products and the long-term consistency of the operation are integral parts of the design of the virtual network. Finally, the *institutional network* has to ensure the political and financial support of the activities. Thus, this network has to include the "owners" of the (physical) infrastructure as well as the "owners" of the services linked to the virtual network, and it has to obtain the necessary long-term mandate from the users. In the case of environmental monitoring, where both the users and the owners might be governmental bodies, a mandate from the relevant political level may be required to support and ensure long-term activities. A given physical network can be associated with a number of virtual networks making use of the same data in different ways and for different applications.

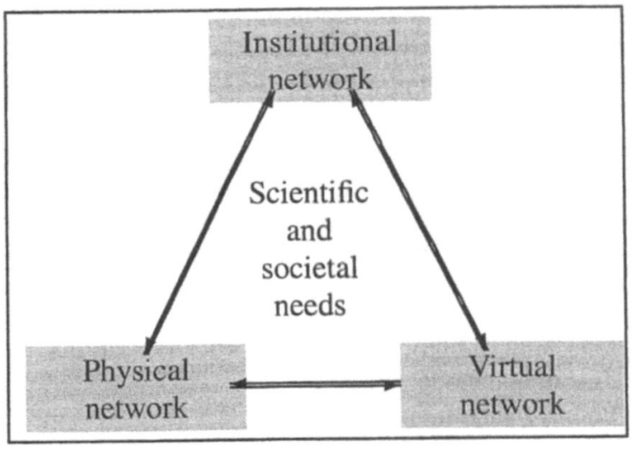

Figure 3: The Network Triangle.
An operational monitoring network should respond to well defined scientific and/or societal needs. Thus, the design of the virtual network should correspond to user requirements derived from these needs. Within the institutional network, the owners of the physical and virtual network have to be linked with the users to provide for the necessary resources and the mandate for maintenance of the network. The physical network needs to comply with the specifications resulting from the virtual network. Thus, the success of the network depends equally on three main aspects, namely, the design of the virtual network, the performance of the physical network, and the efficiency of the institutional structure.

First of all, the design of a virtual network has to be based on a thorough analysis of the "market". As a result, this analysis should identify all potential users (including future ones) of the network and their needs for specific products. A description of the user requirements should clearly specify the required properties of the products in terms of availability, integrity, continuity, consistency, precision and accuracy. The design of the virtual network has to be such as to provide for these properties. This design process will also result in a set of specifications for the performance of the physical network and the quality of the observations provided by this network as well as the communication tools to be used.

The design of a virtual network also depends on the associated institutional network (see below). Thus, the design can be modular or uniform, hierarchical and top-down or more democratic and bottom up. For a network owned by a single institution or an organisation, a hierarchical structure with clear competence for decisions and delegation of work may be appropriate and most efficient. However, in its physical implementation, for regional or global network it is hard to imagine a physical implementation which is not bottom-to-top and to a large extent depending on voluntary contributions of the participants. Such a structure is characterised by relatively high fluctuation of contributors and support, inconsistencies in observations, meta-information, and, eventually, products, relatively high probability of errors, unequal and time-dependent performance levels of different contributors, and a slow and/or insufficient response of contributors to requests from the network. The design of the virtual network has to account for these deficiencies. In particular, the necessity of comprehensive network monitoring and quality control is obvious. Therefore, regional to global monitoring networks should integrate these elements as part of the design of the virtual network.

Within a future global monitoring system, the physical network should be a multi-purpose infrastructure, i.e. there should be more applications for the infrastructure instead of an ever increasing number of sites. Synergy should be exploited as much as possible through co-location.

Over the last two decades, space-geodetic techniques have seen a research state of very rapid development both in terms of accuracy, applicability and availability. Consequently, much of the development has been stimulated by ad-hoc responses to emerging needs and new potentials and capabilities. Within IAG, CSTG has provide some coordination (see e.g., Beutler *et al.*, 1999), but very little was and possibly could have been done to bring forward a clear or even integrated strategy for building networks. Mainly, space-geodetic networks developed in the broader IAG-context were and still are science-driven.

Stimulated by the success of the IGS, science-driven space-geodetic networks are currently in a phase of reorganisation leading to the establishment of several mono-technique services. Similar to the IGS, these new services focus on fostering the specific technique, the quality and availability of the data products, and partly the co-location with other techniques. All of the services connected to the networks are based on voluntary contributions by the participants. They are mostly supported with scientific recognition by IAG and/or IAG commissions. The contributions to the networks depend to a large extent on the good-will of the participants and the availability of budgets of single individuals and their institutions.

It is interesting to note that many of these networks have demonstrated an extreme ability to survive for a long time and to develop a long-term stability. Nevertheless, the establishment of an environmental monitoring systems providing near real-time applications with considerable consequences in case of failure to meet the specifications, might require a different approach.

As stated above, meteorological and some hydrological observations are the very few examples of governmentally-driven observational networks, which have been operational for several decades if not centuries. Only recently, the growing recognition of the likelihood of human-induced climate change and the associated potential hazards as well as the necessity to approach a sustainable development of the global human society on a limited and vulnerable planet has led to the establishment of a rapidly growing number of operational global or regional monitoring systems. All these systems are institutionalised in a different way than the voluntary science-driven networks prevailing in geodesy.

In Figure 4 the structure of a virtual network for an Earth information system is sketched. In such a system, integrated monitoring is an important part resulting in an integrated database. Analysis and modelling draw on this database and aim for both the detection of syndromes and the establishment of an integrated system model. The system is fully driven by the users' need for key variables and assessment of consequences.

7 Conclusions

The emerging IGOS as the umbrella for the G3OS is a very promising development. A basic recognition of IGOS is the need to transform observations presently largely carried out under research programs into operational activities with a long-term perspective.

The integration of space-geodetic networks into a IGGOS is a timely step, which should, in the end, lead to operational monitoring. This should be oriented along the IGOS and integrated into the larger context of a future Global Integrated Earth Monitoring System (GIEMS) as one prerequisite of an Earth information system.

Though IGOS is providing a framework for such an urgently needed GIEMS, it may not be sufficient to promote the integration and adoption over all the disciplines involved in Earth system monitoring and research. In particular, the need for integrating the data sets to enable the production of data products and information useful for both Earth system science and policy makers may require other means to foster the rapid approach towards a user-oriented GIEMS.

Considering the mutual impact of the IPCC on policy makers and the direction of global change research, it may be wise to use a similar instrument to promote integration

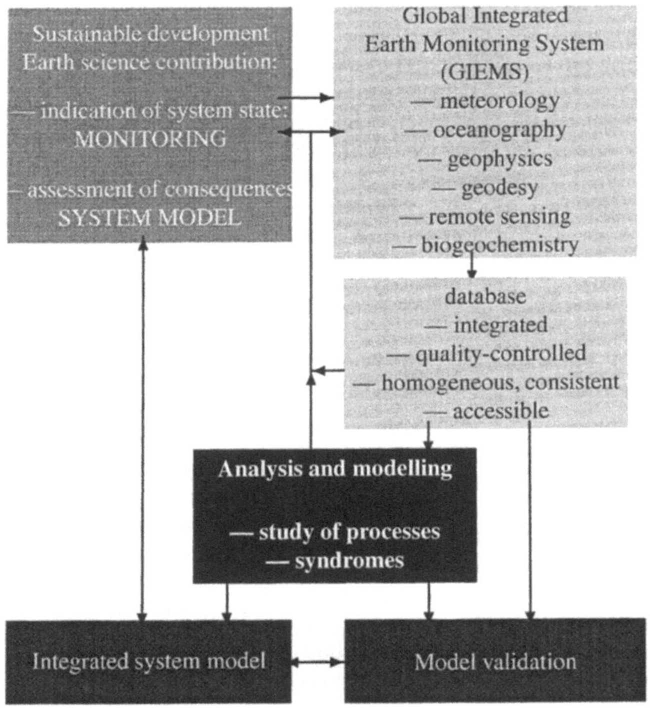

Figure 4: Sketch of a possible integration of monitoring into an Earth information system for sustainability. For a discussion, see text. The contribution of geodesy to the monitoring could be utilised by the IGGOS proposed in Beutler *et al.* (1999).

within Earth monitoring. Therefore, it should be considered to establish an Intergovernmental Panel on Global Monitoring (IPGM). Such a panel would assess the efficiency and quality of Earth system monitoring, specify the user requirements for data products and information, identify gaps in data sets and the dissemination of data sets and information (Figure 5).

This panel could be suggested to IUGG and from there to the United Nations. IAG could take a leading role in initiating this procedure.

While the space or remote-sensing segment of the G3OS appears to be well-developed, there is a clear deficiency in the organisation of the ground-based component leading to a severe lack of in-situ data. The need for an interdisciplinary ground-based observing system is obvious. The Global Geophysical Observing System (GGOS) proposed to the IUGG in 1995 and again in 1999 could provide the means for this organisation.

At the 1995-IUGG meeting in Boulder, the necessity for GGOS was discussed in several plenary sessions and subsequently has been promoted at many occasions. GGOS aims to facilitate (through enhanced communications and co-ordination) the growth of global geophysical observing system real-time networks. In particular, it will promote the co-location of multivariate observing stations for scientific synergy, as well as economy, across all IUGG disciplines and Associations. It derives its motivation from the global nature of the geophysical sciences, technological advances in sensing and telecommunication systems, and the societal imperatives for environmental sustainability and stewardship articulated in Agenda 21. Integral components of GGOS include cultivating the growth of monitoring services, facilitating data dissemination and data quality standards, and fostering interplay between numerical and statistical modelling with data network design, assessment, and utilisation. To be fully successful, GGOS needs to structure a hierarchical approach (national, regional, and global), and to engage research scientists from both the more-developed and the less-developed countries. The IGGOS would not only provide observations to the GGOS but would be a utility for all other parts of the GIEMS.

Acknowledgements

The author would like to thank Christopher N.K. Mooers for stimulating discussions based on his promotion of the GGOS as well as Bjørn Engen and Rune I. Hanssen for clarifying discussions of the network triangle.

References

Ahmad, Y. J., El Serafy, S., and Lutz, E., editors (1989). *Environmental Accounting for Sustainable Developement.* The World Bank, Washington, D.C. 100 pages.

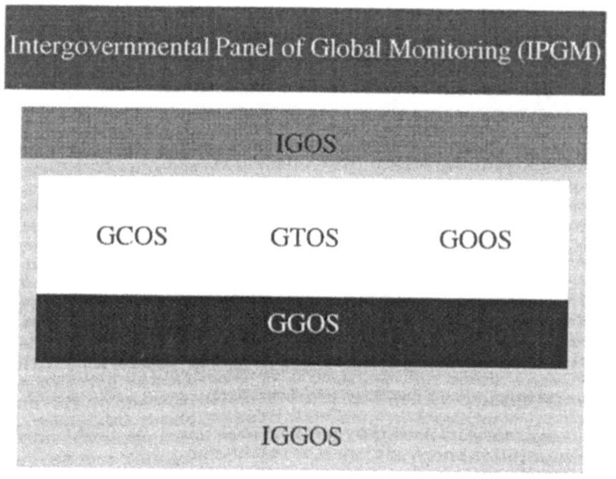

Figure 5: Towards a comprehensive monitoring system. A comprehensive monitoring system would collect all the different networks under a common integrated observing strategy, i.e. IGOS. A solid basis of in-situ data for the three emerging global observing systems GCOS, GTOS and GOOS, which are still strongly focusing on remote sensing, could be provided by a ground-based observing system, the GGOS. The IGGOS would contribute important observations and at the same time be a utility for the other components. Progress of the development in the monitoring should be assessed at a high level, preferably through an IPGM.

Beutler, G., Drewes, H., Reigber, C., and Rummel, R. (1999). Space techniques and their contribution within iag at present and in future. *CSTG Bulletin*, **14**, this issue.

Broeker, W. S. (1987). Unpleasant surprises in the greenhouse? *Nature*, **328**, 123–126.

Dahl, A. L. (1998). IGOS from the perspective of the Global Observing Systems and their sponsors. In *Proceedings for the 27-th Int. Symp. on Remote Sensing of Environment: Information for Sustainability, June 8-12, 1998, Tromsø, Norway*, pages 92–94. Norwegian Space Centre.

Ellsaesser, H. W., MacCracken, M. C., Walton, J. J., and Grotch, S. L. (1986). Global climatic trends as revealed by the recorded data. *Rev. Geophys.*, **24(4)**, 745–792.

Greenland Ice-core Project (GRIP) Members (1993). Climate instability during the last interglacial period recorded in the GRIP ice core. *Nature*, **364**, 203–207.

Houghton, J. T., Meira Filho, L. G., Callander, B. A., Harris, N., Kattenberg, A., and Maskell, K., editors (1996). *Climate Change 1995 - The Science of Climate Change*. Cambridge University Press.

Meadows, D. H., Meadows, D. L., Randers, J., and Behrens, W. W. I. (1972). *The Limits to Growth*. Universe Books, New York.

Nerem, S. M., Rachlin, K. E., and Beckley, B. D. (1997). Characterization of global mean sea level variations observed by TOPEX/POSEIDON using empirical orthogonal functions. *Surveys Geophys.*, **18**, 293–302.

Pfister, C. (1995). Das "'1950er Syndrom"': Die umweltgeschichtliche Epochenschwelle zwischen Industriegesellschaft und Konsumgesellschaft. In C. Pfister, editor, *Das 1950er Syndrom*, pages 51–96. Haupt, Bern. Publikation der Akademischen Kommission der Universität Bern.

Plag, H.-P., Ambrosius, B., Baker, T. F., Beutler, G., Bianco, G., Blewitt, G., Boucher, C., Davis, J. L., Degnan, J. J., Johansson, J. M., Kahle, H.-G., Kumkova, I., Marson, I., Mueller, S., Pavlis, E. C., Pearlman, M. R., Richter, B., Spakman, W., Tatevian, S. K., Tomasi, P., Wilson, P., and Zerbini, S. (1998). Scientific objectives of current and future WEGENER activities. *Tectonophysics*, **294**, 177–223.

Rahmstorf, S. (1995). Bifurcation of the Atlantic thermohaline circulation in response to changes in the hydrological cycle. *Nature*, **378**, 145–149.

Schellnhuber, H.-J. and von Bloh, W. (1993). Homöostasie und Katastrophe: Ein geophysiologischer Zugang zur Klimaentwicklung. In H.-J. Schellnhuber and H. Sterr, editors, *Klimaänderung und Küste: Einblick ins Treibhaus*, pages 11–27. Springer-Verlag Berlin Heidelberg New York.

Schellnhuber, H. J., Block, A., Cassel-Gintz, M., Kropp, J., Lammel, G., Lass, W., Lienenkamp, R., Loose, C., Lüdeke, M. K. B., Moldenhauer, O., Petschel-Held, G., Plöchl, M., and Reusswig, F. (1997). Syndromes of global change. *GAIA*, **6**(1), 19.

Turner II, B. L., Clark, W. C., Kates, R. W., Richards, J. F., Mathews, J. T., and Meyer, W. B., editors (1990). *The Earth as Transformed by Human Action: Global and Regional Changes in the Biosphere Over the Past 300 Years*. University Press, Cambridge. 713 pages.

Williams, D. and Townshend, J. R. G. (1998). The concept of an Integrated Global Observing Strategy. In *Proceedings for the 27-th Int. Symp. on Remote Sensing of Environment: Information for Sustainability, June 8-12, 1998, Tromsø, Norway*, pages 95–98. Norwegian Space Centre.

World Commission on Environment and Development (1987). *Our Common Future*. Oxford University Press, Oxford.

The Integration of Geodetic Measurements into an Earth Science Observing Program: The Example of Glacial Rebound

K. LAMBECK[1]

Introduction

In addressing the issue of integrating geodetic measurements into an Earth Science Observing Program, a convenient starting point may be to look at how the accuracies of geodetic measurements have improved from tens of meters in the 1960's to centimeters and less today, representing a three order of magnitude improvement within three decades. The question foremost is what can we constructively do with such measurements. Not being an engineer, I do not care about these very precise measurements *per se*. What I am ultimately interested in is the evolution of the Earth. What is the planet like today and how did it evolve into this state from its primordial planetary dust cloud? All knowledge about terrestrial resources, hazards, and environmental change follow if we can answer this very basic all-encompassing question. Our particular geodetic question, therefore, is, what can the high precision geodetic measurements contribute to these broader issues. I am not going to try and answer that since many of the papers in this symposium have done that. Instead, I am first going to make some comments on how I think geodesy fits into the broader spectrum and then illustrate this with one example.

There is sometimes a tendency to consider that, because the tools of the trade have a particular bandwidth, the Earth is constrained to behave within this bandwith. But that, of course, is far from the truth, and while the geodetic community may have to limit itself to one part of the spectrum, we should remain aware of the broader question and it is some of these that I wish to raise here: how can one integrate the geodetic data into the much wider spectrum of earth deformation observations? I will illustrate this with the example of crustal rebound and sea-level change in response to time-variant surface loads of ice and water using recent results obtained for Scandinavia and elsewhere. Here, the process is a long-term phenomenon and the geodetic measurements provide only information on the late-stage of the relaxation response, and only by combining these observations with geological data can the phenomenon be fully understood.

The extremes of the earth-deformation spectrum can be illustrated if we ask the question 'what does a radiotelescope have in common with the oldest known terrestrial materials' (Figure 1)? The answer is that both record the history of the Earth but on very different time scales: One measures the displacements on the Earth's surface with millimeter precision on time scales of hours to years, the other records a complex history of earth-deformation processes over a period of 4 billion years.

The particular rock illustrated in Figure 1 is from Archaean terrain in Western Australia and contains the mineral zircon that formed at 4.28 ¥ 109 years ago (NUTMAN et al. 1991). The zircons settled out of molten mantle material at this time when the ambient temperature dropped to about 1200oC. Through complex tectonic processes it became part of the crustal rocks, reached the surface, was eroded and transported away and buried again to depths of more than 20 km. All this before it was 1.2 ¥ 109 years old. It shows that all the geological processes occurring today were already operating in the early Archaean. Then, after about 2.6 ¥ 109 years ago, the rock was returned to near the surface where, more or less, it remained ever since, leaving it for other rocks to pick up the story of the subsequent evolution of the Earth.

At some stage these two seemingly disparate recorders of Earth history, the rock and the telescope, have to come together if we are to understand the full extent of the dynamic behaviour of the planet, including the interactions between the fluid regimes and the more solid parts of the planet. This dynamism is well illustrated in a time-space spectrum of the various processes acting on and within the earth (Figure 2). At very long periods the deformation spectrum is dominated by the mantle-wide convection process, with characteristic wavelengths of the order of the depth of the mantle and characteristic time scales of about 107–108 years. Despite the long time scales of this process, geodetic measurements of the gravity field provide an important input; the long-wavelength gravity field providing an instantaneous but ambiguous snapshot of the convection pattern today. The surface expressions of the convection are seen in plate tectonics, in the tectonic processes at the plate boundaries, with characteristic length scales of 104–103 km and characteristic time scales of 106 years, as well as at the very high frequency end of the spectrum in terms of seismic energy released within the crust. Geodetic measurements are beginning to make important contributions to defining the present-day kinematics of these plate motions and deformations. An important reminder is that not all deformation occurs at the plate margins. Major, largely unpredictable, earthquakes do occur within even

[1] Kurt Lambeck: Research School of Earth Sciences, The Australian National University, Canberra, ACT 0200, Australia, Tel. +61–2–6249 5161; Fax +61–2–6249 5443; Email Kurt.Lambeck@anu.edu.au

the oldest and stablest plates, such as the Proterozoic and Archaean terrains of Australia, and no one part of the Earth can be considered as rigid when viewed on long time scales.

Intermediate processes include the interactions between the core and mantle, operating again on large length scales but on time scales that range from the very short to the order of 103 years. Other intermediate processes include the deformation of the mantle by periodic changes in the surface loads of ice and water, operating on time scales of 105–103 years and on length scales of 104–103 km. I return to this particular problem below.

The geodetic measurements can contribute to processes within one part of the spectrum only, that part in the time interval of hours to years; with the upper limit determined essentially by the attention-span of funding agencies. We can illustrate the position of geo desy in this spectrum further with the specific question of the rheology of the earth. The scientific problem is to understand the rheological properties of the planet so that it becomes possible to predict the response over a very broad frequency band. At high frequencies the response is essentially elastic, at very low frequencies it behaves essentially as a low-viscosity fluid. But how does the transition between the two end-member states occur? At what point does the essentially elastic behaviour become dominated by fluid behaviour? One way of tackling this issue is to measure the Earth's response to known forces over a broad frequency band.

In some problems, the force F is known. It can be external or internal to the Earth. The deformation D or some other attribute of the response, such as gravity or rotational response, is measured, and from the relation $D = F^*R$ the response function R of the Earth is inferred. Examples include the tidal deformation of the Earth, where F can safely be assumed known, D is measured by an array of instrumentation, and R is inferred in term of the Love numbers and the attenuation factor.

A different situation is exemplified by core-mantle coupling processes, in which differential movements between the two parts of the Earth result in a force exerted on the mantle at its base. The D in this case is the rotational response to the torque generated by the force, and the R is some property of the mantle, such as electrical conductivity if the coupling is electromagnetic. In this case, as in most problems, the force is only partly known and the problem is to infer something about both F and R from the observed response.

Geodetic measurements here are important, bridging the gap between the seismic and geological measurements and inferences, the one characteristic of time scales of seconds and minutes, the other characteristic of time scales of thousand years and longer. At higher frequencies we have to call upon the methods of seismology, at lower frequencies upon the methods of geology: geodesy as very-low frequency seismology or as high-frequency geology sums up well the position of this discipline in the larger spectrum of Earth Science.

Now, one could continue with platitudes and talk about how geodetic measurements can be, and should be, integrated into the bigger picture of earth science. Numerous examples of this are given in LAMBECK (1988) but reality is that this integration is still far from successful achievement and I will refrain from further generalisations and attempt to make use of various bits and pieces of information from different disciplines, including geodesy, to address an important problem: the viscosity of the mantle.

Glacial rebound and mantle viscosity: Statement of the problem

The analysis problem is from an observed response D, to infer the response function R of the Earth that is subjected to an only partly known time-dependent ice load F.

R: the response function. The viscosity of the mantle, including the lithosphere, is an important physical property, entering into quantitative evaluations of the thermal history and the convective regime of the mantle. Yet it remains a relatively poorly known property and major unresolved questions remain about its linearity or not, about its depth dependence, and about its lateral variation, particularly within the upper mantle. Laboratory and seismic measurements give some indication of these dependencies at high frequencies, and long-term surface loading problems by volcanos or sediments provide an indication of the response at low frequencies. At intermediate frequencies it is the periodic loading of the Earth's surface by ice sheets that provides the means of estimating the mantle viscosity.

F: the surface load function. The forcing of the earth is the growth and decay of the ice sheets over periods of 105 years and longer. As ice sheets wax, water is withdrawn from the oceans, and as ice sheets wane, the oceans are replenished. Thus through time the solid earth is subjected to a time-dependent surface load that stresses the mantle, inducing flow and deformation. While the general characteristics of the loading cycles are known, major gaps in our knowledge of the ice distribution do occur. How thick was the ice over Europe and North America at the time of the last glacial maximum? Was there a thick ice sheet over the Barents-Kara shelves at the same time? Was the Antarctic ice sheet larger in the past than at present? How did the ice sheets retreat once the climatic amelioration started? In solutions for R, therefore, F be considered to be only partly know.

D: the observed response. Under an expanding ice cap the crust will subside as the load stresses, transmitted through the essentially elastic lithosphere to the mantle, relax. As the ice sheet melts, the crust will rebound as mantle material flows back to the region below the former areas of glaciation. Away from areas of ice loading, smaller adjustments of the crust will occur in response to the concomitant changes in water load. The observed response can be of several complementary kinds: (i) the vertical movement of the crust, measured by geodetic positioning methods, (ii) the movement of the crust relative to sea level, measured by tide gauges or inferred from geological data, (iii) changes in local gravity, and (iv) changes in the earth's rotation and in satellite motion. The analysis problem is how, from these observed responses, to infer both the planet's response function (the mantle viscosity and effective lithospheric thickness) and any unknown aspects of the glacial loads.

Estimating the response and forcing parameters

Observations of vertical movements of the crust provide the most direct measure of the response to the changing surface load. However, such observations have been possible only very recently, using GPS measurements in Fennoscandinavia where the current rates of movement range from 0 to 1 cm/year (SCHERNECK et al. 1999), but the observational record is still too short to provide a significant estimate of mantle rheology. Instead, we will have to rely for the time being on some of the longer geodetic and geological records to solve the rebound problem.

Tide gauge records of sea-level change measure the changing position of the land with respect to the sea surface. In Scandinavia the dominant signal will be of crustal rebound but, at a rate of up to 10 mm/year, parallel changes in ocean volume that result in global sea-level rise of 1 to 2 mm/year are significant and a strategy has to be developed for separating out the two effects. One approach is to use the rebound model to predict where, at the periphery of the former ice sheet, the rebound is zero, to measure sea level there and to attribute any tide-gauge observed change to "global" change. Differences in sea level between these localities and the other Scandinavian sites then determines the rebound signal which can be inverted for mantle rheology. The result is an average eustatic rise of about 1.0-1.1 mm year for the past 100 years (LAMBECK et al. 1998a). Figure 3 illustrates the resulting present-day rates of "absolute" vertical crustal movement inferred from the inversion of the tide gauge (and lake-tilt) data.

But what we see today is the tail-end of the relaxation process and the amounts of vertical movements have been much greater in the past, about 200 m or more in the past 8000 or so years for the central parts of the Gulf of Bothnia. Records of the past movements are contained in shorelines that are now above or below their present position but the observation of the age and height of these features is the sum of two effects: the rebound of the crust and the change in the ocean volume. Inversion of the data for the mantle rheology therefore requires an independent assessment of the changing ocean (hence ice) volumes or a strategy for separating the two kinds of unknowns. This separation can be achieved from an analysis of now-submerged shorelines in regions far from former centers of glaciation, but does introduce uncertainty into the analysis (FLEMING et al. 1998). An initially "local" problem, becomes a global problem when the full solution is sought.

But how good will be the outcomes of the inversions for the mantle rheology. To what extent are the limitations of the ice models mapped into the viscosity estimates. For example, if the model ice sheet is thicker than in reality, the predicted crustal deflection will be excessive and, for a given observed shoreline height-age relation, the inferred mantle viscosity is likely to be too high. Fortunately, the spatial variability of the rebound signal, within and beyond the former ice margins, is quite complex and this permits some separation of the ice and earth-model parameters to be made. One of the outcomes is robust evidence for a degree of radial stratification in the mantle viscosity (LAMBECK et al. 1996) (Figure 4), another is for ice sheets that are generally relatively thin in Lateglacial times. Certainly ice models much more than about 2000 m thick over Scandinavia during the last time of glaciation are quite incompatible with the observed rebound data (LAMBECK et al. 1998b) and Figure 5 illustrates the inferred ice profile across Scandinavia in early Lateglacial time.

One aspect of the analysis is that the geodetic and geological data are quite complementary. The former are for an epoch long after the ice retreated and while the signal is small, the measurements are precise. Also, the rebound predictions are not sensitive to the details of the ice sheet in Lateglacial times, the period for which the ice model is least certain. However, they are sensitive to the total volume of ice within the ice cap. Models in which the ice is distributed differently during the deglaciation phase tend to give the same Earth response function. The earlier part of the geological data corresponds to periods when the ice sheets was still present and the shoreline elevations for this period are strongly dependent on the distribution of ice. Thus the older data is not the most appropriate for inferring mantle viscosity unless one can be certain that the ice model is well constrained. On the other hand, the combination of the geological and geodetic data makes it possible to improve the separation. Recent solutions, based on both data types, produce consistent solutions for mantle viscosity, for Scandinavian ice sheet parameters, and for recent changes in eustatic sea level, with the latter value being of the order 1.0 to 1.2 mm/year averaged over the past 100 years.

An independent data set arises from some of the special circumstances of the Scandinavian ice sheet and the Baltic Basin. During the Younger Dryas, the retreat of the ice came to a momentary stand still with the ice margin extending from east to west very approximately from north of Helsinki, to near Turku, south of Stockholm and towards Oslo. To the south a large periglacial lake developed that was cut of from the Atlantic by a combination of land and ice barriers. At a certain moment, this lake burst through its barrier, lowering the lake level by some 30 m and leaving behind a well developed and preserved shoreline; the last Baltic Ice Lake shoreline. Through subsequent isostatic rebound, this shoreline is now tilted with respect to the present Baltic level, observations of which also provide constraints on the mantle rheology and on the ice model. The analysis of these observations, along with observations of the timing of the various Baltic lake stages, yield further results for the rebound-model parameters that are consistent with the other pieces of data. These models, in return, enable high-resolution reconstructions of the Baltic region and one such result is illustrated in Figure 6 (LAMBECK 1999).

While the solutions for mantle viscosity from the inversion of sea-level data points to a high viscosity for the lower mantle, the precise value of this viscosity, nor the degree of stratification within the lower mantle, is well defined. This arises because it is only the spatially long-wavelength components of the load that stress the lower mantle and these components in the sea-level signal are generally much smaller than the shorter wavelength variations. However, geodetic observations are again important here. In particular the degree 2 component in the ice load induces time-dependent terms of the same degree in the gravitational potential of the Earth, leading to perturbations in both satellite motion and in the Earth's rotation. The amplitudes of these changes are not strongly dependent on the choice of upper-mantle viscosity, the part that is best determined from the sea-level data, but they are quite sensitive to the lower-mantle viscosity. Thus the parameter estimation strategy becomes quite obvious: combine the two data types in a way that weights the upper-mantle structure to the sea-level data and the lower-mantle structure to the satellite and rotation data. The satellite data is, however also sensitive to recent changes in ocean-ice volumes and the viscosity solution will depend on the assumption made about this (see JOHNSTON and LAMBECK 1999). Thus the tide gauge data should also be introduced into the analysis.

One requirement for the rotation analysis is that the global ice volumes and the global rates of melting of the ice sheets is known. This can be constrained with observations of sea-level data from sites far from the centres of glaciation, as previously noted. This becomes another element in the final solution. Finally, the distribution of the ice sources between the two hemispheres remains a matter of importance. The major issue here concerns the water-ice balance over the past 20 000 or so years that I cannot go into detail here. Suffice it to say that the role of Antarctic deglaciation, both now and in the past, remain matters of debate and of importance. Disagreements occur between inferences on past ice volumes drawn from glaciological and sea-level studies and further work is required. Evidence for raised shorelines in Antarctica is substantial and indicates that past ice volumes have exceed those of today (e.g. ZWARTZ et al. 1998). Other than more geological work, geodetic experiments may also play a role here, through the monitoring of present-day rebound.

Conclusion

The specific example discussed here, illustrated with results that are based largely on my own work and that of my colleagues, is one of several that could have been selected to illustrate how only when the geodetic observations are integrated with other observations does it become possible to develope comprehensive solutions to the geophysical problem at issue. In this case, the resultant solution provides insights into the mantle viscosity, into the ice sheet dimensions during the last glaciation, and into changing ice volumes of the present ice sheets and the changes in eustatic sea level. None of the results are yet definitive and the subject will benefit greatly from improved geodetic measurements as well as from re-evaluations of the geological data. Other examples could equally well have been chosen. The study of the Earth's rotation, for example, is one that requires a broad-spectrum approach in order to obtain a complete understanding of the changes that have occurred throughout time. Geodetic observations alone provide a description of only a very narrow part of the total spectrum. Albeit an extremely accurate description. Other inputs come from historical records and from geological records and one of the significant developments for the latter in the interpretation of Late Proterozoic tidal rhythmites by WILLIAMS (1997). Studies of tectonic deformation in complex marginal zones of the major tectonic plates provides another example of where the geodetic data provides only one part of the deformation story, that of the recent deformations. Complementary information here comes from seismic data and from a range of geological and geomorphological indicators of crustal deformation and displacement, as has been well illustrated recently by WALCOTT (1984, 1998) for the deformation of New Zealand lying on the Pacific-Australian plate boundary. In both these and the glacial rebound problem, the geodetic observations become most valuable for understanding the physical processes operating within the Earth only when they are interpreted as part of a very broad bandwidth spectrum of observational evidence.

References

DENTON G.H., HUGHES T.J. (eds.) (1981). *The Last Great Ice Sheets*. Wiley, New York, pp. 484.

FLEMING K., JOHNSTON P., ZWARTZ D., YOKOYAMA Y., LAMBECK K., CHAPPELL J. (1998). *Refining the eustatic sea-level curve since the Last Glacial Maximum using far- and intermediate-field sites*. Earth Planetary Science Letters, Vol. 163, pp. 327-342.

JOHNSTON P., LAMBECK K. (1999). *Postglacial rebound and sea-level contributions to changes in the geoid and the Earth's rotation axis*. Geophysical Journal International, In Press.

LAMBECK K. (1988). *Geophysical Geodesy: The Slow Deformations of the Earth*. Oxford University Press, pp. 718.

LAMBECK K. (1998). *Shoreline displacements in southern-central Sweden and the evolution of the Baltic Sea*. Journal Geological Society, London, In Press.

LAMBECK K., JOHNSTON P., SMITHER C., NAKADA M. (1996). *Glacial rebound of the British Isles - III. Constraints on mantle viscosity*. Geophysical Journal International, Vol. 125, pp. 340-354.

LAMBECK K., SMITHER C., EKMAN M. (1998a). *Tests of glacial rebound models for Fennoscaninavia based on instrumented sea- and lake-level records*. Geophysical Journal International, Vol. 135, pp. 375-387.

LAMBECK K., SMITHER C., JOHNSTON P. (1998b). *Sea-level change, glacial rebound and mantle viscosity for northern Europe*. Geophysical Journal International, Vol. 134, pp. 102-144.

NUTMAN A.P., KINNY P.D., COMPSTON, W., WILLIAMS I.S. (1991). *SHRIMP U-Pb zircon geochronology of the Narryer Gneiss complex, Western Australia*. Precambrian Research, Vol. 52, pp. 275-300.

SCHERNECK et al. (1999). CSTG Bulletin No. #, this volume.

WALCOTT R.I. (1998). *Modes of oblique compression: Late Cenozoic tectonics of the South Island of New Zealand*. Reviews of Geophysics, Vol. 36, pp. 1-26.

WILLIAMS G.E. (1997). *Precambrian length of day and the validity of tidal rhythmite paleotidal values*. Geophysical Research Letters, Vol. 24, pp. 421-424.

WALCOTT R.I. (1984). *The kinematics of the plate boundary zone through New Zealand. A comparison of short- and long-term deformations*. Geophysical Journal, Vol. 79, pp. 613-633.

ZWARTZ D., BIRD M., STONE J., LAMBECK K. (1998). *Holocene sea-level change and ice sheet history in the Vestfold Hills, East Antarctica*. Earth Planetary Science Letters, Vol. 155, pp. 131-145.

a b

Figure 1: Two recorders of Earth deformation (a) an exposed gneissic rock in Western Australia which records a history of crustal deformation from about 4.2 to 2.6 billion years ago, and (b) a radiotelescope which, when used in pairs, determines changes in orientation and scale of baselines, using the techniques of very long baseline interferometry, on time scales of hours to years. In (a) the boulder with the circle is the oldest known terrestrial rock. This has been dated using a high resolution ion microprobe invented by W. Compston, shown on the right.

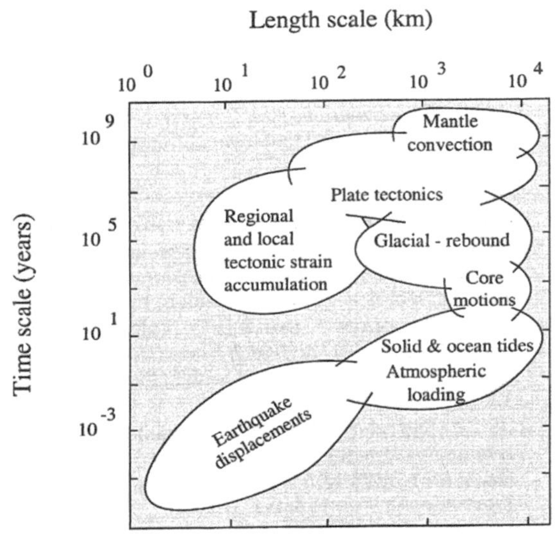

Figure 2: Schematic spectrum on time and space of some of the processes deforming the Earth and for which geodetic measurements provide some constraints.

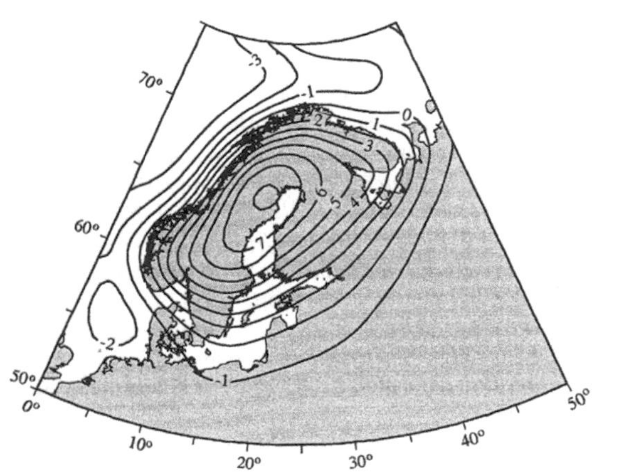

Figure 3: Predicted present-day crustal uplift in northwestern Europe based on tide-gauge and lake-level records (from LAMBECK et al. 1998a).

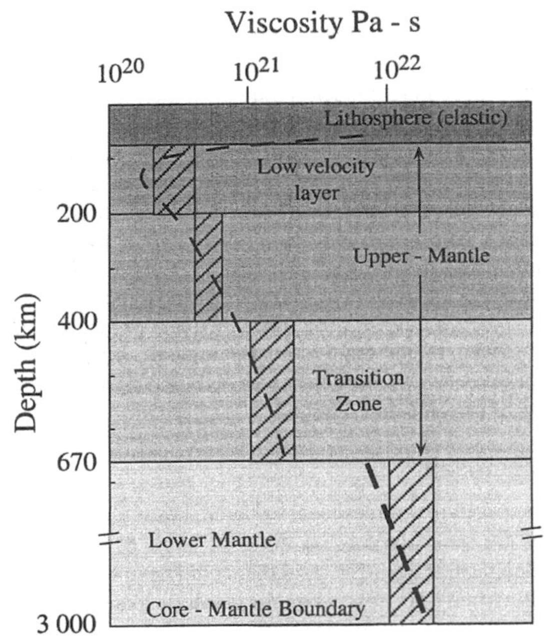

Figure 4: Profile of mantle viscosity inferred from glacial rebound observations from the British Isles. The layering is the result of the parameterization adopted to model the Earth response. The dashed curve is a continuum model more consistent with the temperature and mineralogical zonation of the mantle (from LAMBECK et al. 1996).

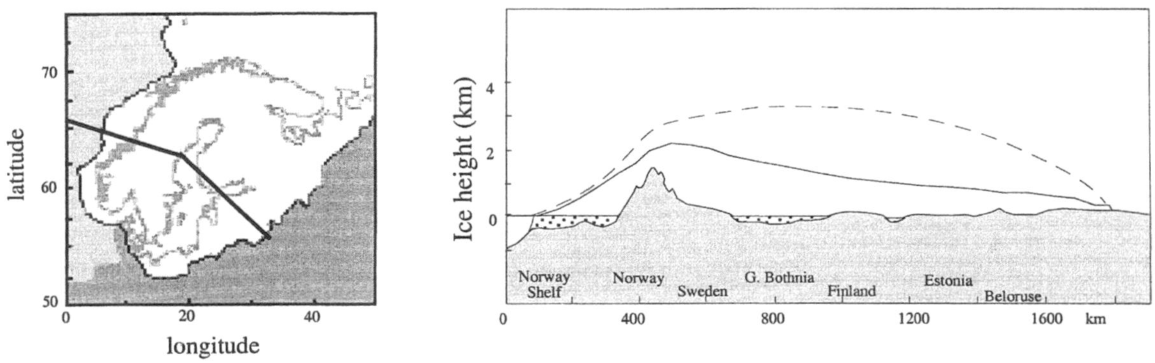

Figure 5: Profile (location in 5a) of ice height over Scandinavia at the time of the Late Glacial Maximum before the onset of the phase of rapid deglaciation (from LAMBECK et al. 1998). The dashed line in (b) refers to a characteristic parabolic ice profile (e.g. DENTON and HUGHES 1981).

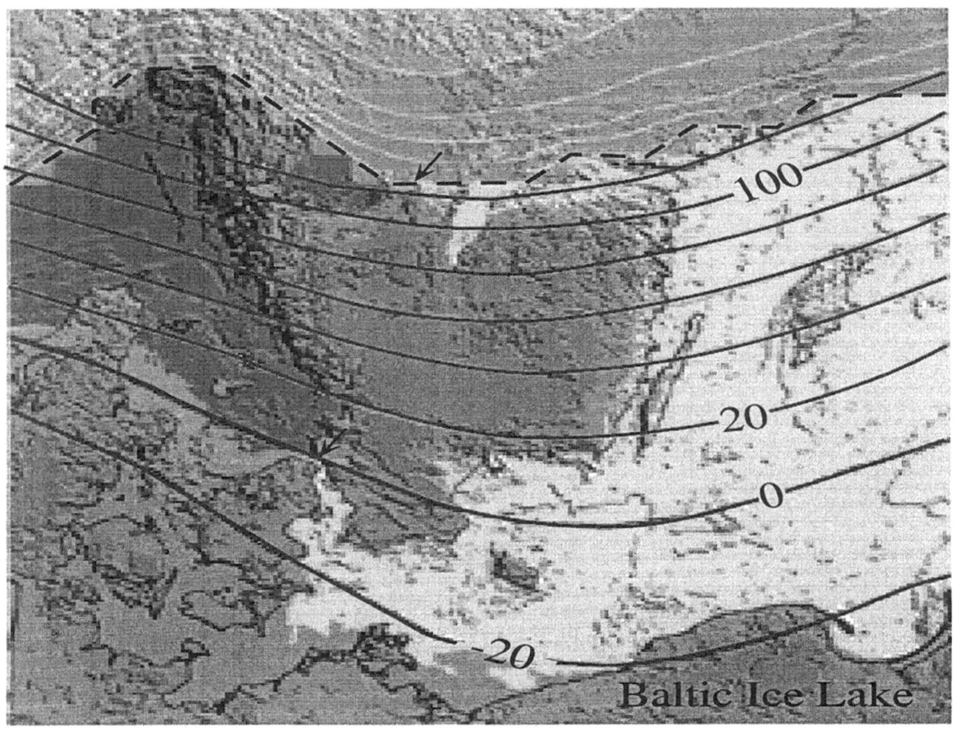

Figure 6: Reconstruction of the Baltic Ice Lake, based on parameters inferred from geodetic and geological observations to describe the ice-and earth-model parameters. The time period is 10 300 years ago and the Lake is held back by the ice pinned on Mt Billingen in Southern Sweden (upper arrow) and by the elevated land barrier in the Danish Straits. The lake overflow occurs through the Öresund (lower arrow) (from LAMBECK 1999). The ice limit is shown by the dashed line and ice height contours (white) are at 200 m interval. The black contours represent sea-level change with respect to present. The dark grey shaded area to the left of southern Sweden is open sea. The pale grey area to the right corresponds to the fresh-water environment.

Posters – Session A

Reference System and Datum Integration

Quality Analysis of Some IGS Weekly Combined Solutions with Respect to ITRF

Z. ALTAMIMI, C. BOUCHER, P. SILLARD[1]

Abstract

IGS has initiated a pilot project in 1995 for ITRF densification. Within this project, the IGS analysis centers produce weekly sets of station positions with full variance-covariance matrices provided in SINEX format. These weekly individual solutions are then combined by 3 associated analysis centers, namely JPL, MIT and NCL. This poster will attempt to analyse some recent IGS weekly combined solutions with respect to ITRF96 used as reference by IGS starting on March 1, 1998. The analysis of these solutions will focus on their quality as well as the fit to ITRF96 in terms of terrestrial reference frame fixing. Moreover, as part of EUREF activities, analogous weekly solutions are also performed by CODE, which are also analysed and presented in this poster.

1. Objective

The main objectives of this paper are twofold:

- Quality assessment of the IGS weekly combined solutions, based on post fit residuals after 7-parameter adjustment w.r.t. ITRF96 (BOUCHER et al, 1998);
- Evaluation of the fit to ITRF96 in terms of Terrestrial Reference Frame Fixing (Origin, Scale, Orientation).

2. Analysed Data

The data used in this analysis are some combined IGS weekly station positions, determined by JPL, MIT, NCL and EUREF/CODE analysis centers and provided in SINEX format. These solutions cover the period of GPS weeks 942 (98:031) to 962 (98:171), except some corrupted SINEX files. Note that IGS started to use ITRF96 on March 1, 1998 (GPS week 947), (KOUBA et al, 1998, ALTAMIMI, 1998).

3. Method

The steps used in this analysis are as follows:

- For MIT, NCL and EUREF/CODE solutions, constraints were removed;
- Orthogonal projection was then applied to MIT and NCL solutions, defining the orientation to be expressed in the same Terrestrial Reference System as that of the constrained solution;
- For JPL free network solutions, only orientation orthogonal projection was applied;
- Comparisons between IGS weekly combined solutions and ITRF96 were performed, estimating 7 transformation parameters at the epoch of each individual weekly solution.;
- Weighted and unweighted RMS "spherical sum of errors" upon ITRF96 and IGS weekly solution were then computed.

4. Results

4.1 Number of common stations with ITRF96

The following figures show the number of common stations between each IGS weekly solution and ITRF96. The 7 transformation parameters as well as the post fit RMS are obviously based on these common stations.

4.2 Weighted and unweighted RMS

The following figures illustrate the weighted and unweighted post fit RMS as result from these comparisons.

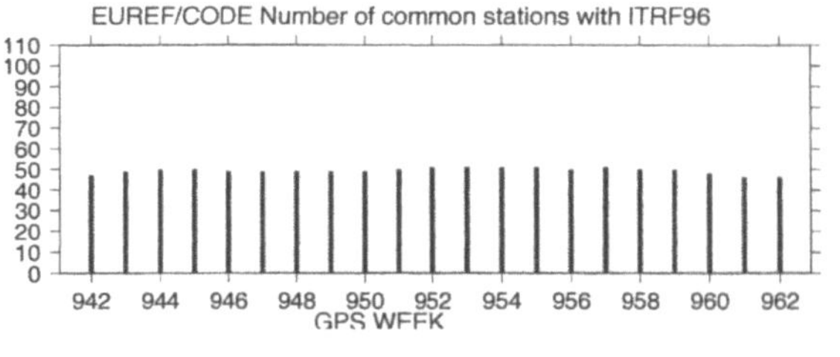

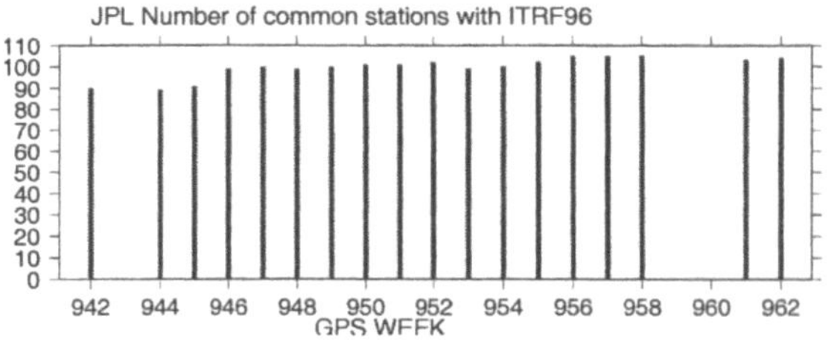

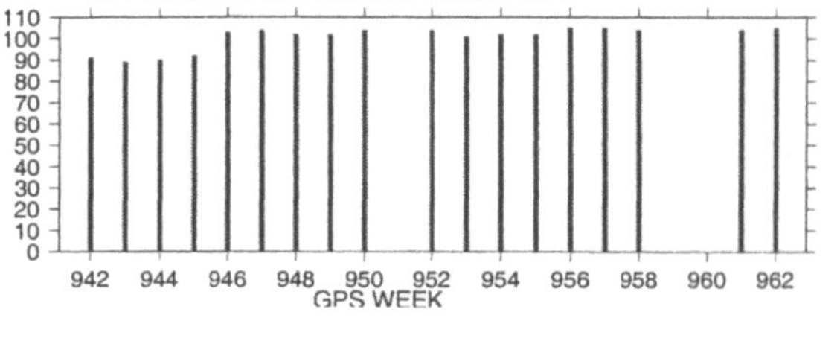

[1] Zuheir Altamimi, Claude Boucher, Patrick Sillard, Institut Géographique National, Ecole Nationale des Sciences Géographiques, 6 et 8 avenue Blaise Pascal, 77455 Marne la Vallée Cedex 2 - France; email: altamim@ensg.ign.fr

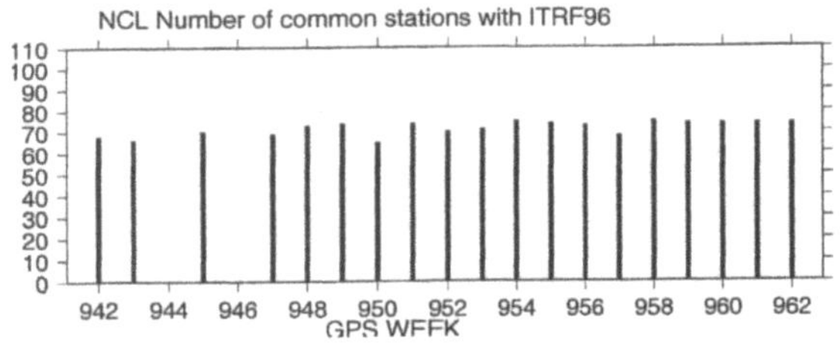

4.2 Weighted and unweighted RMS

The following figures illustrate the weighted and unweighted post fit RMS as result from these comparison.

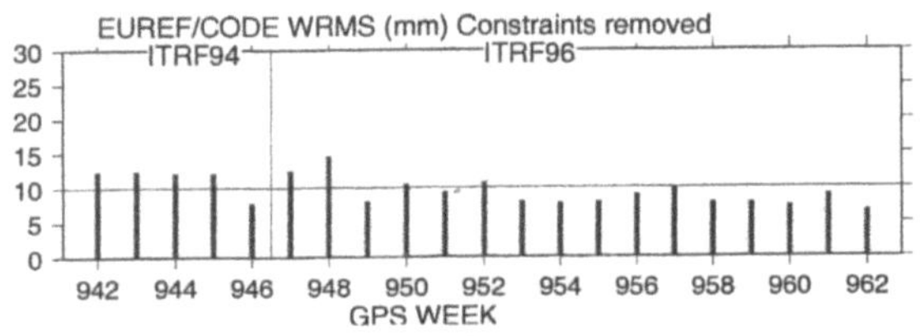

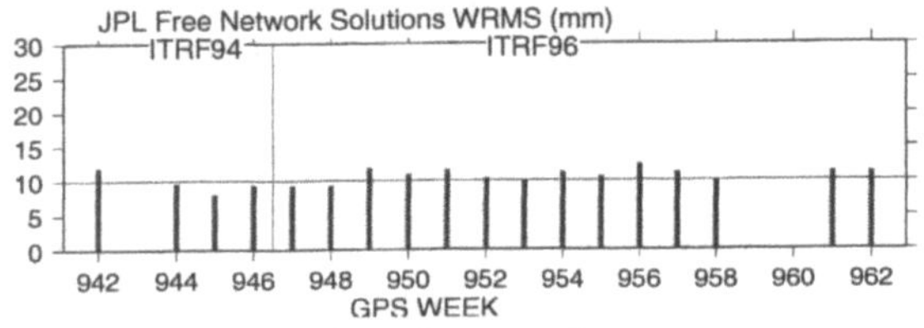

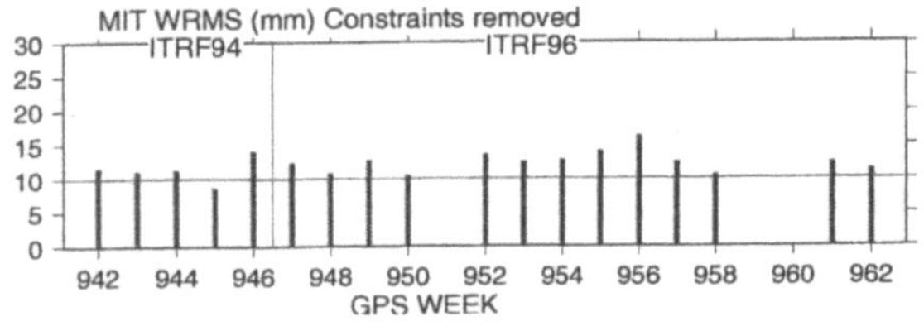

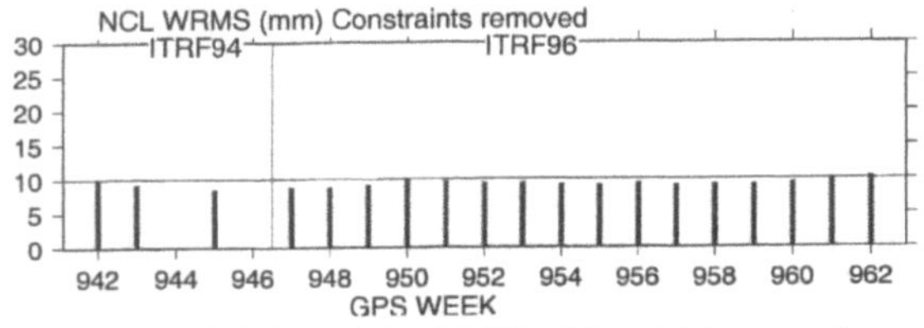

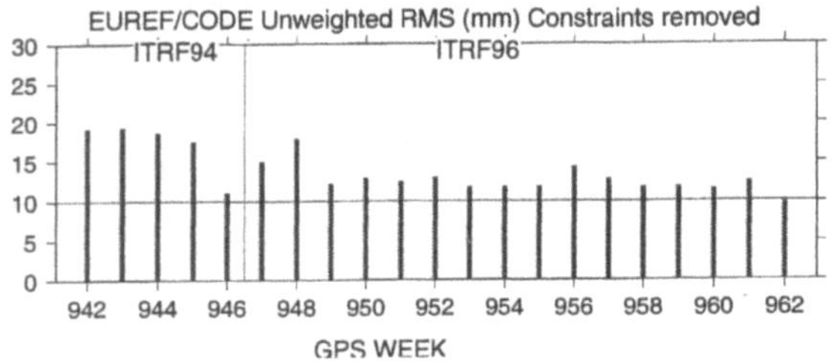

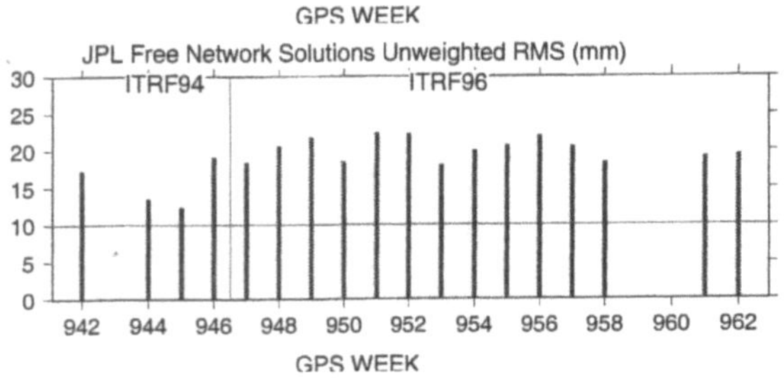

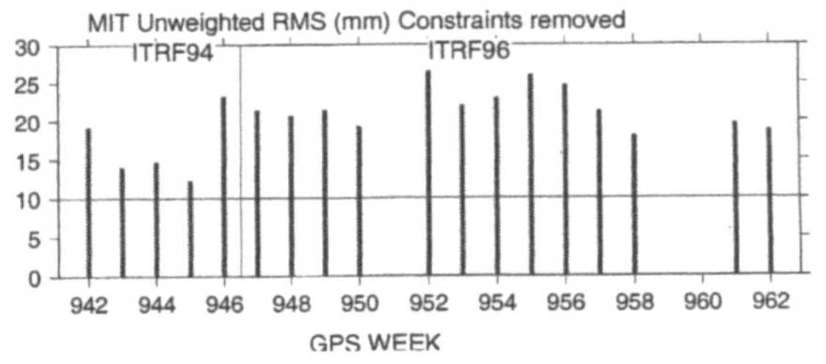

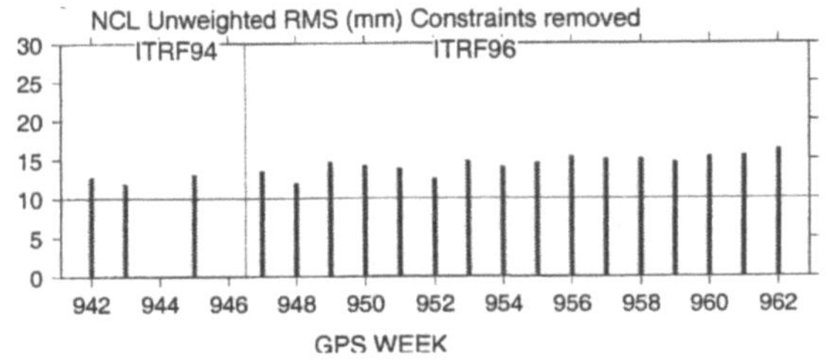

5. Terrestrial Reference Frame Definition

The transformation parameters are plotted in the following figures, illustrating the terrestrial frame definition of each IGS weekly combined solution with respect to ITRF96, in terms of origin, scale and orientation. EUREF/CODE solutions showed translations of several centimeters, which are certainly due to the "regional network configuration".

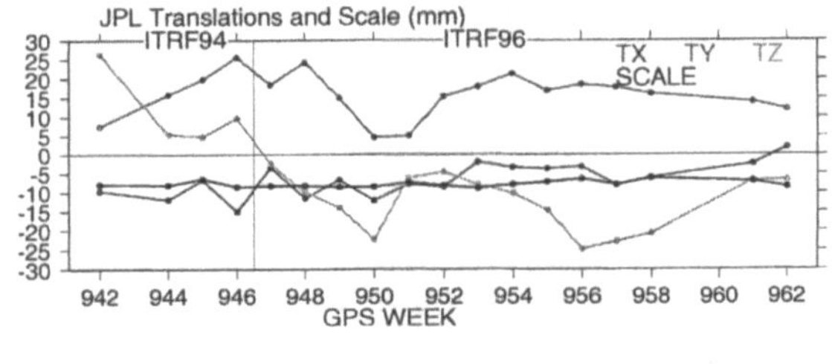

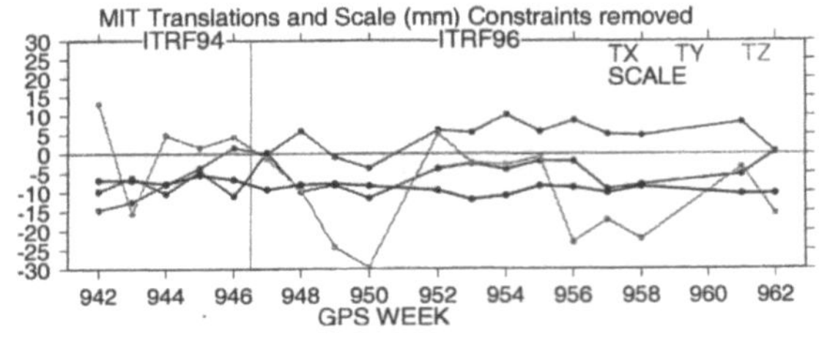

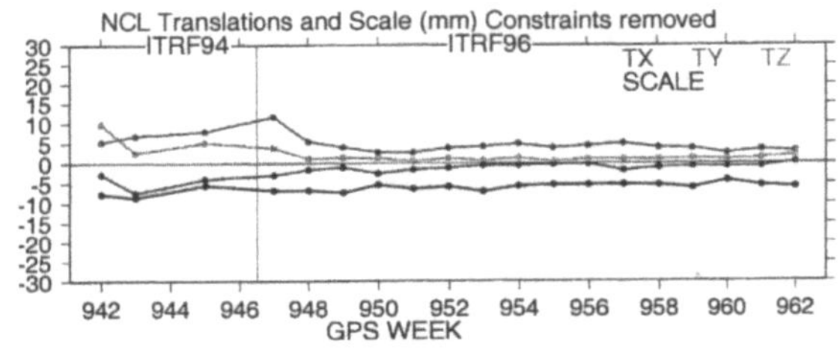

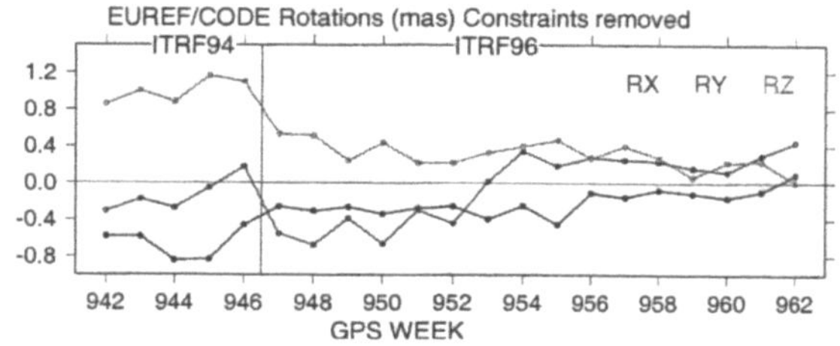

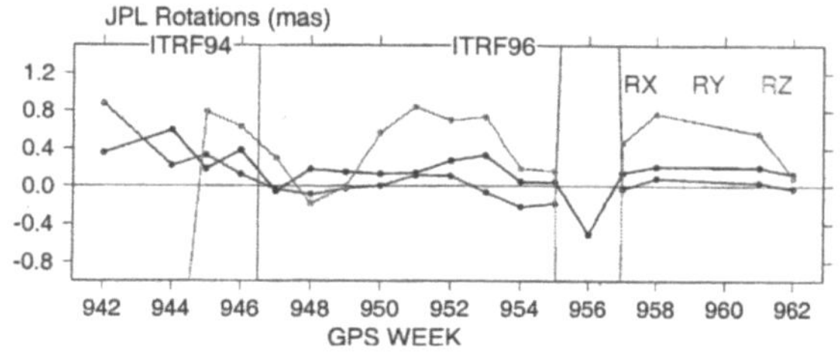

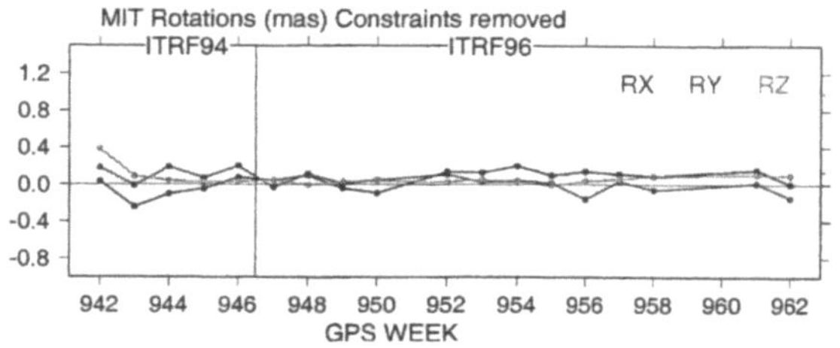

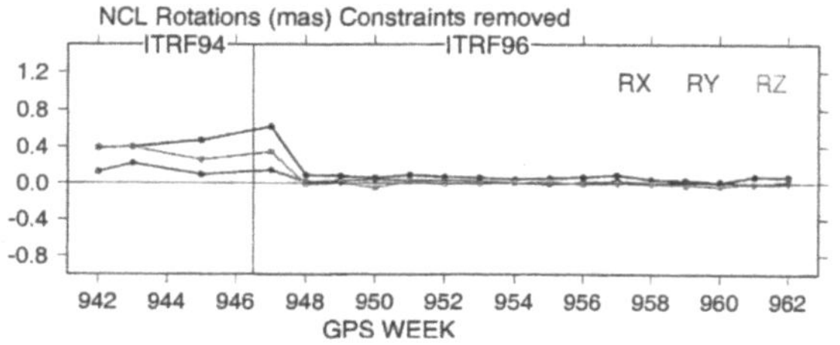

6. Conclusion

From the comparisons of the unconstrained solutions with respect to ITRF96, we can conclude that:

- The **weighted RMS** are roughly 10 mm or better for all the analysed solutions
- The **unweighted RMS** vary from 12 to 25 mm

The following table summarizes the behaviour of the origin, scale and orientation of each IGS weekly combined solution.

Analysis Center	Origin (mm)
JPL	between -25 and 25
MIT	between -20 and 10
NCL	between -5 and 5
Analysis Center	Scale (ppb) (~mm)
JPL	about (-1.4) (-8)
MIT	about (-1.5) (-9)
NCL	about (-1.) (-6)
Analysis Center	Orientation (mas)
EUREF	between -0.4 and 0.4
JPL	between 0 and 0.8 except 3 outliers
MIT	between -0.2 and 0.2
NCL	between 0 and 0.1

References

ALTAMIMI Z., 1998, *IGS reference stations classification based on ITRF96 residual analysis*, Proceedings of 1998 IGS Analysis Center Workshop, Darmstadt, February 9-11, 1998.

BOUCHER C.,ALTAMIMI Z., SILLARD P., 1998, *Results and Analysis of the ITRF96*, IERS Technical Note 24, Observatoire de Paris.

KOUBA J., RAY J., WATKINS M., 1998, *IGS reference frame realisation*, Proceedings of 1998 IGS Analysis Center Workshop, Darmstadt, February 9-11, 1998.

Geosat and ERS-1 Datum Offsets Relative to Topex/Poseidon Estimated Simultaneously with Geopotential Corrections from Multi-Satellite Crossover Altimetry

W. BOSCH[1], J. KLOKOČNIK[2], C.A. WAGNER[3], J. KOSTELECKÝ[4]

Abstract

Crossover residuals, averaged over a sufficient long time period carry the signature of geopotentially based orbit errors. In addition they may exhibit geographical pattern which are inconsistent with orbit dynamics. Tracking system offsets and different orientations, for example, map to the so called "forbidden" harmonics: the degree 1 and the degree 2, order 1 coefficients. Long-term averaged dual satellite crossover residuals among Geosat, ERS-1 and Topex/Poseidon, all with JGM3 based orbits, have been used to solve simultaneously for i) corrections to the gravity field harmonics and ii) parameters which account for a coordinate frame offset relative to Topex/Poseidon. The parameters are estimated by a least squares adjustment. The paper investigates in particular the obstacle of large correlation between the shift parameters and the low degree order one harmonic coefficients. Results of the inversion indicate significant shifts between the coordinate frames of the altimeter missions involved. They must be taken into account, if secular sea level changes are to be investigated.

1. Geopotential Corrections

Crossover differences from altimetry carry the signature of unmodeled or mismodeled physical phenomena (ocean variability, atmosphere, ocean tides, the electromagnetic bias, and the inverted barometer effect). The orbit error, however, is still considered to be dominant, and the two independent observations of the ellipsoidal sea surface height h at a crossover

$$h = (H_{sat}^A + \Delta r^A) - H_{alt}^A = (H_{sat}^D + \Delta r^D) - H_{alt}^D$$

were used to estimate corrections for the radial orbit error components Δr^A of ascending and Δr^D of descending passes. H_{sat} are the computed satellite heights and H_{alt} are the altimeter heights, assumed to be calibrated and fully corrected. If we define crossover residuals by

$$\Delta x^{AD} := \Delta r^A - \Delta r^D$$

the equation above can be rewritten as

$$\Delta x^{AD} = -\left\{\left(H_{sat}^A - H_{alt}^A\right) - \left(H_{sat}^D - H_{alt}^D\right)\right\}$$

showing that crossover residuals can be "observed" by the negative differences of sea surface heights. Kaulas first-order analytical solution of satellite motion establishes a relationship between radial orbit errors and geopotential coefficients. The radial orbit errors of ascending and descending passes

$$\Delta r^A = \Delta\gamma + \Delta\delta \quad \text{and} \quad \Delta r^D = \Delta\gamma - \Delta\delta$$

are composed of a "mean" ($\Delta\gamma$), and a "variable" part ($\Delta\delta$) (ROSBOROUGH 1986), which are related to the errors of the gravity field harmonics dC_{lm}, dS_{lm} by

$$\Delta\gamma = \sum_{l,m} Q_{lm}^C(a,I,\phi)(dC_{lm}\cos m\lambda + dS_{lm}\sin m\lambda)$$

$$\Delta\delta = \sum_{l,m} Q_{lm}^S(a,I,\phi)(dC_{lm}\sin m\lambda - dS_{lm}\cos m\lambda)$$

The functions $Q_{lm}^{C/S}(a,I,\phi)$ depend on the semi-major axis a, the inclination I of the satellite just considered and the geocentric latitude ϕ. For details see ROSBOROUGH (1986) and BOSCH (1997). Combining the equations above for *single* satellite crossover differences (SSC) gives

$$\Delta x^{AD} = \Delta r^A - \Delta r^D = 2\Delta\delta \; .$$

The mean orbit error $\Delta\gamma$ cancels out so that this error component is unobservable from SSC. This changes with crossover residuals between different satellites, indexed by 1 and 2, like ERS-1 and Topex/Poseidon,

$$\Delta x_{12}^{AD} = (\Delta\gamma_1 - \Delta\gamma_2) + (\Delta\delta_1 + \Delta\delta_2) \; .$$

There are three other types of crossovers between different satellites, namely *DA*, *DD* and *AA*. Such *dual* satellite crossover residuals (DSC) are only sensitive to the mean orbit error *differences* (KLOKOČNIK et al. 1993).

2. Datum Offsets

The satellite motion is described in an inertial system: the origin of the coordinate system is the center-of-gravity and moments of inertia for the x- and y-axes vanish. In order to have no center-of-gravity offset all first degree harmonics of the Earth's gravity field must be zero. If the z-axes coincides with the axis of maximum inertia the harmonics of degree 2 and order 1 will also vanish. These terms are called *forbidden* harmonics. Different tracking systems and tracking station distribution or weak calibration conditions, however, may cause differences in the orbital reference system. In the present investigation we consider

- a radial offset dr, to absorb variations in scale or an error of the altimeter bias,

1 Wolfgang Bosch: Deutsches Geodätisches Forschungsinstitut, Marstallplatz 8, D-80539 München, e-mail: bosch@dgfi.badw.de

2 Jaroslav Klokočnik: Astronomical Institute, Czech Academy of Sciences, Cz-251 65 Ondřejov Observatory, Czech Republic

3 C.A. Wagner: NOAA 1305 East-West Highway, Silver Spring, MD 20910, USA

4 Jan Kostelecký: Research Institute of Geodesy, Topography and Cartography, CZ-250 66 Zdiby 98, Fax: +420-02-6 85 70 56, Tel.: +420-02-6 85 72 50, e-mail: kost@fsv.cvut.cz, Czech Republic

- small shifts *dx, dy, dz* of the Earth center-of-gravity.

For both, ascending and descending passes, offset and shifts map into the radial direction by (WAGNER et al. 1997)

$$\Delta\rho = dr$$
$$-\cos\phi\cos\lambda\, dx - \cos\phi\sin\lambda\, dy - \sin\phi\, dz$$

In SSC the radial errors caused by offset and shifts cancel each other and are not visible. Only DSC contain the differential effect of datum transformations for different satellites. However, additional datum parameters are only allowed *for all but one* satellite. Any attempt to estimate simultaneously datum parameters for *all* satellites would generate a singular equation system because arbitrary large datum shifts applied equally to all satellites would have no impact to SSCs and DSCs. Topex/Poseidon has the most reliable tracking support. Therefore datum parameters are estimated only for *other* satellites and they describe datum differences relative to Topex/Poseidon.

3. Stabilization

It was shown that datum offsets and mean orbit error *differences* can only be estimated with DSCs. The capability to estimate the *absolute* mean orbit error depends to a large extent on how much the orbit configurations of the satellites differ. In any case, the least squares estimation of geopotential corrections and datum offsets is an *ill-conditioned problem* that has to be stabilized. We introduced a-priori information by the variance covariance matrix of the JGM3 gravity field, which was used to compute the orbits of the satellites. It was found, however, that this a priori data is still insufficient to obtain a stable estimate of the *mean* orbit error component $\Delta\gamma$. Note, even DSCs are only sensitive to mean orbit error *differences*. Additional a priori information was therefore introduced by so-called "pseudo observations" for the mean orbit error $\Delta\gamma$ computed on a 10° by 10° grid weighted by the standard deviations obtained from a strict error propagation with the variance-covariance matrix of JGM3.

4. Multi Satellite Crossover Data

Eleven data sets with more than 53000 long term averaged crossovers were available from NOAA. These data sets are made as far as possible homogeneous: the satellite orbits are recomputed with the JGM3 gravity field. An empirical one-cycle-per-revolution correction was applied to Geosat and ERS-1 to account for time variable and non-conservative orbit errors. The same models were used to correct for ocean and solid Earth tides. For details see KLOKOČNÍK et al. (1999).

- Topex/Poseidon SSC
- ERS-1 SSC, 35-day repeat
- Geosat, SSC, exact repeat mission
- ERS-1- Topex/Poseidon DSC, synchronous, all four types AA, DD, AD, and DA
- Geosat - Topex/Poseidon DSC, with six years separation, all four types: AA, DD, AD and DA

Except for Topex/Poseidon the ccrossover residuals exhibit a marked geographical pattern that differs considerably from mission to mission. For illustration we show only two of the eleven data sets (see Fig. 1).

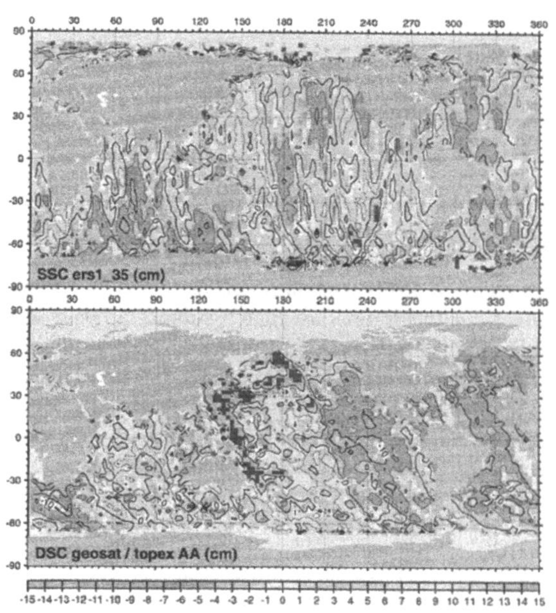

Fig. 1. Geographical distribution of ERS-1 SSC (upper panel) and of DSC between Geosat and Topex/Poseidon (lower panel). Units are cm.

5. Results

We adjusted both datum parameters and harmonic coefficients, complete up to degree and order 30 (excluding forbidden harmonics). The results show remarkable good fits to nearly all crossover data sets (see Tab. 1). Except for Topex/Poseidon SSCs the residual standard deviations ("after adjustment") are roughly 30-50% lower than for the input data sets ("before adjustment").

Tab. 1. Crossover statistic (cm) before and after adjustment

Crossover Data Set	mean ± std. dev. before adjustment	mean ± std. dev. after adjustment
Topex SSC	-0.2 ± 3.2	-0.1 ± 3.3
Geosat SSC	0.0 ± 6.5	0.1 ± 4.1
ERS-1 SSC	-0.1 ± 6.4	-0.2 ± 4.8
Geosat/Topex AA	0.3 ± 8.2	0.0 ± 3.9
Geosat/Topex DD	0.3 ± 7.8	0.0 ± 4.0
Geosat/Topex AD	0.5 ± 9.3	0.4 ± 7.7
Geosat/Topex DA	-2.0 ± 8.7	-0.5 ± 6.9
ERS1/Topex AA	0.1 ± 5.0	-0.1 ± 3.1
ERS1/Topex DD	0.0 ± 5.1	0.1 ± 2.9
ERS1/Topex AD	0.5 ± 5.7	0.4 ± 3.5
ERS1/Topex DA	-0.4 ± 5.1	-0.4 ± 3.2

The estimated datum parameters for ERS-1 and Geosat relative to Topex/Poseidon are compiled in Tab. 2. There are significant offsets and shifts, in particular *dx, dy*-shifts of 6.0 and -5.9 cm for Geosat, and a *dx*-shift of 3.4 cm for ERS-1. It must be emphasized, however, that *dr*-offsets and *dz*-shifts are highly correlated with each other.

Tab. 2. Datum parameters (cm), their standard deviation and the number of significant digits for ERS-1 and Geosat

	ERS-1	digits	Geosat	digits
dr	- 0.8 ± 0.1	1.1	- 0.1 ± 0.1	0.1
dx	+ 3.4 ± 0.2	1.2	+ 6.0 ± 0.3	1.2
dy	- 1.1 ± 0.2	0.7	- 5.9 ± 0.3	1.2
dz	+ 0.4 ± 0.1	0.5	- 1.4 ± 0.2	0.9

This can be seen from a subset of the correlation matrix shown in Tab. 3. The negative correlations are most likely caused by the dominance of the southern hemisphere ocean which lets a positive *dz*-shift appear as a global negative offset *-dr*. In addition, the matrix subset in Tab. 3 shows rather high correlations between *dx*- and *dy*-shifts and the low odd degree order one coefficients. It is known that *dx*, *dy*-shifts translate to dC_{31}, dS_{31}, see e.g. KLEUSBERG (1986). Because JGM3 already has high correlation among the odd degree order one coefficients it appears to be in general difficult to separate shifts (*dx, dy*) and coefficients (dC_{31} dC_{51} dC_{71}, dS_{31} dS_{51} dS_{71}).

Tab. 3. Subset of correlation matrix (scaled by 1000) for datum parameters of ERS-1 and Geosat and the odd degree, order one harmonic coefficients

dr	dx	dy	dz	<< (ERS-1)				
1000								
35	1000							
-50	-19	1000						
-751	-37	49	1000	dr	dx	dy	dz	<< (Geosat)
3	7	-4	-2	1000				
49	78	1	-46	-58	1000			
4	1	138	-3	-80	-3	1000		
-2	-6	7	1	**-523**	34	83	1000	
7	**546**	0	-7	1	**342**	-1	-1	C(3,1)
42	**635**	-8	-39	8	**422**	-4	-6	C(5,1)
27	**476**	-9	-25	9	**408**	-3	-6	C(7,1)
0	0	**567**	1	-1	-1	**343**	1	S(3,1)
-6	-6	**630**	6	-6	-3	**424**	8	S(5,1)
-4	-8	**447**	4	-6	-3	**404**	8	S(7,1)

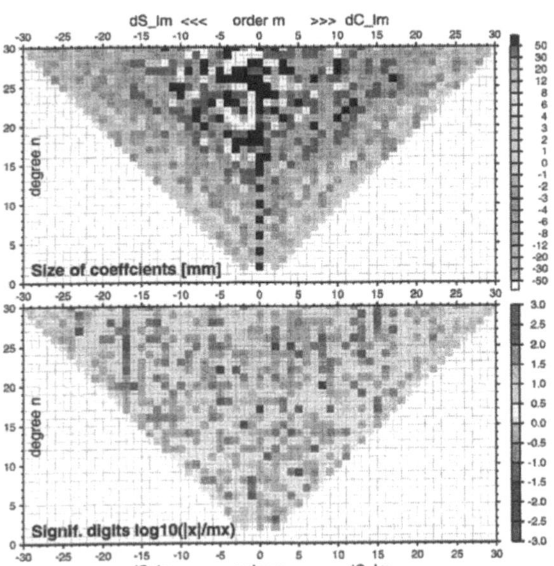

Fig. 2. Estimates of harmonic coefficients. The size (upper panel) is given in mm by scaling with R_{Earth}.

The estimated harmonic coefficients and their significant digits are shown in fig. 2. For low order and higher degree the size of the coefficients tend to oscillate. This may be caused by a non optimal balance between crossover residuals and a priori information. In summary, however, the coefficients lead to reasonable estimates of variable and mean orbit errors for *all* satellites. As an example fig. 3 shows the orbit errors of ERS-1.

We summarize that the simultaneous estimation of datum parameters and harmonic coefficients is feasible if sufficient a priori information is introduced to stabilize the system to be solved. Due to high correlations the separation of datum parameters and low odd degree order one coefficients is difficult. The solution demonstrates the capability to get reasonable estimates of the mean orbit error of altimeter satellites which is otherwise by no way visible.

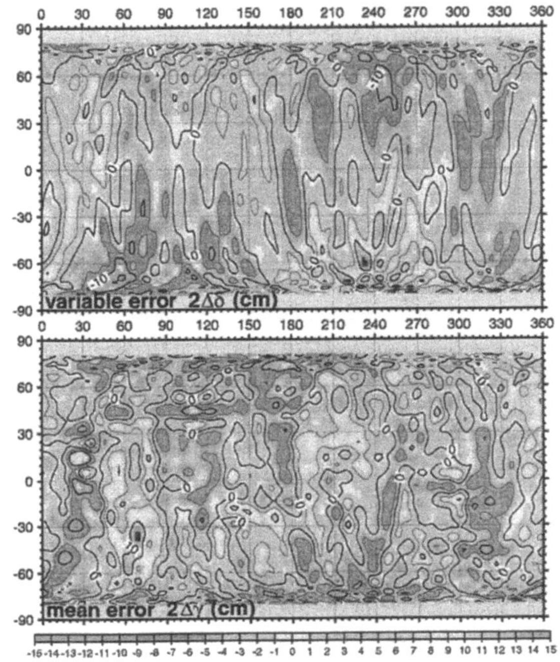

Fig. 3. Spatial pattern of variable and mean orbit error components computed for ERS-1 from the geopotential corrections. The variable part approximates the ERS-1 SSC as shown in the upper panel of fig. 1.

References

BOSCH, W. (1997): *Geoid and orbit corrections from crossover satellite altimetry.* DGFI Report No. 63, Deutsches Geodätisches Forschungsinstitut, München

KLEUSBERG A. (1980): *The Similarity Transformation of the Gravitational Potential Close to the Identity*, manuscripta geodaetica, Vol. 5, 241-256

KLOKOČNÍK J., KOSTELECKY J., JANDOVÁ M. (1995): *Altimetry with dual satellite crossovers.* manuscripta geodaetica 20: 22-95

KLOKOČNÍK J., WAGNER C.A., KOSTELECKY J. (1999): *Spectral accuracy of Jgm3 from satellite crossover altimetry*, J. Geodesy, 73, 138-146

WAGNER C., KLOKOČNÍK J., KOSTELECKY J (1997): *Dual-satellite crossover lumped coefficients, their use in geodesy and oceanography.* J. Geod 71: 603-616

ROSBOROUGH G.W. (1986): *Satellite Orbit Perturbation due to the Geopotential.* Report CSR-86-1, Univ. of Texas at Austin, Center for Space Research.

The Integration of Brazilian Geodetic Network into SIRGAS - Preliminary Results -

S. M. A. COSTA[1], D. S. BEATTIE [2], K. D. PEREIRA[3]

Abstract

Prior to the advent of satellite positioning, geodetic systems were based on a mathematic model of the Earth which best fits over a specific area. The surveys were developed for mapping requirements where precision of a meter was considered satisfactory. With the adaptation of space positioning techniques by the scientific community, the geodetic systems assumed a global character where centimetric precision could be obtained by these systems. As a consequence of the new technologies, the change from local datum to geocentric reference systems is required, in order to make the information compatible on an international level. To this end Brazil has been involved in the definition of the South American Geocentric Reference System (SIRGAS), whose purpose is to establish a geodetic system for the whole of South America. Brazil will complete the next step to accomplish this change.

The integration of Brazilian Geodetic Network to SIRGAS will be carried out through a simultaneous adjustment of the network using the software *Geodetic Adjustment using Helmert Blocking of Space and Terrestrial data* (GHOST). In the adjustment, the GRS80 ellipsoid will be adopted and all Brazilian Network will be constrained by 11 SIRGAS stations within the national area. Terrestrial and spatial data will be combined to generate coordinates for about 5000 stations. Some comparison studies about geoid model quantities to be adopted in the adjustment will be presented as well as some preliminary adjustment results.

1. Introduction

Being responsible for geodesy and mapping in Brazil, the Brazilian Institute of Geography and Statistics (IBGE) started the establishment of the Brazilian geodetic network in the 1940's. Since this time, the survey methods and computations have gone through great transformations resulting in better accuracy.

The first global and simultaneous adjustment of Brazilian net was carried out in 1996. The adjustment results showed some distortions between new realization and the original establishment of the South American Datum 1969 (SAD69) [IBGE, 1996].

Nowadays, the application of spatial positioning techniques requires the choice of a geocentric datum as a reference system. With this in aim, Brazil has been an active participant in the SIRGAS Project, through technical cooperation, as a project data center and by applying the project recommendation about the integration of geodetic network into the SIRGAS reference system [SIRGAS, 1997]. The SIRGAS reference system being the most accurate realization of ITRF in South America.

Brazil is a country of continental proportions and has advantage of an existing classical network, therefore the decision to establish a totally new GPS network could not be supported. The choice was the establishment of an Active Control GPS Network, named Brazilian Network for Continuous Monitoring of GPS - RBMC [FORTES, 1997], with 9 stations (the same as the ones occupied during the SIRGAS Project) together with regional and local GPS networks.

For the integration of the Brazilian network into the SIRGAS System, 11 SIRGAS stations within the country were connected to existing geodetic network stations.

Initially an adjustment of all GPS campaigns up to 1998 was performed, the structure thus formed is known as Brazilian GPS Network. The combination of this network with the classical network in a simultaneous adjustment was possible using Helmert Blocking technique and tridimensional modelling. The classical observations have been previously validated and stored in the last adjustment. There are 56 stations in common between classical and GPS networks.

This paper presents the adjustment data used and the methodology, as well as preliminary results of the integration of the Brazilian geodetic network into SIRGAS system.

2. Terrestrial Data

All observations for the classical network used in the global adjustment in SAD69 carried out in 1996, were also used in the SIRGAS integration and considering that validation and storage steps had been previously developed, nothing was added for this adjustment.

3. Spatial Data

During the period 1973-1991, IBGE made use of the TRANSIT satellite geodetic positioning system, especially in areas where classical fieldwork was difficult, e.g.

[1] Sonia M. A. Costa, Brazilian Institute of Geography and Statistics (IBGE); Av. Brasil 15671 - Parada de Lucas - Rio de Janeiro - RJ - CEP: 21241-051, Brazil; e-mail: soniamaria@ibge.gov.br

[2] Don S. Beattie, Geomatics Canada, Geodetic Survey Devision, e-mail: beattie@geod.emr.ca

[3] Katia D. Pereira: Brazilian Institute of Geography and Statistics (IBGE); Av. Brasil 15671 - Parada de Lucas - Rio de Janeiro - RJ - CEP: 21241-051; e-mail: katiadu@ibge.gov.br

Amazon region. For station determination the observational method was point positioning. Program GEODOP V along with precise ephemeris was used in postprocessing in order to get better results from DOPPLER observations. For this adjustment only 179 DOPPLER stations, which are coincident with classical stations, were included.

GPS technology was first used in 1991 and our first experience was the international project GIG91. Nowadays, Brazilian GPS network is composed by all GPS campaigns performed until middle of 1998, which involves 655 stations. The processing of these data was not all the same. Until 1994 a total of 187 GPS stations were processed with TRIMVEC Plus (version D), using broadcast ephemeris, this means the results referred to the original definition of WGS84. Later on data was processed with Bernese version 4.0 [ROTHACHER and MERVAT, 1996] with precise ephemeris, meaning the results referred to IRTFyy. In both cases, the station coordinates and its variance-covariance matrix (VCM) were generated in the final processing stage. The mathematical correlations were handled correctly for the data processed with Bernese software. The data of each individual campaign was transformed to GHOST format and a minimum constraint adjustment was performed. The variance factor obtained from this adjustment was used to scale the VCM. For each campaign this procedure was repeated before it was included in the final adjustment file. For SIRGAS integration we have used a total of 1757 GPS baselines.

4. Geoid Undulations and Deflections

The mathematical model of the adjustment software requires that every occupied network station in the horizontal database must have an associated geoid undulation and components of the deflection of the vertical. Previously, deflections were available for only 5% of the network and the last geoid model version for South America was based on geopotential model GEMT2. Considering the facts previously reported, it was decided to generate all the geoid information using two geopotential models, OSU91A and EGM96. Program TSCHERN was used for OSU91 and program F477 was used for EGM96, in the both cases geoid entities were based on the set of fully-normalized spherical harmonic coefficients of gravitational potential to degree and order 360. The values of the geoid components estimated by the two programs were compared to geoid undulations computed from GPS and DOPPLER observations on 316 stations in Brazil. The statistical analysis of this comparison is listed on Table 1. Although, statistics parameters are very similar, EGM96 was chosen because of smaller differences in maximum and minimum values.

Table 1: Comparison between observed geoidal heights with GPS/DOPPLER and geopotential models OSU91A and EGM96.

	DOP/GPS X OSU91A	DOP/GPS X EGM96
Average	-0.1	0.26
Std. Dev.	2.01	2.02
Minimum	-6.26	-4.97
Maximum	5.24	4.96

5. SIRGAS Integration

The GHOST software was chosen for the adjustment of the Brazilian geodetic network because of these features:

(1) Use of Helmert Blocking method for partitioning large geodetic networks [BEATTIE, 1987];

(2) Tridimensional mathematical model with fixed heights, allowing the combination of classical and GPS networks.

Note: Canada used the GHOST software in the NAD83 adjustment)

(3) GHOST software was installed at IBGE on the DOS/PC and UNIX/Hp platforms.

The defined strategy for block partitioning is based on the Cholesky algorithm for normal equation reduction and solution. Appropriate selection of a minimum number of junctions stations between blocks, minimizes the computational effort in the adjustment by isolating zero from non zero areas in the matrix.

As can be seen in Figure 1, the geodetic network was divided into 8 lower level blocks. Pairs of blocks were joined by a junction level in three different levels. Following is summary of the Terrestrial and Spatial observations that were combined:

Nr. of Directions :	16913
Nr. of Distances :	1534
Nr. of Azimuths :	389
Nr. of Positions Equations :	179
Nr. of Positions Difference Equations :	1757
Nr. of Partial Reduced Normal Eqs. :	12

The unknowns of adjustment include the coordinates of 5381 stations as well as the 15 auxiliary parameters described bellow:

- 1 orientation for azimuths in FCK5;
- 7 scale parameters for distances of classical network;
- 7 transformation parameters between WGS84 and ITRFyy.

Studies are in progress to decide if it is necessary to transform different ITRFyy' solutions to SIRGAS epoch (1995.4). The 11 SIRGAS station formed the integration by introducing overly constrained VCM along with the SIRGAS coordinates.

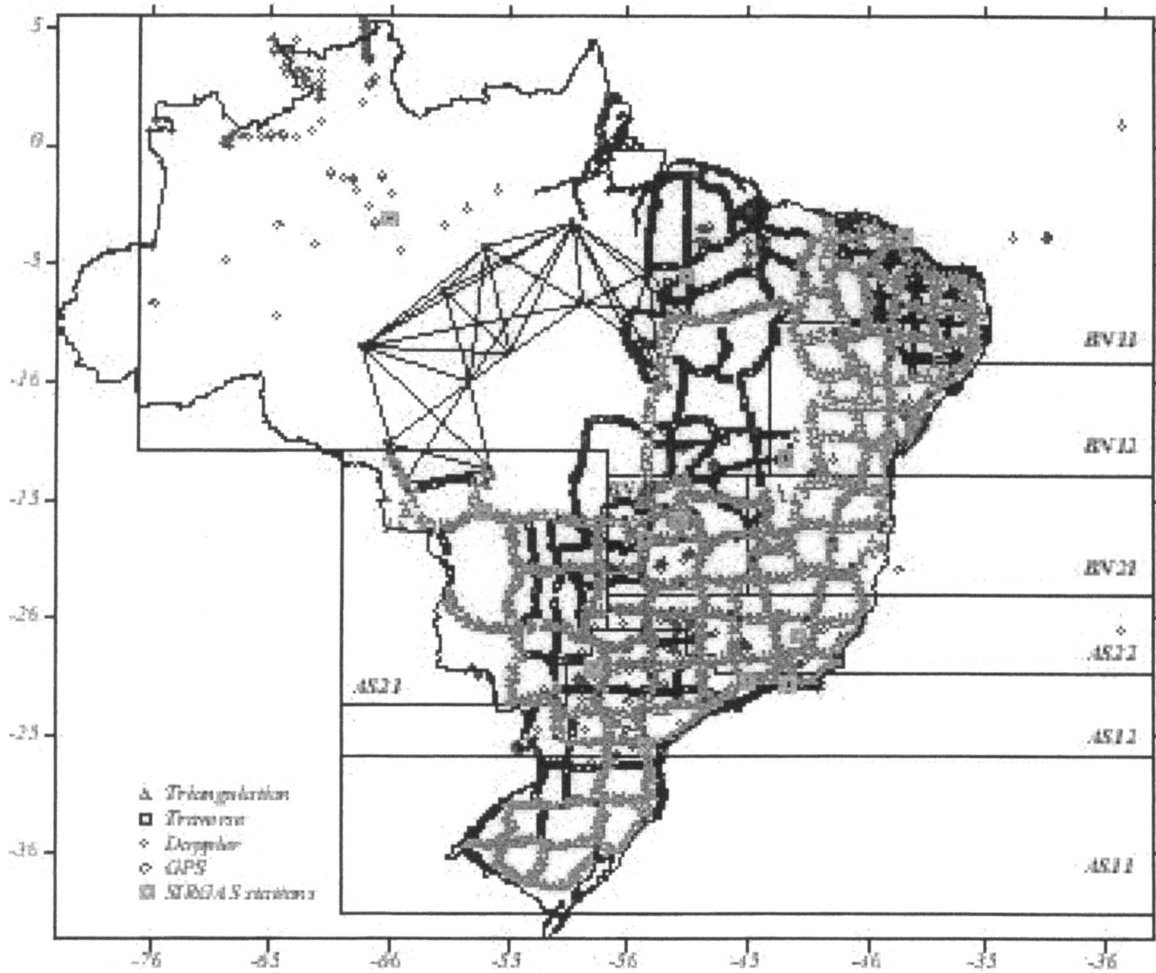

Figure 1: Brazilian Geodetic Network and Helmert Block Division

Tables 2.b and 2.c: Absolute and Relative errors at a significance level of 95%

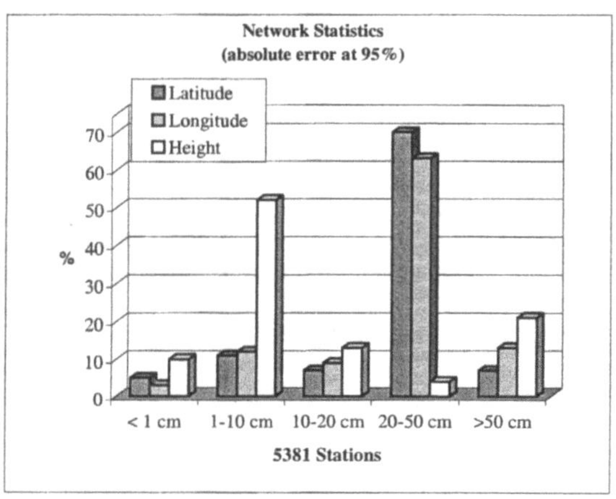

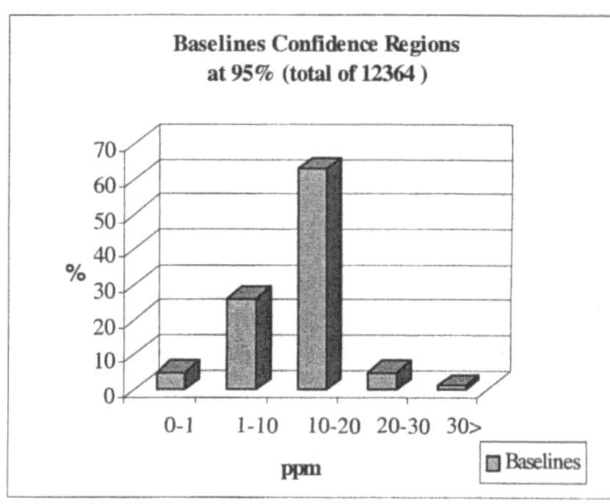

A significance level of 95% was chosen for accuracy analysis of adjustment for both absolute and relative errors. Tables 2.b and 2.c show that results was as expected for a classical network. For GPS network, 80% of the absolute errors are between 1 to 10 cm. Analyzing 1219 baselines, 60% of the relative errors are between 0.1 to 1 PPM.

6. Relation of SAD-69 to SIRGAS

Established at the end of 1960's, the SAD-69 is the official geodetic datum adopted by Brazil. Considering the different satellite positioning techniques used, it becomes necessary to analyze these differences by solving for transformation parameters between various geocentric reference systems, as NWL-10D, NSWC-9Z2 and WGS-84. As well the agreement between SAD-69 and SIRGAS is required. In 1997 with SIRGAS results a set of transformation parameters between SAD69 and SIRGAS were computed from 4 stations (Curitiba, Cachoeira Paulista, Brasília, Presidente Prudente), being:

- Translation in x : 67.327 ± 0.036 m
- Translation in y : -3.899 ± 0.036 m
- Translation in z : 38.292 ± 0.036 m

A set of 4 transformation parameters (translations and translations/scale) in each Helmert Block was solved for in the adjustment. Table 3 shows the values obtained for translations and scale parameters as well as the standard deviations computed in the new adjustment.

Table 3: Transformation parameters between SAD69 and SIRGAS for each Helmert Block area.

Block Code	No. of St.	DX (m)	Std.Dev.	DY (m)	Std.Dev.	DZ (m)	Std.Dev.	Scale (ppm)	Std.Dev.	Variance factor
AS 11	530	-66.031	0.520	2.879	0.669	-39.073	0.460	-0.266	0.151	0.8
AS 12	824	-65.421	0.356	2.004	0.410	-39.457	0.233	-0.326	0.092	1.6
AS 21	596	-65.607	0.368	2.526	0.505	-39.281	0.199	-0.409	0.102	0.9
AS 22	503	-66.356	0.552	2.984	0.550	-38.866	0.292	-0.051	0.130	0.8
BN 11	1267	-65.208	0.209	2.551	0.192	-38.890	0.043	-0.320	0.045	1.0
BN 12	474	-67.759	0.749	4.091	0.626	-38.530	0.186	-0.030	0.155	0.6
BN 21	408	-66.671	0.847	3.230	0.769	-38.756	0.338	-0.277	0.186	0.6
BN 22	592	-65.785	0.747	2.312	0.830	-39.083	0.330	-0.392	0.182	0.9
TOTAL	5194									

7. Conclusion

The results of final adjustment system integrated into SIRGAS will be available to the community at the beginning of 1999. Some studies are being made for adoption of SIRGAS system for mapping. The first step is the estimation of an adequate set of transformation parameters in order to reduce the digital map coordinates to the SIRGAS system in a more automatic fashion. The conversion of analogue maps to digital format is a slow process requiring much work, but efforts are underway to finish this process as soon as possible.

References

BEATTIE D.S. (1987): *Program GHOST User Documentation*, Geodetic Survey of Canada, Ottawa.

CADESS H. et al (1991): *Inter American Geodetic Integration*, PAIGH.

FORTES L.P.S. (1997): *Operacionalização da Rede Brasileira de Monitoramento Contínuo do Sistema GPS (RBMC)*, master thesis, Instituto Militar de Engenharia.

IBGE: *Ajustamento da Rede Planimétrica do Sistema Geodésico Brasileiro - Relatório (1996)*, Departamento de Geodésia, Rio de Janeiro, Brasil.

NGS (1983): *North American Datum of 1983*, NOAA Professional Paper NOS2, National Geodetic Information Center, NOAA.

ROTHACHER M., MERVART L. (1996): *Bernese GPS Software Version 4.0*, Astronomical Institute University of Berne, Berne.

SIRGAS Relatório Final (1997): IBGE, Rio de Janeiro.

Construction of the New Japan Datum Using Space Geodetic Technologies

S. OGI, M. MURAKAMI[1]

Abstract

The Geographical Survey Institute, Japan is constructing a new geodetic framework, JGD2000 (Japanese Geodetic Datum 2000), for the next century. JGD2000 is being built by space geodetic technologies such as GPS and VLBI based on ITRF94 and GRS80.

JGD2000 is connected at the Kashima VLBI station to ITRF94 by international VLBI experiments. By domestic VLBI experiments, several very long baselines are precisely determined. These baselines and 595 GPS stations fixed the backbone of JGD2000. The coordinates of the first through third order triangulation stations are recalculated. Those of the fourth order triangulation stations will be mapped by interpolation. Height system is also changed from normal orthometric heights to Helmert orthometric heights. JGD2000 is planned to be put in force on 1 April 2000.

1. Background

The current Japanese geodetic reference system is called Tokyo Datum. Tokyo Datum was constructed by Meiji government in late 19 century. The datum origin was determined by the series of astronomical observations at the Tokyo Astronomical Observatory and by tidal observations at Tokyo Bay. It has been working only with small revision such as correction for longitude and the azimuth angle, local re-adjustment after some large earthquakes and the introduction of the fourth order triangular stations for the cadastral survey.

Bessel 1841 ellipsoid is employed as the reference ellipsoid. Vertical value is described in normal orthometric height above the mean sea level at Tokyo Bay. Most of horizontal control survey was done by triangulation survey with theodolite.

In these two decades, EDM and GPS are rapidly introduced into the local geodetic survey. Accuracy of EDM is about 10^{-6} and GPS is about 10^{-7}. Results by GPS survey are sometimes connected with the geocentric reference system. It is observed by these techniques that Tokyo Datum has a large internal distortion and the center of its reference ellipsoid is eccentric from the geocenter.

The crustal activities around Japan are very high, so that crustal movements and earthquakes cause large internal distortion in geodetic network. Neglecting geoidal undulation also caused a large part of the distortion. The limitation of the computational resources and observation instruments decreased the strength of the geodetic network. Crustal movements also cause the elevation changes. In some areas large ground subsidence are observed. Internal distortion is shown in fig. 1.

2. Concept of JGD2000

We have started constructing a new Japanese datum using space geodetic technologies to solve these problems. A code name of the new datum is JGD2000 (Japanese Geodetic Datum 2000). We have decided that JGD2000 should be based on ITRF94/GRS80 and Helmert orthometric height should be adopted. Backbone of JGD2000 is VLBI baselines and 947 permanent GPS stations (GEONET).

To determine the ITRF94/GRS80 coordinates of exiting triangular stations, an ideal Bessel ellipsoid that has the same datum origin as Tokyo datum but has a different orientation is introduced. Re-survey data and past survey results were adjusted by projection method onto the Bessel holding coordinates of GPS stations fixed.

Fig.1: Correction vector from Tokyo Datum to Adjusted Tokyo Datum

[1] Shoichi Ogi, M. Murakami, Second Geodetic Division, Geodetic Department, Geographical Survey Institute, Ministry of Construction, JAPAN; Kitasato-1, Tsukuba, Ibaraki, 305-0811 JAPAN; Tel.: +81-298-64-4801, Fax: +81-298-64-1802, E-mail: ogi@gsi-mc.go.jp

3. VLBI Experiments

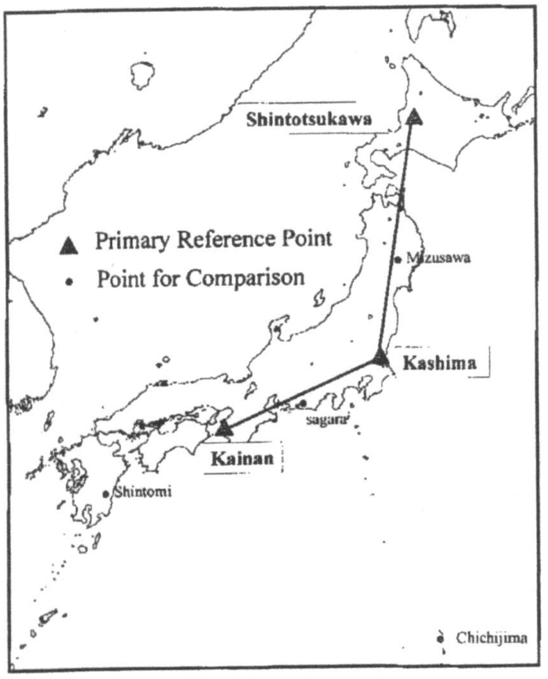

Fig. 2: Distributions of VLBI stations

To take the place of the conventional origin, we have to settle for a start point that is precisely connected with the geocenter. GSI has been participating in the international VLBI experiments with the Kashima VLBI Station. The ITRF94 coordinates of Kashima are precisely determined through more than ten years of observation as

(ITRF94 Epoch 1997.0 at Kashima VLBI Station)

X=-3,997,892.2752m

Y= 3,276,581.2676m

Z= 3,724,118.2304m

During 1986 to 1997, we had some domestic VLBI experiments using mobile and permanent VLBI stations. After assessment of the effect of crustal movements, three VLBI stations were selected as primary reference points and the three others were not selected as reference but used for comparison. The distributions of these domestic VLBI stations are shown in fig. 2.

4. Permanent GPS Stations

Geographical Survey Institute operates 947 permanent GPS stations (GEONET). We tried to fix a new geodetic framework at once. GPS campaign was held during 27 December 1996 to 5 January 1997. At that period the number of the GEONET stations was 616.

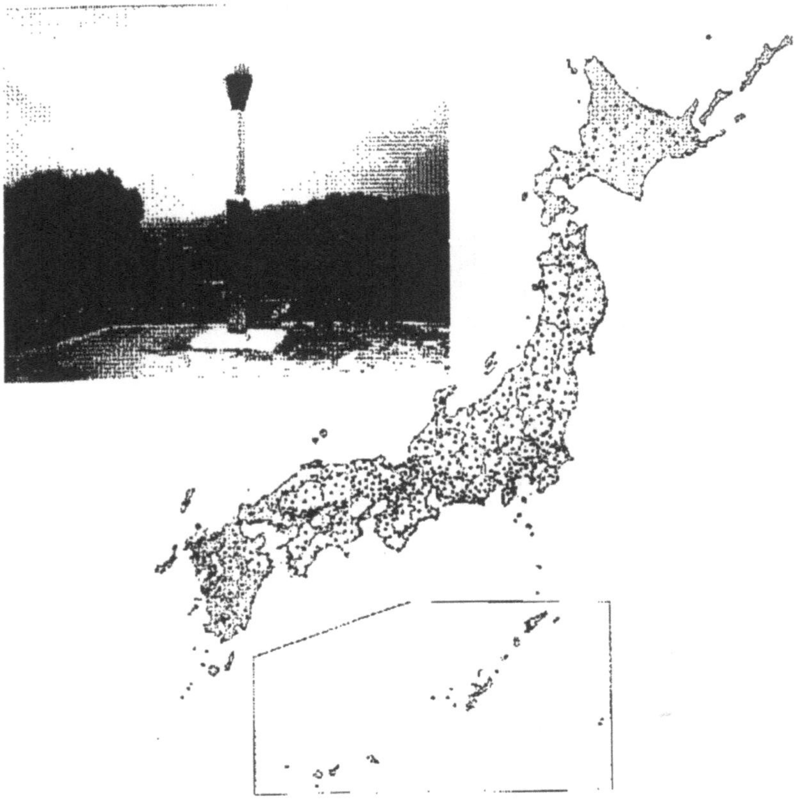

Fig.3: Distributions of the permanent GPS Stations

Total of 595 stable stations was selected from 616 stations. Observed GPS data was collected by public telecommunication lines. The data was analyzed with GAMIT and adjusted with GLOBK. The Three VLBI stations were tied to GEONET and their coordinates were held fixed in the net adjustment. Accuracy assessments have been done with the coordinates at VLBI stations for comparison. Baseline analysis was also done with Bernese to check the results.

The typical separation between permanent GPS stations is about 25km. VLBI baselines and GEONET consists of a backbone of JGD2000 (see fig. 3).

5. Re-Survey and Net Adjustment of the Triangulation stations

An ideal Bessel ellipsoid that has the same datum origin as Tokyo Datum but has a different orientation is introduced to project the coordinates of GEONET. Three axes of the Bessel are exactly parallel with ITRF94.

By GPS campaign to connect the origin of Tokyo Datum to the Kashima VLBI station, we determined shift vector between the center of the Bessel and origin of ITRF94 as

$$\begin{pmatrix} X \\ Y \\ Z \end{pmatrix}_{ITRF} = \begin{pmatrix} X \\ Y \\ Z \end{pmatrix}_{Bessel} + \begin{pmatrix} -146.414m \\ 507.337 \\ 680.507 \end{pmatrix}$$

Triangulation Stations are divided into three classes. Five hundred Stations of Class-A were connected with permanent GPS stations by GPS survey and adjusted by the three-dimensional net adjustment. The Cartesian coordinates in ITRF94 were transformed to longitude, latitude and ellipsoidal height in the Bessel. The other 3000 stations of Class-A were re-surveyed by GPS or EDM (1979-1997) and were adjusted by the projection method onto the Bessel ellipsoid. Through this adjustment, coordinates of the permanent GPS stations and 500 station of Class-A were held fixed and the effect of the local geoid was included.

Method for Class-B is the same as Class-A. Coordinates of the stations of Class-A were held fixed. The net adjustment was block net adjustment by the projection method onto the Bessel ellipsoid.

Triangulation stations of Class-C (except the fourth order) were not re-surveyed. Using the triangulation survey record, block net adjustment was done by projection method onto the Bessel ellipsoid. Vertical deflections were not corrected because they can be neglected in this block adjustment.

After adjustment, data set will be converted to Cartesian coordinates in ITRF94 and then to longitude, latitude and ellipsoidal height in GRS80.

The number of the fourth triangulation point is more than 60000. A mapping method will be introduced for transformation. Firstly, shift values on regular 30-second grid will be calculated by Kriging interpolation from shift values at adjusted stations. Then shift values for respective forth order triangulation stations will be obtained by bilinear interpolation using this grid data set. We are planning to distribute this grid data set and bilinear interpolation program among local governments and private sector to translate their own control stations and cartographic data.

6. Vertical Control

At the same time elevation data and geoid height data will be changed. Orthometric height is adopted for JGD2000. Leveling data (1987-1996) will be corrected by Helmert Height correction using gravity data and adjusted by net adjustment (see fig. 4).

Geoid model will be re-calculated with this orthometric height.

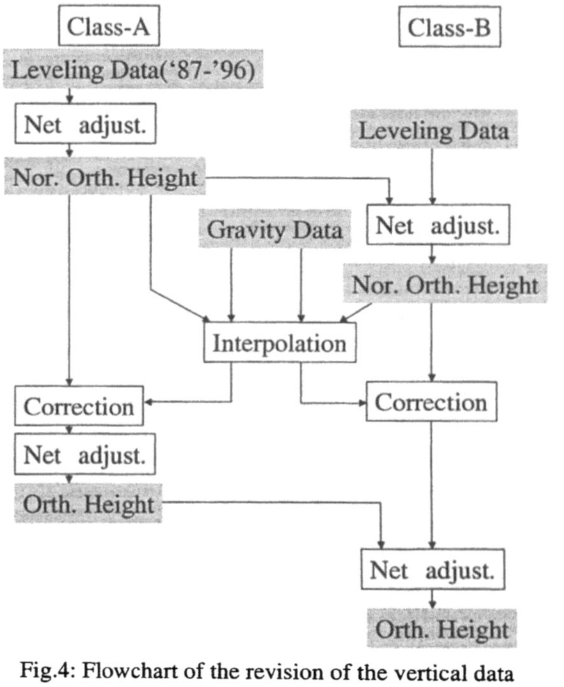

Fig.4: Flowchart of the revision of the vertical data

7. Time schedule

Almost all of the re-survey campaign has been finished except in isolated islands. Now we are carrying out the net adjustments. The horizontal net adjustment is planned to be completed by the end of 1998. The vertical adjustment will be finished by the middle of fiscal 1999. Since the Japanese geodetic datum is defined in the Survey Act, its amendment is prepared to submit the Diet.

Deviation of Japanese Vertical Datum and Reference Surface with Respect to the Global Geoid

Y. KUROISHI[1]

Abstract

The deviation of the Japanese vertical reference surface from the geoid is discussed. The analysis method is based on RAPP (1994); GPS-derived geoid heights at benchmarks (BMs) are compared with the geoid undulations calculated from a global geopotential model, EGM96. The Geographical Survey Institute (GSI) conducted GPS observation at some 800 BMs throughout Japan and determined the ellipsoidal heights with reference to a nationwide continuous GPS array of Japan. The published normal orthometric heights at BMs were adjusted in 1969/71 with local corrections for ground subsidence afterwards. Since the Japanese islands are located in an island and arc region, we have to be careful in handling the orthometric heights because of the existence of vertical crustal movements during the leveling and GPS observation periods.

The mean differences in areas of about 1.5 degree width in both longitude and latitude show geographically gradual variations of the peak to peak amplitude of around 60 cm. With evaluation of height changes estimated from preliminary adjustment results of the latest leveling survey in the period from 1987 to 1995, the p-p amplitude is slightly reduced to 50 cm. When the area of approximately 300 km by 300 km around the vertical datum station is considered, the mean deviation is derived as 14 cm and 18 cm below the EGM96 geoid with and without height change corrections, respectively.

1. Method

Three-dimensional geometric positioning with an accuracy of 1 cm or better becomes practical with space geodetic techniques and local/regional geoid modeling is about to enter a realm of 1 cm precision. Then the unification of national vertical datums and the absolute determination of global, regional and local geoid are highly expected to be realized in the same accuracy.

Fundamental data for geoid determination, such as surface gravity, digital elevation model, and GPS observation with leveled heights, however, are based on respective local height systems, whose absolute values are referred to locally defined datums. The evaluation of national vertical datums with such accuracy is, therefore, difficult at present, but would be possible in the decimeter level with respect to latest geopotential models.

As an initial evaluation of the Japanese vertical datum and reference surface (in the meaning of null height surface realized by national leveling survey at BMs) a simple method of geoid undulation comparison based on RAPP (1994) is applied by using ellipsoidal heights from nationwide GPS observations at BMs and the latest geopotential model complete to degree and order 360, EGM96 (LEMOINE et al., 1997). GRS80 (MORITZ, 1980) and ITRF89 (BOUCHER and ALTAMIMI, 1991) are employed as the normal earth model/ellipsoid and the reference frame in the geoid computation, respectively.

EGM96 geoid undulations, N_{EGM} are recovered from the coefficients in consideration of differences in semi-majors and geocentric gravitational constants between GRS80 and EGM96 (SMITH and MILBERT, 1997). GPS/Leveling geoid heights, N_{GPS} are obtained from GPS-derived ellipsoidal heights, h_{GPS} and orthometric heights, H by leveling;

$$N_{GPS} = h_{GPS} - H \qquad (1)$$

We consider that both undulations are referred to the tide-free system (POUTANEN et al., 1996). The bias, B between EGM96 and a local vertical datum can be estimated as;

$$B = N_{EGM} - N_{GPS} \qquad (2)$$

Then, we can evaluate the bias for a specific datum if we have well distributed GPS/leveling data throughout the region.

2. Data

GSI carried out GPS observations at about 800 first-order BMs throughout the country in 1995 (KURODA et al., 1997). Three-dimensional geocentric coordinates were determined with reference to the nationwide continuous GPS array of Japan.

Normal orthometric height has been employed as the Japanese height system. The official height values were adjusted in 1969/71 from national releveling data with holding fixed one and only vertical datum in Tokyo. Corrections for ground subsidence afterwards were partially applied from local releveling results.

The normal gravity field is considered in the normal orthometric height reduction. Real gravity perturbations produce significant errors in normal orthometric height in higher elevations in principle. Shorter wavelength geoid undulations are also expected to be significant in higher elevations in general. Those variations induce substantial errors in the bias estimation.

Mean differences of geoid undulations between EGM96 and GPS/leveling data are evaluated in 100 m height sections (i.e. height sections of 0 – 100 m, 100 – 200 m, and so on) in Fig. 1. The thick solid line with triangles shows the mean differences for the sections, the thin line with squares and the broken line with circles give the

[1] Yuki Kuroishi; Space Geodetic Laboratory, Geographical Survey Institute; 1 Kitasato, Tsukuba, Ibaraki 305-0811, Japan; Tel. +81-298-64-5968; Fax +81-298-64-2655; E-mail: yuki@geos.gsi-mc.go.jp

numbers of data and the standard deviations (SDs) of differences, respectively. It is seen that mean differences are significantly in contrast between the groups below and above 400 m height.

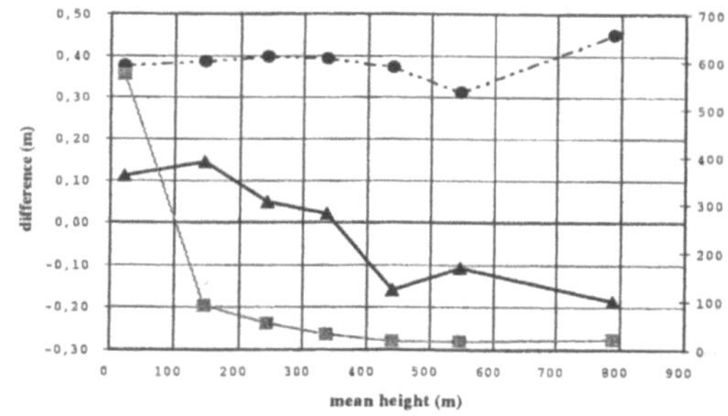

Fig. 1: Mean differences between EGM96 height anomaly and GPS/leveling geoid heights in each 100m-height section.

Japanese islands are located in an island and arc region and have large crustal movements vertically as well as horizontally. Accordingly, we have to be careful in handling the official heights of BMs because of the existence of vertical crustal movements occurred during the leveling and GPS observation periods.

The eighth national releveling at the first-order BMs had been conducted by GSI in the period from 1987 to 1995 and its preliminary adjustment results came out. After removing data points with spike-like values in height changes (spot change points) manually, gridding by continuous curvature splines in tension (SMITH and WESSEL, 1990) is applied to height changes (newly adjusted heights minus the published ones) on 10 by 10 minutes lattice. The geographical distribution is shown in Fig. 2. The changes range from about -30 cm to +30 cm.

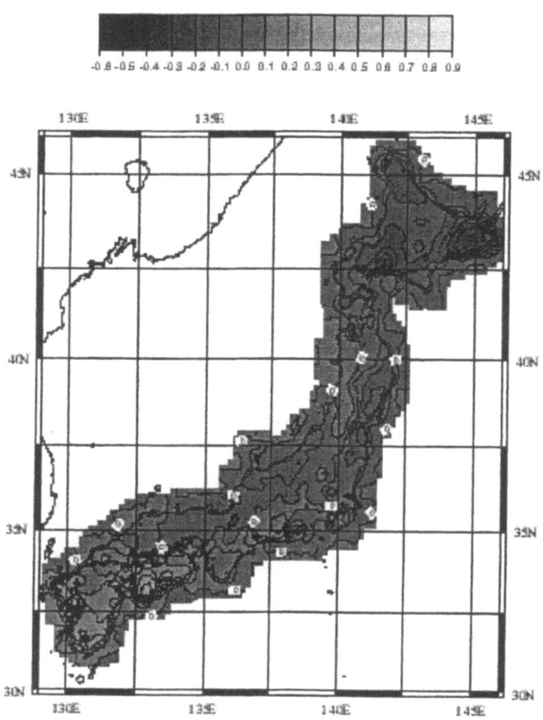

Fig. 2: Height changes until the 8-th releveling. Unit in m.

Although the newly adjusted heights were not finalized and might be revised, the reliability of GPS/leveling geoid undulation data shall be improved by correcting height changes from the published leveling heights to the new heights. Some GPS observations were made in eccentric locations from BMs. We apply the height corrections to GPS data where newly adjusted heights were given on the same positions or within the distance of 2km. Eighty-eight out of all 806 GPS sites have no such correspondence.

3. Results

In the estimation of the bias we analyze three cases for GPS/leveling data:

①– at all 806 sites with the official heights;

②– at 665 sites with the official heights; and

③– at 665 sites with corrected heights.

In the latter two cases additional 53 sites of height exceeding 400 m are ignored. Intercomparison among the three cases provides information on the effects of height discrimination and height change correction.

To evaluate geographical variations of bias, the entire region is divided into seven blocks, each of which has an area of about 1.5 by 1.5 degree in longitude and latitude. The 665 GPS sites and the block boundaries are shown in Fig. 3.

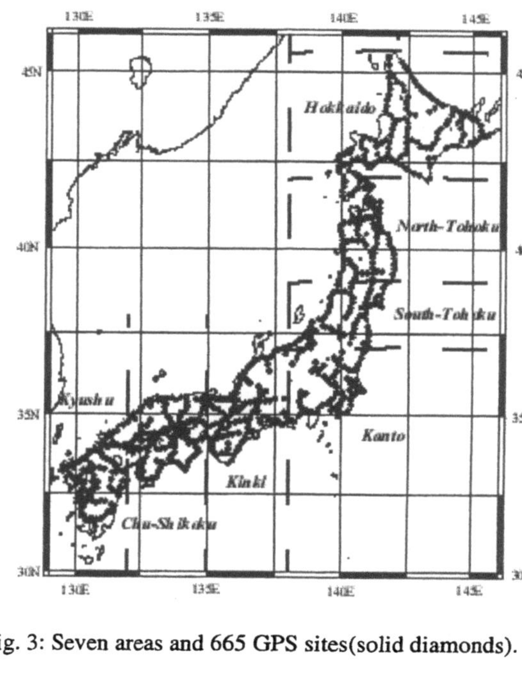

Fig. 3: Seven areas and 665 GPS sites(solid diamonds).

Geoid undulation differences between EGM96 and GPS/leveling data for the case ③ are presented in Fig. 4. Note that the differences are taken in the opposite sign to the bias. Existence of short wavelength variations are conspicuous as expected because of the limit of the EGM96 geoid undulation resolution. Moreover geographical perturbation is also found in medium wavelength.

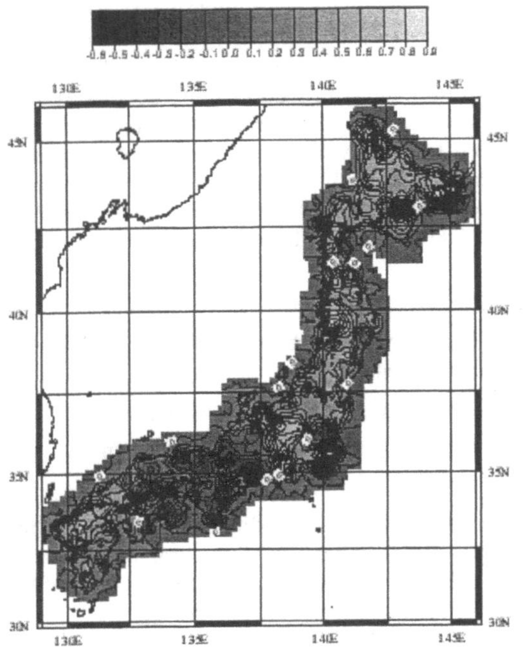

Fig. 4: Geoid undulation differences between EGM96 and GPS/leveling values from the 665 sites with height change corrections. Unit in meter.

Statistics of geoid undulation differences in each block and the whole country are summarized in Table 1. The geographical distribution of mean differences are given in Fig. 5. The results for the whole country suggest that the rejection of GPS sites of height exceeding 400 m improves SD by 2.5% and the height correction does by about 7.5%.

The blocks can be divided into two groups, each of which has almost the same value of mean differences. Chu-Shikoku and Kinki blocks belong to one group and the remainder to the other; block mean values are different by about 40 cm between the groups. Although the estimated bias is comparable to the SD values, we think that the geographical deviation of mean differences is significant.

This features can be attribute to the medium wavelength errors in EGM96 as one source since deep trenches run east off the coast from north to south.

The peak-to-peak range of block means is reduced to 50 cm in the case ③ from 60 cm in the case ①. In addition to the improvement in SD, this shows the enhancement of reliability in the bias estimation. For the evaluation of bias at the vertical datum station, we take the mean difference in Kanto block where the station is located; 14 cm and 18 cm below the EGM96 geoid with and without height change corrections, respectively.

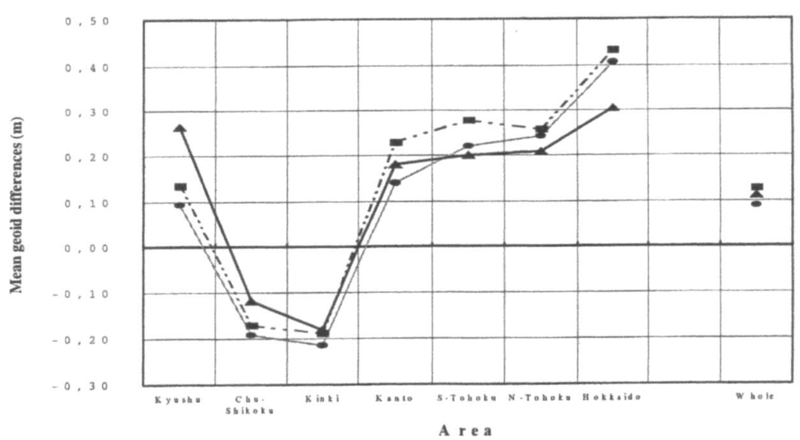

Fig. 5: Mean geoid differences in geographic blocks. Thin solid line with circles, broken line with squares, and thick solid line with triangles correspond to the case ①, ②, and ③ in that order.

Existence of large signals in geoid undulations in short wavelength, of course, results in the large SD for bias estimation. We may consider to use locally improved geoid models with reference to EGM96.

KUROISHI (1998) developed the latest gravimetric geoid model for Japan in high precision. The geographical distribution of geoid undulation differences showed that the short wavelength features are drastically reduced. But the existence of geoid model errors in medium wavelength was pointed out partly due to the errors in the network-adjusted ship gravity data used. Since the bias estimation requires the accurate information on geoid undulations, we cannot employ the local geoid model as the reference.

Tab. 1: Statistics of geoid undulation differences. Unit in meter.

Area	All 806 stations with 1969/71 heights			665 stations with 1969/71 heights			665 stations with height corrections		
	mean	SD	Range	mean	SD	Range	mean	SD	Range
Kyushu	0.09	0.230	1.112	0.136	0.204	0.998	0.263	0.188	0.874
Chu- Shikoku	-0.19	0.244	1.340	-0.171	0.242	1.295	-0.118	0.252	1.395
Kinki	-0.22	0.298	1.836	-0.188	0.277	1.783	-0.182	0.254	1.618
Kanto	0.141	0.448	1.869	0.228	0.435	1.869	0.180	0.433	1.844
South Tohoku	0.22	0.250	1.110	0.278	0.224	1.044	0.200	0.225	1.060
North Tohoku	0.241	0.280	1.322	0.257	0.273	1.295	0.208	0.288	1.299
Hokkaido	0.407	0.367	1.971	0.433	0.360	1.904	0.305	0.361	1.979
Whole	0.09	0.385	2.537	0.128	0.375	2.485	0.113	0.348	2.297

4. Conclusion

The separation of the Japanese vertical reference surface is evaluated with respect to the EGM96 global geoid. Mean deviations show geographically gradual variations of the peak to peak amplitude of 60 cm. When corrections for height changes between the time of adjustment of the official heights and that of the latest re-leveling are included, the amplitude is reduced to 50 cm with smaller SD and smaller range of deviations.

The mean deviation around the datum station is estimated at 14 cm and 18 cm below the EGM96 geoid with and without height change corrections, respectively.

Acknowledgment

GPS/leveling data are compiled and provided by the First Geodetic Division of the Geographical Survey Institute. Some figures are produced with GMT software (WESSEL and SMITH, 1991).

References

BOUCHER C., ALTAMIMI Z. (1991): *ITRF89 and other realizations of the IERS Terrestrial Reference System for 1989.* IERS Technical Note 6

KURODA J., TAKABATAKE J., MATSUSHIMA M., FUKUDA Y. (1997): *Integration of Gravimetric Geoid and GPS/ leveling Survey by Least Square Collocation.* Journal of the Geographical Survey Institute, Vol. 87, pp. 1-3 (in Japanese)

KUROISHI Y. (1998): *Improvement of Japanese geoid with 1D-FFT method and its comparison with altimetry-derived geoid.* Abstracts, 2nd Joint Meeting of the International Gravity Commission and the International Geoid Commission, p. 82

LEMOINE F. G., SMITH D., SMITH R., KUNZ L., PAVLIS E., PAVLIS N., KLOSKO S., CHINN M., TORRENCE M., WILLIAMSON R., COX C., RACHLIN K., WANG Y., KENYON S., SALMAN R., TRIMMER R., RAPP R., NEREM S. (1997): *The development of the NASA GSFC and DMA joint geopotential model.* IAG Symposia, Vol. 117, Segawa et al. (ed.), Gravity, Geoid and Marine Geodesy, Springer, pp. 461-469

MORITZ H. (1980): *Geodetic Reference System 1980.* Bulletin Géodésique, Vole 54, pp. 395-405

POUTANEN M., VERMEER M., MAKINEN J. (1996): *The permanent tide in GPS positioning.* Journal of Geodesy, Vole 70, pp. 499-504

RAPP R. G. (1994): *Separation between reference surfaces of selected vertical datums.* Bulletin Géodésique, Vol. 69, pp. 26-31

SMITH, D. A., MILBERT D. G. (1997): *Evaluation of Preliminary Models of the Geopotential in the United States.* International Geoid Service Bulletin no.6, D.I.I.A.R. Politecnico di Milano, Milano, Italy, pp. 7-32

SMITH, W. H. F., Wessel P. (1990): *Gridding with continuous curvature splines in tension.* Geophysics, Vol. 55, pp. 293-305

Wessel P., SMITH, W. H. F. (1991): *Free software helps map and display data.* EOS Trans. Amer. Geophys. U., Vol. 72, pp. 441, 445-446

Reference Systems
Used in Global Navigation Satellite Systems

U. ROSSBACH[1]

Abstract

The global navigation satellite systems used today employ different reference systems to express ECEF coordinates of their satellites: GPS employs WGS84, whereas GLONASS makes use of PZ-90. Although the definitions of these two reference systems are similar, in practical realization there are offsets in both origin and orientation between these reference frames. This affects those users of satellite navigation systems, who want to make use of both systems for all the advantages that this combined use provides.

This poster looks at these two reference frames, their definitions, realizations and development. Furthermore, it looks at the importance of using a unified datum in satellite geodesy. It provides an overview of the efforts to integrate the WGS84 and PZ-90 reference frames and shows possible transformation parameters between these two frames.

1. Reference Systems in Use

The two global navigation satellite systems currently in use worldwide employ different reference frames for the expression of coordinates. The American Global Positioning System (GPS) uses the World Geodetic System 1984 (WGS84), whereas the Russian Global Navigation Satellite System (GLONASS) employs the system Parametry Zemli (Parameters of the Earth) 1990 (PZ-90, sometimes also referred to as PE-90).

The GPS Interface Control Document [3] defines the WGS84 coordinate frame in the following way:

- origin is Earth's center of mass
- z-axis is parallel to the direction of the Conventional International Origin (CIO) for polar motion, as defined by the Bureau International de l'Heure (BIH) on the basis of the latitudes adopted for the BIH stations
- x-axis is the intersection of the WGS84 reference meridian plane and the plane of the mean astronomic equator, the reference meridian being parallel to the zero meridian defined by the BIH on the basis of the longitudes adopted for the BIH stations
- y-axis completes a right-handed Earth-centered, Earth-fixed orthogonal system

The PZ-90 reference frame is defined in the GLONASS Interface Control Document [2] as:

- Origin is Earth's center of mass
- The z-axis is parallel to the direction of the mean North pole according to the mean epoch 1900 - 1905 as defined by the International Astronomical Union and the International Association of Geodesy
- The x-axis is parallel to the direction of the Earth's equator for the epoch 1900 - 1905, with the XOZ plane being parallel to the average Greenwich meridian, defining the position of the origin of the adopted longitude system
- The y-axis completes the geocentric rectangular coordinate system as a right-handed system

Both WGS84 and PZ-90 are realized by means of a set of stations with defined coordinates in the specific system. For the WGS84 system, these stations originally were surveyed by means of Doppler observations to US Navy TRANSIT satellites. The coordinates of these stations later were refined gradually using SLR and VLBI observations. The internal accuracy of WGS84 can be considered to be 1 m horizontally (longitude, latitude) and 2 m vertically (height), each one sigma.

Analogously, the stations used to realize the PZ-90 reference frame originally were surveyed using Doppler measurements to TSIKADA satellites, the Soviet counterpart of the US TRANSIT system. Likewise, their coordinates were refined later on by means of SLR/VLBI observations. The internal accuracy of PZ-90 also is specified to be 1m in the horizontal coordinates and 2 m in height.

The set of stations used to define PZ-90 differs from that used for the realization of WGS84. Thus, although the definitions of WGS84 and PZ-90 sound similar, due to these different realizations, there are differences in origin, orientation and scale. The WGS84 coordinates of an arbitrary point therefore may not be identical to its coordinates in PZ-90.

2. Combining GPS and GLONASS

Many users of the GPS system, both in navigation and geodesy, meanwhile recognize the advantages of combining GPS with GLONASS satellite observations: The increase in the number of available satellites improves the accuracy of the positioning determination by improving the geometry of the tracked satellites. In mountainous or urban areas with parts of the sky blocked, only the additional satellite observations may make a position fix possible at all. Even the currently diminished GLONASS constellation is likely

[1] Udo Rossbach; Institute of Geodesy and Navigation (IfEN), University FAF Munich; D-85577 Neubiberg, Germany

to be useful for this purpose, since only observations to two GLONASS satellites are required for this. (One GLONASS observation is needed to compensate for the different GPS and GLONASS time frames.)

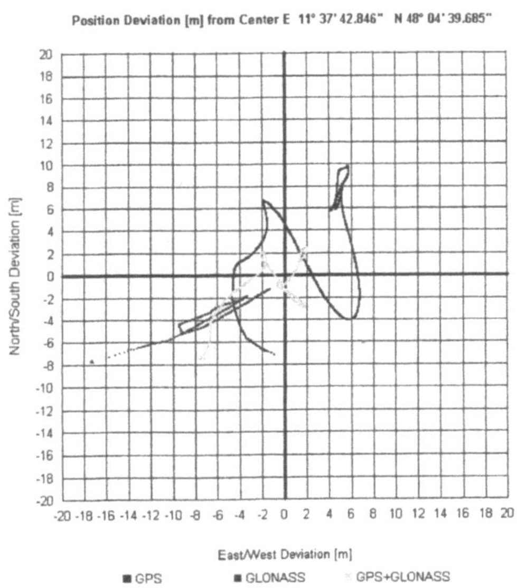

Figure 1: Combined GPS/GLONASS positioning without coordinate transformation (example).

Including observations to a different, independent satellite system will also improve the integrity of the navigation system. This is especially important for use of GPS in aviation and other safety critical applications.

However, in GPS-only positioning solutions, the satellite positions are given in WGS84 and thus the user position is in WGS84. In GLONASS-only solutions, satellite positions in PZ-90 are obtained from the ephemeris data, thus the user position is in PZ-90. Due to the differences in reference frame realizations, in a combined positioning solution, with some of the satellite coordinates in WGS84 and some of them in PZ-90, the coordinate frame of the calculated user position is undefined.

Figure 1 shows an example of such processing. GPS and GLONASS range measurements were processed without regard of different coordinate frames. Receiver position was determined in a Kalman-filtered epoch-by-epoch solution. Time span is approximately half an hour, with a data rate of 1 Hz. On display are the deviations in the local plane of the computed coordinates with respect to an arbitrarily chosen center. Please note that the centers of the GPS, GLONASS and combined GPS/GLONASS position distributions, where the Kalman filter finally converges, differ by some meters. They are aligned roughly along a line from North-East (GPS) to South-West (GLONASS).

Therefore, when combining GPS and GLONASS measurements to improve the position determination, the differences between the coordinate frames employed by these two systems must be accounted for. A straight-forward way to accomplish this is to transform the obtained satellite coordinates at the time of signal transmission from one coordinate frame to another, before forming the design matrix and calculating the user position.

Since GPS navigation has become the standard in Western countries and WGS84 therefore is more widely spread and better known than PZ-90, it is considered best to transform GLONASS satellite positions from PZ-90 to WGS84, thus obtaining the user position also in WGS84.

WGS84 and PZ-90 are different realizations of very similarly defined coordinate systems. Therefore, the differences in orientation between the two frames can be assumed to be small. This enables the application of a seven parameter Helmert transformation to transform coordinates from one frame into another. This seven parameter transformation reads:

$$\begin{pmatrix} x \\ y \\ z \end{pmatrix}_w = \begin{pmatrix} \Delta x \\ \Delta y \\ \Delta z \end{pmatrix} + (1+\delta s) \cdot \begin{pmatrix} 1 & \delta\omega & -\delta\psi \\ -\delta\omega & 1 & \delta\varepsilon \\ \delta\psi & -\delta\varepsilon & 1 \end{pmatrix} \cdot \begin{pmatrix} x \\ y \\ z \end{pmatrix}_p$$

with

Dx, Dy, Dz coordinates of the origin of frame *p* in frame *w*

de, dy, dw differential rotations around the x-, y- and z-axes of the *p* frame, respectively, to establish parallelism with the *w* frame

ds differential scale change

Neither of the ICDs ([2], [3]) defines a set of parameters to be used in transforming satellite coordinates. These transformation parameters must be determined separately. This currently is one of the most intensely investigated topics in GPS/GLONASS combination. Different groups of researchers have determined their own sets of parameters that differ from one another like the methods to determine these parameters.

3. Techniques for Determination of Transformation Parameters

There are several possible methods to determine the transformation parameters from PZ-90 to WGS84. Common to all these methods is the necessity to measure or otherwise obtain the coordinates of a given set of points. Coordinates of these points must be determined in both coordinate frames, PZ-90 and WGS84. Afterwards, a set of transformation parameters is calculated that brings the coordinates into coincidence when applied to the coordinates of one of the coordinate frames.

To determine the seven parameters of the coordinate transformation as introduced above, at least seven point coordinates must be known in both frames to obtain seven equations for solving seven unknowns. Since each point in space supplies three coordinates (one each for the x-, y- and z-axes), measuring three points is mathematically sufficient to calculate the desired transformation parameters. However, to have a good quality of the obtained parameters, one is desired to have coordinates of as much points as possible for reasons of redundancy in the equations. In addition, these points should be globally distributed to extend the validity area of the derived parameters. Otherwise, translational and rotational parameters may not be sufficiently separated from each other. This will result in a set of transformation parameters that is only valid in a specific area of the earth.

Possible methods of parameter determination can be distinguished by the location of the points whose coordinates are obtained in both systems:

- Ground-based techniques: Coordinates of points on the surface of the earth are made known in both coordinate frames. Usually, either a set of points known in WGS84 is occupied and measured in PZ-90 or the other way round.
- Space-based techniques: Coordinates of satellites at a specified epoch in time are made known in both coordinate frames. Usually, coordinates of GLONASS satellites are obtained from their ephemerides (in PZ-90) and from ground tracking from sites known in WGS84.

Each of the ground-based techniques suffers from a disadvantage with respect to the space-based techniques: There are no known points with coordinates known in PZ-90 outside the territory of the former Soviet Union, making it nearly impossible to occupy these points with GPS receivers, determine their coordinates in WGS84 and derive a globally valid set of transformation parameters. On the other hand, there is plenty of points with coordinates known in WGS84 all around the world. But until recently, there was only a few geodetic quality GLONASS receivers available. These were too few to occupy these points, determine their coordinates in PZ-90 and derive a globally valid set of transformation parameters. Only in 1998, a sufficient number of Ashtech Z-18 L1/L2 GPS/GLONASS receivers became available to be used in a global observation campaign, IGEX-98. One objective of this campaign is the determination of a set of globally valid transformation parameters [4], [8].

Regarding this, the space-based techniques have one major advantage: With only a few GLONASS receivers, broadcast ephemeris data (in the PZ-90 frame) of all GLONASS satellites all around the world can be received, providing global coverage. However, getting GLONASS orbit data in the WGS84 frame can be expensive. These can be obtained by radar or SLR tracking of the satellites, both requiring a large infrastructure, if global coverage is to be obtained. Therefore, each group of scientists that determined transformation parameters using a space-based technique cooperated closely or was sponsored by an organization that can provide such an infrastructure, e.g. NASA.

A different possibility to determine orbits of GLONASS satellites in the WGS84 frame is to track the satellites using a network of receivers located at sites surveyed in WGS84 and then compute the satellite orbits from the range measurements, like IGS does to obtain precise ephemerides of GPS satellites. But again, this requires a sufficient number of globally distributed GLONASS receivers.

4. Estimated Transformation Parameters

One of the first estimations of the transformation parameters between PZ-90 and WGS84 was performed by Russian scientists [1]. They determined the following set of transformation parameters:

$Dz = 1.5$ m

$dw = -0.076''$

$Dx = Dy = de = dy = ds = 0$

Extensive work on the determination of transformation parameters between PZ-90 and WGS84 was conducted at the Lincoln Laboratory of the Massachusetts Institute of Technology [5]. They estimated these parameters by comparing the coordinates of GLONASS satellites in PZ-90 (from the broadcast ephemerides) and in WGS84 (from radar tracking). Parameters for the transformation from PZ-90 coordinates to WGS84 were published in 1996:

$Dy = 2.5$ m

$dw = -0.4''$

$Dx = Dz = de = dy = ds = 0$

A terrestrial observation campaign to determine transformation parameters between PZ-90 and WGS84 was carried out by three German institutes (the Institute of Geodesy and Navigation of the University FAF Munich, the Institute of Applied Geodesy in Frankfurt on Main and the German Aerospace Research Establishment, Remote Sensing Ground Station Neustrelitz) in May 1996 [7]. Six GLONASS P-Code receivers were set up at IGS stations, distributed over Europe, at known coordinates in the WGS84 frame. By means of GLONASS satellite observations, the coordinates of these stations were determined in the PZ-90 frame. Transformation parameters were derived from these two sets of coordinates:

$dw = -0.33''$

$Dx = Dy = Dz = de = dy = ds = 0$

In 1998, another group of Russian scientists presented a different estimation of the transformation parameters [6]. They derived their estimation of the parameters by comparing the orbits of GLONASS satellites obtained from the GLONASS control center (given in PZ-90) with orbits determined by means of SLR tracking of these satellites from stations given in WGS84. From these data, they determined the following set of transformation parameters:

$Dx = -0.47$ m

$Dy = -0.51$ m

$Dz = -2.00$ m

$de = -0.002''$

$dy = -0.001''$

$dw = -0.356''$

$ds = 22\times10^{-9}$

5. Applying the Transformation Parameters

As stated above, when combining GPS and GLONASS measurements to calculate the position of a satellite navigation systems user, the satellite coordinates at the time of signal transmission (obtained in WGS84 and PZ-90 frames, respectively) have to be converted to a single coordinate frame before forming the design matrix and calculating the user position. Usually, the coordinates of GPS satellites will be left unchanged, and the coordinates of GLONASS satellites will be transformed from PZ-90 to WGS84. To do so, a seven parameter Helmert transformation is applied, using one of the above mentioned sets of transformation parameters.

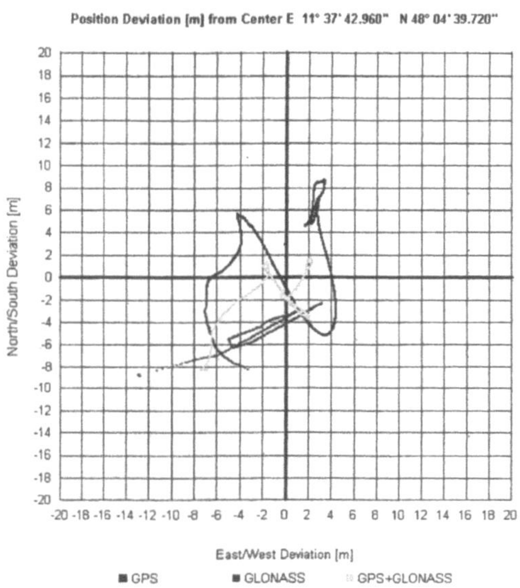

Figure 2: GPS/GLONASS positioning with coordinate transformation [7] applied.

Figure 2 shows an example of such processing. The data used in Figure 1 were processed in the same way, but this time the calculated coordinates of the GLONASS satellites at the time of signal transmission were transformed to WGS84 before the user position was calculated. To convert PZ-90 coordinates to WGS84, the set of transformation parameters as proposed in [7] was employed.

It can be clearly seen that now the centers of the GPS, GLONASS and combined GPS/GLONASS position distributions, where the Kalman filter finally converges, are much closer together than without the coordinate transformation. Since the coordinate transformation according to [7] contains only a rotation around the z-axis, the z-coordinates (and therewith the latitudes) of the computed positions are not influenced by this transformation. Therefore, centers of the position distributions are now aligned almost exactly along a line from North to South.

References

[1] Boykov, V. V. et al.: *Experimental Compiling the Geocentric System of Coordinates PZ-90*, in: Geodezia i Kartografia, No. 11, November 1993, pp. 17-21 (in Russian Language)

[2] *GLONASS Interface Control Document (Rev. 1995)*, Coordinational Scientific Information Center, Russian Space Forces, Moscow, 1995

[3] *GPS Interface Control Document (ICD-GPS-200)*, ARINC Research Corporation, July 1991

[4] IGEX-98 WWW Home Page, http://lareg.ensg.ign.fr/IGEX/

[5] Misra, P. M., R. I. Abbot, E. M. Gaposchkin: *Integrated Use of GPS and GLONASS: PZ-90 – WGS84 Transformation*, Proceedings of ION Satellite Division GPS-96 International Technical Meeting, Kansas City, Missouri, September 17-20, 1996

[6] Mitrikas, V. V., S. G. Revnivykh, E. V. Bykhanov: *WGS84/PZ-90 Transformation Parameters Determination Based on Laser and Ephemeris Long-Term GLONASS Orbital Data Processing*, Proceedings of ION Satellite Division GPS-98 International Technical Meeting, Nashville, Tennessee, September 15-18, 1998

[7] Rossbach, U., H. Habrich, N. Zarraoa: *Transformation Parameters Between PZ-90 and WGS84*, Proceedings of ION Satellite Division GPS-96 International Technical Meeting, Kansas City, Missouri, September 17-20, 1996

[8] Willis, P., G. Beutler, W. Gurtner, G. W. Hein, R. Neilan, J. Slater: *The International GLONASS Experiment 1998 (IGEX-98)*, International Call for Participation, May 1998.

Recent Contribution to the ITRF and its Realization in Canada

R. FERLAND, P TÉTREAULT, C. HUOT, D. HUTCHISON, J. KOUBA[1]

Abstract

Natural Resources Canada (NRCan) is processing data daily from a subset of at least eight Canadian automated GPS tracking stations, the Active Control Points (ACP), and from at least twenty five of the International GPS Service (IGS) stations using a modified version of JPL GIPSY-OASIS II software. NRCan contributes to IGS solutions for satellite orbit, clock corrections, earth orientation parameters (EOPs), station coordinates and more recently tropospheric delays and ionospheric grids.

In the daily estimation process, station coordinates are constrained by a subset of station positions of the current epoch in of the International Terrestrial Reference Frame (ITRF). Removing a priori constraints is required for the rigorous station coordinate and velocity estimation.

In the most recent NRCan contribution to IERS, four years of daily station coordinates and EOPs were combined. A consistent set of station coordinates, velocities and EOPs were obtained by back substituting the final station coordinates in the daily solutions. The NRCan and ITRF station coordinate and velocity solutions were compared, as well as the NRCan station velocity estimates were compared to the NUVEL1A plate motion model. Analysis and comparison between the time series of residuals are also presented.

This approach, through the Canadian Active Control System, provides a convenient globally consistent realization of ITRF in Canada.

Introduction

NRCan yearly contribution to IERS includes a consistent set of station coordinates, velocities and EOP estimates for the maintenance of the ITRF and EOP time series. The NRCan contribution is obtained by combining daily station coordinates and EOP estimates. The estimates are extracted from the daily NRCan GPS orbit processing performed in support of the Canadian Active Control System (CACS) and the International GPS Service (IGS). A brief description of the processing and combination strategies are given in section 2 and 3 respectively. Combination residual statistics, comparisons with the ITRF96 solution and the NUVEL1A plate motion model are given in section 4.

1. Processing Strategy

The daily processing of the GPS tracking data at NRCan requires the estimation of several parameters, which can be divided in 3 groups: station related, satellite related and EOPs. Ionospheric-free (L3) combinations of code and phase measurements every 7.5 minutes are used in the processing. Station coordinates, initial phase ambiguities, clock offsets and wet tropospheric delays are estimated at each station; state vectors, radiation pressure and clock offsets are estimated for each satellite. The estimated EOP parameters include the polar motion and UT1-UTC as well as their rates (TÉTREAULT et al. 1996). A priori information for a number of parameters is used to constrain the estimation process. Station coordinates and their uncertainties are provided in the current ITRF. The tropospheric delay at each station is initialized using the previous day solution. Initial state vector for each satellite is also taken from the previous day solution and is given a nominal uncertainty. Radiation pressure is initialized with a moving average of the previous 4 days. A priori clock parameters are unconstrained subject only to the reference clock.

2. Daily Stations and EOPs Processing

The station coordinates and EOPs, along with their appropriate variance-covariance information, are extracted from the complete daily solutions and stored using the Software INdependent EXchange format (SINEX). A number of software modules developed at NRCan are used to combine the station coordinates and EOPs. The combination process can be divided into daily, weekly, annually or multi-year solutions. Unconstrained solutions are used in the combination process. This removes the small discontinuities introduced when changing ITRF realization as well as the small inconsistencies present in ITRF station positions. To avoid singularity in orientation of the unconstrained daily solutions, small orientation constraints at 1 to 10 arc-seconds are left in the "unconstrained" solutions. They prevent numerical problems while having no adverse consequence on the processing. The implicit geocenter can also affect the daily and combined solutions. In this analysis, the effect of the implicit geocenter is removed from the unconstrained daily solutions. This introduces a singularity in position. A small 1m constraint is left on the geocenter in the "unconstrained" daily solutions for the same reason as for the orientation. The unconstrained daily solutions are aligned in rotation and translation to ITRF96 of the epoch using the common stations. This removes small rotations and translations that occur when removing the a priori constraints and the geocenter effect. The daily unconstrained solutions are tested for abnormal behaviour of the parameters and appropriate corrections are applied whenever required. Variations between each station solution are caused by the measurements noise, the changes in the network geometry, the processing strategy and also by the local conditions at a site.

[1] R. Ferland, P Tétreault, C. Huot, D. Hutchison, J. Kouba, Geodetic Survey of Canada, Geomatics Canada, Natural Resources Canada

3. Stations and EOPs Combination

The combination of station coordinates and EOPs are treated differently. The station coordinates have in general a smooth linear time evolution and, therefore, both position and velocity are estimated for each station. All the available stations are included in the combination. But, in this analysis, only the stations (Fig. 1) with daily solutions spanning at least six months are used. Several of the excluded stations have only a few days of observations. With very short observation time span, velocity estimates are unreliable. The EOPs have a more complicated time evolution and therefore, their velocity is not modeled. The daily estimates are kept independent from day to day. To prevent the number of parameters from becoming unreasonably large due to the daily addition of new EOP parameters, a solution without the EOP is first estimated.

Fig. 1: Stations in the NRCan contribution to ITRF.

The unconstrained daily solutions are combined into weekly solutions without velocity estimation. The weekly solutions are then combined into annual solutions with station velocity estimation. All annual solutions have been estimated at the reference epoch 1998.0. The annual solutions are then combined to form the unconstrained 1994-1997 cumulative solution. The 4-years cumulative solution is then aligned in position and velocity to ITRF96 using a 14-parameter transformation (3 rotations, 3 translations, 1 scale and their first derivative). The transformation is determined using all stations common to the solution and ITRF96 at the reference epoch. Outliers are either corrected if the problems are known (e.g.: incorrect antenna height) or rejected.

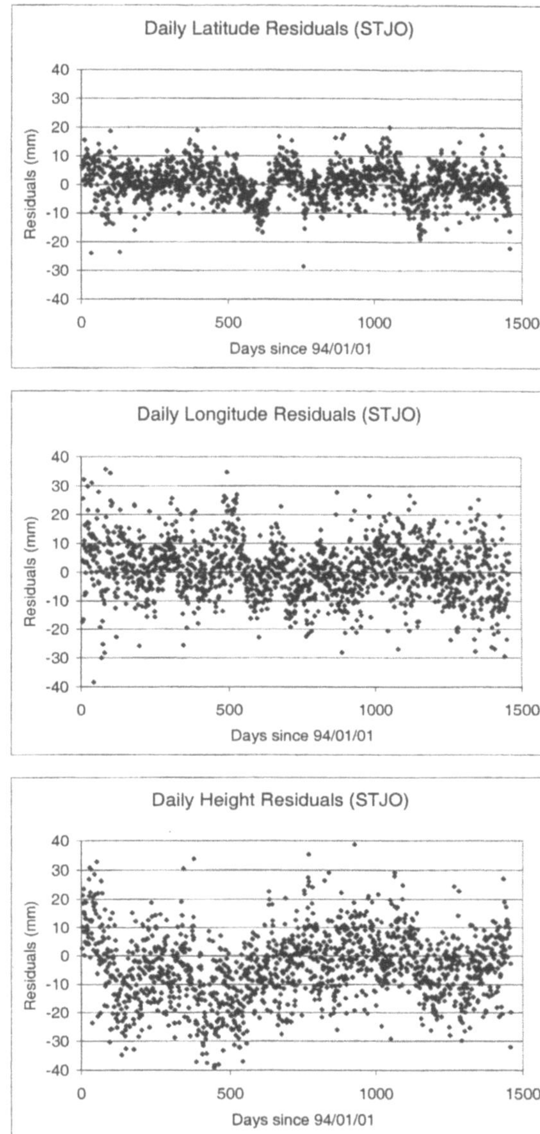

Fig. 2: Daily Residuals in Latitude, Longitude and Height at station STJO.

Tab. 1: Yearly and 4-year daily station residuals R.M.S.

Year	ϕ (mm)	λ (mm)	h (mm)	n
94	11.6	17.0	23.8	6956
95	8.9	12.8	20.3	8997
96	7.6	12.0	18.3	10093
97	8.7	14.4	18.7	9926
94-97	9.1	14.0	20.1	35972

Tab. 2: The NRCan station daily residuals R.M.S. and the velocity solution differences with respect to NUVEL1A.

Stns	NRCan - ITRF96 (1998.0)					
	Position Differences			Velocity Differences		
	dN mm	dE mm	dU mm	dVN mm/y	dVE mm/y	dVU mm/y
ALBH	5.9	8.8	11.4	4.7	7	-0.6
ALGO	5.6	8.0	13.6	-2.1	1.1	0.1
AREQ	9.8	18.3	20.8	3.2	13.9	-0.6
CAS1	7.4	10.6	15.8	-6.9	-5.9	3.6
CHUR	5.1	6.7	10.2	-1.4	2.1	12
DAV1	10.4	15.6	22.5	-9.3	-4.4	2.9
DRAO	5.7	8.5	11.3	1.6	3.3	3.3
DUBO	3.9	5.6	11.7	-1.2	-5.5	-15.6
FAIR	7.2	10.2	18.3	-2.3	2.5	0.1
FLIN	4.5	6.3	6.5	-5.9	4.5	2.1
FORT	12.0	19.2	33.0	0.3	4.2	3.4
GOLD	6.6	10.0	15.0	6.8	-2.7	0.2
HART	12.7	20.6	33.4	-0.6	-4.3	2.3
IRKT	8.1	12.4	13.6	3.7	4.1	0.4
KERG	8.2	15.0	21.9	-5.6	-4.9	6.4
KIT3	11.0	16.1	17.9	5.9	5.8	3.3
KOKB	9.4	19.3	28.1	0.8	-2.8	-1.5
KOSG	8.8	11.4	13.5	4.6	-2.6	4.3
LHAS	7.8	14.4	17.5	23.5	18.7	2.1
MADR	11.4	14.1	17.4	-1.8	1.5	3.8
MALI	9.0	23.5	25.4	-2.7	22.4	-17.6
MCM4	10.8	12.0	22.4	-1.4	8	11.3
NRC1	5.7	7.8	11.4	-0.6	2.7	8.8
PAMA	14.7	25.6	35.6	2.2	19.5	-0.2
RCM5	6.8	13.8	20.9	-1.6	-3.3	-4.7
SANT	13.5	20.2	24.9	2.6	18.8	6.9
SCHE	5.2	6.6	11.4	0	-0.3	20.5
STJO	6.2	10.2	13.2	-4.2	-1.6	4.5
TAIW	13.4	16.8	23.2	1.4	18.9	-1.4
TIDB	11.1	15.9	23.8	1.6	6	-4
TROM	9.9	10.8	19.8	3.5	-4.3	-0.6
TSKB	9.6	13.3	18.0	6.5	-19.1	-2.8
WETT	9.9	10.9	15.1	2.8	1.5	5.6
WHIT	5.5	8.4	10.8	3	0.9	-4.5
WILL	5.5	7.5	10.6	0.9	1.8	3.7
WTZR	8.7	13.4	14.8	-0.1	0.2	-2.3
YAR1	10.6	17.1	24.2	-4	3.5	-0.3
YELL	5.6	9.1	13.9	-0.7	2	0.2

Tab. 3: Station position and velocity differences between the NRCan solution and ITRF96 at epoch 1998.0.

Stns	NRCan - ITRF96 (1998.0)					
	Position Differences			Velocity Differences		
	dN mm	dE mm	dU mm	dVN mm/y	dVE mm/y	dVU mm/y
ALBH	2.6	-3.7	-7.6	-0.3	-1.5	-3.5
ALGO	4.1	-1.3	3.3	0.1	0.0	0.6
AREQ	2.0	3.3	0.2	0.1	-0.7	0.6
CAS1	12.7	-18.9	-5.9	0.4	-8.2	-10.3
CHUR	2.1	0.8	22.0	-0.9	1.1	13.0
DAV1	14.0	7.3	2.8	-1.3	-0.3	1.6
DRAO	3.1	0.3	0.5	0.2	0.6	2.1
DUBO	1.9	5.7	11.4	-3.4	4.4	6.3
FAIR	2.3	-5.1	-6.6	0.2	0.2	0.2
FLIN	8.4	-0.4	4.4	4.4	0.2	3.3
FORT	5.8	-7.7	-7.2	0.4	0.9	1.2
GOLD	3.1	0.6	-5.8	-0.1	-0.1	0.0
HART	7.8	5.5	-5.3	0.6	-0.2	0.8
IRKT	-0.8	-1.6	-4.8	0.5	1.8	0.5
KERG	10.9	14.9	2.7	0.1	1.7	-0.5
KIT3	3.2	-4.1	-2.8	2.1	2.0	1.4
KOKB	0.1	6.5	-1.6	0.0	-0.3	0.1
KOSG	5.2	-6.4	3.0	3.9	-1.7	3.5
LHAS	-2.6	2.2	-2.3	-0.4	-0.8	0.2
MADR	-6.4	-2.8	-4.6	-1.2	0.5	-0.1
MALI	7.6	25.0	-26.9	2.9	26.6	-20.3
MCM4	-5.2	-19.5	-23.3	-1.8	-1.5	-16.1
NRC1	1.5	0.8	17.9	-1.5	0.9	11.8
PAMA	-1.5	35.5	-27.7	1.7	15.4	-5.5
RCM5	2.2	-18.8	-15.8	-1.2	-5.2	-10.5
SANT	-1.5	7.0	3.2	-1.6	-0.3	-1.2
SCHE	4.0	1.6	10.7	-0.1	0.4	5.3
STJO	1.1	-1.4	17.0	-1.8	-1.3	5.7
TAIW	-5.5	4.2	-0.9	-0.6	4.8	4.0
TIDB	-4.7	2.0	3.8	-0.7	0.9	-0.2
TROM	-0.8	-1.0	-4.6	0.3	0.0	0.1
TSKB	-0.3	-2.8	-4.9	1.5	1.7	2.5
WETT	2.6	-2.9	5.6	2.9	1.3	7.9
WHIT	6.5	-3.6	-4.3	4.4	-2.8	-3.1
WILL	3.4	-1.8	-2.4	0.2	-0.3	-0.8
WTZR	2.8	-4.4	-0.6	0.0	0.0	0.0
YAR1	-0.2	-0.4	-2.9	-0.6	-0.1	0.3
YELL	3.4	-1.0	1.5	0.2	0.3	-0.5

The cumulative solution is then back-substituted in the daily solutions to estimate the EOPs and optionally the explicit daily geocenter. After the back-substitution, the station coordinates/velocities, EOPs and optionally the geocenter are available in a consistent reference frame. The final residuals are the differences between the unconstrained daily solutions and the cumulative solution where a 6-parameter (3 rotations and 3 translations) transformation has been applied to align each daily solution to the cumulative solution of the epoch.

Daily residual time series for station STJO, shown in Fig. 2, are fairly typical. The annual and total R.M.S. of the daily residuals for all stations is summarized in Tab. 1. A 10% annual R.M.S. improvement can be seen between 1994 and 1996, while there is a slight degradation for 1997. The number of stations/days processed has also increased significantly between 1994 and 1996. The number of daily solutions available at each station as well as the time span of the series is summarized in Tab. 4.

The weekly combination residuals R.M.S. is reduced by a factor of about 2 when compared to the daily residuals which is less than the expected 2.6 ($\sqrt{7}$). This indicates potential non-random behaviour of the residuals. The first 3 columns of Tab. 2 show the R.M.S. for 38 stations. The R.M.S. varies significantly from station to station.

Tab. 4: Daily solution time span and the number of daily solutions at each station

Stns	Time span dT y	# of daily Soln's	Stns	Time span dT y	# of daily Soln's
ALBH	4.0	1239	MADR	4.0	1376
ALGO	4.0	1444	MALI	1.6	346
AREQ	3.6	1049	MCM4	2.9	1034
CAS1	0.6	100	NRC1	3.7	1278
CHUR	2.7	411	PAMA	3.2	410
DAV1	3.1	638	RCM5	2.7	908
DRAO	4.0	1445	SANT	4.0	1209
DUBO	1.0	101	SCHE	1.4	191
FAIR	4.0	1399	STJO	4.0	1428
FLIN	1.1	128	TAIW	2.9	1077
FORT	4.0	1121	TIDB	4.0	1379
GOLD	4.0	1374	TROM	3.4	1016
HART	4.0	1101	TSKB	3.0	1076
IRKT	2.0	562	WETT	2.1	743
KERG	3.0	965	WHIT	1.4	492
KIT3	3.2	848	WILL	2.7	870
KOKB	4.0	1327	WTZR	1.9	674
KOSG	3.9	1346	YAR1	4.0	1259
LHAS	1.2	315	YELL	4.0	1451

A spectral analysis of the residuals reveals a significant annual period at a number of stations. Height amplitude of the annual period approaching the cm level can be found at some stations; horizontal amplitude is significantly smaller. Seasonal variations at some stations could explain part of this annual signal. However, seasonal variations cannot explain the significant differences in R.M.S. between the stations. They range in height (Tab. 2) from 11mm to over 35mm. The R.M.S. variations are influenced by local environment and receiver performance variations between stations and to a lesser degree the processing strategy. Large station residual R.M.S. is probably not a reliable indication of a poor station coordinate and velocity estimate. Several stations with large RMS have an excellent agreement with ITRF96.

The NRCan aligned and unconstrained 4-year solution is compared to ITRF96 (1998.0). Tab. 3 lists the position differences (dN, dE and dU) and the velocity differences (dVN, dVE, dVU) for each station. Position and velocity differences R.M.S. are 6.8mm, 12.1mm, 15.0mm and 2.8mm/y, 6.4 mm/y, 8.3 mm/y in the north, east and vertical components, respectively. The agreement with ITRF96 in the north position and velocity component is significantly better than in the east and vertical by factors of about 2 and 3, respectively. Stations with longer historical information tend to have better agreement with ITRF96.

The NRCan 4-year solution velocity was also compared to the NUVEL1A plate motion model. The difference with NUVEL1A is contained in the last 3 columns of Tab. 2. Velocity differences R.M.S. for the north, east and vertical are 5.2 mm/y, 8.7 mm/y and 6.7 mm/y respectively. NUVEL1A models the horizontal rigid plates motion only; whereas, the vertical R.M.S. is with respect to an assumed zero vertical motion.

4. Conclusion

The NRCan solution covering a period from 1994 to 1997 includes 38 stations with a minimum of 6 months of data. Daily residuals R.M.S. are 9.1mm, 14.0mm and 20.1mm for the north, east and height components respectively. The R.M.S. decreases at an annual rate of about 10% between 1994 and 1996. The station coordinate and velocity solution was compared to ITRF96. Position and velocity differences R.M.S. are 6.8mm, 12.1mm, 15.0 mm; and 2.8mm/y, 6.4mm/y and 8.3 mm/y for the north, east and height components, respectively. The stations velocity solution was compared to the plate motion model NUVEL1A and the velocity differences R.M.S. are 5.2mm/y, 8.7mm/y and 6.7 mm/y for the north, east and height components, respectively.

References

TÉTREAULT P., HUOT C., FERLAND R., HUCHITSON D., KOUBA J., POPELAR J.(1996). *NRCan Analysis Center Annual Report for 1996*, in International GPS Service for Geodynamics, 1996 Annual Report, IGS Central Bureau, JPL, Pasadena, Cal., USA, 1997.

Realization and Unification of NAD83 in Canada and the U.S. via the ITRF

M. CRAYMER[1], R. FERLAND[2], R. SNAY[3]

Abstract

A more unified realization of the North American Datum of 1983 (NAD83) was recently implemented by Canada and the U.S. tö provide more accurate and convenient access to this reference system. In both countries, NAD83 is the adopted datum for spatial positioning. It was initially realized in the mid-1980s by adjusting primarily classical geodetic observations that connected a network of monumented control points spanning North America. Although the new NAD83 realization is compatible with the original one within the accuracies of those classical observations, the wide use of high precision GPS surveys since then have required a more rigorous definition of NAD83 in terms of network scale, ellipsoidal heights and crustal motion. In an effort to provide a more precise realization of a 3D NAD83 common to both countries, and to comply with IAG resolutions recommending that reference systems be tied to the ITRS, Canada and the U.S. have collaborated in the determination of a common transformation between the ITRF and the fundamental 3Ddatum of NAD83. This conformal similarity transformation is based upon ITRF96, the latest realization of ITRF. Through the use of continuously operating GPS stations (CACS in Canada and CORS in the US) forming part of the ITRF, this transformation also provides more accurate and convenient access to the NAD83 reference frame. It also allows for the conversion of IGS precise GPS ephemerides from ITRF to NAD83, thereby enabling high accuracy GPS surveys to be performed entirely within NAD83.

1. Background

The adopted datum or reference system for spatial positioning in Canada and the US is the North American Datum of 1983 (NAD83). The original 1986 realization of this datum was defined to be compatible with the BIH Terrestrial System BTS84, which was the best available global reference frame at the time. The origin of BTS84 was defined largely by satellite laser ranging (SLR), while its orientation was defined by astronomic observations and its scale by both SLR and very long baseline interferometry (VLBI).

NAD83, on the other hand, was realized mainly through Doppler observations using internationally accepted transformations from the Doppler reference frame to BTS84. VLBI stations were also included to provide an accurate connection to other reference frames.

Although NAD83 is defined by a three dimensional reference frame, access to it by end users for general spatial positioning needs was provided through a dense network of horizontal control monuments. This network was established mainly by conventional horizontal control methods, such as triangulation, trilateration and astronomic azimuths, with some GPS baselines used in later years. The adjustment of these horizontal observations, together with several hundred observed Doppler positions, provided a practical realization of NAD83 fixed to the North American tectonic plate.

2. Problems

At the time of its initial realization, NAD83 was intended to be a geocentric system and was compatible with the other geocentric systems of the time, such as the original realization of WGS84. However, due to the use of more accurate techniques, it is now known that NAD83 is offset by about 2 m from the true geocenter. The best geocentric reference frame currently available is the International Terrestrial Reference Frame (ITRF).

Another limitation of the original realization of NAD83 was that access to it was provide mainly through a horizontal control network. Today, many applications of GPS require convenient access to the more fundamental 3D definition of the NAD83 reference system.

The accuracy of horizontal control networks based on the original NAD83 is limited by the accumulation of errors in both the observations and methods of network integration. Because Doppler observations provided essentially the only measure of absolute position, errors in the original NAD83 realization primarily reflect the metre level uncertainty in these observations. Figure 1 illustrates these errors at selected points across Canada by comparing the adopted NAD83 horizontal control to high accuracy GPS surveys tied almost directly to the fundamental reference frame of NAD83 as defined by the VLBI stations.

3. Solution

At the XXth General Assembly of the International Union of Geodesy and Geophysics in 1992, the International Association of Geodesy (IAG) made the following recommendations in Resolution No. 1 [IAG, 1992]:

[1] M. Craymer, Geodetic Survey Division, Natural Resources Canada, 615 Booth Street, Ottawa, Ontario K1A 0E9, Canada; Tel. 613-947-1829, Fax. 613-992-6628, Email: craymer@nrcan.gc.ca

[2] R. Ferland, Geodetic Survey Division, Natural Resources Canada, 615 Booth Street, Ottawa, Ontario K1A 0E9, Canada; Tel. 613-995-4002, Fax. 613-995-3215, Email: rferland@nrcan.gc.ca

[3] R. Snay, National Geodetic Survey, N/NGS5, National Oceanic and Atmospheric Administration, 1315 East-West Highway, Silver Spring, Maryland 202910-3282, U.S.A.; Tel. 301-713-3205x155, Fax. 301-713-4327, Email: rich@ngs.noaa.gov

"1) that groups making highly accurate geodetic, geodynamic or oceanographic analysis should either use the ITRF directly or carefully tie their own systems to it"

"4) that for high accuracy in continental areas, a system moving with a rigid [tectonic] plate may be used to eliminate unnecessary velocities provided it coincides exactly with the ITRS at a specific epoch"

Considering recommendation (1), it is assumed that recommendation (4) also allows for the use of other systems providing they are carefully tied to the ITRS. Note that in ITRF the motions of the individual tectonic plates will cause horizontal coordinates to constantly change in time. It is therefore necessary specify which epoch ITRF coordinates refer to and to account for tectonic motion when changing epochs.

To comply with the IAG recommendations, ties between ITRF and NAD83 were made via the common VLBI stations, allowing for the accurate determination of a conformal 3D similarity (Helmert) transformation between the two. This transformation effectively provides a more accurate realization of the fundamental 3D NAD83 reference system via the ITRF that is common to both Canada and the U.S. It also provides convenient and nearly direct access to the 3D reference frame of NAD83 for the first time.

The first ITRF-NAD83 transformation adopted by both Canada and the U.S. was determined with respect to ITRF89 in the early 1990s [SOLER et al., 1992]. The scale of ITRF was adopted for compatibility with more recent versions of WGS84 by setting the estimated scale parameter to zero. This transformation was used in both Canada and the U.S. to derive NAD83 coordinates for the continuously operating GPS stations (CACS in Canada and CORS in the U.S.) that contribute to the definition of the ITRF reference frame.

Unfortunately, there was no agreement on how to update the transformation to new versions of the ITRF. Canada updated using the incremental transformations between different versions of ITRF as published by the International Earth Rotation Service (IERS). The U.S., on the other hand, recomputed the transformation for each new ITRF, adopting the scale of each. Consequently, the transformation between NAD83 and the more recent ITRFs differed slightly, mainly due to changes in scale. Although these scale changes were small and had little affect on the horizontal NAD83 coordinates, it significantly affected the definition of the vertical (ellipsoidal heights), resulting in a bias of about 5 cm between ITRF89 and ITRF96.

4. ITRF96-NAD83 Transformation

In order to reconcile the different realizations of NAD83 based on different ITRF-NAD83 transformations, a new common NAD83 transformation was derived with respect the most recent ITRF96.

The data used in determining the transformation were NAD83 and ITRF96 coordinates at 12 VLBI stations in both Canada and the U.S. (see in Figure 2). These are the only fundamental points in the original definition of NAD83 with 3D coordinates in both NAD83 and ITRF96. The ITRF96 coordinates at epoch 1997.0, together with their associated covariance matrix were obtained from the IERS. The NAD83 coordinates and covariance matrix were obtained from ITRF89 using the original ITRF89-NAD83 transformation. This realization of NAD83 is referred to here as NAD83(ITRF89).

The ITRF96-NAD83 transformation was modeled using a weighted 7-parameter similarity (Helmert) transformation. The variance factor was estimated only for the covariance matrix of the NAD83 coordinates (obtained from ITRF89). The ITRF96 covariance matrix was considered known.

The estimated transformation parameters from ITRF96 at epoch 1997.0 to NAD83(ITRF89) are:

$$
\begin{aligned}
TX(1997.0) &= 0.9910 \text{ m} \\
TY(1997.0) &= -1.9072 \text{ m} \\
TZ(1997.0) &= -0.5129 \text{ m} \\
RX(1997.0) &= -25.79 \text{ mas} \\
RY(1997.0) &= -9.65 \text{ mas} \\
RZ(1997.0) &= -11.66 \text{ mas} \\
DS(1997.0) &= 6.62 \text{ ppb (not used, set to 0)}
\end{aligned}
$$

where the rotations are clockwise following the unconventional IERS notation in McCarthy [1996].

Note that the scale of ITRF96 was adopted for this realization of NAD83 by setting the scale change parameter (DS) to zero after estimation. This ensures the scale of NAD83 will be compatible with the more accurate scale defined by ITRF96 and used by other systems such as WGS84 (i.e., GPS).

In order to correctly account for the tectonic motion of the North American tectonic plate when transforming from/to ITRF96 positions at any arbitrary epoch, the NUVEL-1A crustal motion model was adopted [DEMETS et al., 1996]. Larson et al. [1997] have shown NUVEL-1A to be in good agreement with velocities estimated from GPS in North America. The effect of this motion can be treated as additional clockwise rotations of the reference frame. For the North American plate these rotations are [MCCARTHY, 1996, Table 3.2]

$$
\begin{aligned}
dRX &= 0.0532 \text{ mas/y} \\
dRY &= -0.7423 \text{ mas/y} \\
dRZ &= -0.0316 \text{ mas/y}
\end{aligned}
$$

The adopted transformation from ITRF96 at epoch t to NAD83 is then given by

$$
\begin{bmatrix} X_{NAD83} \\ Y_{NAD83} \\ Z_{NAD83} \end{bmatrix} = \begin{bmatrix} TX \\ TY \\ TZ \end{bmatrix} + \begin{bmatrix} 1 & -RZ(t) & RY(t) \\ RZ(t) & 1 & -RX(t) \\ -RY(t) & RX(t) & 1 \end{bmatrix} \begin{bmatrix} X_{ITRF96} \\ Y_{ITRF96} \\ Z_{ITRF96} \end{bmatrix}
$$

where TX, TY and TZ are the translations estimated above, and the combined rotations from ITRF96 (epoch t) to ITRF96 (epoch 1997.0) to NAD83 are

$$
\begin{aligned}
RX(t) &= RX(1997.0) + dRX \cdot (1997.0\text{-}t) \\
RY(t) &= RY(1997.0) + dRY \cdot (1997.0\text{-}t) \\
RZ(t) &= RZ(1997.0) + dRZ \cdot (1997.0\text{-}t)
\end{aligned}
$$

The inverse transformation from NAD83 to ITRF96 is obtained by reversing the signs on the above transformation parameters.

5. Improved Realization of NAD83

The above transformation is used to define an improved realization of a 3D NAD83 reference frame. This NAD83 reference frame is therefore essentially defined in terms of ITRF96, following IAG recommendations. The NUVEL-1A plate motion model is also adopted to transform ITRF96 positions at any epoch.

To ensure a consistent application of the transformation to other versions of ITRF, both Canada and the U.S. also agreed to adopt the most current IERS Conventions [McCarthy, 1996] for transforming from other ITRF reference frames to ITRF96 prior to using the adopted ITRF96-NAD83 transformation.

In the U.S., this improved realization of NAD83 is referred to as NAD83(NSRS), where NSRS is an acronym for the National Spatial Reference System. In Canada, it is referred to as NAD83(CSRS), where CSRS is an acronym for the Canadian Spatial Reference System.

The main advantage of this improved realization of NAD83 is that it provides almost direct access to the NAD83 reference frame through ties to CACS and CORS, the continuously operating GPS stations. CACS and CORS stations are part of the fundamental global network defining the ITRF and have accurate ties to the NAD83 VLBI stations. These stations effectively act as both ITRF and NAD83 datum points for spatial positioning in Canada and the U.S., enabling the more accurate integration of surveys into NAD83 (less accumulation of error) as well as more convenient access (effectively acting as extra receivers on NAD83 datum points). The adopted transformation also enables the dissemination of IGS-compatible precise GPS ephemerides in both ITRF and NAD83 reference frames, allowing users to perform precise GPS surveys entirely within NAD83.

Figure 1 displays the differences between (i) NAD83 coordinates derived from high accuracy GPS surveys using direct ties to CACS stations in NAD83(CSRS) and (ii) those from the NAD83 horizontal control network based on conventional methods and indirect ties to the NAD83 datum points. The differences are up to the meter level and reflect primarily the distortions present in the horizontal control network due to the accumulation of errors in the observations and methods of network integration.

It is important to realize that these horizontal differences do not represent changes in the NAD83 datum. The effect of differences in the realizations of the NAD83 datum are very small. Horizontal coordinate shifts are generally within the accuracy of positioning techniques and are insignificant for most positioning applications.

The only significant difference between realizations of NAD83 is a change in scale. Although the scale difference between the original 1986 NAD83 and NAD83(ITRF89) is about 87 ppb or 9 mm per 100 km, it has little practical effect on most local and regional positioning applications. It's main effect is a shift in ellipsoidal heights of about 55 cm. Because the original NAD83 control network was mainly horizontal, this effect was of no practical consequence.

The change in scale between NAD83(ITRF89) and NAD83(CSRS/NSRS) is much smaller, amounting to only 7 ppb. Although the effect on the horizontal coordinates is negligible, it results in a change of 5-7 cm in ellipsoidal heights. However, this has little impact because of the limited use of NAD83 ellipsoidal heights. As noted earlier, NAD83(ITRF89) has been used only for CACS stations in Canada as well as for some special purpose networks.

In the U.S., realizations of NAD83 based on transformations from ITRF93 and ITRF94 were used for the CORS network which in turn was used to integrate the High Accuracy Reference Networks (HARNs). Because these ITRFs are not very different from ITRF96, the differences between these realizations and NAD83(CSRS/NSRS) are negligible, both horizontally and vertically.

References

DeMets C., Gordon R.G., Argus D.F., Stein S. (1994). *Effect of recent revisions to the geomagnetic reversal time scale on estimates of current plate motions.* Geophysical Research Letters, Vol. 21, No. 20, pp. 2191-2194.

International Association of Geodesy (1992). *IAG Resolutions adopted at the XXth IUGG General Assembly in Vienna.* The Geodesist's Handbook. Bulletin Geodesique, Vol. 66, No.2, pp. 132-133.

Larson K.M., Freymueller J.T., Philipsen S.(1997). *Global plate velocities from the Global Positioning System.* Journal of Geophysical Research, Vol. 102, No. B5, pp. 9961-9982.

McCarthy D.D. (ed.) (1996). *IERS Conventions (1996).* Technical Note 21, International Earth Rotation Service, Paris Observatory, Paris.

Soler T., Love J.D., Hall L.W., Foote R.H.(1992). *GPS Results from Statewide High Precision Networks in the United States.* Proceedings of the 6th International Geodetic Symposium on Satellite Positioning, The Ohio State University, Columbus, OH, March 17-20, pp. 573-582.

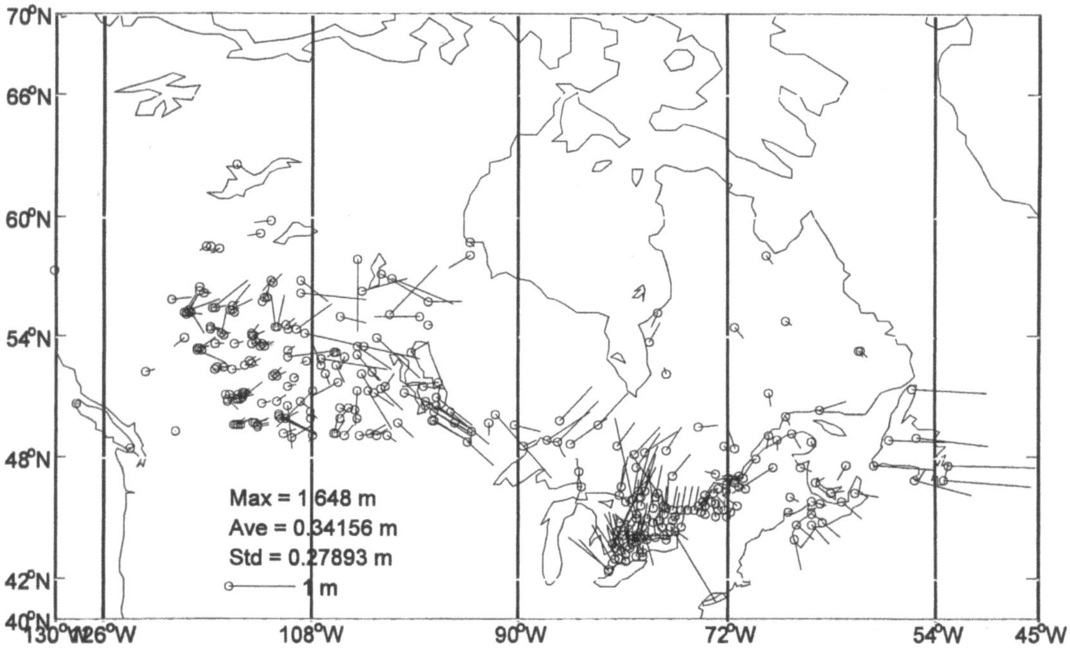

Fig. 1: Errors in the NAD83 horizontal control network in Canada as represented by differences with respect to high accuracy GPS surveys tied almost directly to the fundamental reference frame of NAD83 defined by the VLBI stations.

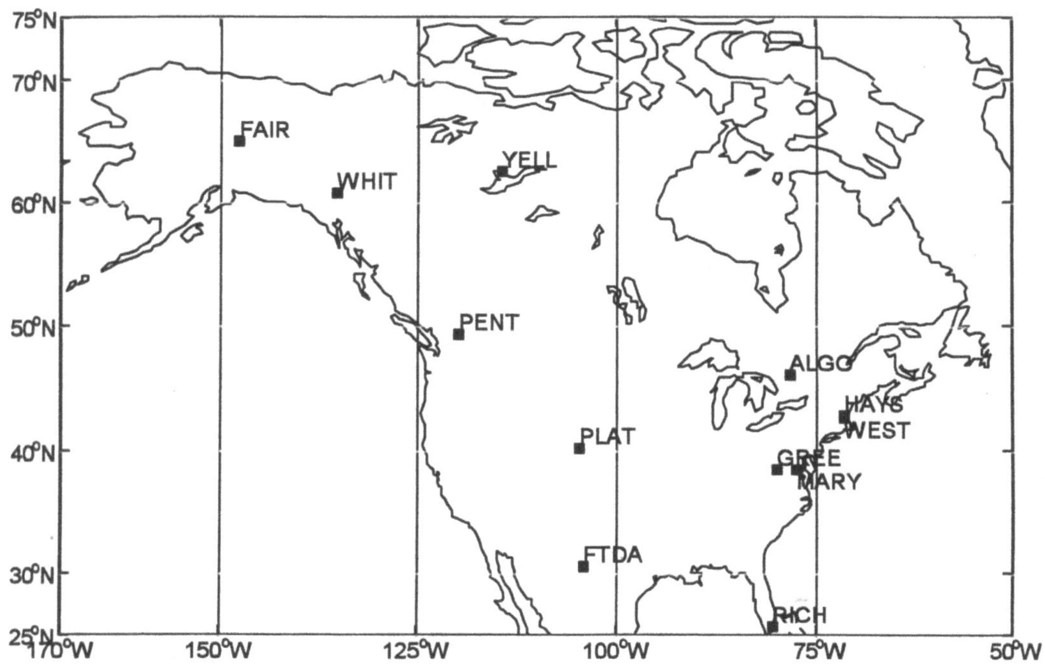

Fig. 2: VLBI stations used to derive the new ITRF96-NAD83 transformation.

Geodetic Datum Definition of the SAGA Network

D. ANGERMANN[1,2], J. KLOTZ[1], CH. REIGBER[1]

Abstract

In cooperation with various organizations in the host countries, the GFZ Potsdam established the large-scale GPS network SAGA (South American Geodynamic Activities) covering Chile and parts of Argentina. The network consists of 215 sites including two sites on the islands Robinson Crusoe and San Felix that are part of the Nazca plate.

The GPS data were processed with the GFZ Software EPOS resulting in a station coordinates repeatability of 1 - 3 mm horizontally and 5 mm vertically. We processed the SAGA data simultaneously with data of the IGS station network to transform the SAGA network into the ITRF96 frame. Doing this, we computed global solutions with about 20 globally distributed IGS stations and regional solutions with IGS stations in South America. The mean station coordinate residuals for the globally distributed IGS stations between the SAGA solution and the ITRF96 coordinates are 1cm horizontally and 1 - 2 cm vertically. With respect to the South American reference frame the mean station coordinates residuals are in the range of 2-5 mm horizontally and below 1 cm for the height. The residual motions for five IGS stations located on the stable part of South America are 1 - 3 mm/yr.

1. Introduction

During the past decade, the scope and accuracy of space-geodetic techniques have expanded greatly. In addition to VLBI, SLR, DORIS and PRARE stations, a large number of GPS sites started operation during the last years, leading to a fairly homogeneous distribution of tracking stations on the globe. The data of these GPS stations provide an excellent basis for many geodynamic applications, such as the observation of global plate motions and intraplate deformations. These tasks require an appropriate definition of the reference system. The latest and most precise realization of the International Terrestrial Reference System (ITRS) is the ITRF96 [BOUCHER et al., 1998], [SILLARD et al., 1998], which was realized by the International Earth Rotation Service (IERS). The position parameters are produced from a combination of sets of station coordinates and velocities determined by means of the most accurate space-geodetic techniques: Very Long Baseline Interferometry (VLBI), Satellite Laser Ranging (SLR), Global Positioning System (GPS) and Doppler Orbit Determination and Radio Positioning Integrated by Satellite (DORIS).

Within the SAGA project the GFZ Potsdam established in cooperation with many partner institutions of the host countries in 1993/94 a network of 215 GPS stations covering the whole territory of Chile and the western part of Argentina [KLOTZ et al., 1996]. Since 1993 this SAGA network was periodically re-observed in three GPS campaigns.

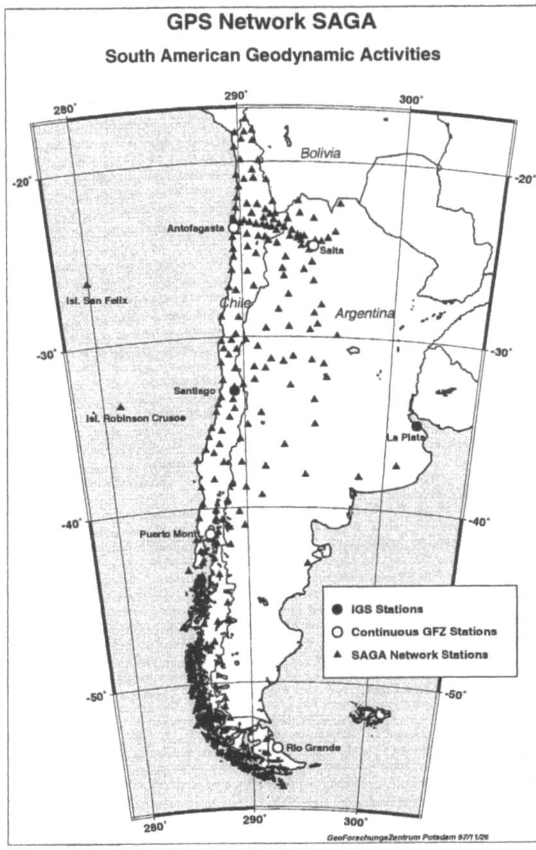

Fig. 1: GPS Network SAGA

This paper describes the strategy for the GPS data analysis using the GFZ Software EPOS (Earth Parameter and Orbit System) and deals with the datum definition of the SAGA network in the ITRF96 frame. Our main focus in this paper is the realization of a reference frame that represents the stable part of the South American continent. The second focus is on the motions of the IGS reference stations with respect to stable South America. Geodynamic results of our periodically-observed SAGA stations will be discussed in separate papers [e.g. KLOTZ et al., 1999].

1. Data and Processing Strategy

In the framework of the SAGA project we conducted four large GPS campaigns in cooperation with various organizations of the host countries. The first observation of the entire

1 Detlef Angermann (till June 30, 1999), Jürgen Klotz, Christoph Reigber, GeoForschungsZentrum Potsdam, Div. 1, Sect. 1.1, D-14473 Potsdam, Germany, Tel: +49 331 288 1113, Fax: +49 331 288 1169, email: dang@gfz-potsdam.de

2 Detlef Angermann (since July 1, 1999), Deutsches Geodätisches Forschungsinstitut, Marstallplatz 8, D-80539 München, Germany, Tel: +49 89 23031109, Fax: +49 89 23031240, email: angerman@dgfi.badw.de

GPS network was carried out between October 1993 and March 1994. The first re-observation of the northern part of the network between Antofagasta and Salta was realized in October and November 1995. The major part of the SAGA network was re-observed in November and December 1996. The second re-observation of the northern network took place in October and November 1997. In all campaigns we used Trimble 4000 SSE receivers and the same kind of antennas, thereby producing a homogeneous data set. The average acquired observation time per site was 70 hours. In order to connect all SAGA observations with the ITRF96 frame we included data of the IGS station network. In case of a global network solution we used the data of about 20 globally distributed IGS stations. For a regional solution we included data of IGS stations located in South America.

The GPS data were processed with the GFZ Software EPOS [ANGERMANN et al., 1997]. The processing chain of this software package is based on undifferenced phase and pseudo range observations. In EPOS the parameters are estimated in a rigorous least squares adjustment. The parameters solved for are station coordinates, ambiguities, receiver and satellite clock parameters for each observation epoch and tropospheric parameters for each station. Optional is the processing of orbit and earth orientation parameters (EOP) in case of a dynamic solution. The network computation is done in the form of daily solutions with ionosphere-free linear combinations of phases and pseudo ranges. In this study we used an elevation cut-off angle of 15 degree and a sampling rate of 4 minutes. The combined IGS orbits and the corresponding EOP's were fixed and all station coordinates were solved without any constraints (fiducial-free solution). The ambiguities resulting from the ionosphere-free linear combination of phases were estimated as real values (float solution). The tropospheric effects were modelled with the CfA mapping function [DAVIS et al., 1985] using meteorological data from a standard atmosphere. The tropospheric zenith path delay was estimated for each station at 4 hour intervals. To account for the antenna phase center variations, the IGS correction tables were applied [ROTHACHER and MADER, 1996].

2. Reference Frame

As described above, we computed the SAGA solutions with fixed combined IGS orbits and EOP's and solved the station coordinates without any constraints. Hence the reference frame of these fiducial-free solutions is defined by the IGS orbits and EOP's. In a second step we transformed our SAGA solutions to the ITRF96 station coordinates by estimating a seven parameter transformation. We used different IGS station constellations for this Helmert-transformations.

(A) Transformation to a global frame using about 20 globally distributed IGS stations.

(B) Transformation to a regional or continental South American frame using 6-7 IGS stations in South America.

We define the reference frame for station velocities in two ways. For studying plate motions at a global scale we estimate the station velocities directly from method A. For the interpretation of our SAGA results we focus on regional tectonics and relative plate motions. To achieve this goal we define the datum in the South American frame (method B) leading to station velocities with respect to the stable part of the South American continent. This is done in two steps:

(1) We derive a best fitting Euler vector (pole and rotation rate) for the South American plate in a least squares adjustment using the ITRF96 velocities of the IGS stations Fortaleza, Kourou, Brazilia, La Plata and Ascension Island. These stations are assumed to be located on stable South America.

(2) The ITRF96-based Euler vector of step 1 is then used to transform the SAGA results as well as the ITRF96 station velocities to the South American plate. Doing this, we make sure that the SAGA-derived velocities are related to the same reference frame as the ITRF96 velocities.

3. Results and Discussion

To assess the quality of the SAGA network two methods are used: (1) The daily repeatability is estimated using the residuals of a Helmert-transformation between mean and daily network solutions. The mean station coordinates residuals are in the order of 1-3 mm in the horizontal and 5 to 6 mm in the vertical directions. (2) To estimate the accuracy we compare our SAGA network solutions with the ITRF96 station coordinates. Table 1 shows the residuals after a Helmert-transformation between our global SAGA 95 solution to the ITRF96 station coordinates (19 IGS Stations).

Tab. 1: Residuals of Helmert-Transformation between global SAGA 95 solution and ITRF96

Stations	North mm	East mm	Height mm
Kootwijk, Netherlands	1.6	4.3	-23.4
Madrid, Spain	-2.5	-1.6	-25.9
Tromsoe, Norway	8.0	11.1	-32.3
Wettzell, Germany	-2.4	3.2	-20.3
Tsukuba, Japan	-4.7	-3.1	20.2
Taipei, Taiwan	3.8	3.8	-11.0
Kitab, Uzbekistan	-12.3	9.9	24.7
Tidbinbilla, Australia	2.0	-13.0	-19.6
Yaragadee, Australia	-9.2	-14.1	-10.0
Algonquin, Canada	-1.2	3.0	15.0
Fairbanks, USA	11.7	0.2	21.6
Yellowknife, Canada	11.0	7.6	14.2
Santiago, Chile	-3.3	12.7	-4.8
Kourou, French Guiana	-12.4	-13.6	21.4
Fortaleza, Brazil	-12.5	-17.3	11.2
Arequipa, Peru	-9.1	8.2	14.3
La Plata, Argentina	1.6	10.0	2.5
Kokee Park, USA	11.2	-4.0	26.3
O'Higgins, Antarctica	1.8	20.0	-24.2
RMS	7.8	10.1	19.6

The global network accuracy is 1 cm horizontally and 2 cm in the vertical directions. We assess the regional network accuracy with respect to the South American reference frame by comparing our regional SAGA solutions with the ITRF96 coordinates of the IGS stations in South America. Table 2 shows the mean residuals for the IGS stations Santiago, Kourou, Fortaleza, Arequipa, La Plata and O'Higgins. The accuracy of the regional SAGA solutions with respect to the South American reference frame is in the order of 2-5 mm horizontally and better than 1 cm vertically.

Tab. 2: Residuals of Helmert-Transformation between regional SAGA 95 solution and ITRF96

Stations	North mm	East mm	Height mm
Santiago, Chile	1.1	1.1	-4.4
Kourou, Fr. Guiana	5.7	-1.8	0.4
Fortaleza, Brazil	-8.6	-1.4	-3.1
Arequipa, Peru	2.7	0.5	3.6
La Plata, Argentina	0.9	1.0	4.0
O'Higgins, Antarctica	-2.3	2.1	-0.5
RMS	4.5	1.4	3.1

The results of Table 1 and 2 show that the mean residuals for the South American IGS stations are significantly smaller in the case of a regional Helmert-transformation. In the case of a global transformation a significant amount of spatially coherent noise may affect station coordinates in specific regions. This is mainly due to the fact, that both the SAGA computations as well as the realization of the ITRF96 reference frame may be influenced by errors in the underlying physical and geometrical models, (e.g. plate motion models, solid earth tides, ocean and atmospheric loading, gravity models, satellite orbits, tropospheric and ionospheric models), which may affect the station coordinates systematically. The mean residuals of the global transformation show a systematic behaviour especially in the vertical directions (see Table 1). The height residuals for the IGS stations in Europe are negative (-2.5 cm in the average), whereas the station height residuals for the IGS stations in South America show a positive trend. The results of Table 2 indicate that in case of the regional transformation, the mean residuals for the South American stations are reduced significantly. Consequently we prefer a regional Helmert-transformation if we focus on regional tectonic studies.

As described above, we defined the station velocities of our SAGA network with respect to the stable part of South America. In the first step we computed the Euler vector for the South American plate using the ITRF96 velocities of the IGS stations Kourou, Fortaleza, Brazilia, La Plata and Ascension Island. The resulting ITRF96-based Euler pole is located at -6.7 ± 2.8° N, 213.0 ± 10.7° E with a rotation rate of 0.10 ± 0.01°/Ma. The estimations of the Actual Plate Kinematic Model (APKIM 8.80) [DREWES, 1997] of -19.4 ± 5.1° N, 210.1 ± 12.3° E and 0.127 ± 0.010°/Ma are for the longitude of the rotation pole and the rotation rate in good agreement with our estimations. The difference of 12.7° for the latitude is not significant (3 σ). The corresponding NUVEL-1A predictions are -25.4°N, 235.6°W and 0.12°/Ma [DEMETS et al., 1990], [ARGUS et al., 1991], [DEMETS et al., 1994]. The NUVEL-1A rotation rate agrees with both space geodetic estimations and the difference of more than 20° for the longitude of the pole is not significant. A significant difference, however, exists for the latitude of the pole between our solution and NUVEL-1A.

The residuals of the ITRF96 velocities of the stations Kourou (KOUR), Fortaleza (FORT), Brazilia (BRAZ), La Plata (LPGS) and Ascension Island (ASC1) with respect to the ITRF96-based Euler vector are shown in Tab. 3. The residuals for the horizontal velocities are in the range of 1-3 mm/yr being below their one sigma standard deviations. These results indicate that those five stations represent stable South America.

Tab. 3: Station velocity residuals with respect to the ITRF96-based Euler vector for the South American plate

Stations	North mm/yr	East mm/yr	Height mm/yr
KOUR	-0.1 ± 0.8	0.5 ± 2.6	0.4 ± 2.7
FORT	0.7 ± 0.7	-0.5 ± 1.6	2.2 ± 1.6
BRAZ	-2.3± 2.6	-3.7 ± 5.1	-10.1 ± 5.0
LPGS	-0.9 ± 1.6	0.6 ± 2.1	1.1 ± 1.7
ASC1	-0.9 ± 2.7	3.0 ± 5.0	2.1 ± 2.0

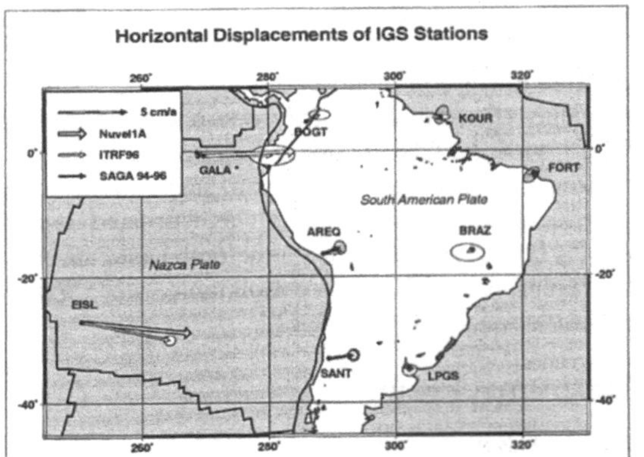

Fig. 2: ITRF96 and SAGA-derived horizontal displacements with respect to stable South America along with their 95% confidence ellipses, including the NUVEL-1A velocities of the IGS stations Galapagos and Easter Island located on the Nazca plate. For the IGS station Galapagos we used the velocity estimation of the IGS Analysis Center in Berne (CODE) [ROTHACHER et al., 1998], since the ITRF96 velocity of station GALA is poor estimated.

In the second step we used our ITRF96-based Euler vector to transform the ITRF96 and the SAGA-derived velocities to stable South America. Fig. 2 show the horizontal station velocities for 7 IGS stations in South America and the IGS stations Galapagos and Easter Island of the Nazca plate along with their 95% confidence ellipses. In addition the velocities of the SAGA94 and 96solutions are displayed. The SAGA 94/96 velocities and the corresponding ITRF96 velocities agree within 1-3 mm/yr. The differences are not significant, as indicated by their 95% error ellipses.

The ITRF96 velocities of the IGS stations Bogota, Arequipa, Santiago, Galapagos and Easter Island, which do not belong to the stable part of South America are displayed in Table 4. The stations Bogota, Arequipa and Santiago, which are located in the plate boundary zone, show significant motions of 1-2 cm/yr with respect to the South American reference frame. The velocities of the IGS stations Easter Island (66.1 mm/yr at 102.2°) and Galapagos Island (5.12 mm/yr at 90.1°) are significantly slower than the NUVEL-1A prediction of 80.2 mm/yr at 96.4° and 6.76 mm/yr at 87.5° respectively, indicating that the present-day convergence rate between the Nazcaplate and the South American plate is significantly slower than the average motion over the last millions of years (NUVEL-1A).

Due to the ongoing extension and densification of the IGS station network and due to the activities within the SIRGAS project [SIRGAS, 1997], [SEEMÜLLER and DREWES, 1998] further improvements for the definition and realization of a reference system for South America are expected in the near future.

Tab. 4: ITRF96 station velocities with respect to stable South America.

Stations	North mm/yr	East mm/yr	Height mm/yr
SANT	2.5 ± 1.5	17.8 ± 1.4	8.1 ± 1.7
AREQ	1.3 ± 0.6	11.7 ± 0.8	-1.2 ± 0.8
BOGT	5.2 ± 1.5	9.9 ± 3.1	0.1 ± 1.5
EISL	14.0 ±1.5	64.6 ±1.6	1.8 ± 1.8
GALA	0.0 ±3.1	51.3 ± 6.8	0.0 ± 2.4

Acknowledgements

This work was supported by the 'GeoForschungsZentrum Potsdam' and by the SFB 267 'Deformationsprozesse in den Anden'. We thank the Instituto Geografico Militar (IGM) Chile for supporting the GPS observations in Chile as well as the IGS and the IERS communities for providing GPS data, IGS products and ITRF96 station coordinates and velocities.

References

ANGERMANN D., BAUSTERT G., KLOTZ J., REINKING J., ZHU S.Y.: *The Impact of IGS on the Analysis of Regional GPS-Networks*. IAG Series No. 115: GPS Trends in Precise Terrestrial, Airborne, Space Applications, 35-39, Springer Verlag, 1996.

ANGERMANN D., BAUSTERT G., GALAS R., ZHU S.Y.: *EPOS.P.V3 (Earth Parameter and Orbit System)*, Software User Manual for GPS Data Processing. Scientific Technical Report STR97/14, 52pp., GeoForschungsZentrum Potsdam, 1997.

ARGUS D. F., GORDON R. G.: *No-net-rotation model of current plate velocities incorporating plate motion model NUVEL-1*, Geophys. Res. Lett., 18, 2038-2042, 1991.

BOUCHER C., ALTAMIMI Z., FEISSEL M., SILLARD P.: *Results and Analysis of the ITRF94*, IERS Tech. Note, 20,198 pp., 1996.

BOUCHER C., ALTAMIMI Z., SILLARD P.: *Results and Analysis of the ITRF96*, IERS Tech. Note, 24, 166 pp., 1998.

DAVIS J. L., HERRING T. A., SHAPIRO I. I., ROGERS A. E.: *Geodesy by radio interferometry: Effects of atmospheric modeling on estimates of baseline length*, Radio Sci. 20, 1593-1607, 1985.

DEMETS C., GORDON R. G., ARGUS D. F., STEIN S.: *Current plate motions*, Geophysical J. Int., 101,425-478, 1990.

DEMETS C., GORDON R. G., ARGUS D. F., STEIN S.: *Effect of recent revisions on to the geomagnetic reversal time scale on estimates of current plate motions*, Geophys. Res. Lett., 21, 2191-2194, 1994.

DREWES H.: *Combination of VLBI, SLR and GPS determined station velocities for actual plate kinematic and crustal deformation models*. IAG Symp. 119, eds. R. Forsberg, M. Feissel, R. Dietrich, 377-382, 1997.

International GPS Service for Geodynamics, 1996 Annual Report, eds. J. F. Zumberge, D. E. Fulton, R. E. Neilan, IGS Central Bureau, Jet Propulsion Laboratory, California Institute of Technology, Pasadena, California, USA, 1997.

KLOTZ J., REINKING J., ANGERMANN D.: *Die Vermessung der Deformation der Erdoberfläche*, Geowissenschaften, 14, 389-394, 1996.

KLOTZ J., ANGERMANN D., MICHEL G.W., PORTH R., REIGBER CH., REINKING J., VIRAMONTE J., PERDOMO R., RIOS V.H., BARIENTOS S., BARRIGA R., CIFUENTES O.: *GPS-derived Deformation of the Central Andes Including the 1995 Antofagasta Mw=8.0 Earthquake*, Pure and Appl. Geophys., in press, 1999.

REIGBER CH., XIA Y., MICHEL G., KLOTZ J., ANGERMANN D.: *The Antofagasta 1995 Earthquake: Crustal Deformation Pattern as observed by GPS and D-INSAR*. Proceedings of 3. ERS Symposium, 507-513, Florence, 1997.

ROTHACHER, M., MADER G.: *Combination of Antenna Phase Center Offsets and Variations: Antenna Calibration set IGS 01*, IGS Central Bureau / University of Berne, Berne, Switzerland, 1996.

ROTHACHER M., BEUTLER G., SPRINGER T.A., SCHAER S.: *IGS-Analysis Center Solution SSC(CODE)98P01*, University of Berne, Berne, Switzerland, 1998.

SEEMÜLLER W., DREWES H.: *Annual Report 1997 of the RNAAC SIRGAS*. IGS 1997 Technical Reports, 173-174, Pasadena, USA,1998.

SILLARD P., ALTAMIMI Z., BOUCHER C.: *The ITRF96 realization and its associated velocity field*, Geophys. Res. Lett., 25, 3223-3226, 1998.

SIRGAS: *SIRGAS Final Report, Working Groups I and II*. Ed. by Inst. Brasileiro de Geografia e Estatistica (IBGE), Rio de Janeiro, 1997.

Weak Gravitational Lensing and Instability of the Celestial Reference Frame

V.E. Zharov[1], M. V. Sazhin[1], T.A.Kalinina[1]

Abstract

The fluctuation of the angular positions of reference extra galactic radio sources caused by the gravitational lensing on visible stars and dark bodies of our Galaxy is considered. It is shown that these angular fluctuations range from a few up to hundreds of microarcseconds. The random proper motions of the stars and the dark bodies lead to a small rotation of the celestial reference frame. It limits the accuracy of realization of the inertial reference frame. It is shown that the gravitational lensing does not allow to align the axes of the different reference frames better than a few microarcseconds.

1. Weak microlensing

Here, we consider the effect of weak microlensing on the fluctuation of the angular positions of reference extra galactic radio sources, though this effect is general, and will also be relevant, in principle, to the reference optical stars. We calculate the values of the position variations of the selected radio sources due to this effect, and determine the influence of the gravitational field of visible Galactic stars and dark bodies on the stability of the celestial reference frame.

Let us consider the simple gravitational lens: a point-like extra galactic source S of photons, a point-like massive body B with mass M_b that bends the light, and an observer O. The direction of the incident light does not coincide with the straight line connecting the extra galactic source and the observer. Therefore, the apparent celestial position of the extra galactic source does not coincide with its true position.

The effect is characterized by the Einstein radius, which has the value

$$\varphi_E = \sqrt{\frac{4GM_b}{c^2}\frac{L_{sb}}{L_{ob}\left(L_{sb}-L_{ob}\right)}},$$

where L_{ob} and L_{sb} are the distances from the observer O to the body B and from source S to the body B, respectively. In the microlensing effect, two images appear, which have angular distances from the body B:

$$\varphi_1 = \frac{\varphi}{2}+\frac{1}{2}\sqrt{\varphi^2+4\varphi_E^2}$$

and

$$\varphi_2 = \frac{\varphi}{2}-\frac{1}{2}\sqrt{\varphi^2+4\varphi_E^2}$$

where φ is the angular distance from B to the true position of the source S, and φ_1 and φ_2 are the angular distances of the two images relative to B. The brightness of the second image φ_2 is inversely proportional to the fourth power of φ, and in the case $\varphi >> \varphi_E$, which we will call weak microlensing, the observer cannot see the second image due to its weakness.

If the true right ascension and declination of S are α_t, δ_t, the apparent coordinates of S are α_a, δ_a, and the coordinates of B are α_b, δ_b, then the fluctuation of the angular position is defined by equations:

$$\Delta\alpha = \left(\frac{\varphi_E}{\varphi_a}\right)^2 \frac{\cos\delta_b}{\cos\delta_a}\sin\left(\alpha_a-\alpha_b\right), \qquad (1)$$

$$\Delta\delta = -\left(\frac{\varphi_E}{\varphi_a}\right)^2 \frac{\sin\delta_b-\cos\varphi_a\sin\delta_a}{\cos\delta_a}. \qquad (2)$$

One can see from (1, 2) that the difference between the true and apparent angular positions is proportional to the mass of the body-lens and inversely proportional to the second power of the angular distance φ_a It means that the effect is defined mainly by the location of the stars relative to the radio sources but not the masses of the stars. So the high precision positional measurements can be probe for study of the distribution of the massive compact halo objects (Machos) in the vicinity of the Sun.

2. Variations of the coordinates of reference sources

Two catalogues of the extra galactic radio sources positions were used to test the effect of gravitational lensing on their stability. The first catalogue RSC(IERS)95 C02 was published in the International Earth Rotation Service (IERS) Annual Report (IERS, 1996). The second catalogue RSC(WGRF)95 R01 was created by the IAU Working Group on Reference Frame (WGRF) (Ma et al., 1998). Each of them realizes the International Celestial Reference Frame (ICRF) as a set of relative positions oriented to the axes of the International Celestial Reference System, ICRS (Arias et al., 1995).

[1] Vladimir Zharov, M.V.Sazhin, T.A. Kalinina, Sternberg State Astronomical Institute, Moscow, 119899, Universitetsky Prospekt, 13, Russia; Tel.: +7 095 9395024; Fax: +7 095 9328841, Email: zharov @ sai.msu.su

The extra galactic frame RSC(WGRF)95 R01 was obtained by putting the radio source coordinates from the WGRF solution in ICRS via comparison to RSC(IERS)95 C02 on the base of common sources. From the 212 defining sources selected by WGRF, 117 were defining sources in RSC(IERS)95 C02. The "defining" sources define the axes of the ICRF as accurately as possible. So they are the most compact and best observed. The remaining sources are included to obtain a better distribution of sources on the sky. The stability of the frame is based on the assumption that the defining sources have no proper motion. Gravitational lensing leads to changes in the coordinates of *all* sources with time or "proper" motions, and to a small rotation of the frame.

Here, we calculate the effect of gravitational lensing on visible stars at first. Then we offer a hypothesis that could explain the negative parallaxes of some stars by the transverse motions of the massive dark bodies and faint stars relative to the line-of-sight from observer to the source. It gives possibility to find the distribution of the dark bodies in the vicinity of the Sun and estimate the rotation of the celestial reference frame due to the lensing on the dark bodies. We used the Guide Star Catalogue (LASKER et al., 1990) to find stars whose angular distances from the extra galactic reference sources do not exceed 1'. A total of 25 radio sources from the catalogue RSC (IERS)95 C02 have a total of 30 nearby stars. 58 radio sources from RSC (WGRF)95 R01 have 65 neighbouring stars in a circle with radius 1'. For most of these stars, we know neither their mass nor their distance.

To estimate the rotation of the true ICRF relative to the observed ICRF, we used only the defining sources. The rotation vector $\Theta = (\theta_1, \theta_2, \theta_3)$ is defined by the matrix

$$r_i = \begin{pmatrix} 1 & -\theta_3 & \theta_2 \\ \theta_3 & 1 & -\theta_1 \\ -\theta_2 & \theta_1 & 1 \end{pmatrix} \mathbf{r}_i \qquad (3)$$

where $\mathbf{r}_i$ and $\mathbf{r}_i'$ are the unit vector of the *i*th source before and after the deflection of the light from this source. We estimated the distance to the star or the dark body L_{ob} using the equation (SAZHIN et al., 1998):

$$\lg L_{ob.} + 0.2aL_{ob} = 0.2(m + 0.26) + 1.7 \lg \frac{M_b}{M}$$

where M. is the mass of the Sun and $a = 1.9\ \text{kpc}^{-1}$ is the total absorption in the vicinity of the Sun. The mass of each star was determined in the range from $1M_\odot$ to $3M_\odot$ using the Salpeter mass function $dn / dM_b \sim M_b^{-2.35}$. We find that $\Theta = (-0.08 \pm 0.02, -0.47 \pm 0.03, -0.36 \pm 0.03)$ μas for various realizations of the Salpeter mass distribution (the variances of Θ correspond to the ranges for various mass distributions) for the catalogue RSC(IERS)95 C02 and $\Theta = (-0.07 \pm 0.05, -1.19 \pm 0.10, -0.01 \pm 0.01)$ μas for the catalogue RSC (WGRF)95 R01. It means that two reference frames can not be aligned on the microarcsecond level of accuracy. Different sets of stars as the gravitational lenses lead to different rotation of these frames.

Really the obtained values of Θ give the lower estimation of the effect of the gravitational lensing on the stability of the ICRF. It is necessary to correct the results for the lensing on the dark bodies.

We suppose that the negative parallaxes of some stars from the Hipparcos catalogue (ESA, 1997) can be explained by the motion of the dark bodies. In this case, the distribution of these stars could be connected with the distribution of the dark bodies. One can estimate the number of the radio sources that will be affected by the gravitational lensing, and the value of the coordinate fluctuations for calculation of the instability of ICRF.

The Fig.1, 2 help to understand how the effect of the gravitational lensing changes the coordinates of the fictitious radio source that have no proper motion and one of the Hipparcos star (N20814), respectively.

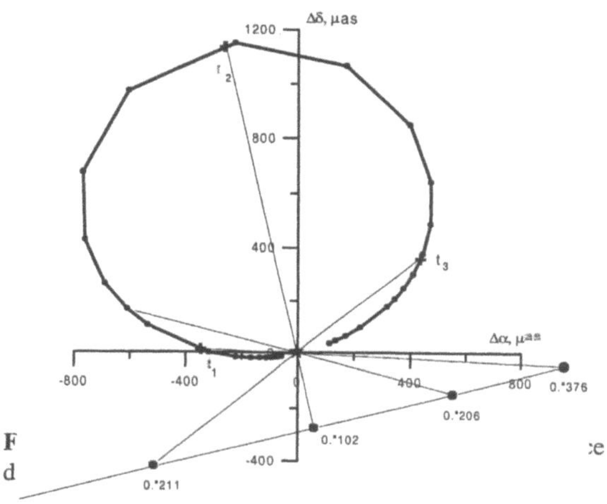

Fig. 1: The variation of the coordinates of the radio source due to proper motion of the dark body.

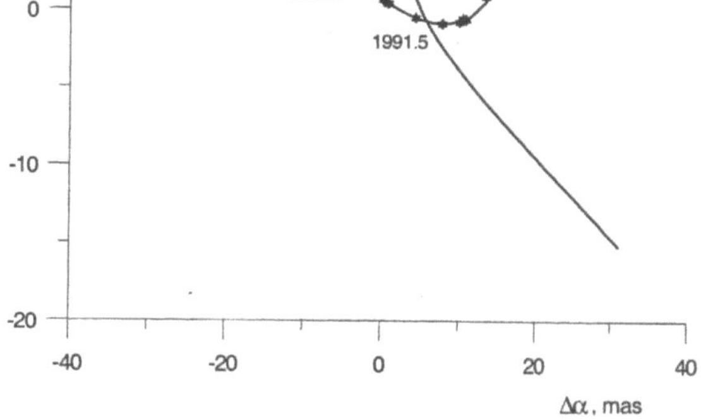

Fig. 2: The variation of the coordinates of the star 20814 due to proper motion of the dark body.

In the first case, the apparent motion is the circular motion, but in the second case, we see that the star moves along a complicated curve.

One can show that

$$\mu_\alpha^c \cos\delta_a = \mu'_\alpha \cos\delta_a + \pi_s F_1 + \left(\frac{\varphi_E}{\varphi_a}\right)^2 G_1, \qquad (4)$$

$$\mu_\delta^c = \mu'_\delta + \pi_s F_2 + \left(\frac{\varphi_E}{\varphi_a}\right)^2 G_2 \qquad (5)$$

where $\mu_\alpha^c, \mu_\delta^c$ are the components of proper motion of the star in the heliocentric system (they are shown in the catalogue), $\mu'_\alpha, \mu'_\delta, \mu''_\alpha, \mu''_\delta$ are the components of proper motion of the star and the dark body in the geocentric system (they are observed), and F_1, F_2, G_1, G_2, are the functions of the coordinates of the Sun $\alpha_\odot, \delta_\odot$, of the star and of the dark body. The equations (4,5) describe the transformation between the geocentric and heliocentric systems. Here we do not take into account the difference between heliocentric and baricentric systems.

It is obviously that effect of the gravitational lensing will change the values $\mu_\alpha^c, \mu_\delta^c, \pi_s$, significantly, if the last terms in (4,5) will of the same order as the sum of two first terms. The system of equations (4,5) was solved for arbitrary moments of observations that were shown by asterisks on Fig.2. The coordinates of the HIP20814 star are $\alpha = 4^h27^m35.^s6946$, $\delta = 48^0 17' 51.''173$,

$\mu_\alpha^c = -2.39 \pm 2.03$ mas/year, $\mu_\delta^c = -1.44 \pm 1.06$ mas/ year $\pi_s = -2.34 \pm 1.41$ mas. We supposed that really the star has positive parallax (π_s =5 mas) and $\mu'_\alpha = -8$ mas/year, $\mu'_\delta = 4$ mas/year. If there is the dark body with coordinates $\alpha = 4^h27^m35.^s696$, $\delta = 48^0 17' 51.''195$ and mass $M_b = 0.7 M_\odot$, $\mu''_\alpha = -77$ mas/year, $\mu''_\delta = 47$ mas/year, $\pi_b = 50$ mas, then the calculated values $\mu_\alpha^c, \mu_\delta^c, \pi_s$ are equal to -2.2 ± 3.1 mas/year, -1.6 ± 2.1 mas/year, -0.3 ± 0.5 mas/year.

This example shows that the negative parallaxes of some Hipparcos stars can be explained by the gravitational lensing.

We chose a total of 1326 stars with negative parallaxes, which have not flags indicating the double or multiple system (field H59) and absence of variability (field H52) (ESA, 1997). The distribution of these stars, normalized to 1326, per galactic latitude b is shown in Fig.3 by thin line.

It is seen that it differ significantly from the distribution of all Hipparcos stars, normalized to the total number of stars (thick line). If our hypothesis is true, their relation will be the distribution function of the Machos per galactic latitude. This distribution is approximately uniform when $|b|>15°$, but in the plane of the Galaxy ($|b|<15°$) the density of the Machos three times more (Fig.4).

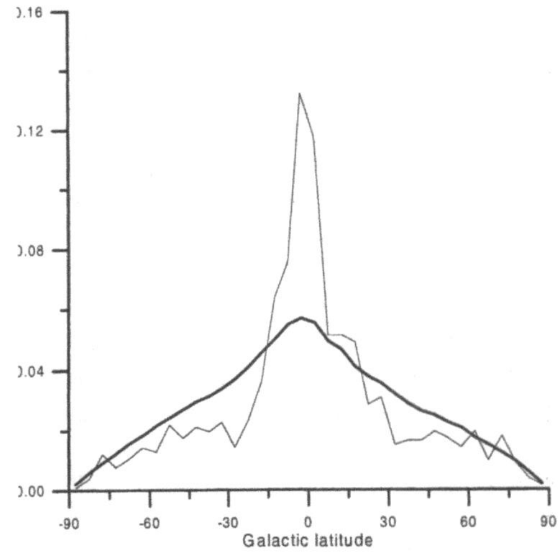

Fig. 3: The normalized numbers of the stars in the Hipparcos catalogue (thick line) and stars with negative parallaxes (thin line) per galactic latitude.

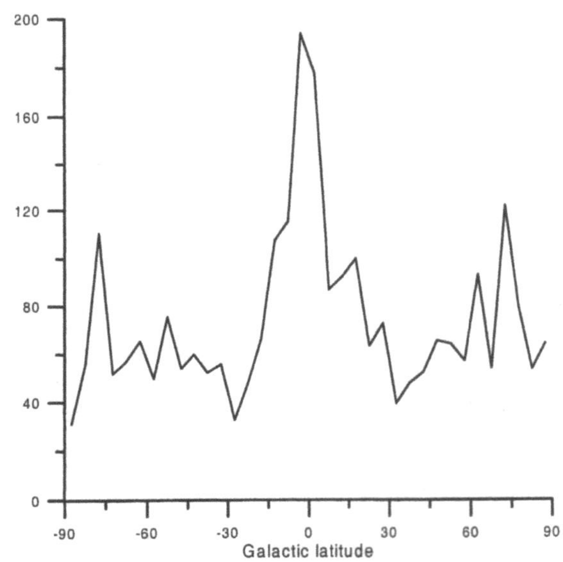

Fig. 4: The distribution function of the dark bodies per galactic latitude.

The median precision of the astrometric parameters in the Hipparcos catalogue is of order of 1 mas (KOVALEVSKY, 1998). It means that almost all Hipparcos stars are located within 1 kpc of the Sun. Therefore, it could be an estimation of the Machos distribution near the Sun.

Really, some stars can be located further than 1 kpc from the Sun. In this case, the negative parallaxes can be explained by the uncertainties of observations. So this distribution has to be corrected: distant stars have to be excluded but some "double" stars have to be added.

One can see from Fig.2 that the apparent motion of the star on time interval 1990.5-1991.5 is very similar to the motion of a component in the double system. It is known (KOVALEVSKY, 1998), that over 10000 stars in the Hipparcos catalogue are suspected to be non-single but were not resolved, generally because the components are too close or one of them is too faint. If one of the components is invisible and period of orbital motion is close to the observation period, then we can not maintain that the star is double. Only after repeated observations one can answer question: was the star double or was the gravitational lensing event observed?

In order to take into account the Machos effect on the stability of the ICRF, we note:

1. The stars with negative parallaxes make up approximately 1% fraction of the Hipparcos stars.
2. There is strict limitation to the angular distance φ_a. The angle φ_a has to be less then 0.1" in order to detect the coordinate fluctuation, if precision of the observations of 1 mas.
3. The period of the coordinate fluctuation has to be of a few years.
4. The probability of the lensing event for the radio source lying in the Galaxy plane is three times more then for the high latitude source.

For estimation of the rotation of the ICRF the coordinates of two arbitrary radio sources with $b \approx 0°$ were corrected on 1 mas. It corresponds the close passage ($\varphi_a \approx 0.1"$) of the dark body with $M_b=0.2M_\odot$ (Fig.1). The coordinates of two other sources with large b were shifted on 0.3 mas. It can be explained by the passage ($\varphi_a \approx 0.2"$) of the dark body with $M_b=0.5M_\odot$.

Using (3) we find that $|\Theta|$ can reach $5 \div 10\ \mu$as. It means that effect of the Machos on the stability of the ICRF is more important.

5. Conclusion

The effect of weak microlensing leads to a small rotation of the celestial reference frame. It will limit the accuracy of the extra galactic reference frame. So this frame can be considered inertial and homogeneous only to microarcsecond accuracy.

The effect considered here is defined mainly by the Machos. Therefore, the high precision astrometric catalogues could be powerful tools for studying of the density of microlensing objects. At the same time homogeneity of these catalogues will be broken by the temporal variations of the coordinates of the reference sources. We can expect that coordinates of 1% of the sources are fluctuated with amplitudes ≈ 1 mas and with periods of order of $1 \div 3$ years. More significant fraction of the stars will be bent by the Machos too, but the fluctuations will be less, and period will be longer.

Acknowledgements

Authors thank the Russian Foundation of Fundamental Research (grants 97-05-64342, 98-05-64797), "Russian Universities" (grant 2-5547) and program "Integration" (K0641) for support of the work.

V.E. ZHAROV gratefully acknowledges travel support from the IAG. M. V. SAZHIN acknowledges for financial support of Centro Volta-Landau Network during the writing this paper.

References

ARIAS E.F., CHARLOT P., FEISSEL M., LESTRADE J.-F. (1995). *The exrtagalactic reference system of the International Earth Rotation Sernice, ICRS.* Astronomy and Astrophysics, Vol.303, pp.604-608.

ESA. (1997). *The Hipparcos and Tycho Catalogues.* ESA Publication SP-1200.

IERS. (1996). *1995 IERS Annual report.* Observatoire de Paris, Paris.

KOVALEVSKY J. (1998). *The new astrometry.* Reports of Progress Physics, Vol. 61, pp.77-115.

LASKER B.M., STRUCH C.R., MCLEAN B.J., RUSSEL J.L., JENKNER H., SHARA M.M. (1990) *The Guide Star Catalogue. 1. Astronomical foundations and image processing.* The Astronomical Journal, Vol. 99, pp.2019.

MA C., ARIAS E.F., EUBANKS T.M., FEY A.L. *et al.* (1998). *The international Celestial Reference Frame as realized by Very Long Baseline Interferometry.* The Astronomical Journal, Vol.116, pp.516-546.

SAZHIN M.V., ZHAROV V.E., VOLYNKIN A.V., KALININA T.A. (1998) *Microarcseconds instability of the celestial reference frame.* MNRAS, Vol.300, pp.287-301.

Height Time Series of the Italian GPS Fiducial Network compared with VLBI and SLR Estimations

R. Devoti[1], M. Fermi[1], C. Ferraro[2], R. Lanotte[2], V. Luceri[2], A. Nardi[1], R. Pacione[2], P. Rutigliano[2], C. Sciarretta[1], G. Bianco[3], F. Vespe[3]

Introduction

In this work, we present the results of the determinations of vertical motions obtained from the analysis of data coming from different space geodetic techniques, Global Positioning System (GPS), Satellite Laser Ranging (SLR), Very Long Baseline Interferometry (VLBI). All the geodetic solutions have been computed at the Centre of Space Geodesy (CGS) at Matera.

The vertical motions are the least well determined in the geodetic analysis even if the determination and monitoring of the vertical kinematics behaviour of points on the Earth's surface provides important information useful in constructing dynamic model of tectonic. In some sites of the analysed network are collocated different space geodetic techniques; in these sites the vertical velocities obtained by the analysis of different data are compared to strengthen our knowledge of the tectonic motions. The analysed network is shown in fig. 1.

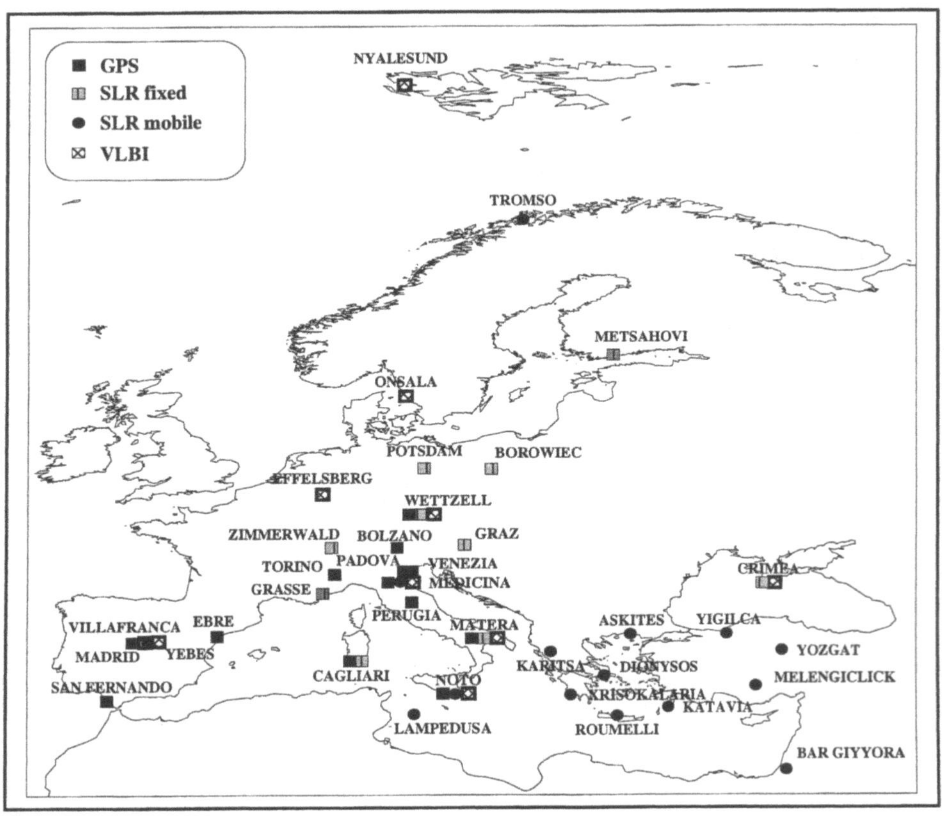

Fig.1 Analysed network

[1] Roberto Devoti, Marco Fermi, Antonio Nardi, Cecilia Sciarretta, Telespazio, S.p.A., Via Triburtina, 965-00156 Roma, Italy, Tel.: +39 - 06 - 4079-3329 /-3862 /-3349 /-3864, Fax: +39 - 06 - 4079-6202, e-mail: roberto@caronte.mt.asi.it, marco@sidus.mt.asi.it, antonio@sidus.mt.asi.it, cecilia@hp735.mt.asi.it

[2] Carmela Ferraro, Roberto Lanotte, Vincenza Luceri, Rosa Pacione, Paolo Rutigliano, Telespazio, S.p.A., Centro di Geodesia Spaziale „G. Colombo", C.da Terlecchia, P.O.Box - ADP, 75100 Matera, Italy, Tel.: +39 - 0835 - 3772-269, /-238, /-231, /-224, /-200, Fax: +39 - 0835 - 334951, e-mail: ferraro@asi.it, lanotte@asi.it, lucieri@asi.it, pacione@asi.it, pacione@asi.it, rutigliano@asi.it

[3] Giuseppe Bianco, Francesco Vespe, ASI - Agenzia Spaziale Italiana, c/o Centro di Geodesia Spaziale „G. Colombo" P.O. Box . ADP, 75100 Matera, Italy, tel.: +39 - 0835 - 3772-209, /-211, Fax:+39 - 0835 - 339005, e-mail: bianco@asi.it, vespe@asi.it

1 Data analysis

Table 1 summarises the data statistics, the adopted models and strategies for the SLR, VLBI and GPS solutions. The station coordinates obtained by each solution have been expressed in the ITRF96 coordinates system to allow comparison of the results. The station velocities have been obtained estimating linear slopes versus time from the estimated coordinates.

Tab. 1: GPS, SLR and VLBI solutions

Solutions	GPS_Net	GPS_pp	SLR	VLBI
Software used	MicroCosm	Gipsy - Oasis II	Geodyn II/Solve	Calc/Solve
Data	Feb.95-Sep.98, daily sessions	Feb.95-Sep.98, daily sessions	Jan86 - Dec.97 (Lageos - I) / Nov.92 - Dec.97 (Lageos-II)	37 observing sessions from 1990 since 1998
Celestial References Frame	Satellite ephemerides ad EOP from IGS	JPL "fiducial free" satellite ephemerides ad EOP	Estimated Lageos I & II orbits	Source coordinate from IAU95 / EOP and nutation series from the IERS series 97C04
Terrestrial References Frame	Position and velocity of Wettzell fixed at ITRF96 values	JPL parameters from fiducial free solution to ITRF96	Position and velocity of Herstmonceaux ad Washington fixed at ITRF96 values	Position and velocity of Wettzell fixed at ITRF96 values
Geodetic Estimated Parameters	Site positions	Site positions	Satellite monthly state vector / Site positions every 2 months / Daily EOP	Site positions

The SLR solution is a global solution based on the reduction of the whole available data set for Lageos II and for the data since 01/01/1985 for Lageos I satellites. Details about the analysis procedure are available in DEVOTI et al. (1996); the main differences are that in this analysis the station coordinates have been estimated bimonthly starting from ITRF96 a priori stations position and that we performed a last square linear fit of the coordinates to estimate the velocities.

The VLBI solution is a very preliminary solution obtained using observing sessions from the EUROPE campaign. The a priori station positions are ITRF96, Wettzell coordinates and velocity has been fixed to ITRF96 values. The analysis procedure is described in LANOTTE et al (1996).

GPS data from the network reported in fig.1 are routinely analysed using MicroCosm software with a Fiducial Approach, in order to provide weekly solution to EUREF as Local Analysis Center. The details about the solution obtained with the Fiducial Approach are in FERMI et al (1997). The data from two stations, Medicina and Padova, have been analysed also with a different strategy, the Precise Point Positioning using Gipsy-Oasis II software (ZUMBERGE et al., 1997). For both the GPS solutions, the data have been analysed on a daily basis to provide the 3-D determinations of site positions in the ITRF96 reference frame. This allows comparing the results from two GPS independent analysis between them and with the results derived from VLBI and SLR analysis. Hereafter we call the GPS solution obtained with the Fiducial Approach as GPS_net solution and the other one as GPS_pp solution.

2 Results

Many stations of the network shown in fig. 1 have a long history of tracking and good quality of data, so this leads to accurate velocity estimation.

In table 2 are reported the Up component of the motions only for these stations, compared with the value provided by ITRF96. For the other stations of the network the results are not reliable due to short data history or noisy data.

This happens for the mobile SLR stations and for some VLBI (Yebes, Crimea, Ny Alesund) and GPS stations (Ebre, S. Fernando, Villafranca, Bolzano, Perugia, Torino)

Among SLR stations, Graz seems to let down with a velocity of 2.0 ± 0.5 mm/yr; the results from Grasse, Potsdam and Zimmerwald lead us to conclude that no uplift motions are ongoing in this area, all these stations show 0 or sub-millimetric movements at 1-σ level,.

For Wettzell we estimate two different velocities due to change of SLR system in 1991. So we identify these two systems with their own IERS ID., Wettzell 7834 for the former, Wettzell 8834 for the latter. For Wettzell 8834 we obtained an doubtful 5.8 ± 1.3 mm/yr (ITRF96 value is -2.3 $\pm$ 0.1 mm/yr); this result need of further investigation. Wettzell has been keep fixed in the GPS_net and VLBI solutions, so no comparison is possible. The solution for Wettzell 7834 is in agreement with the previous tectonic interpretation; the estimated velocity -2.1 ± 2.4 mm/yr seems to confirm that there are no significant vertical motion in this area.

The results of VLBI confirm the previous interpretation. No vertical motions are detected at Effelsberg and Onsala stations.

Madrid seems to be affected by vertical movements. This result confirm the conclusion of a previous VLBI analysis (LANOTTE et al, 1996) GPS result for Madrid could confirm the uplift motion but the data available for this station are quite noisy and so this result needs of further confirmation.

Tab. 2: Up velocity estimated from GPS, SLR and VLBI analysis compared with the ITRF96 results

STATION	DATASET	SLR mm/yr	GPS mm/yr	VLBI mm/yr	ITRF96 mm/yr
Matera	SLR From 01/01/1986 To 31/12/1997 GPS From 05/02/1995 To 05/09/1998 VLBI From 01/01/1991 To 28/02/1998	0.7±0.8	-0.9±0.2	-1.2±1.3	-0.7±0.3
Cagliari	SLR From 01/06/1989 To 31/12/1994 GPS From 09/04/ 1995 To 05/09/1998	-7.8±2.0	-4.6±0.3		1.2±0.2
Effelsberg	VLBI From 01/08/1994 To 28/02/1998			2.5±2.1	-3.7±1.9
Grasse	SLR From 01/07/1986 To 31/12/1997	-1.4±0.9			0.3±0.3
Graz	SLR From 01/01/1986 To 31/12/1997	-2.0±0.5			0.9±0.3
Madrid	GPS From 05/02/1995 To 05/09/1998 VLBI From 01/01/1990 To 28/02/1998		1.2±0.6	4.9±1.9	3.9±0.3
Medicina	GPS From 28/07/1995 To 05/09/1998 VLBI From 01/01/1990 To 28/02/1998		-3.8±0.3	-4.7±1.3	1.4±0.2
Noto	SLR Mobile station occupation 1991-93 GPS From 05/02/1995 To 05/09/1998 VLBI From 01/01/1990 To 28/02/1998	-9.2±6.3	-5.1±0.3	-5.2±1.7	0.8±0.4
Onsala	VLBI From 01/01/1990 To 28/02/1998			-0.2±1.6	0.2±0.2
Potsdam	SLR From 01/01/1993 To 31/12/1997	-2.5±2.0			4.3±2.1
Zimmerwald	SLR From 01/01/1986 To 31/12/1997	0.9±1.1			3.2±0.3
Wettzell (7834)	SLR From 01/05/1986 To 28/02/1991	-2.1±2.4			-2.3±0.1
Wettzell (8834)	SLR From 01/03/1991 To 31/12/1997	5.8±1.3			-2.3±0.1

In the following, we report the time series of the Up component for the Italian stations in which are available data from different geodetic techniques. For Matera are shown the results from SLR, GPS and VLBI analysis, while for Medicina and Noto are reported the results of VLBI and GPS_net analysis. We obtained a GPS_pp solution for Padova but the data span analysed with the GPS_net solution is to short to allow comparison.

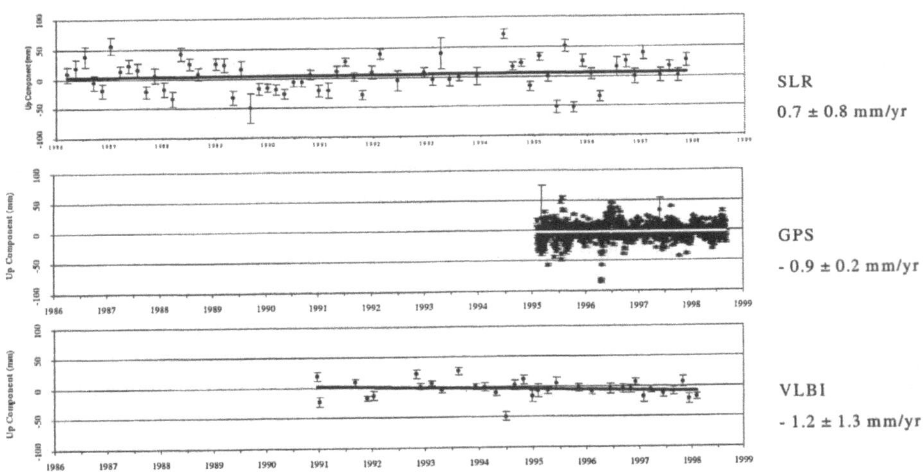

Fig. 2: Up velocities for Matera station obtained with SLR, GPS_net and VLBI analysis

These results, in good agreement among them, lead us to conclude that no vertical motions are going on at Matera. They are also in agreement with the value of the ITRF96 solution (– 0.7 ± 0.3 mm/yr.) that at 3σ level confirm the previous interpretation.

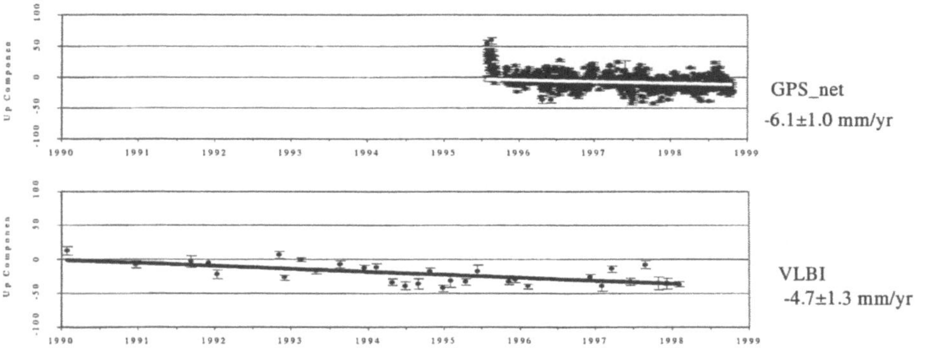

Fig. 3: Up velocities from GPS and VLBI data for Medicina station.

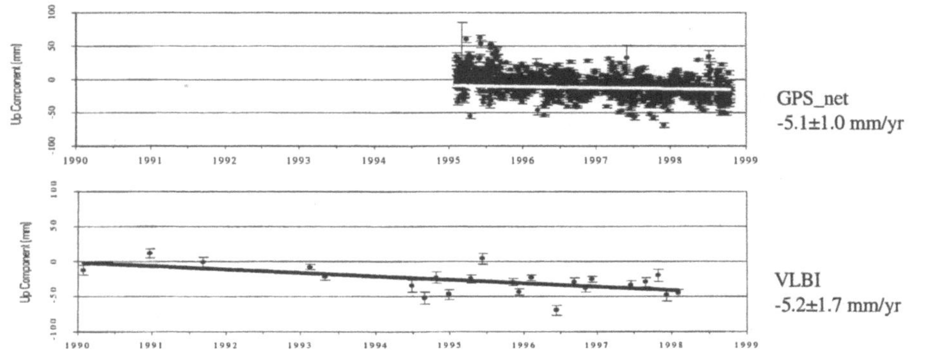

Fig. 4: Up velocities from GPS and VLBI data at Noto station

The results of the GPS_net and VLBI solutions show that Medicina is subject to subsidence phenomena with a velocity of ~ 4 mm/yr. The velocity estimated with GPS_pp solution is almost zero. These differences could be ascribed to a different realisation of the reference system obtained with Precise Point Positioning technique with respect to the other analysis. Therefore, the results obtained could be not directly comparable.

For Noto the results of GPS_net and VLBI are in good agreement and show that this area is letting down with a velocity of ~ 5 mm/y

3 Conclusion and remarks

The main critical point in the vertical movement determination is in the realisation of the terrestrial reference frame: even in the case the same ITRF has been referenced to, the different sites selected as constraints affect the results. For instance, the choice of ITRF96 Wettzell position and velocity values, both in the case of GPS and VLBI solutions, brings very consistent height velocity results for collocated sites, but maybe 'dragged' down by the behaviour of Wettzell height, pointed as -2.3+/-0.1 mm/yr by ITRF96 (otherwise indicating, for the majority of sites of the European network, 0 height movement at the 3-s level).

This suspicion comes also from the results obtained from GPS point-positioning technique for Medicina: height shows 0 movement at the 3-s level, but in this case the realisation of the ITRF96 is very different, having been used the JPL transformation parameters from free network to ITRF96, derived from a global network of sites. Probably, having geophysical information about heights at disposal may help in selecting the reliable sites to realise the terrestrial reference frame. The growing of collocated stations and their use in the realisation of reference frame in the analysis of the different space geodesy data are of fundamental importance: they surely bring a better consistency among the results.

Bibliography

DEVOTI R., FERMI M., LUCERI V., RUTIGLIANO P., SCIARRETTA C. BIANCO G.: *A Contribution in the estimation of tectonic motion in crucial areas: the CGS96 SLR Solution.* Tectonophysics, Proceedings of "7th WEGNER Meeting, Porto, June 1996" In Press.

FERMI M., FERRARO C., NARDI A., BIANCO G., VESPE F. (1997): *First Results of the Italian GPS Fiducial Network,* Proceedings of the 4th International Seminary "GPS In Central Europe", Penc 7-9 May 1997, Warsaw University of Technology N. 4 -27.

LANOTTE R., BIANCO G., FERMI M., RUTIGLIANO P., VERDONE G.R. (1996): *The CGS VLBI EUR96 Geodetic Solution,* Proceedings of the Eleventh Working Meeting VLBI for Geodesy and Astronometry.

ZUMBERGE J.F., HEFLIN M.B., JEFFERSON D.C., WATKINS M.M., WEBB F.H. (1997): *Precise Point positioning for the efficient and robust analysis of GPS data from large networks,* J.Geophys. Res., 102, 5005-5017.

Monitoring the Continental Reference Frame in South America

H. DREWES[1], K. KANIUTH[1], W. SEEMÜLLER[1], K. STUBER[1], H. TREMEL[1],
N. HERNANDEZ[2], M. HOYER[3], E. WILDERMANN[3], L. P. FORTES[4], K. D. PEREIRA[4]

Abstract

The South American Terrestrial Reference System (SIRGAS) is defined according to the IERS conventions. The reference frame is realized by a set of three-dimensional geocentric coordinates of about 60 stations derived from a continental GPS campaign in 1995 (epoch 1995.4). The time evolution of station coordinates is defined by the true station motions, in a linear approximation given by the station velocities. These velocities may either be determined from continuous monitoring or from repeated GPS campaigns. The actual continuous monitoring within the IGS Regional Network for South America includes 16 stations on the continent. The linear station velocities derived from two years of observations are presented. As an example of irregular deformations, results from repeated GPS campaigns in a boundary zone along the Caribbean and South American plates are shown. Linear station velocities are estimated as well as unsteady station motions as a consequence of an earthquake. The adequate consideration of these motions in a global geodetic observing system is an important issue.

Introduction

The South American Geocentric Reference System (Sistema de Referencia Geocéntrico para América del Sur, SIRGAS 1997, HOYER et al. 1998, KANIUTH et al. 1998a) is defined by the standards of the IERS Terrestrial Reference System (ITRS). It is realized by a reference frame of 58 sites on the South American mainland and some surrounding areas (see fig. 1). The coordinates of the reference stations were determined by the analysis of a GPS campaign in May/June 1995 in connection with ITRF94 and refer to the epoch 1995.4. The main problem of the maintenance of the reference frame is the time evolution of the station coordinates, i.e., the determination of station velocities which are caused by recent crustal movements (plate motions and deformations) and which form part of the realization of the reference system (DREWES 1998). This is the topic of the present paper.

State of the Project

The overall feature of recent crustal movements in South America is given by the velocities derived from permanently observing GPS stations within the IGS Regional Network of South America. The network comprises a total of 24 stations, 16 of which are located on the South American mainland, and 8 in the surrounding areas. All the observation data of these stations are weekly processed by the IGS Regional Network Associate Analysis Center for SIRGAS (RNAAC-SIR) at Deutsches Geodätisches Forschungsinstitut, München (DGFI), Germany (SEEMÜLLER and DREWES 1998). The results of the processing are forwarded as SINEX files (Software Independent Exchange Format) to the IGS global data centers. They are then combined with the global solutions of the IGS analysis centers to the polyhedron solutions. These solutions provide on a weekly basis the coordinates of all included stations in the actual IERS Terrestrial Reference Frame (ITRF) at the observation epoch.

Results

From the weekly polyhedron solutions computed by the IGS Global Network Associate Analysis Center (GNAAC) at MIT, Boston, USA (HERRING 1997), we computed linear station velocities of the individual South American stations. In table 1 we compare numerically the epoch 1995.4 coordinates of the SIRGAS stations with those of the RNAAC SIR and of ITRF96, respectively. The latter coordinates were extrapolated from their actual reference epoch (weekly or 1997.0, resp.) to epoch 1995.4 using the individual station velocities of the corresponding solution. As these are not well determined for RNAAC SIR stations, observing only since a short time, there occur some discrepancies due to the extrapolations (also to be seen in the large standard deviations). This fact demands for more efforts in extending the observations in time. In addition more permanent stations should be established.

Table 1 includes also the comparison of the RNAAC SIR velocities and the ITRF96 velocities. Figure 2 shows these velocities graphically. Interpreting the discrepancies one has to consider the short observation period (some stations less than one year) included in the RNAAC SIR solutions.

1 Hermann Drewes, Klaus Kaniuth, Wolfgang Seemüller, Klaus Stuber, Herbert Tremel: Deutsches Geodätisches Forschungsinstitut, Marstallplatz 8, 80539 München, Germany,Tel.: +49 - 89 - 230 31 -106, /-114, /-109, /-119, /-119, Fax: +49 - 89 - 240, e-mail: name@dgfi.badw.de

2 José Napoléon Hernandez: Servicio Autónomo de Geografía y Cartografía National, Av. Este 6, C.S.B., Esq. Camejo Edif. Camejo, Piso 2, Caracas, 1010, Venezuela, Tel.: +58 - 2 - 408 17 73, Fax: +58 - 2 - 542 03 74

3 Melvin Hoyer, Eugen Wildermann: Escuela de Ingeniería Geodésica, Facultad de Ingeniería, LUZ, Apartado Postal 10311, Maracaibo / Edo. Zulia, Venezuela, Tel.: +58 - 61 - 59 87 58, Fax: +58 - 61 - 51 22 14

4 Luiz Paulo Souto Fortes, Katia Duarte Pereira: Instituto Brasileiro de Geografia e Estatistica, Departamento de Geodesia, Av. Brasil 15671, Parada de Lucas, Rio de Janeiro CEP 21241-051, Brasil, Tel.: +55 - 21 - 391 82 17, Fax: +55 - 21 - 481 27 47

Regional Networks

In addition to the linear station velocities determination, the maintenance of the reference frame requires the detection of episodic motions as caused, e.g., by earthquakes (REIGBER et al. 1997, RUEGG et al. 1996). There are several geodynamic projects going on in South American regions of high seismic risk, e.g., the Central and South America (CASA) project in the northern part of the continent (FREYMUELLER et al. 1993, KANIUTH et al. 1998b, KELLOGG and VEGA 1995), the Central Andes Project (CAP, KIENDRICK et al. 1999, LEFFLER et al. 1997, NORABUENA et al. 1998) in the central part, the South American Geodynamics Activities (SAGA, ANGERMANN et al. 1997) in the southern part. As an example we show in figure 3 the eastern part of the CASA network with linear station velocities derived from GPS campaigns in 1988, 1993, and 1996, and the displacements in some stations after the Cariaco July 09, 1997 earthquake. The coordinate changes are up to 15 cm and they have, of course, to be reduced from the reference epoch coordinates and from the linear velocities. This is a principal problem in the monitoring of a terrestrial reference frame.

Conclusion

The monitoring of the continental reference frame in South America has to include the linear station coordinate variations (velocities) as well as the episodic motions. This should preferably be done by permanent observation stations. If repeated measurement campaigns are used for the monitoring, it's problematic to separate linear variations (e.g., rigid plate motions) from episodic ones (e.g., caused by earthquakes). Therefore it is recommended to all countries to join with permanent observation stations the IGS Regional SIRGAS Network being regularly processed by the RNAAC SIR.

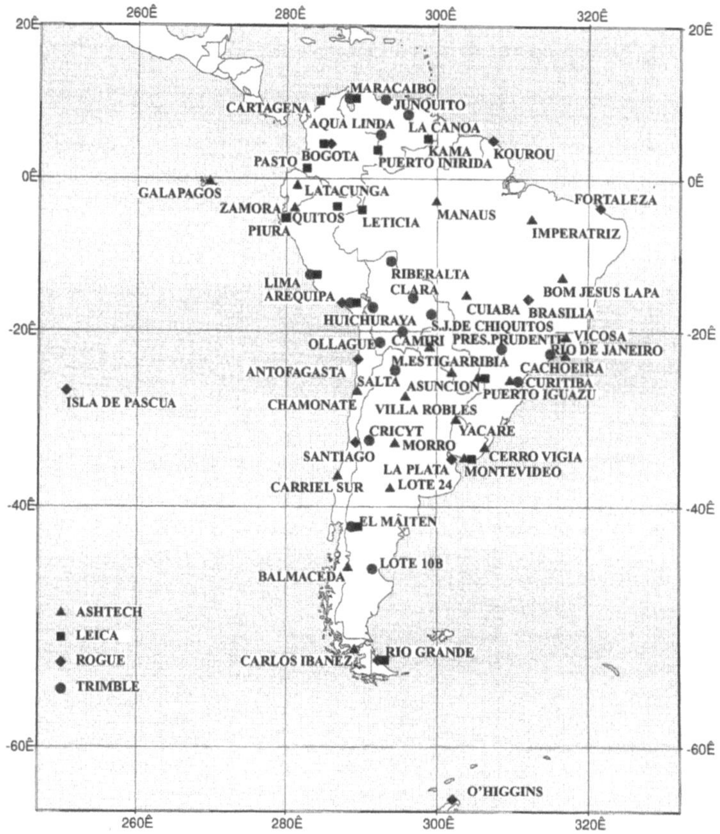

Fig. 1: Stations realizing the „Sistema de Referencia Geocéntrico para América del Sur (SIRGAS)" observed during the May/June 1995 GPS campaign.

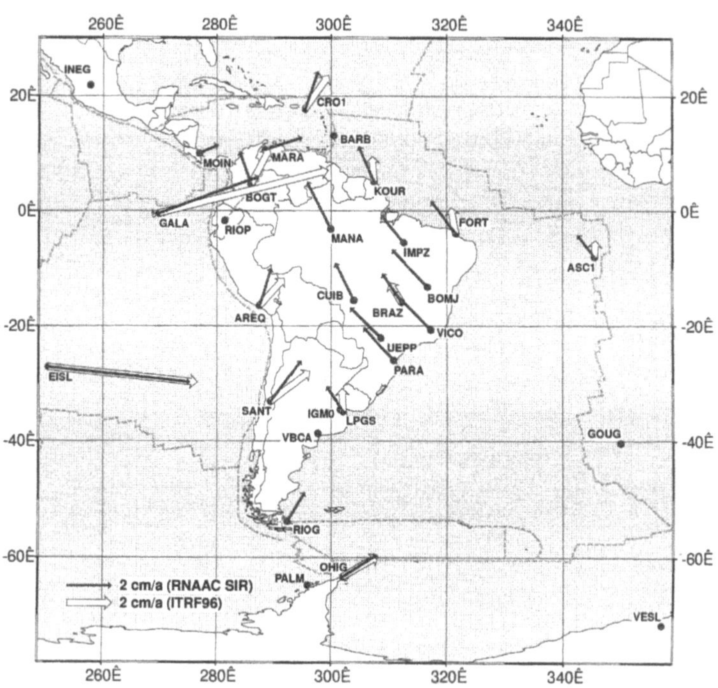

Fig. 2: Velocities of stations included in the RNAAC SIR network (status January 1999)

Tab. 1: Station coordinates from different solutions at epoch 1995.4 (extrapolated with velocities) [m]

Station	IGS RNAAC-SIR		ITRF96		SIRGAS		velocities [m/a]			
(I d)	X Y Z	sigma	X Y Z	sigma	X Y Z	sigma	RNAAC	sigma	ITRF96	sigma
Arequipa	1942826.727	.002	.713	.003	.727	.004	.0067	.0009	.0114	.0006
(AREQ)	-5804070.237	.003	.248	.004	.249	.003	-.0035	.0012	.0013	.0006
	-1796894.006	.002	.997	.003	.994	.004	.0161	.0007	.0123	.0006
Bogotá	1744399.136	.005	.100	.007	.102	.005	-.0131	.0020	.0076	.0032
(BOGT)	-6116037.848	.009	.824	.005	.852	.005	.0271	.0034	.0034	.0012
	512731.594	.003	.597	.004	.597	.005	.0124	.0012	.0161	.0015
Bom Jesus	4510195.854	.005			.859	.004	-.0025	.0019		
(BOMJ)	-4268322.243	.006			.294	.005	-.0194	.0022		
	-1453035.385	.003			.370	.004	.0147	.0011		
Brasilia	4115014.099	.002	.104	.009	.106	.004	-.0032	.0010	-.0089	.0049
(BRAZ)	-4550641.500	.003	.528	.010	.513	.004	-.0093	.0012	.0017	.0053
	-1741444.094	.002	.092	.005	.086	.004	.0124	.0007	.0115	.0023
St. Croix	2607771.192	.002	.181	.003			-.0056	.0008	.0079	.0012
(CRO1)	-5488076.836	.004	.843	.004			.0263	.0014	.0085	.0011
	1932767.663	.002	.675	.003			.0095	.0008	.0145	.0012
Cuiaba	3430711.420	.005			.398	.003	-.0024	.0017		
(CUIB)	-5099641.518	.007			.528	.004	-.0112	.0023		
	-1699433.016	.003			.996	.003	.0155	.0011		
Easter I	-1884951.850	.003	.862	.005	.849	.006	.0616	.0010	.0643	.0016
(EISL).	-5357595.845	.003	.837	.006	.833	.007	-.0186	.0013	-.0210	.0020
	-2892890.498	.002	.496	.004	.495	.006	-.0062	.0008	-.0070	.0014
Fortaleza	4985386.646	.003	.633	.004	.661	.005	-.0054	.0011	.0010	.0017
(FORT)	-3954998.555	.002	.573	.004	.585	.005	-.0102	.0009	-.0036	.0016
	-428426.522	.002	.513	.003	.515	.005	.0143	.0006	.0114	.0007
Galapagos	-33796.174	.011	.214	.346			.0457	.0035	.0772	.1968
(GALA)	-6377522.643	.014	.620	.697			-.0015	.0045	-.0240	.3954
	-82120.934	.004	.932	.080			.0166	.0015	.0191	.0453
Imperatriz	4289656.448	.010			.485	.004	.0075	.0031		
(IMPZ)	-4680884.870	.012			.940	.004	-.0229	.0037		
	-606347.402	.014			.402	.004	.0103	.0011		
Kourou	3839591.441	.017	.430	.006	.438	.005	-.0020	.0067	-.0010	.0025
(KOUR)	-5059567.525	.012	.540	.008	.532	.005	-.0084	.0046	.0001	.0029
	579956.892	.002	.906	.003	.906	.005	.0165	.0009	.0114	.0008
La Plata	2780102.984	.002	.969	.005	.979	.004	-.0048	.0008	.0034	.0022
(LPGS)	-4437418.844	.002	.850	.005	.845	.004	-.0060	.0008	-.0061	.0017
	-3629404.614	.002	.610	.004	.604	.004	.0121	.0008	.0080	.0015
Manaus	3179009.397	.008			.385	.004	-.0032	.0028		
(MANA)	-5518662.079	.016			.101	.004	-.0160	.0054		
	-344401.918	.003			.888	.004	.0207	.0009		
O'Higgins	1525872.444	.002	.449	.006	.458	.006	.0206	.0009	.0179	.0021
(OHIG)	-2432481.288	.003	.300	.006	.304	.006	-.0043	.0011	.0006	.0019
	-5676146.074	.003	.099	.005	.102	.006	-.0068	.0013	.0025	.0012
Curitiba	3763751.655	.004			.639	.004	-.0041	.0013		
(PARA)	-4365113.727	.005			.768	.004	-.0159	.0019		
	-2724404.763	.003			.755	.004	.0119	.0011		
Rio Grande	1429907.686	.004			.729	.005	.0163	.0014		
(RIOG)	-3495354.694	.006			.729	.005	-.0185	.0019		
	-5122698.648	.007			.707	.005	-.0096	.0025		
Santiago	1769693.315	.002	.304	.004	.314	.004	.0186	.0007	.0219	.0013
(SANT)	-5044574.119	.002	.134	.005	.146	.004	-.0107	.0010	-.0074	.0019
	-3468321.076	.002	.073	.004	.077	.003	.0114	.0006	.0070	.0013
Pres.Prud.	3687624.332	.004			.310	.003	-.0063	.0012		
(UEPP)	-4620818.539	.005			.571	.004	-.0140	.0017		
	-2386880.416	.003			.407	.004	.0119	.0011		
Vicosa	43732 83.338	.005			.325	.004	-.0090	.0017		
(VICO)	-4059638.979	.005			.008	.005	-.0140	.0018		
	-2246959.817	.003			.796	.004	.0174	.0012		

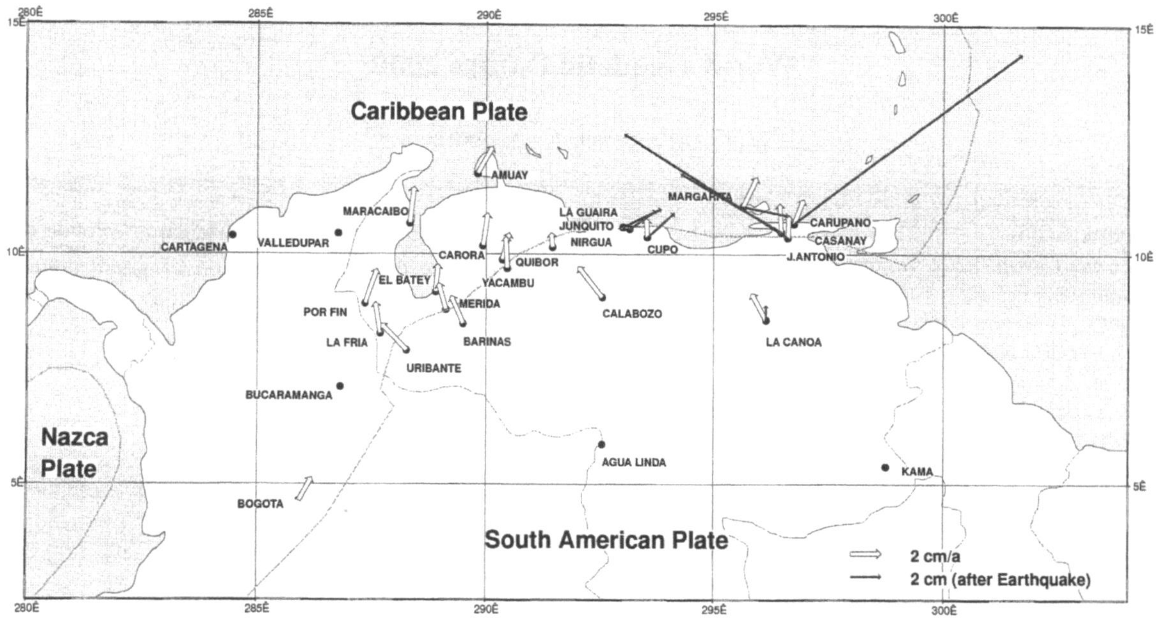

Fig. 3: Eastern part of the CASA network: Linear station velocities 1988...1996 and displacements after the Cariaco July 9, 1997 earthquake

Literature

ANGERMANN D., KLOTZ J., MICHEL G., REIGBER CH., REINKING J.: *Großräumige GPS-Netze zur Bestimmung der Krustenkinematik in Zentralasien und Südamerika.* DVW Schriftenreihe (28) 79-93, 1997.

DREWES H.: *Time evolution of the SIRGAS reference frame.* Springer, IAG Symposia, No. 118, 174-179, 1998.

FREYMUELLER J.T., KELLOGG J.N., VEGA V.: *Plate motions in the North Andes region.* J. Geophys. Res. (98) 21853-21863, 1993.

HERRING T.A.: MIT T2 Associate Analysis Center Report. IGS 1996 Annual Report, 255-267, IGS Central Bureau, JPL, Pasadena 1997.

HOYER M., ARCINIEGAS S., PEREIRA K., FAGARD H., MATURANA R., TORCHETTI R., DREWES H., KUMAR M., SEEBER G.: *The definition and realization of the reference system in the SIRGAS project.* Springer, IAG Symposia, No. 118, 167-173, 1998

KANIUTH K., DREWES H., STUBER K., TREMEL H., MOIRANO J.: *Results of the SIRGAS 95 GPS network processing at DGFI/I.* Springer, IAG Symposia, No. 118, 180-186, 1998.

KANIUTH K., DREWES H., STUBER K., TREMEL H., HERNANDEZ N., HOYER M., WILDERMANN E., KAHLE H.-G., STRAUB C.: *Crustal deformation along the Caribbean - South American plate boundary derived from the CASA GPS project.* Springer, IAG Symposia, No. 118, 417-422, 1998.

KELLOGG J.N., VEGA V.: *Tectonic development of Panama, Costa Rica and the Colombian Andes: Constraints from Global Positioning System geodetic studies and gravity.* Geol. Soc. Am. Special Paper 295, 1995.

KENDRICK E.C., BEVIS M., SMALLEY JR. R.F., CIFUENTES O., GALBAN F.: *Current rates of convergence across the Central Andes: Estimates from continuous GPS observations.* Geophys. Res. Lett. (26) 541-544, 1999.

LEFFLER L., STEIN S., MAO A., DIXON T., ELLIS M.A., OCOLA L., SACKS I.S.: *Constraints on present-day shortening rate across the central eastern Andes from GPS data.* Geophys. Res. Lett. (24) 1031-1034, 1997.

NORABUENA E., LEFFLER-GRIFFIN L., MAO A., DIXON T., STEIN S., SACKS I.S., OCOLA L., ELLIS M.: *Space geodetic observations of Nazca-South America convergence across the Central Andes.* Science (279) 358-362, 1998.

REIGBER CH., XIA Y., MICHEL G.W., KLOTZ J., ANGERMANN D.: *The Antofagasta 1995 earthquake: Crustal deformation pattern as observed by GPS and D-INSAR.* Proc. 3rd ERS Symp. on Space at the service of our environment (ESA SP-414) 507-513, 1997.

RUEGG J.C., and 14 others: *The M = 8.1 Antofagasta (North Chile) earthquake of July 30, 1995: First results from teleseismic and geodetic data.* Geophys. Res. Lett. (23) 917-920, 1996.

SEEMÜLLER W., DREWES H.: *The IGS regional associate analysis center for South America at DGFI/I.* Springer, IAG Symposia, No. 118, 211-215, 1998.

SIRGAS Final Report, Working Groups I and II. IBGE Rio de Janeiro 1997.

World Geodetic Datum 2000

E. W. GRAFAREND, A. A. ARDALAN[1]

Abstract

Based on the current best estimates of the fundamental geodetic parameters, i.e., W_0 (*Grafarend and Ardalan, Journal of Geodesy* 71 *(1997) 673-679)*, *GM* (*Ries et al., Geophys. Res. Letters* 19 *(1992) 529-531)*, J_2 (*Lemoine et al., GRAGEOMAR 1996, International Association of Geodesy, Symposia* 117, *(1996) 461-469)* and Ω (given in *Internal Communications of IAG/IUGG Special Commission 3, Darmstadt (1997))*, the *form parameters* of a *Somigliana-Pizetti level ellipsoid*, namely the semi-major axis a and semi-minor axis b (or equivalently the linear eccentricity $\varepsilon=\sqrt{a^2-b^2}$) are computed. There are *six* parameters namely the four fundamental geodetic parameters $\{W_0, GM, J_2, \Omega\}$ and the two form parameters $\{a, b\}$ *or* $\{a,\varepsilon\}$, which determine the ellipsoidal reference gravity field of Somigliana-*Pizetti* type constraint to two non-linear condition equations. Their iterative solution leads to best estimates $a=(6378136.572 \ \pm 0.053)m$, $b=(6356751.920 \ \pm 0.052)m$, $\varepsilon = (521853.580 \ \pm 0.013)m$ for the *tide-free geoid of reference* and $a=(6378136.602 \ \pm 0.053)m$, $b=(6356751.860 \ \pm 0.052)m$, $\varepsilon=(521854.674 \ \pm 0.015)m$ *for the zero-frequency tide geoid of reference.* The best estimates of the form parameters of a *Somigliana-Pizetti* level ellipsoid, $\{a, b\}$, differ significantly by -0.398m, -0.454m, respectively, from the data of the Geodetic Reference System 1980.

1 Spheroidal Coordinates, spheroidal gravity field

We define spheroidal coordinates $\{\lambda,\phi,u\}$ through their relations with Cartesian coordinates $\{x,y,z\}$

$$\begin{cases} x=\sqrt{u^2+\varepsilon^2}\cos\phi\cos\lambda, \\ y=\sqrt{u^2+\varepsilon^2}\cos\phi\sin\lambda\ , \\ z=u\sin\phi. \end{cases} \tag{1}$$

In equation (1), $\varepsilon=\sqrt{a^2-b^2}$, is the linear eccentricity. As it is shown in Fig. 1, in terms of spheroidal coordinates, a point in space can be identified as the intersection point of three coordinate surfaces, each of which depends on one of the parameters $\{\lambda,\phi,u\}$.

For a static, uniformly rotating earth, the gravity potential field $W(\lambda,\phi,u)$ can be additively decomposed into the gravitational potential field $U(\lambda,\phi,u)$ and centrifugal potential field $V(\lambda,\phi,u)$, namely

$$W(\lambda,\phi,u)=U(\lambda,\phi,u)+V(\lambda,\phi,u). \tag{2}$$

The spheroidal eigenvalue-eigenfunction solution of the 3-D Laplace differential equation of gravitational potential in the space outside the ellipsoid of reference $E^2_{a,b}$ reads

$$U(\lambda,\phi,u)=\sum_{n=0}^{\infty}\sum_{m=-n}^{+n} u_{nm}\frac{Q^*_{n|m|}(\frac{u}{\varepsilon})}{Q^*_{n|m|}(\frac{b}{\varepsilon})}e_{nm}(\lambda,\phi), \tag{3}$$

where, $e_{nm}(\lambda,\phi)$ is the surface spheroidal harmonic

$$e_{nm}(\lambda,\phi)=P^*_{n|m|}(\sin\phi)\begin{bmatrix}\cos m\lambda & \forall m\geq 0\\ \sin|m|\lambda & \forall m<0\end{bmatrix} \tag{4}$$

$P^*_{nm}(\sin\phi)$ is the normalised associated Legendre function of the first kind, and $Q^*_{nm}(u/\varepsilon)$ is the associated Legendre function of the second kind.

Representation of centrifugal potential in (i) Cartesian coordinates, (ii) spheroidal coordinates, and (iii) surface spheroidal harmonics is

$$\begin{aligned} V(\phi,u)&=\frac{1}{2}\Omega^2(x^2+y^2)\\ &=\frac{1}{2}\Omega^2(u^2+\varepsilon^2)\cos^2\phi\\ &=\frac{1}{3}\Omega^2(u^2+\varepsilon^2)e_{00}-\frac{1}{3\sqrt{5}}\Omega^2(u^2+\varepsilon^2)e_{20}\end{aligned} \tag{5}$$

where, Ω is the angular velocity of the earth.

2 The Somigliana-Pizetti gravity field of a level ellipsoid, the two constrains

If the spheroidal gravity potential field (2), (3), (5) are specified to level ellipsoid *(i)* $u=b$ and *(ii)* $w_0=W_0$ constant, where W_0 is the gravity potential of the geoid, then the eigenvalues of the three dimensional Laplace partial differential equation are restricted accordingly, to

[1] E. W. Grafarend, A. A. Ardalan: Department of Geodesy and Geoinformatics, Stuttgart University, Geschwister-Scholl-Str. 24 D, D-70174 Stuttgart, Germany; Tel: +49+711+1213390, Fax: +49+711+1213285

$$
\begin{aligned}
W(\lambda,\phi,b) &= (u_{00} + \frac{1}{3}\Omega^2 a^2)e_{00} \\
&+ (u_{20} - \frac{1}{3\sqrt{5}}\Omega^2 a^2)e_{20}(\lambda,\phi) \\
&+ \sum_{n=1}^{\infty} \sum_{\substack{m=-n \\ (n,m)\neq(2,0)}}^{+n} u_{nm} e_{nm}(\lambda,\phi) \\
&= w_0 = \text{const.}
\end{aligned} \tag{6}
$$

Since only the first term on the left-hand side of (6) is constant, the following relations must hold

$$u_{00} + \frac{1}{3}\Omega^2 a^2 = w_0, \tag{7}$$

$$u_{20} - \frac{1}{3\sqrt{5}}\Omega^2 a^2 = 0, \tag{8}$$

$$u_{nm} = 0 \ \ \forall n \geq 1, (n,m) \neq (2,0). \tag{9}$$

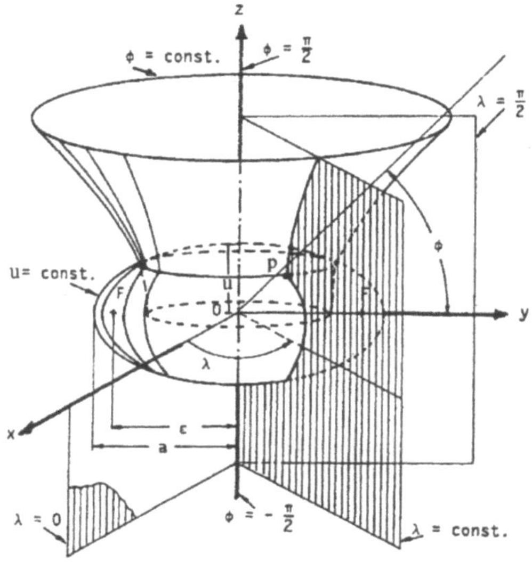

Fig. 1: Oblate spheroidal coordinates. The coordinate surfaces are spheroids (u=const), a half hyperboloids of one sheet (ϕ=const), and half planes (λ=const).

It can be shown that the spheroidal harmonic coefficients u_{00} and u_{20} are related to GM and J_2 as follows

$$u_{00} = \frac{GM}{\varepsilon}\cot^{-1}(\frac{b}{\varepsilon}), \tag{10}$$

$$
\begin{aligned}
u_{20} = \frac{\sqrt{5}}{4}\frac{GM}{\varepsilon}&\left[(3\frac{b^2}{\varepsilon^2}+1)\cot^{-1}(\frac{b}{\varepsilon}) - 3\frac{b}{\varepsilon}\right] \\
&\times\left[3\frac{a^2}{\varepsilon^2}J_2 + 1\right],
\end{aligned} \tag{11}
$$

where J_2 is the second zonal spherical harmonic coefficient. J_2 can be converted to second normalised zonal spherical harmonic coefficient J_2^* according to $J_2 = \sqrt{5}J_2^*$. Upon substituting (10) and (11) into (2), (3), and (5) we reach to two constrains of Somigliana-Pizetti gravity field which will be discussed in *section 3*.

3 Newton iteration and error propagation relations

The constrains, (7) and (8) are non-linear functional relations between fundamental geodetic parameters $\{W_0, GM, J_2, \Omega\} = \{y_1, y_2, y_3, y_4\}$, called pseudo-observations, and the two form parameters $\{a,b\} = \{x_1, x_2\}$.

$$f_1(x_1,x_2,y_1,y_2,y_3,y_4) =$$

$$\frac{y_2}{\sqrt{x_1^2 - x_2^2}}\cot^{-1}(\frac{x_2}{\sqrt{x_1^2 - x_2^2}}) + \frac{1}{3}y_4^2 x_1^2 - y_1 = 0 \tag{12}$$

$$f_2(x_1,x_2,y_1,y_2,y_3,y_4) =$$

$$
\begin{aligned}
&\frac{1}{4}\frac{y_2}{\sqrt{x_1^2 - x_2^2}}(15y_3\frac{x_1^2}{x_1^2 + x_2^2} + \sqrt{5})[(3\frac{x_2^2}{x_1^2 + x_2^2} \\
&+1)\cot^{-1}(\frac{x_2}{\sqrt{x_1^2 - x_2^2}}) - 3\frac{x_2}{\sqrt{x_1^2 - x_2^2}}] \\
&-\frac{1}{3\sqrt{5}}y_4^2 x_1^2 = 0.
\end{aligned} \tag{13}
$$

Equations (12) and (13) can be linearized and solved for $\{a,b\} = \{x_1, x_2\}$ in an iterative way, namely

$$\mathbf{x} - \mathbf{x}_0 = \Delta\mathbf{x} = (\mathbf{J}(\mathbf{x}_0))^{-1}(\mathbf{f} - \mathbf{f}_0) \tag{14}$$

$$
\begin{aligned}
&\mathbf{x} - \mathbf{x}_0 = \mathbf{J}^{-1}(\mathbf{f} - \mathbf{f}_0) \\
\Rightarrow\ &\mathbf{x}_1 = \mathbf{x}_0 + \mathbf{J}_0^{-1}(\mathbf{f} - \mathbf{f}_0) \\
\Rightarrow\ &\mathbf{x}_2 = \mathbf{x}_1 + \mathbf{J}_1^{-1}(\mathbf{f} - \mathbf{f}_1) \\
\Rightarrow\ &\cdots \Rightarrow \mathbf{x}_n = \mathbf{x}_{n-1}
\end{aligned} \tag{15}
$$

where

$$\mathbf{x} := \begin{bmatrix} x_1 \\ x_2 \end{bmatrix}, \tag{16}$$

$$\mathbf{J} := \begin{bmatrix} \frac{\partial f_1}{\partial x_1} & \frac{\partial f_1}{\partial x_2} \\ \frac{\partial f_2}{\partial x_1} & \frac{\partial f_2}{\partial x_2} \end{bmatrix}, \tag{17}$$

and

$$\mathbf{f} := \begin{bmatrix} f_1(x_1,x_2,y_1,y_2,y_3,y_4) \\ f_2(x_1,x_2,y_1,y_2,y_3,y_4) \end{bmatrix}. \tag{18}$$

The propagation of random errors of pseudo observations into derived parameters $\{x_1, x_2\}$ are derived according to

$$\mathbf{\Sigma_x} = \mathbf{J_x^{-1}J_y\Sigma_y(J_y)'(J_x^{-1})'}, \tag{19}$$

where, $\mathbf{\Sigma_y} = \mathrm{Diag}(\sigma_1^2\ \sigma_2^2\ \sigma_3^2\ \sigma_4^2)$ is covariance matrix of pseudo-observations, $\mathbf{\Sigma_x}$ is covariance matrix of parameters, and $\mathbf{J_y}$ is the *Jacobian* matrix of partial derivatives of **f** with respect to (y_1, y_2, y_3, y_4).

$$\mathbf{J_y} = \begin{bmatrix} \frac{\partial f_1}{\partial y_1} & \frac{\partial f_1}{\partial y_2} & \frac{\partial f_1}{\partial y_3} & \frac{\partial f_1}{\partial y_4} \\ \frac{\partial f_2}{\partial y_1} & \frac{\partial f_2}{\partial y_2} & \frac{\partial f_2}{\partial y_3} & \frac{\partial f_2}{\partial y_4} \end{bmatrix}. \tag{20}$$

4 Analysis of the ellipsoidal parameters of the Somigliana-Pizetti level ellipsoid

Table 1 refers to the data $\{W_0, GM, J_2, \Omega\}$ we applied within our computation. Our calculation of the two form parameters $\{a, b\}$ of the level ellipsoid of Somigliana-Pizetti type in zero-frequency-tide reference system are collected in Table 2, while Table 3 is devoted to the results in the tide-free reference system. In addition based upon error propagation, the variance-covariance matrix of $\{a, b\}$ with respect to the best estimates of $\{W_0, GM, J_2, \Omega\}$ is outlined in Table 4.

5 Conclusions

From Tables 1-4 the following conclusions can be made: The change of the *GM* value from TDT to SI units produces a millimetre variation in the form parameters $\{a, b\}$ of $E^2_{a,b}$ which is below the level of the calculated root-mean-square error values. J_2^* data from the geopotential models JGM3, EGM96 and OSU91A result in almost the same values of the form parameters $\{a, b\}$ of $E^2_{a,b}$. The root-mean-square errors of the form parameters $\{a, b\}$ of $E^2_{a,b}$ are directly proportional to the root-mean-square error of the gauge value of the potential W_0, the ruling parameter for the variance budget. Indeed, we have documented that the *tidal potential which is constant in time* in its indirect effect with a properly chosen secular Love number ("zero-frequency tidal reference system") flattens *the reference ellipsoid* as a level ellipsoid of Somigliana-Pizetti type. The abstract contains the final data set of the form parameters $\{a, b\}$ of level ellipsoid-of-revolution of Somigliana-*Pizetti* type (*tide-free* as well as zero-*frequency tide reference*).

Tab. 1: Best estimates of $\{W_0, GM, J_2, \Omega\}$.

Author	Parameter	Value	Sources
Burša et al., 1997a	W_0	$(62636855.72 \pm 0.5)\ m^2/s^2$	Satellite altimetric data, gauge stations values
Burša et al., 1997b	W_0	$(62636855.80 \pm 0.5)\ m^2/s^2$	Satellite altimetric data, gauge stations values
Grafarend and Ardalan, 1997	W_0	$(62636855.80 \pm 3.6)\ m^2/s^2$	Gauge station data, GPS, Baltic Sea Level Project
Ries et al., 1992	GM	$(398600.4418 \pm 0.0008)\ km^3/s^2$	Satellite Laser Ranging, satellite altimetric data; in SI units, including mass of the earth's atmosphere
Ries et al., 1992	GM	$(398600.4415 \pm 0.0008)\ km^3/s^2$	Satellite Laser Ranging, satellite altimetric data; in TDT units, including mass of the earth's atmosphere
Tapley et al., 1996	J_2^*	$-4.8416954845647 \times 10^{-4}$ $\pm 4.66 \times 10^{-11}$	JGM3 geopotential model; in zero frequency tide system
Lemoine et at., 1996	J_2^*	$-4.84165371736 \times 10^{-4}$ $\pm 3.56 \times 10^{-11}$	EGM96 geopotential model; in tide-free system
Rapp et al., 1991	J_2^*	$-4.84165532804 \times 10^{-4}$ $\pm 4.47 \times 10^{-11}$	OSU91A geopotential model; in tide-free system
Groten, 1997	Ω	$(7.292115 \times 10^{-5} \pm 10^{-12})$ rad/s	

Tab. 2: Form parameters $\{a, b\}$ of the Somigliana-Pizetti level ellipsoid in zero-frequency tide system.

a (m)	B (m)	$a\text{-}a_{GRS80}$ (m)	$b\text{-}b_{GRS80}$ (m)	W_0 (m^2/s^2)	GM (km^3/s^2)	J_2^*
6378136.610 ±0.053	6356751.868 ±0.052	-0.390	-0.446	62636855.72$^{(i)}$ ±0.5	398600.4418$^{(iv)}$ ±0.0008	$-4.8416954845647 \times 10^{-4(vi)}$ $\pm 4.66 \times 10^{-11}$
6378136.605 ±0.053	6356751.863 ±0.052	-0.395	-0.451	62636855.72$^{(i)}$ ±0.5	398600.4415$^{(v)}$ ±0.0008	$-4.8416954845647 \times 10^{-4(vi)}$ $\pm 4.66 \times 10^{-11}$
6378136.602 ±0.053	6356751.860 ±0.052	-0.398	-0.454	62636855.80$^{(ii)}$ ±0.5	398600.4418$^{(iv)}$ ±0.0008	$-4.8416954845647 \times 10^{-4(vi)}$ $\pm 4.66 \times 10^{-11}$
6378136.602 ±0.369	6356751.860 ±0.366	-0.398	-0.454	62636855.80$^{(iii)}$ ±3.6	398600.4418$^{(iv)}$ ±0.0008	$-4.8416954845647 \times 10^{-4(vi)}$ $\pm 4.66 \times 10^{-11}$

i: Burša et al., 1997a	iv: Ries et al., 1992 (in SI units)	$\Omega = 7.292115 \times 10^{-5} \pm 10^{-12} (rad/s)$
ii: Burša et al., 1997b	v: Ries et al., 1992 (in TDT units)	(Groten, 1997)
iii: Grafarend and Ardalan, 1997	vi: Tapley et al., 1996	

Tab. 3: Form parameters $\{a, b\}$ of the Somigliana-Pizetti level ellipsoid in tide-free system.

a (m)	B (m)	a-a_{GRS80} (m)	b-b_{GRS80} (m)	W_0 (m^2/s^2)	GM (km^3/s^2)	J_2^*
6378136.580	6356751.928	-0.420	-0.386	$62636855.72^{(i)}$	$398600.4418^{(iv)}$	$-4.84165371736\times10^{-4(vi)}$
±0.053	±0.052			±0.5	±0.0008	$\pm3.561\times10^{-11}$
6378136.575	6356751.923	-0.425	-0.391	$62636855.72^{(i)}$	$398600.4415^{(v)}$	$-4.84165371736\times10^{-4(vi)}$
±0.053	±0.052			±0.5	±0.0008	$\pm3.561\times10^{-11}$
6378136.572	6356751.920	-0.428	-0.394	$62636855.80^{(ii)}$	$398600.4418^{(iv)}$	$-4.84165371736\times10^{-4(vi)}$
±0.053	±0.052			±0.5	±0.0008	$\pm3.561\times10^{-11}$
6378136.572	6356751.920	-0.428	-0.394	$62636855.80^{(iii)}$	$398600.4418^{(iv)}$	$-4.84165371736\times10^{-4(vi)}$
±0.369	±0.366			±3.6	±0.0008	$\pm3.561\times10^{-11}$
6378136.581	6356751.926	-0.419	-0.388	$62636855.72^{(i)}$	$398600.4418^{(iv)}$	$-4.84165532804\times10^{-4(vii)}$
±0.053	±0.052			±0.5	±0.0008	$\pm4.472\times10^{-11}$
6378136.573	6356751.917	-0.427	-0.397	$62636855.80^{(iii)}$	$398600.4418^{(iv)}$	$-4.84165532804\times10^{-4(vii)}$
±0.369	±0.366			±3.6	±0.0008	$\pm4.472\times10^{-11}$

i: Burša et al., 1997a
ii: Burša et al., 1997b
iii: Grafarend and Ardalan, 1997
iv: Ries et al., 1992 (in SI units)
v: Ries et al., 1992 (in TDT units)
vi: Tapley et al., 1996
vii: Rapp et al., 1991
$\Omega = 7.292115\times10^{-5}\pm10^{-12}(rad/s)$ (Groten, 1997)

Tab. 4: Variance-covariance matrix of optimal $\{a, b\}$ of $E^2_{a,b}$, Somigliana-Pizetti type, via error propagation from variance data of $\{W_0, GM, J_2, \Omega\}$.

σ_{W_0} (m^2/s^2)	σ_{GM} km^3/s^2	σ_{J_2}	σ_Ω (rad/s)	$\hat{\sigma}_a$ (m)	$\hat{\sigma}_b$ (m)	$\hat{\sigma}_{ab}$	$\rho=\frac{\hat{\sigma}_{ab}}{\hat{\sigma}_a\hat{\sigma}_b}$
0.5	0.0008	4.5×10^{-11}	10^{-12}	0.053	0.052	2.738×10^{-3}	0.99984

References:

BURŠA M, KOUBA J, RADEJ K, TRUE SA, VATRT V, VJTÍŠKOVÁ M (1997a): *Monitoring geoidal potential on the basis of TOPEX/POSEIDON altimeter data and EGM96.* Paper presented at scientific assembly of IAG, Rio de Janeiro 1997.

BURŠA M, RADEJ K, ŠÍMA Z, TRUE SA, VATRT V (1997b): *Determination of the geopotential scale factor from TOPEX/POSEIDON satellite altimetery.* Stud Geoph Geod 14: 203-216

GRAFAREND EW, ARDALAN AA (1997): *W_0 - an estimate in the Finnish Height Datum N60, epoch 1993.4, from twenty-five GPS points of the Baltic Sea Level Project.* J Geod 71:673-679

GROTEN E (1997): *Current best estimates of the parameters of common relevance to astronomy, geodesy, and geodynamics.* Internal Communications of IAG/IUGG Special Commission 3, Darmstadt 1997

RAPP RH, WANG YM, PAVLIS NK (1991): *The Ohio State 1991 geopotential and sea surface topography harmonic coefficient models.* Report 410, Ohio State University Department of Geodetic Science and Surveying, Columbus, Ohio

RIES JC, EANES RJ, SHUM CK, WATKINS MM (1992): *Progress in the determination of the gravitational coefficient of the Earth.* Geophys Res Lett 19: 529-531

TAPLEY BD, WATKINS MM, RIES JC, DAVIS GW, EANES RJ, POOLE SR, RIM HJ, SCHUTZ BE, SHUM CK, NEREM RS, LERCH FJ, MARSHALL JA, KLOSKO SM, PAVLIS NK, WILLIAMSON RG (1996): *The JGM3 gravity model.* J Geophys Res 101: 28029-28049

Posters – Session B

Strengths and Weaknesses of Space Techniques

Validation of TOPEX/POSEIDON Measurements in the Southern Baltic Sea

G. LIEBSCH, R. DIETRICH[1]

Abstract

In the present investigation altimetric sea-level heights from TOPEX/POSEIDON are compared with measurements of 8 coastal tide gauges in the southern Baltic Sea. The zero points of the tide gauges were determined by GPS measurements and by precise spirit levellings. Therefore, the instantaneous sea level above the ellipsoid and above an equipotential surface of the earth gravity field can be estimated only by means of tide gauge measurements.

Because tide gauge and altimeter measurements refer not to the same location it is necessary to know the differences of the ellipsoidal sea-level heights. For this purpose

- different geoid models, which were fitted to the GPS-levelling points, were used;
- the instantaneous sea-level height above the geoid was extrapolated to the altimeter footprints by empirical functions.

The comparison includes data from more than 5 years (Cycle 1, September 1992 to Cycle 188, October 1997). The method described here can in principle be used for the calibration of altimeter measurements. Special attention is paid to the different error sources concerning the sea-level height determination by altimetry respectively tide gauge measurements.

1. Introduction

Satellite altimetry as well as tide gauges can be used to determine ellipsoidal sea level heights, taking into account the satellite orbit and various environmental corrections on the one hand and GPS measurements at the tide gauge stations on the other hand. Both measuring techniques record the sea-level signal independently from each other with approximately the same accuracy. Therefore, a comparison of both techniques can be used to look for possible systematic inconsistencies, which is of special importance for the determination of long-term sea-level changes.

For this purpose the southern Baltic Sea is a suitable area for the following reasons:

1. There is a small tidal signal in the Baltic Sea.
2. For the comparison with altimetric sea-surface heights the tide gauge stations should have long-term sea-level records. This allows a better understanding of the sea-level signals, a consistency check of the land motion signals and enables the estimation of long-term sea-level changes independently from the altimeter data (MITCHUM, 1997a). Corresponding investigations using sea-level records with an observation period of more than 100 years are available in the southern Baltic Sea (LIEBSCH, 1997a).
3. Globally distributed sites should be used for the calibration of altimeter measurements in order to monitor a possible geographical dependence of instrumental effects (NEREM, 1997). From that point of view the Baltic Sea is a good supplement to the existing calibration sites (MITCHUM, 1997a; CHRISTENSEN, 1994; MENARD, 1994; WHITE, 1994; MORRIS, 1994).

In the present paper further improvements of an earlier investigation (LIEBSCH and DIETRICH, 1997) concerning the comparison of TOPEX/POSEIDON measurements and tide gauge measurements in the southern Baltic Sea will be presented. An important progress could be made due to the availability of GPS measurements at the tide gauge stations as well as better geoid models.

2. Data

In this investigation altimetric sea-level heights from TOPEX and measurements of 8 permanent coastal tide gauge stations in the southern Baltic Sea (figure 1) were used. The zero points of all tide gauge stations were determined by GPS measurements and by spirit levellings. The accuracy of the levellings is better than 1 mm/km. That means, the height differences between the tide gauges have an uncertainty of less than 1 cm, taking into account the extension of the network of about 50 km by 50 km.

The GPS measurements at the tide gauges have been carried out during the measurements of the European United Vertical Network (EUVN) GPS campaign and lasted about two days. The resulting accuracy of the ellipsoidal heights is also below the 1 cm level. Therefore, the instantaneous sea level at the tide gauge stations above the ellipsoid as well as with respect to an equipotential surface of the earth gravity field can be estimated with an accuracy of about 2 cm. The sampling interval of the tide gauge measurements is 1 hour.

The ascending altimeter track No. 111 crosses the region of the tide gauges network. The comparison includes altimeter data (merged TOPEX/POSEIDON Geophysical Data Records, version C (AVISO 1996) of a latitude between 54.2° and 54.6° and a 5-years time span (1992-1997, up to cycle 188). Corrections due to the tropospheric

[1] Gunter Liebsch, Reinhard Dietrich: Technische Universität Dresden, Institut für Planetare Geodäsie, Mommsenstraße 13, D - 01062 Dresden, Germany; Tel. +49 - 0351 - 463 46 52, Fax: +49 - 0351 - 463 70 63, E-mail: dietrich@ipg.geo.tu-dresden.de

and ionospheric refraction, the sea state bias as well as the solid earth tides were applied. Because some measurements were corrupted a careful outlier detection was necessary.

3. Comparison of the sea-level heights

The tide gauge measurements and the altimetric sea surface heights do not refer to the same reference surface and not to the same location. The comparison will be done at the altimeter footprints. For this purpose one has to

- relate the altimetric sea surface height above the ellipsoid to the geoid
- extrapolate the instantaneous sea surface above the geoid from the tide gauge stations to the altimeter footprints.

The geoid undulations at the altimeter footprints were determined using two geoid models (EGG97 and NKG96). These models were fitted to about 20 GPS/levelling points. The resulting residuals are below the 1 cm level at the majority of the stations. In general both geoid models are in good agreement. The differences between the geoid models along the altimeter track are in the order of a few centimeter.

The sea-level records of the different tide gauges have a very high correlation (above 0.92). Therefore, the extrapolation of the instantaneous sea-level heights from the tide gauges to the altimeter footprints within a certain distance to the tide gauge network may be realized by a simple linear model for each altimeter cycle. A special handling of the tides is not necessary because of the very small tidal signal in the test area (LIEBSCH, 1997a; LIEBSCH, 1997b).

The resulting sea level heights have a correlation of 0.94. A comparison in the time domain shows, that the altimetrically derived sea-level heights have a slightly lower trend compared to sea-level heights extrapolated from the coastal tide gauge data (figure 2). The accuracy may be evaluated representing the differences of the two kinds of sea-level heights in the time domain as well as in the space domain (figure 3). The overall rms of the differences is 63 mm, whereas most of the values are within a ±20 cm band. The mean value of all differences (comparable with an absolute offset) is 12 ±2.0 mm. The differences have a temporal trend of -4 mm/a with a random error of ±1.4 mm/a (comparable with an altimeter drift). An additional uncertainty of 1... 2 mm/a has to be taken into account due to the sensitivity of the trend estimation to the outlier procedure. The sign of the trend indicates, that the observed altimetric sea-level heights tend to fall respectively the TOPEX altimeter is measuring longer with time. This is in general agreement to the result of other groups (NEREM 1997).

In the space domain the differences show some systematic variability in the order of few centimeter. These variations can be explained probably by errors of the geoid model. Therefore, one has to be aware of the limited accuracy of the absolute offset.

4. Interpretation

The trend difference shown in figure 3 may be caused by several effects. On the one hand the tide gauges measure relative sea-level changes, that means sea-level changes with respect to a local benchmark. Therefore, the sea-level trend may be influenced by recent crustal movements at the tide gauge stations. The monitoring of these crustal movements may be realized by permanent GPS measurements.

On the other hand altimetrically derived sea-level trends depend on the stability of the internal oscillator, but also on systematic effects in the satellite orbit and the environmental corrections. Figure 4 illustrates the differences of the TOPEX/POSEIDON ellipsoidal orbit heights from NASA and CNES for all measurements used in the present study. The orbits show a very good agreement in the order of ± 2 cm. Concerning the long-term behaviour, the orbit differences have a trend of about 2 mm/a, which may be caused by a different datum realization. Similar effects, which may influence the mean value and seasonal variations can be obtained comparing the dual frequency ionospheric corrections from TOPEX and DORIS as well as the tropospheric corrections of the on board radiometer and a weather model.

5. Conclusions and Outlook

The comparison of altimetric sea-level heights and sea-level heights from coastal tide gauge data show a random noise level of about ± 6 cm. This value represents the accuracy for a single difference of the sea-level heights. As the main contributions to the overall error budget can be identified:

- the altimetric sea surface heights (FU et al., 1994) 4...5 cm
- geoid models including the fit to the GPS/levelling points 2...3 cm
- tide gauge measurements including the extrapolation of the instantaneous sea-level heights 3...4 cm

A possible long-term drift of the altimeter bias can be estimated with a standard deviation of about 1.4 mm/a for the 5 year interval. The estimated drift of -4 mm/a may partly be caused by recent crustal movements at the tide gauge stations and the datum realization of the TOPEX/POSEIDON satellite orbit. The absolute offset between the altimetric sea-level heights and the sea-level heights from coastal tide gauge stations was determined by 12 mm ± 2 mm.

The results in the present investigation are limited due to the sensitivity of the trend estimation to the outlier detection problem. The absolute offset may be influenced by possible errors of the geoid models. Therefore, further improvements of the method may be achieved by a better configuration of the tide gauge network and an improved geoid modelling.

Acknowledgment

In this study the recent geoid models EGG97 and NKG96 were used. We thank W. TORGE, H. DENKER and R. FORSBERG for their kind support.

References

AVISO, (1996). *AVISO} user handbook: merged TOPEX/ POSEIDON products.* AVI-NT-02-101-CN, 3. edition.

CHRISTENSEN E., HAINES B., KEIHM S., MORRIS C., NORMAN R., PURCELL G., WILLIAMS B., WILSON B., BORN G., PARKE M., GILL S., SHUM C., TAPLEY B., KOLENKIEWICZ R., NEREM R. (1994).: *Calibration of TOPEX/POSEIDON at platform harvest.* JGR, 99(C12):24465-24485.

FU L., CHRISTENSEN E., YAMARONE C., LEFEBVRE M., MENARD Y., DORRER M., ESCUDIER P. (1994).: *TOPEX/ POSEIDON mission overview.* JGR, 99(C12):24369-24381.

LIEBSCH G. (1997).: *Aufbereitung und Nutzung von Pegelmessungen für geodätische und geodynamische Zielstellungen.* Deutsche Geodätische Kommission, Reihe C, Heft Nr. 485.

LIEBSCH G., DIETRICH R. (1997).: *Sea level variations measured by topex/poseidon and by tide gauges: a validation study in the southern baltic sea.* In Forsberg, R., Feissel, M., Dietrich, R., editors, IAG Symposium 119: Geodesy on the Move, pages 138-143. Springer Verlag, Berlin Heidelberg.

MENARD Y., JEANSOU E., VINCENT P. (1994).: *Calibration of the TOPEX/POSEIDON altimeters at Lampedusa: Additional results at Harvest.* JGR, 99(C12):24487-24504.

MITCHUM G. (1997).: *A tide gauge network for altimeter calibration.* In Neilan, R., Van~Scoy, P., Woodworth, P., editors, Workshop on methods for monitoring sea level, pages 45-55. IGS Central Bureau, JPL, Pasadena, California, USA.

MORRIS C., Gill S. (1994).: *Evaluation of the TOPEX/ POSEIDON altimeter system over the Great Lakes.* JGR, 99(C12):24527 - 24539.

NEREM R. (1997).: *The contribution of satellite altimetry to measurering long-trem sea-level changes.* In Neilan, R., Van~Scoy, P., Woodworth, P., editors, Workshop on methods for monitoring sea level, pages 25-37. IGS Central Bureau, JPL, Pasadena, California, USA.

WHITE N., COLEMAN R., CURCH J., MORGAN P., WALKER S. (1994).: *A southern hemisphere verification for the TOPEX/POSEIDON satellite altimeter mission.* JGR, 99(C12):24505-24516.

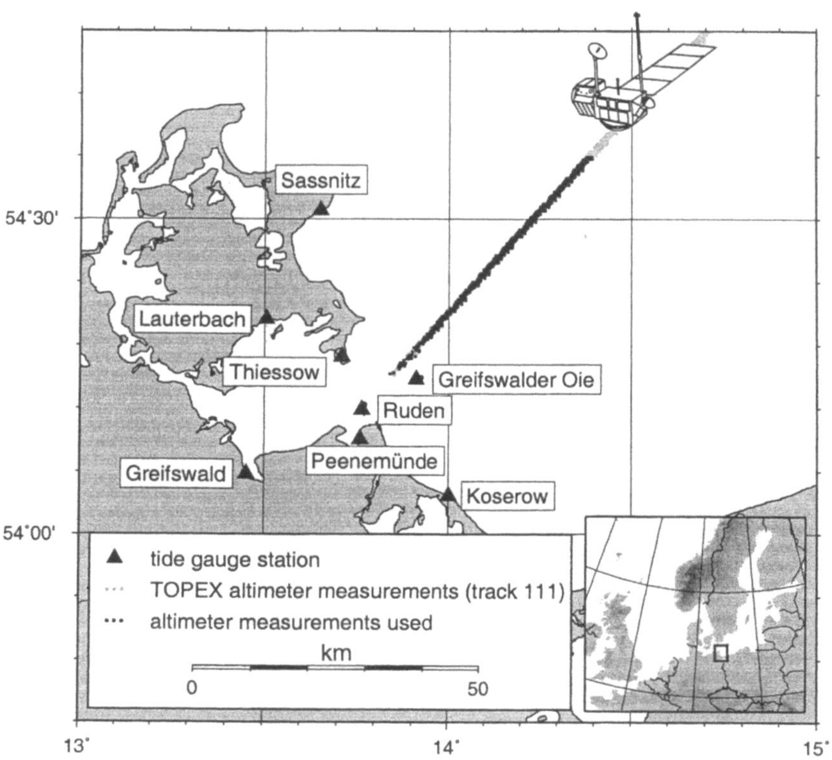

Fig. 1: Test Area in the Southern Baltic Sea for the comparison of sea-level heights derived from TOPEX measurements and from tide gauge measurements

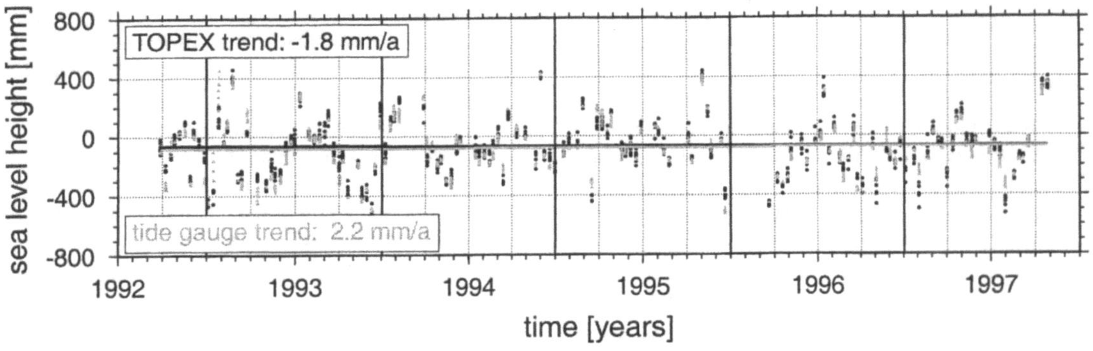

Fig. 2: Comparison of sea-level heights derived from TOPEX measurements (black) and from tide gauge measurements (grey)

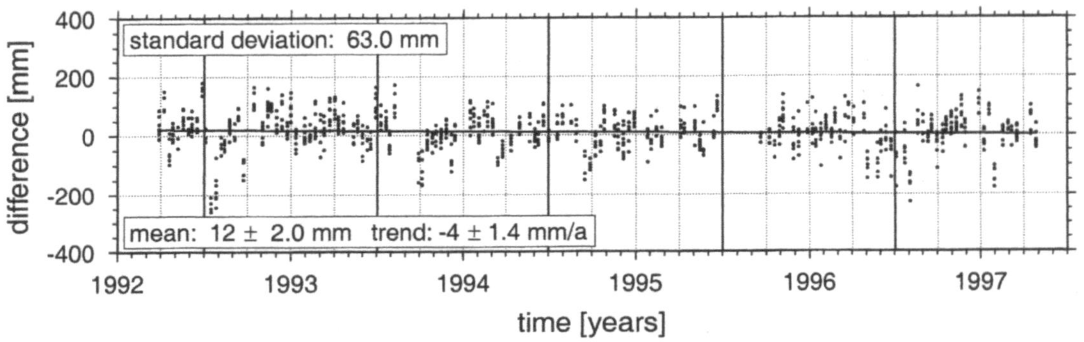

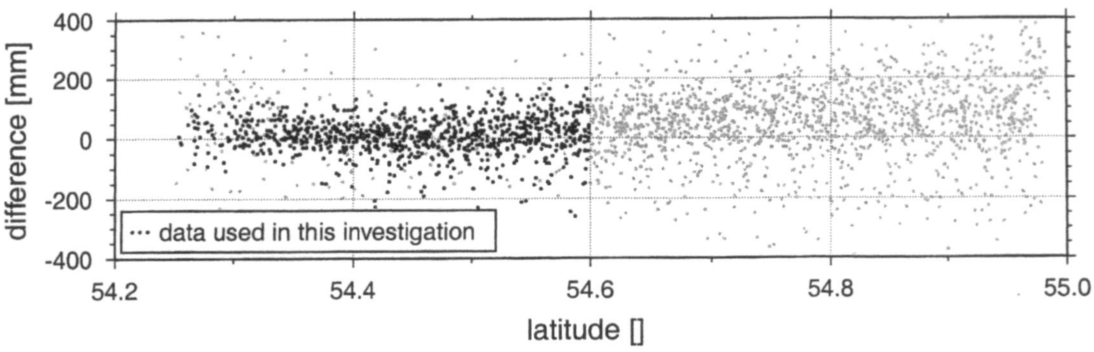

Fig. 3: Differences of sea-level heights derived from TOPEX measurements and tide gauge measurements as a function of time (top) and latitude (bottom) sign: sea-level heights from TOPEX measurements minus sea-level heights from tide gauge measurements

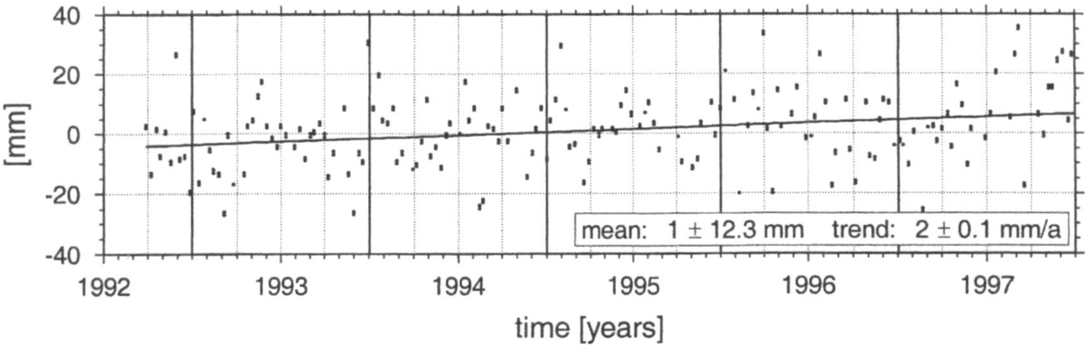

Fig. 4: Differences of the TOPEX/POSEIDON ellipsoidal orbit heights (NASA minus CNES) of the measurements used in this investigation

DORIS : Contribution of an Orbit Determination System to Accurate and Long Term Altimetry from Space

M. COSTES[1], P. VINCENT[2]

Abstract

The DORIS system (Doppler Orbitography and Radiopositioning Integrated by Satellite) was developed by CNES (Centre National d'Etudes Spatiales), IGN (Institut Géographique National) and GRGS (Groupe de Recherche en Géodésie Spatiale) to meet science and operational user requirements in very precise orbit determination and high accuracy location of ground beacons for point positioning.

First of all, an overview of the overall DORIS system is performed, including the space and ground segments. Emphasis is made on the worldwide homogeneous beacon network and on the upgrade of a multi-mission ground segment to control the DORIS instruments on-board various satellites and to process the data.

The second objective is to introduce and illustrate several important science questions dealing with the role of DORIS in a unified reference system for space oceanography. Indeed, DORIS is the kernel of most altimeter systems through orbitography (TOPEX/POSEIDON, JASON, ENVISAT). Because oceanographers have proved that very tiny signals such as the secular sea-level drift (of the order of 1.5 mm/year) may be detected thanks to altimetry, it is now clear that the future challenge will be to decorrelate the contribution of the long term variations of the reference system from the "true" ocean signal.

Then, problems of stability of the reference system on a multi-year time scale, as well as problems of comparing tide gauge measurements to altimetry are discussed.

1. The DORIS system

The DORIS system was designed and optimized to provide high precision orbit determination and beacon positioning. It was developed in the frame of the TOPEX/POSEIDON oceanographic altimetry mission. It is operational since 1990, when SPOT2 satellite was launched with the first DORIS receiver on-board. It provides a wide variety of services to different users.

DORIS is an uplink radio system based on the Doppler principle. It measures relative velocity between the orbiting satellite and a dense, permanent network of orbit determination beacons.

The core of the system is the beacon network distributed homogeneously over the earth. The bi-frequency signals at 400 MHz and 2 GHz emitted by the beacons are used by the receivers on-board the various satellites to perform Doppler measurements. The DORIS data can then be used:

- on-board, to perform autonomous orbit determination,
- on ground, after storage and downlink to perform:
  - precise orbit determination,
  - beacon precise positioning,
  - atmosphere modelling (troposphere, ionosphere),
  - earth gravity modelling.

1.1 The beacon network

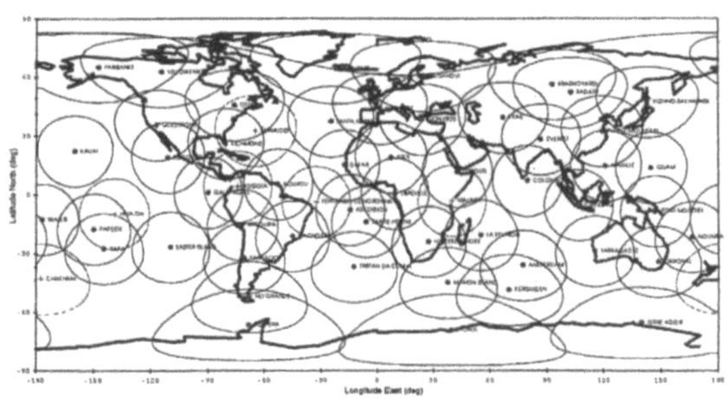

Fig.1 : DORIS permanent beacon network

The DORIS permanent network includes 52 beacons hosted by institutes from more than 30 different countries.

This network is managed by the DORIS Beacon Installation and Maintenance Service (SIMB), an IGN department. More than 20 beacons are colocated with other precise positioning techniques to allow intercalibration.

Each site is equipped with a beacon package including:

- a bi-frequency 400 MHz and 2 GHz transmitter, including a USO (Ultra Stable Oscillator),
- an omni-directional bi-frequency antenna,
- a battery pack, to provide autonomy versus supply,
- a meteorological package providing temperature, pressure and humidity measurements, used for tropospheric correction.

The beacons emit a narrow band ultra stable signal plus auxiliary data:

[1] Murielle Costes: DORIS System Manager,CNES, 18 Avenue Edouard Berlin, 31401 Toulouse Cedex 4, France, Tel.: +33 - 5 - 61 28 22 60, Fax: +33 - 5 - 61 28 25 95 email: murielle.costes@cnes.fr

[2] Patrick Vincent: Altimetry Manager,CNES, 18 Avenue Edouard Berlin, 31401 Toulouse Cedex 4, France, Tel.: +33 - 5 - 61 27 44 69, Fax: +33 - 5 - 61 28 25 95, email: patrick.vincent@cnes.fr

- beacon identifier,
- housekeeping data,
- meteorological data,
- time tagging reference data.

Presently, two beacons, one in Toulouse, the other in Kourou, named Master Beacons, are connected to the control center to allow data upload to the on-board package and are linked to atomic clock to allow synchronization of the DORIS system with international reference time.

1.2 On-board package

The DORIS on-board package includes:

- a receiver performing Doppler measurements and receiving auxiliary data from the beacons,
- a bi-frequency omni-directional antenna,
- a USO (Ultra Stable Oscillator).

There may be an unlimited number of DORIS on-board packages.

Presently, there are different generations of on-board instrumentation:

- 1 channel receivers (on-board SPOT2, T/P, SPOT3)
- 1 channel receiver including DIODE (real time on-board orbit determination software) presently working on-board SPOT4.
- 2 channel receivers (as a baseline, DIODE is included in the DORIS software; on-board ENVISAT to be launched in 2000).
- 2 channel miniaturized receivers are under development and will be on-board JASON and SPOT5. A mono-frequency receiver could be developed from this light receiver to perform the navigation of satellite constellations.

The bi-frequency receivers are used for high accuracy applications; they allow ionospheric correction. The bi-channels receivers allow to track 2 beacons simultaneously. Starting with SPOT4, all DORIS receivers include the real time on-board orbit determination software DIODE.

For lower accuracy applications, a monofrequency mono-channel smaller package of 3 kg (patch antenna and clock included) could be developed from the 2 channels miniaturized receiver.

1.3 Control and processing center

The control and processing center is located in Toulouse, France. It is in charge of:

- beacon network monitoring,
- on-board package monitoring and programming,
- science telemetry acquisition and pre-processing,
- technological archive,
- precise orbit determination,
- beacon positioning.

This center is included in the SSALTO (Orbitography and Altimetry Multi-mission Center), that is a facility devoted to DORIS system and to altimetry mission control and/or processing.

User archives collect measurements from the control and processing center, archive and distribute them. Since DORIS was included in IERS, two user archives are in operation, one in Europe managed by IGN, the other IN USA under NASA/CDDIS management.

1.4 Performances

Following is a summary of system performances.

characte-ristics	first generation (lG)	second generation (2G)	second generation miniaturized (2GM)	second generation (2GMM2)
applications	SPOT2; SPOT3; TOPEX-POSEIDON; SPOT4 (DIODE)	ENVISAT	JASON; SPOT5	Navigator
perfor-mances				
precise orbit	< 3 cm/ radial	< 3 cm/ radial	< 3 cm/ radial	
real time orbit	5 m/3 axes (SPOT4)	1 m/3 axes	30 cm/ radial; 1 m/ other axes	qq m/3 axes
time transfer	3 μs	3 μs	3 μs	10 μs
mass	18 kg	11 kg	5,6 kg	1,5 kg
volume	20 l	15 l	5,6 l	1,5 l
supply	13 W	18 W	17 W	10 W

Positioning performances:

data gathering duration	accuracy (1 satellite)	accuracy (2 satellites)
1 hour	1 m	0.5 m
1 day	20 cm	15 cm
5 days	10 cm	7 cm
26 day	3 cm	1-2 cm

2. DORIS: a geodetic contributor to climate change studies

Within the climate change research framework, the rate of present-day global sea-level change is a crucial topic. Traditional measurement techniques rely on tide gauges, a small number of which has been measuring sea level changes on an effective manner for a few decades. From the analysis of edited tide gauge collections of measurements, several authors already issued a 1.5 mm/yr figure for the rate of sea-level change. Telling if such a figure is fully representative of the global Earth is difficult as the geographic distribution of reliable tide gauges is not regular and somewhat sparse. Also, separating the contributions of all potential causes: thermal expansion of the oceans, water transfer from continental glaciers, ice sheets, others, is still more challenging but will of course not be the subject of the following sections. More important in the frame of geodesy, is to concentrate on the way classical tide gauges and its space symmetric that is space altimetry have to be used for a proper estimation of global sea-level change. In such a framework, we will demonstrate: 1. the importance of a very good knowledge of a terrestrial reference system, 2. the growing importance of accurately monitoring such a terrestrial reference frame along time, 3. DORIS is well suited to be a predominant contributor to such a reference frame.

2.1 TOPEX/POSEIDON: a unique and still perfectible contributor to sea-level change estimate

An important consideration in the success of the TOPEX/POSEIDON (T/P) mission is the ability to define and maintain the terrestrial reference frame throughout the duration of the mission. This was indeed extremely important as one of the objectives of the mission was explicitly to help in performing decadal studies of ocean surface change.

The best example is provided by the ability of T/P to allow the first measurement of interannual variations of the global mean sea level with a precision better than 1 mm/yr: NEREM et al., 1997, CAZENAVE et al., 1998, for instance). Using 4.5 years of T/P data, Cazenave et al. report a 1.2 +/- 0.15 mm/yr global mean sea-level rise. Nevertheless, the formal error estimate accompanying the 1.2 mm/yr figure may be optimistic, since it is now known that the T/P microwave radiometer whose data are used to correct altimeter measurements for wet tropospheric effect is probably drifting probably due to ageing. And, in addition such a kind of error source, other sources have been neglected up to now and should surely be taken into account to reach the few tenths of mm accuracy that is desired to make a large step in our knowledge of mean sea level change. Among the potential error sources are those issued from geodetic aspects of the altimetry system.

2.2 Sea-level change, orbits and reference frame

One remarkable feature of altimetry is that the observation is performed at a global scale in a center of mass-fixed reference frame. The positions of the stations tracking the satellite define the orbit reference frame, and consequently, the ability to precisely determine their locations within a coordinate system whose origin is located at the Earth's center of mass is of considerable importance. It is widely agreed that the international network of satellite laser ranging systems is an important contributor to T/P reference frame definition. The permanent orbitography network of DORIS beacons is the other major contributor. Indeed, since T/P launch, the knowledge of the coordinates of the ground beacons has greatly improved, which allowed to introduce the DORIS system in the IERS reference frame computations, and then to reprocess T/P orbits in a consolidated geodetic reference frame.

The T/P POD (Precise Orbit Determination) efforts have been continuous and exceptional since launch and have helped to reach the better than 3 cm accuracy rms that is generally agreed. DORIS is probably the major contributor to such a low figure. Indeed, it should be first mentioned that the weight of the DORIS system in the JGM-3 gravitational potential used in T/P orbit computations is very high. Also, besides the effort concerning gravitational forces, efforts on other non gravitational surface forces have also been done, which lead to the drastic reduction of the corresponding item of the orbit error budget. Another error source should now be looked at very cautiously: it lies in the effect of horizontal and vertical motions of the DORIS stations used to compute the T/P orbits. The best example of the problem (at least for horizontal motion) is brought through the present T/P data sets. Indeed, the present T/P data sets as distributed by AVISO include 2 different estimates of the satellite precise orbit: one is issued from NASA/GSFC computational process, the other being generated by the french Service d'Orbitographie DORIS. Since 1996, the 2 entities decided to apply 2 different strategies in terms of reference frame: the former was taking into account horizontal velocities to allow station positions evolving with time, while the latter chose to work with a fixed reference frame during a given period (with of course the open possibility to change the reference frame at dates to be decided in accordance with oceanographic requirements, when results prove the necessity). An explanation for working with a fixed frame was that, beginning of 1996, one of the team considered that horizontal velocities of the DORIS beacons were not sufficiently well known to take their preliminary estimates into account. Recently, using the two orbits to derive estimates of sea-level change during the El Nino period extending over 1997-1998 showed a 5 mm/yr discrepancy. And it was evidenced that the essential part of the discrepancy was effectively due to differences in reference frame strategies. Even if such reference frame effects do not affect most of the ocean studies performed by T/P investigators, it has thus appeared important, for long term sea-level change studies that involve tiny figures of the order of 1 mm/yr, to manage the evolution of the reference frame through time at least by taking into account station horizontal velocities in the orbit computation process.

Notice that the impact of vertical velocities of tracking stations is not yet so well estimated. In addition to absolute positions and horizontal velocities of DORIS beacons, CRETAUX et al. (1997) also computed the vertical velocities of DORIS stations, based on the analysis of 4 years of DORIS data from 3 satellites. It appears that uncertainties are still at the level of 1 mm/yr. Use of the 2nd generation of DORIS instruments onboard SPOT4, Jason and ENVISAT will allow to continue the DORIS station motion analysis, with an even better accuracy, since the instrumental noise of the DORIS receiver will be of the 0.1 mm/yr order of magnitude to be compared with the 0.3 mm/yr noise for T/P like receivers. T/P orbits taking into account both the horizontal and the vertical motions of the DORIS beacons should now be computed to look for a potential trend between sea-levels computed in the 2 cases (A. CAZENAVE, personal communication); then the rate of sea-level change from T/P may be updated as well as the associated error budget.

Last, one other subject of study should also be the stability of the reference frame in which sea-levels are computed. It is known that geocenter variations are related to the adopted reference system, and in particular to its origin. Sea-surface heights are related to the Earth's center of mass since the satellite orbit is defined in an inertial reference frame which center is the center of mass of the Earth. In practice, tracking data involved in the orbit computation are collected by stations that are distributed at the Earth surface, which define an earth fixed reference frame which origin is the center of figure. Hence motions of the center of figure relative to the center of mass should be taken into account in the global mean sea-level problem.

2.3 Tide gauges are the required complement to space altimetry

It has to be emphasized that the tide gauge / altimetry comparison process that has been initiated with great success by G. MITCHUM (1997) is now generally accepted as a vital contribution to the assessment of the quality of any altimetry system. Fixing tide gauges geodetically is now recognized of very high priority. Indeed, it is well known that in addition to the irregular geographical distribution of high quality tide gauges over the world, another difficulty in getting a reliable estimate of sea-level change from such a technique is to accurately separate the absolute part of sea-level change from vertical land movements such as post glacial rebound, seismic and tectonic deformations. Preliminary results by CAZENAVE et al. (pers. comm.) using data from tide gauge equipped with DORIS beacons showed extremely convincing features in terms of decorrelation of land movements in sea level measurements, and a very fruitful comparison to altimetry analysis. To the knowledge of the authors, such a work of comparing tide gauge data equipped with DORIS beacons to altimeter measurements will be continued in the frame of Jason and ENVISAT missions by A. Cazenave's team. And the results should be better and better with series of more than 10 year long coupled tide gauge/DORIS stations.

3. Conclusion

From its probatory status onboard SPOT 2, the DORIS system has evolved to become a major contributor to many geodetic applications. Continuous performance improvements of the onboard and ground instruments, improvement of the onboard (DIODE real time navigation system) and ground system capabilities, intend to install the DORIS system as a major permanent actor in geodesy: In particular, it is now quite mature to be a key element of sea-level studies, which is very high performance demanding in the frame of altimetry missions. And the fact is that the DORIS system is now the core of many altimeter missions: T/P, but also the coming Jason and ENVISAT missions.

The T/P experience has strongly proved that, for interdacadal ocean surface change studies, there is a crucial importance in knowing and monitoring very well the terrestrial reference system in which orbit computation is performed. The continuous maintenance and development of the DORIS system, as well as the exploitation of the data by several groups worldwide, will make such a task feasible. Ultimately as far as sea-level studies are concerned, DORIS will be a quite well suited companion to tide gauges to allow investigators continue their sea-level work by comparing data from tide gauge instruments and future altimetry missions when both systems have their stations in the same reference frame.

References

- CAZENAVE A. et al., 1998, *Global mean sea level changes observed by Topex-Poseidon and ERS1*, Physics and Chemistry of the Earth, In Press
- CRÉTAUX J.F. et al, 1997, *Global tectonic motions from the DORIS space system*, J. Geophys. Res.
- MITCHUM G., 1997, *Monitoring the stability of satellite altimeters with tide gauges*, J. ATMOS. And Oceanic Tech..
- NEREM S. et al., 1997, *Improved determination of global mean sea-level variations using TOPEX/POSEIDON data*, Geophys. Res. Lett. , 24, 1331-1334.

Acknowledgments

A. CAZENAVE (CNES/LEGOS) and her team are acknowledged for all fruitful discussions and interactions we had about sea-level issues from tide gauges. J.P. BERTHIAS (CNES) and J. DORANDEU (CLS) are also thanked for their contribution in TOPEX/POSEIDON data analysis especially within the frame of the AVISO activities.

IGS Combinations of Polar Motion, Length of Day and Universal Time

Y. MIREAULT, J. KOUBA[1]

Abstract

Since March 1997, the International GPS Service (IGS) has officially introduced new Length Of Day (IGS LOD) and Universal Time (IGS UT) combination products based on a weighted average of available Analysis Centre's (AC) LOD solutions. Each AC LOD series is calibrated with respect to the IERS Bulletin A over a 21-day period. IGS UT is estimated by integrating the IGS combined LODs from a chosen IERS Bulletin A reference point (also called the anchor point). Several processing and combination strategies were tested and implemented in the Rapid (available 22 hours after the last observation) and the Final (available 10 days after the last observation) IGS LOD/UT combinations. IGS LOD and UT precision is at or below the 100 and 30 µsec levels respectively for both the Rapid and Final combinations when compared to the IERS Bulletin A. The IGS has also been combining Polar Motion (PM) x and y since May 1995 and PM rates since June 1996. Steady improvements were made throughout the years resulting in an IGS PM precision, when compared to the IERS Bulletin A, at or below the 0.3 and 0.1 mas levels for the Rapid and Final combinations respectively. The PM and LOD/UT combinations are described and results are presented.

1. Introduction

The first IGS PM x and y combination was performed on GPS Wk 803 (IGS Rapid products only) and was oriented to ITRF93 (Kouba and Mireault, 1996). Starting on GPS Wk 860, the EOP combination for the Final IGS products was initiated and all IGS products including EOPs were oriented to ITFR94 (Kouba and Mireault, 1997). The last reference frame orientation change happened on GPS Wk 947 when all IGS products were oriented to ITRF96 (KOUBA and MIREAULT, 1998). EOP related changes and enhancements since GPS Wk 803 are summarized in Tab. 1.

2. IGS EOP Combination: Method Description

Two different procedures for the EOP combination are used: one for PM x, y, x_{rate} and y_{rate} (group I) and another for LOD/UT (group II). Both procedures are described below.

Group I: PM x, y, x_{rate} and y_{rate}

The parameters in the first group are combined as a straight-forward weighted average of AC (Cent) available values using AC orbit weights (eqn. 1).

$$\overline{ERP}_{Comb} = \sum_{Cent}^{Ncent} ERP_{Cent} \cdot W_{Cent}^{Orb} \tag{1}$$

where: W_{Cent}^{Orb} is the AC orbit weight (Kouba et al., 1995); $\overline{ERP}_{Comb}$ is the combined IGS value of one of the parameters of group I.

Group II: LOD/UT

Each AC LOD series is first calibrated with respect to the IERS Bulletin A. This is done by estimating a 21-day calibration (ncal) bias between IERS Bulletin A and each AC individually (eqn. 2). AC LOD values from day "Start" (or *anchor* point) and up to the day required (nday total) are then properly corrected (eqn. 3) before combining them to form IGS combined LOD values (eqn. 4). AC LOD weights are based on the bias calibration RMS from (eqn. 2) and given explicitly in (eqn. 5 and 6). Finally, IGS combined LODs from (eqn. 4) are integrated into IGS UTs (eqn. 7) from the *anchor* point up to the day required. Figure 1 illustrates the general procedure.

Table 1. History of IGS EOP Changes and Enhancements

Wk/Day	Products	Changes
803/0 860/0	Rapid Final	IGS PM x and y combinations from available AC PM values using AC orbit weights.
857/0 860/0	Rapid Final	IGS PM x and y rate combinations from available AC PM rate values using AC orbit weights.
894/0 895/0	Final Rapid	IGS LOD combination/integration officially implemented using IERS Bulletin A UT1 values that only included **24h VLBI data**.
896/0 898/0	Final Rapid	IGS LOD combination/integration using **up to the most current** IERS Bulletin A UT1 **observed** values.
907/0 908/3	Final Rapid	IGS LOD combination/integration using **up to five days prior** to the most current IERS Bulletin A UT1 **observed** values.
917/0 919/0	Final Rapid	New EOP comparison table implemented in the IGS/IGR summary report.
929/0 931/0	Final Rapid	New IGS LOD weighting scheme based on the AC LOD RMS bias calibration ($1/RMS^2$) with respect to IERS Bulletin A.
933/0	Rapid only	Use of EMR UT-derived LOD values in the IGS Rapid LOD combination/integration.

[1] Yves Mireault, Jan Kouba: Geodetic Survey Division, NRCan, 615 Booth Street, Rm 468, Ottawa, Ontario, Canada, K1A 0E9, email: mireault@geod.emr,ca

$$\text{Bias}_{\text{Cent}} = \frac{\sum_{\text{ical}}^{\text{ncal}} (\text{LOD}_{\text{Bull.A}} - \text{LOD}_{\text{Cent}})}{\text{ncal}} \tag{2}$$

$$\text{LOD}'_{\text{Cent}_{\text{iday}}} = \text{LOD}_{\text{Cent}_{\text{iday}}} + \text{Bias}_{\text{Cent}} \qquad \textit{iday} = 1, \textit{nday} \tag{3}$$

$$\overline{\text{LOD}}\text{comb}_{\text{iday}} = \sum_{\text{Cent}}^{\text{Ncent}} W_{\text{LOD}_{\text{Cent}}} \cdot \text{LOD}'_{\text{Cent}_{\text{iday}}} \qquad \textit{iday} = 1, \textit{nday} \tag{4}$$

where:

$$W_{\text{LOD}_{\text{Cent}}} = \frac{1}{\text{RMS}^2_{\text{Bias Cent}}} \tag{5}$$

$$\text{RMS}_{\text{Bias}_{\text{Cent}}} = \sqrt{\frac{\sum_{\text{ical}}^{\text{ncal}} (\text{LOD}_{\text{Bull.A}} - \text{LOD}_{\text{Cent}})^2}{\text{ncal}}} \tag{6}$$

Finally,

$$\overline{\text{UT}}\text{comb}_{\text{iday}} = \text{UT}_{\text{Bull.A}_{\text{Start}}} - \sum_{i}^{\text{iday}} \overline{\text{LOD}}\text{Comb}_i \qquad \textit{iday} = 1, \textit{nday} \tag{7}$$

where:

"Start" is 5 days prior to the IERS Bulletin A last observed UT1,

$\text{LOD}_{\text{Bull.A}}$ is IERS Bulletin A UT1-derived LOD,

LOD_{Cent} is the AC LOD or AC UT-derived LOD (in the case of EMR),

$\text{Bias}_{\text{Cent}}$ is the AC LOD calibration bias,

$\text{LOD}'_{\text{Cent}}$ is the corrected AC LOD or AC UT-derived LOD (e.g. EMR),

$\overline{\text{LOD}}_{\text{Comb}}$ is the estimated IGS combined LOD,

"ncal" is the number of calibration days (maximum 21),

"nday" is the number of combination/integration days,

$\text{UT}_{\text{Bull.A}_{\text{Start}}}$ is the IERS Bulletin A UT1 starting or *anchor* value, and,

$\overline{\text{UT}}_{\text{Comb}}$ is the estimated IGS combined UT.

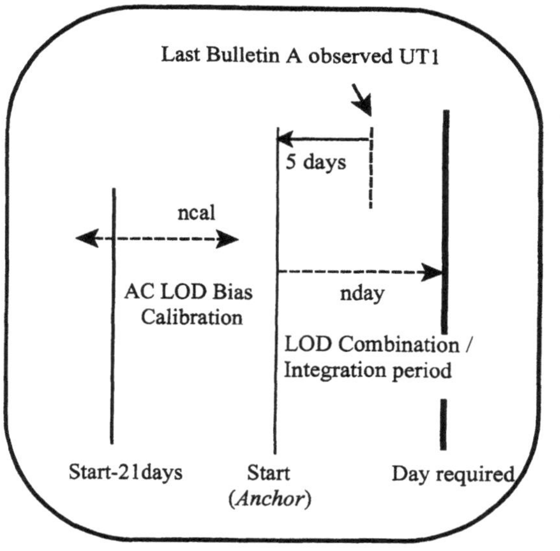

Fig. 1. LOD combination and integration procedure.

3. IGS EOP Combination Results

Initially, only the IERS Bulletin A UT1 values containing 24h VLBI data were used for the LOD calibration and integration. The strategy was modified a few weeks later to also include all observed, i.e. non-predicted values. After extensive testing, it was shown that using IERS Bulletin A UT1 values ending five days prior to the last observed UT1 value would give more optimal and reliable results (MIREAULT et al., 1997).

The UT integration period is generally no more than 6-10 days for the Rapid (IGR) combinations and 1-4 days for the Final (IGS) combinations. However, with the recent introduction of daily IERS Bulletin A updates, these periods were cut down to 6 and 1 days respectively.

At first, as in the IGS Polar Motion (PM) combinations, AC orbit weights were used for AC LOD weights. Other weighting strategies based on AC LOD calibration standard deviation ($1/\sigma^2$) and calibration RMS ($1/RMS^2$) were tested. The $1/RMS^2$ weighting scheme produced the best results overall (when compared to IERS Bulletin A) and was therefore implemented. Another improvement (as already noted by (Ray, 1996)) was realized when EMR UT-derived LOD values were used instead of its estimated LOD. Tests were conclusive and the new strategy was immediately implemented (strategy IGR-C-3* in Table 2). Table 2 contains a sample of the results from some of the LOD tests performed between GPS Wks 921 and 937.

Tab. 2: Comparison of strategies of IGS LOD/UT combination and IERS Bulletin A

Product tested: IGS Rapid (IGR)

Period covered: Wk 921/day 0 to Wk 937/day 6

Strategy used	UT Comparison wrt IERS Bulletin A. RMS (μsec)	Improvement of RMS (%)
IERS Bulletin A (originally)	377	---
IGR-A-1	162	57
IGR-B-1	158	58
IGR-C-1	137	64
IGR-C-2	127	66
IGR-C-3	121	68
IGR-C-3*	***96***	***75***

Description of IGR LOD combination/integration; strategies used:

IERS Bulletin A (originally):UT values taken directly from IERS Bulletin A (at the time of the combination);

(A): IGR LOD combination with AC LOD bias calibration based on IERS Bulletin A UT1 values that included 24h VLBI data;

(B): IGR LOD combination with AC LOD bias calibration based on all observed (i.e. non predicted) IERS Bulletin A UT1;

(C): IGR LOD combination with AC LOD bias calibration based on IERS Bulletin A UT1 values ending five days prior to its last observed UT1 value;

(1): AC LOD weights from AC orbit combinations;

(2): AC LOD weights based on the standard deviation of AC LOD bias calibration wrt IERS Bulletin A ($1/\sigma^2$);

(3): AC LOD weights based on the RMS of AC LOD bias calibration with respect to IERS Bulletin A ($1/RMS^2$);

(*): Use of EMR UT-derived LOD and use of AC LOD for all others.

Figure 2 shows a comparison of IGS Rapid and Final EOP values with respect to IERS Bulletin A since GPS Wk 860. PM x, y, x_{rate}, y_{rate} and LOD differences for the Rapid products are shown on the top graphic while the middle graphic shows the same differences for the IGS Final products. Each EOP series are offset by a value of 200. Units are 0.01 mas for PM x, y, x_{rate}, y_{rate} and μsec for LOD. The bottom graphic compares both the IGS Rapid and Final UT with IERS Bulletin A. On all graphics, the x-axis is divided into two periods: GPS Wk 860-946 (oriented to ITRF94) and GPS Wk 947-973 (oriented to ITRF96). Means and standard deviations are also given for each series. Since the introduction of ITRF96, an improvement in precision for all IGS/IGR EOP series can be noticed.

4. Conclusion

IGS has been combining PM x and y since May 1995, PM x_{rate} and y_{rate} since June 1996 and LOD/UT since March 1997. In this paper, the EOP combination methods were described in detail. It was shown that LOD/UT combination/integration was not as straightforward as the Polar Motion since it involves an AC LOD calibration with respect to IERS Bulletin A and a choice of an UT1 anchor value from IERS Bulletin A. Tests regarding the LOD/UT combinations were summarized and IGS/IGR EOP combined products since GPS Wk 860 were compared to IERS Bulletin A. Improvement throughout the years for all EOP series is noticeable. The change of reference frame on GPS Wk 947 (ITRF96) clearly improved IGS/IGR EOP precision.

References

KOUBA J., MIREAULT Y, LAHAYE F. (1995). *Appendix to the 1994 IGS Analysis Coordinator Report*, in International GPS Service for Geodynamics, Annual Report, IGS Central Bureau, JPL, Pasadena, Cal., USA, pp. 70-94.

KOUBA J., MIREAULT Y. (1996). *1995 IGS Analysis Coordinator Report*, in International GPS Service for Geodynamics, Annual Report, IGS Central Bureau, JPL, Pasadena, Cal., USA, pp. 45-76.

KOUBA J., MIREAULT Y. (1997). *1996 IGS Analysis Coordinator Report*, in International GPS Service for Geodynamics, Annual Report, IGS Central Bureau, JPL, Pasadena, Cal., USA, pp. 55-100.

KOUBA J., MIREAULT Y. (1998). *1997 IGS Analysis Coordinator Report*, in International GPS Service for Geodynamics, Annual Report, IGS Central Bureau, JPL, Pasadena, Cal., USA, in print.

MIREAULT, Y., KOUBA, J., RAY J.R. (1997). *Use of GPS for Near Real Time Estimates of UT1*, EOS Trans. AGU, 78(17), Spring Meeting Suppl., S109.

RAY, J.R. (1996). *Measurements of Length of Day using the Global Positioning System.* Journal of Geophysical Research, Vol. 102, No. B9, pp. 20141-20149.

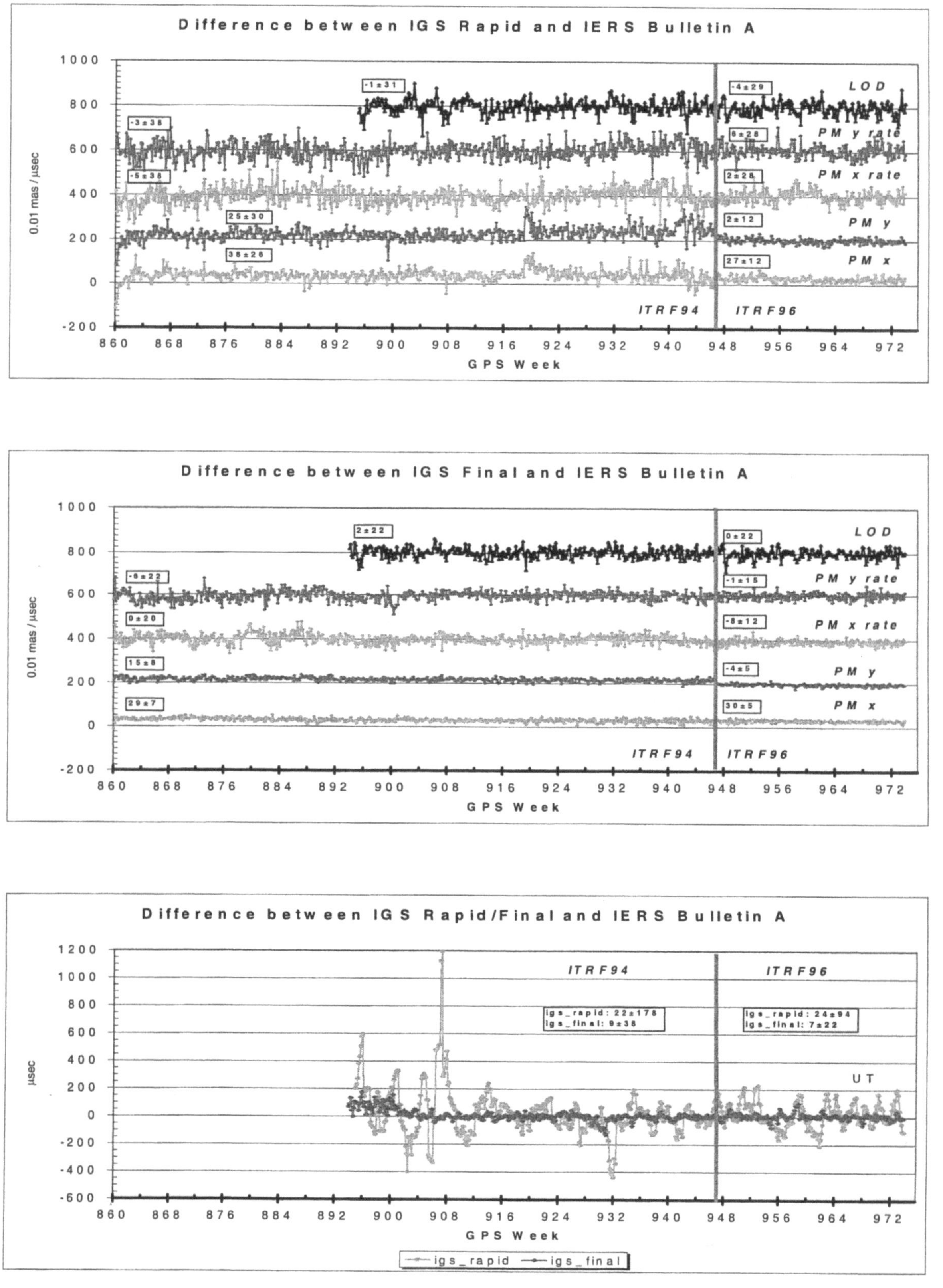

Figure 2. Difference between IGS EOP Combinations and IERS Bulletin A

Comparison of Site Velocities Measured by VLBI and GPS in the Key Stone Project Network

Y. KOYAMA[1], R. ICHIKAWA[1], T. KONDO[1], N. KURIHARA[1], Y. TAKAHASHI[2], T. YOSHINO[2], K. SEBATA[2], M. FURUYA[2]

1. Introduction

Communications Research Laboratory has established four space geodetic observation sites in and around Tokyo, Japan under the Key Stone Project (KSP) [KOYAMA et al, 1997]. At each of the four sites (i.e. Koganei, Kashima, Miura, and Tateyama), an 11-m antenna VLBI system, a 75-cm telescope SLR system, and a dual-frequency geodetic GPS receiver system are collocated closely with each other. The relative positions of the three reference points are repeatedly measured by ground survey measurements. The comparisons of the site coordinates and velocities estimated from three different space geodetic techniques are considered to be quite important to understand error sources of each measurement technique and to evaluate consistency of the coordinates and velocities defined in a conventional terrestrial reference frame. In this paper, site velocities estimated from VLBI and GPS observations are compared.

2. Observations and data analysis

The first VLBI observation with the KSP facility was performed with the Koganei-Kashima baseline on August 29, 1994. VLBI facilities at Miura became operational in May 1995, and at Tateyama in September 1996. Regular VLBI observations and data processing began on the daily basis from December 1, 1995, and the duration of an experiment was expanded from about 6 hours daily to about 24 hours once-every-other-day since September 30, 1997. The observed data are transferred to the data processing facility at Koganei by using high speed (2.4 Gbps) digital communication optical fibres in Asynchronous Transfer Mode (ATM) and they are processed by the real-time VLBI correlator system simultaneously with the observations. After all the observations and correlation processing in one experiment complete, the obtained data are analyzed by using the automated data analysis system which employs CALC version 8.1 which has been developed by a group in the Goddard Space Flight Center of National Aeronautics and Space Administration. In the data analysis, the position of the KSP Kashima VLBI station is fixed to the coordinates determined on the ITRF96 reference frame and the positions of the other three VLBI stations are estimated. The accurate ITRF96 coordinates of the KSP Kashima VLBI station were estimated through a series of joint VLBI experiments with the 34 m antenna VLBI station at Kashima and the four KSP VLBI stations. In the estimation, the velocity of the KSP Kashima VLBI station is assumed to be same as the velocity of the Kashima 34 m antenna station defined in the ITRF96. With the current VLBI observation mode of 128 MHz total bandwidth (80 MHz for X-band and 48 MHz for S-band), about 2-mm of the repeatability for the baseline length estimations are regularly achieved [KONDO et al, 1997]. By using the site coordinates estimated from VLBI observations, velocities of each sites can be estimated with formal errors less than 0.2 mm/year for the horizontal components and less than 1.0 mm/year for the vertical component.

GPS measurements at four KSP observation sites began in July, 1997. With about one year of the accumulated GPS data, precise estimations of the site velocities from the GPS data are becoming feasible. GPS observation data are collected at 30 seconds of intervals for 24 hours a day, and the observed data stored in a PC system are transferred to a GPS data analysis workstation through the computer network configured for the VLBI observation system. Observed data are analyzed by using Bernese Version 4.0 software which has been developed in Bern University. In the data analysis, the position of a GPS site at Tsukuba is strongly constrained to the ITRF96 coordinates and the positions of the four GPS sites in the KSP network are adjusted. The GPS site at Tsukuba is one of the globally distributed International GPS Service (IGS) sites and its position is considered to be accurately defined on the ITRF96 reference frame.

3. Discussions

The estimated site velocities determined from VLBI and GPS observations are given in Table 1. These velocities are estimated in the ITRF96 reference frame and then converted to the local Cartesian coordinate system. The estimated one-sigma uncertainties are attached in the table except for the Kashima VLBI station's velocity since it is provided by the ITRF96 site velocity for the 34m antenna VLBI station at Kashima.

[1] Yasuhiro Koyama, Ryuichi Ichikawa, Tetsuro Kondo, Noriyuki Kurihara: Communications Research Laboratory, Kashima Space Research Center, 893-1 Hirai, Kashima, Ibaraki 314-0012 Japan; Tel.: +81 - 299 - 847-143/-142/-137/-141, Fax:+81 - 299-159, email: koyama@crl.go.jp, richi@crl.go.jp, kondo@crl.go.jp, kurihara@crl.go.jp

[2] Yukio Takahashi, Taizoh Yoshino, Kouichi Sebata, Masato Furuya: Communications Research Laboratory, 4-2-1 Nukuikita, Koganei, 184-8795 Japan; Tel.:+81 - 42 - 3277-565/-560/-904, 3275392, Fax: +81 - 42 - 327-4584/-6077, email:takahashi@crl.go.jp, yosh@crl.go.jp, sebs@crl.go.jp, mf@crl.go.jp

Tab. 1: Estimated VLBI and GPS site velocities expressed in the local Cartesian coordinate system.

Site		Eastward Component (mm/year)	Northward Component (mm/year)	Upward Component (mm/year)
Koganei	VLBI	0.12±0.10	-8.06±0.09	-2.23±0.51
	GPS	-0.63±0.09	-11.50±0.09	-6.83±0.74
Kashima	VLBI	-1.03	-11.93	-5.53
	GPS	-4.00±0.08	-14.46±0.08	-26.42±0.71
Miura	VLBI	-7.38±0.12	1.34±0.14	-11.53±0.64
	GPS	-10.36±0.08	4.27±0.12	-26.87±0.78
Tateyama	VLBI	-7.42±0.13	4.56±0.17	-8.72±0.73
	GPS	-12.69±0.08	9.94±0.14	-16.37±0.79

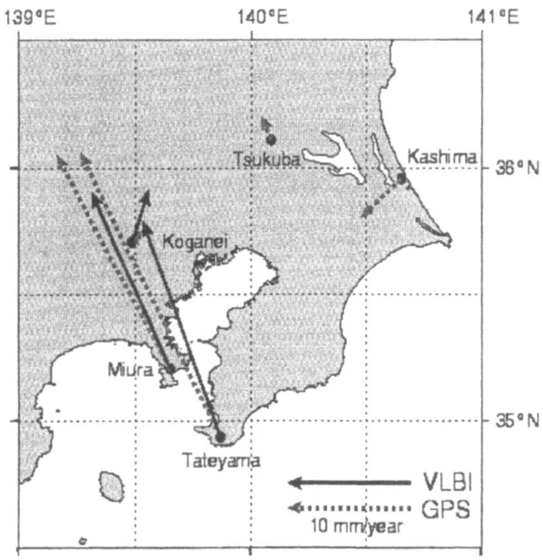

Fig. 1: Site velocities determined from VLBI and GPS observations. ITRF96 site velocity of Kashima VLBI station is subtracted from all the other site velocities. Site velocity of Tsukuba GPS site (TSKB) is given from ITRF96 velocity.

Horizontal site velocities estimated from VLBI and GPS observations are compared in the Figure 1. A discrepancy between VLBI and GPS site velocities ranges from 4.6 mm/year to 20.9 mm/year in the vertical component, and from 3.5 mm/year to 7.5 mm/year in the horizontal plane. These discrepancies are significantly larger than the estimated uncertainties. Common part of the discrepancies can be considered as the inconsistency between GPS site velocity at Tsukuba and VLBI site velocity at Kashima defined in the ITRF96 reference frame, and the rest can be considered as the estimation errors. But, since the accumulation period of the GPS observation data is only about a year, more accumulation of the GPS observation data is considered necessary for more detailed comparison.

Figure 2 shows the VLBI site velocities in the horizontal plane with respect to the stable North American plate. Kashima's velocity was evaluated from the NUVEL-1a Plate motion model [DEMETS et al, 1994] and the Kashima's velocity with respect to the Eurasian plate determined by HEKI [1996]. On the other hand, Figure 3 shows the motions of three tectonic plates around Japan with respect to the North American plate evaluated by the NUVEL-1a model and the Philippine sea plate motion model by SENO et al [1994].

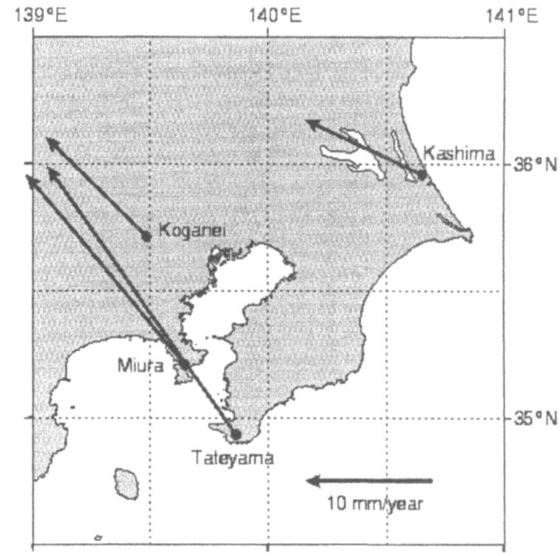

Fig. 2: Site velocities determined from VLBI. Kashima's velocity is assumed to be -9.21 mm/year (East) and 3.98 mm/year (North) with respect to the North American Plate [HEKI, 1996].

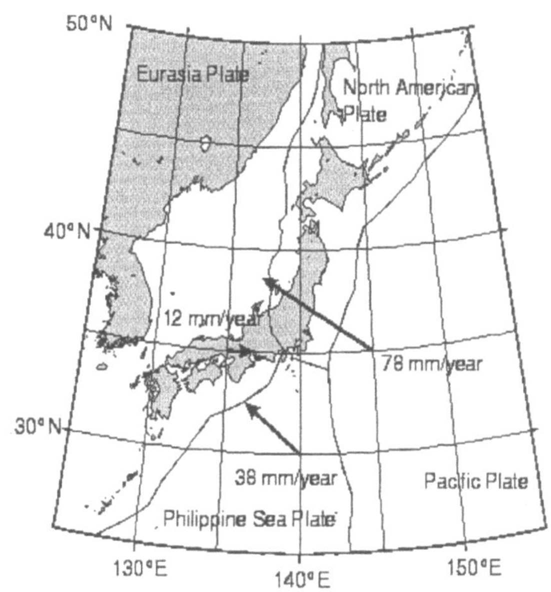

Fig. 3: Four tectonic plates around Japan and their relative motions seen from the stable North American plate. Plate motions are evaluated from the NUVEL-1a [DEMETS 1994] plate motion model and the Philippine sea plate motion model proposed by SENO et al [1993].

Directions of Kashima and Koganei site velocities agree very well with the direction of the Pacific plate motion shown in the figure 3. On the other hand, directions of site velocities of Miura and Tateyama are very close to the direction of the Philippine sea plate motion shown in the figure 3. From these results, the northern region of the KSP network is moved by the Pacific plate whereas the southern region of the KSP network is compressed by the Philippine sea plate.

4. Concluding remarks

Site velocities of the KSP space geodetic network observation sites were estimated from VLBI and GPS observations and the results were compared. Significant discrepancies were shown, but the results are still preliminary and longer span of the observations will be required for detailed discussions. SLR observations using the KSP facilities just began on a regular basis and the estimation of the SLR site velocities will become possible soon. Comparisons of site velocities and site coordinates determined from three different space geodetic techniques will give us important information to construct a reliable and consistent terrestrial reference frame.

Acknowledgments

The authors would like to express deep appreciation to colleagues in Geographical Survey Institute who allowed us to use GPS data observed at Tsukuba.

References

DEMETS C., GORDON R. G., ARGUS D. F., STEIN S.(1994), *Effect of recent revisions to the geomagnetic reversal time scale on estimates of current plate motions*, Geophys. Res. Let., Vol. 21, pp. 2191-2194

HEKI K. (1996), *Horizontal and vertical crustal movements from three-dimensional very long baseline interferometry kinematic reference frame: Implication for the reversal timescale revision*, J. Geophys. Res., Vol. 101, No. B2, pp. 3187-3198

KONDO T., KURIHARA N., KOYAMA Y., SEKIDO M., ICHIKAWA R., YOSHINO T., AMAGAI J., SEBATA K., FURUYA M., TAKAHASHI Y., KIUCHI H., KANEKO A. (1998), *Evaluation of repeatability of baseline lengths in the VLBI network around the Tokyo metropolitan area*, Geophys. Res. Let., Vol. 25, pp. 1047-1050

KOYAMA Y.,ICHIKAWA R. Gotoh T., SEKIDO M., KONDO T., KURIHARA N., TAKAHASHI F., AMAGAI J., OTSUBO T., NOJIRI H., SEBATA K., KUNIMORI H., KIUCHI H., KANEKO A., TAKAHASHI Y., HAMA S., HANADO Y., IMAE M., MIKI C., HOSOKAWA M., YOSHINOT. (1997), *VLBI, SLR and GPS observations in the Key Stone Project*, proceedings of the IAG Scientific Assembly, Rio de Janeiro, Sep. 1997, pp. 394-399

SENO T., STEIN S., GRIPP A. E. (1993), *A model for the motion of the Philippine sea plate consistent with NUVEL-1 and geological data*, J. Geophys. Res., Vol. 98, pp. 17941-17948

Large Lunar Laser Ranging at its Best

J. MÜLLER[1], D. EGGER[2], B. REICHHOFF[2], M. SCHNEIDER[2], U. SCHREIBER[2]

Abstract

Lunar Laser Ranging (LLR) data can be used to determine parameters of the Earth-Moon system (e.g. lunar gravity, tidal parameters) and quantities parametrizing relativistic effects in the solar system. Moreover, LLR contributes to the realisation of the International Terrestrial Reference Frame (ITRF) and provides a set of Earth Orientation Parameters (EOP) like UT0 or nutation coefficients.

Among the relativistic parameters are those representing secular effects like $\dot{G}/G$ and periodic effects such as violation of the equivalence principle (EP). Altogether 10 relativistic parameters have been determined, e.g. metric parameters β and γ, the geodetic precession of the lunar orbit, Yukawa coupling constant, the preferred frame parameters $\alpha 1$ and $\alpha 2$ and others. Here, we give an overview over the prospects of LLR today and its impact for Geodesy and Relativity.

1. Introduction

Lunar Laser Ranging (LLR) is one of the oldest space geodetic techniques. Observations have been started shortly after the deployment of the first retro-reflector during the first manned lunar mission Apollo 11 in the year 1969. Since that time many LLR observations have been performed, inhomogeneously distributed but continued up to now.

Figures 1 and 2 show the uneven data distribution as a function of the synodic angle D and of the sidereal angle S. The observations are collected in 36 bins of 10 degrees width.

In Figure 1, the data gaps at full Moon (180 degrees) and new Moon (0 resp. 360 degrees) and the asymmetry about quarter Moon (at 90 and 270 degrees) are caused by operational restrictions, such as difficulties to target laser reflector arrays on the Moon when they are in darkness or there is excessive background solar illumination noise.

In Figure 2, the uneven distribution is caused by the absence of observatories at the southern hemisphere, because observations from northern hemisphere laser stations are less frequent when the Moon is in the southern sky.

The parameter determination is affected badly, e.g. the missing observations near new Moon weaken the quality of the fit for the equivalence principle η.

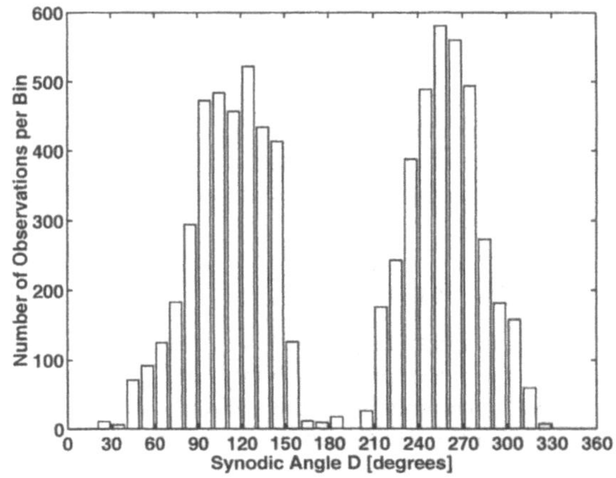

Fig. 1: Data distribution as a function of the synodic angle D

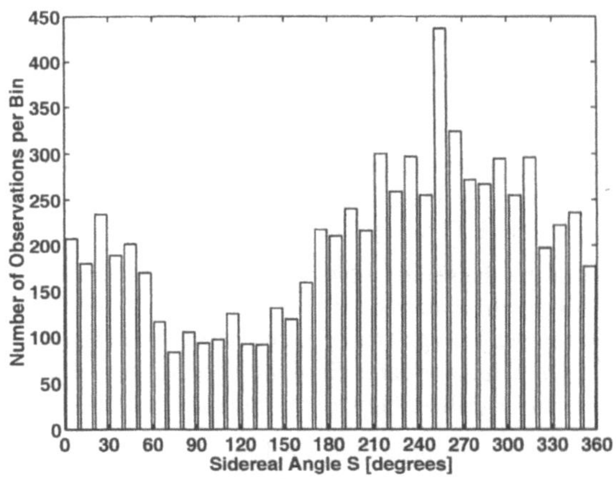

Fig. 2: Data distribution as a function of the sidereal angle S

As an optical technique, LLR is strongly dependent of weather conditions which lead to further inhomogeneities in the distribution of measurements.

Due to the large distance the energy balance for each laser observation looks very bad: only one photon out of 10^{19} transmitted ones finds its way back to the receiver. Nevertheless a total number of 12200 normal points have been obtained which have an accuracy of less than 1.5 cm (nowadays).

1 Jürgen Müller: Institute for Astronomical and Physical Geodesy; Technische Universität München, D-80290 Munich, Germany; Tel.: +49 - 89 - 289-23198, Fax: +49 - 89 - 289-23178, email: jxmx@alpha.fesg.tu-muenchen.de

2 Dieter Egger, Burkard Reichhoff, Manfred Schneider, Ulrich Schreiber: Research Facility for Space Geodesy, Technische Universität München, D-80290 Munich, Germany; Tel.: +49 - 89 - 289-23191, Fax: +49 - 89 - 289-23178, email: dieter@alpha.fesg.tu-muenchen.de

On the other hand, LLR benefits from the very long period of observations (almost 30 years) which allows to solve for the largest periods in the nutation series (i.e. the nodal drift of the lunar orbit: 18.6 years).

While some parameters benefit from the very long period of lunar observations, others benefit from the increased measurement precision in recent years, e.g. the Earth-Moon mass parameter *GM* or Earth rotation UT0.

In the Seventies LLR was the only space technique (beside classical astronomy) which was able to determine Earth orientation parameters. Today LLR can still compete with other space geodetic techniques and contributes to combined EOP solutions computed by the International Earth Rotation Service (IERS).

2. LLR Model

Based upon EINSTEIN'S theory of gravitation a model complete up to the first post Newtonian level has been developed to compute the values of the LLR observables - the round trip travel times of laser pulses between stations on the Earth and passive reflectors on the Moon (see e.g. MÜLLER et al., 1996 or MÜLLER & NORDTVEDT, 1998 and the references therein).

During the last year the model has been improved to better include effects like ocean loading, solar radiation pressure, precession rate of the equator in longitude and obliquity, while other parts of the model have been updated (better description of solid Earth tides or effects of asteroids). Useful suggestions can be found in the IERS Conventions (1996), e.g. for the effect of tidally driven diurnal and semidiurnal UT1 variations or solid Earth tides.

With such a theoretical model many parameters (about 170) describing the dynamics of the Earth-Moon system can be investigated. Two groups of parameters are determined by a weighted least-squares fit of the observations. The first group are so-called ‚Newtonian' quantities like

- geocentric coordinates of four stations (including Wettzell where only a few observations are available), optional drift rates of the corresponding continental plates;
- a set of Earth's orientation parameters (luni-solar precession constant, nutation coefficients of the 18.6 years period, Earth's rotation UT0 and variation of latitude by polar motion);
- selenocentric coordinates of four retroreflectors;
- rotation of the Moon for one initial epoch (physical librations);
- orbit (position and velocity) of the Moon for this epoch;
- orbit of the Earth-Moon system about the Sun for one epoch;
- mass of the Earth-Moon system times the gravitational constant;
- the lowest mass multipole moments of the Moon;
- lunar Love number and a rotational energy dissipation parameter;
- lag angle indicating the lunar tidal acceleration responsible for the increase of the Earth-Moon distance (and the slow down of Earth's angular velocity);
- quadrupole moment of the Sun $J_2^{\odot}$ which could also be added to the second group because it contributes to the anomalous perihelion advance of Mercury.

The second group includes possible modifications of general relativity (values in EINSTEIN'S theory are given in brackets):

- geodetic precession Ω_{GP} ($\approx$1.92"/cy) of the lunar orbit;
- metric parameters γ (= 1) (space curvature) and β (= 1) (non-linearity), included in the Einstein-Infeld-Hoffmann equations and in the light time equation;
- EP parameter η(= 0), i.e. test of the strong equivalence principle;
- time variation of the gravitational constant $\dot{G}/G$ (= 0 yr^{-1}) which is important for the unification of the fundamental interactions;
- coupling constant α (=~0) of Yukawa potential for the Earth-Moon distance which corresponds to a test of NEWTON'S inverse square law;
- combination of parameters $\zeta_1 - \zeta_0 - 1$ (= 0) derived in the MANSOURI and SEXL (1977) formalism indicating a violation of special relativity (there: LORENTZ contraction parameter ζ_1 = 1/2, time dilation parameter ζ_0 = -1/2);
- violation of the equivalence principle in the coupling to dark matter assumed in the center of the galaxy;
- α_1 (= 0) and α_2 (= 0) parametrizing preferred frame effects within metric gravity.

Most relativistic effects cause periodic perturbations of the lunar orbit

$$\Delta r_{EM} = \sum_{i=1}^{n} A_i \cos(\omega_i \Delta t + \Phi_i).$$

A_i and ω_i are the amplitude and frequency of the perturbation, Φ_i the corresponding phase angle. Typical periods of some (relativistic) parameters are given in Table 1.

Tab. 1: Typical periods of some relativistic quantities.

Parameter	Typical Periods
η	synodic ($29^d\ 12^h\ 44^m\ 2.9^s$
ζ_1-ζ_0-1	annual (365.25^d)
δg_c	sidereal ($27^d\ 7^h\ 43^m\ 11.5^s$)
α_1	sidereal, annual, sidereal - 2*annual, anomalistic ($27^d\ 13^h\ 18^m\ 33.2^s$) ± annual, synodic
α_2	2*sidereal, 2*sidereal - anomalistic, nodal (6798^d)
$\frac{\dot{G}}{G}$	secular

These properties (relativistic phenomena show up with typical periods) can be used to identify and separate the different effects and to determine corresponding parameters.

3. Results

The global adjustment gives improved values for the estimated parameters and their formal standard errors. More realistic errors have been derived by covariance analysis. As an example we describe the error analysis for the amplitude $A_D = 12.8\ \eta$[m] of the EP effect

$\Delta r_{em} = A_D \cos(D)$

more detailed. The formal fit of $A_D = 1.0 \pm 0.2$ cm indicates some unmodelled synodic signal in the data, but not necessarily of relativistic origin. Insufficiently modelled effects such as solid Earth tides, ocean loading or geocenter motion and uncertainties in values of fixed model parameters lead to a total realistic error of σ_{AD} (realistic) = 1.1 cm resp. $\eta \leq 9 \cdot 10^{-4}$. If this result is interpreted as a test of the weak EP (i.e. the silicate Moon and the iron-cored Earth fall differently in the field of the Sun) one has

$$\eta_{weak} = \eta \frac{U_G}{Mc^2} \frac{M_{mantle}}{M_{core}} \approx 1 \cdot 10^{-12}$$

where $U_G \approx 4.6 \cdot 10^{-10}$ is the gravitating self-energy of the Earth-Moon system, M its mass and M_{mantle}/M_{core} corrects for the different compositions of the Earth mantle and core.

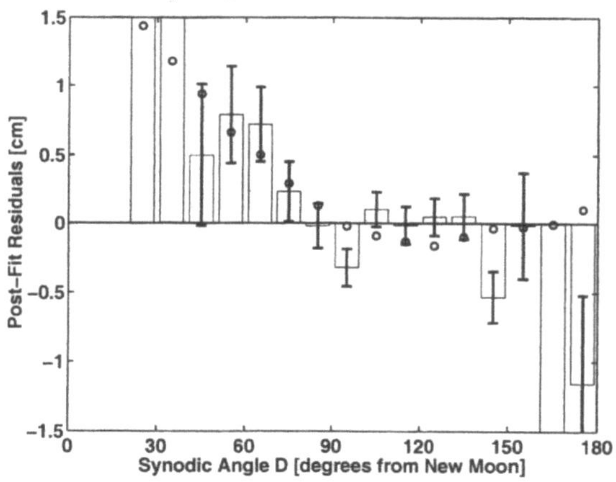

Fig. 3: Synodic bins. The 'o' show the effect of '1cm cos(D)' signal.

These results have been checked by constructing and plotting 'synodic bin-averaged' post-fit residuals as a function of synodic angle D (measured from nearest new Moon). As shown in Figure 3, each bin bar represents the average of hundreds of post-fit range residuals whose synodic angles fall within the bins' 10 degree widths. The plotted bin errors reflect only the observer-supplied errors of the original normal points. The curve of 'o' markings indicates the changes of these bin averages which resulted when a signal '1cm cos(D)' was added to the actual LLR residuals. The marked curve appears compatible with the actual post-fit residuals, supporting our 'realistic' fit for this amplitude of about 1.0 ± 1.1 cm (see also MÜLLER & NORDTVEDT, 1998).

Tab. 2: Determined values for the relativistic quantities and their realistic errors.}

parameter	results
diff. geod. prec. $\Omega_{GP} - \Omega_{deSit}$ ["/cy]	$(-2 \pm 10) \cdot 10^{-3}$
metric par. $\gamma - 1$	$(4 \pm 5) \cdot 10^{-3}$
metric par. $\beta - 1$	$(-1 \pm 4) \cdot 10^{-3}$
equiv. principle par. η	$(8 \pm 9) \cdot 10^{-4}$
time var. grav. const. $\dot{G}/G$ [yr^{-1}]	$(3 \pm 5) \cdot 10^{-12}$
Yuk. coupl. const$_{\alpha\lambda=4\cdot10^5\ km}$	$(2 \pm 2) \cdot 10^{-11}$
spec. relativi. $\zeta_1 - \zeta_0 - 1$	$(-5 \pm 12) \cdot 10^{-5}$
infl. of dark matter δg_c[cm/s^2]	$(4 \pm 4) \cdot 10^{-14}$
preferred frame effect α_1	$(-8 \pm 9) \cdot 10^{-5}$
preferred frame effect α_2	$(2 \pm 2.5) \cdot 10^{-5}$

Similar investigations have been performed for the other relativistic parameters. Final results are shown in Table 2. The realistic errors are comparable with those derived in other recent investigations (see e.g. WILLIAMS et al., 1996).

Although we concentrate our investigations to determine relativistic quantities, LLR data are used to compute new solutions of EOPs (UT0 and variation of latitude by polar motion) each year covering the whole period of observations since 1970. The accuracy for UT0 is about 50 μs nowadays which is in agreement with other space geodetic techniques.

The results are submitted to the IERS where they are combined to a global solution. The same is valid for a set of station coordinates which are estimated simultaneously and submitted to the IERS to contribute to the realisation of the international terrestrial reference frame ITRF.

Also for the 'Newtonian' parameters, a similar error analysis can be performed as for the relativistic ones. Therefore given errors represent always realistic estimations (Actual values can be received on request).

In 1998 the International Laser Ranging Service (ILRS) has been established to improve the international cooperation of the various laser centers and to better pre-process the laser-related products for the scientific community. Our group contributes to the ILRS as Lunar Analysis Center.

4. Conclusion

For the IERS, LLR contributes to the realisation of the International Terrestrial Reference Frame and to combined solutions of Earth Orientation Parameters. Moreover, LLR permits a variety of relativistic quantities to be determined with high accuracy. No definitive violation of general relativity is found, though evidence of modelling inadequacies is present in the synodic residuals. The (weak and strong) equivalence principle is verified, NEWTON'S inverse square law tested, and preferred frame effects (within special and general relativity) are investigated as well. Therefore LLR is the most important tool to test EINSTEIN'S theory of gravitation in the solar system. These results can be further improved if LLR observations are continued into the future and the basic modelling of the LLR experiment is extended down to the millimeter level of accuracy.

References

ERS Conventions (1996), IERS Technical Note 21 (ed. D.D. MCCARTHY).

MANSOURI R.M., SEXL R.U. (1977) *A test theory of Special Relativity*. General Relativity and Gravitation, 8, No. 7, 497-513 (Part I); No. 7, 515-524 (Part II); No. 10, 809-814 (Part III).

MÜLLER J., NORDTVEDT K. (1998): *Lunar laser ranging and the equivalence principle signal*. Physical Review D, 58, 062001.

MÜLLER J., NORDTVEDT K., VOKROUHLICKÝ D. (1996): *Improved constraint on the α_1 PPN parameter from lunar motion*. Physical Review D, 54, R5927-R5930.

WILLIAMS J.G., NEWHALL X.X., DICKEY D.O. (1996): *Relativity parameters determined from lunar laser ranging*. Physical Review D, 53, 6730.

Acknowledgement

We express our thanks to KEN NORDTVEDT and DAVID VOKROUHLICKÝ for many fruitful discussions in this context over the last years.

Combining GPS and CCD to Modernise Astronomical Levelling

C. FOSU, B. EISSFELLER, G. W. HEIN[1]

Abstract

The advent of GPS has substantially increased the demand for local geoid determination. The local geoid is needed to convert the very accurate geometric co-ordinates provided by GPS to physically meaningful co-ordinates required for many applications. In order words local geoid determination is vital for most GPS users to fully exploit its tremendous and ever increasing potential. There are various methods for determining the local geoid. However these existing methods are either insufficiently accurate, inefficient slow and expensive. Theoretically, astronomical levelling is probably the most accurate and computationally simple solution to the problem of local geoid determination. The idea here is to determine astronomic co-ordinates, and compare to geodetic co-ordinates now obtained from GPS. The difference gives deflection of the vertical which can subsequently be used for precise or local geoid determination. However this method has never been of any practical importance. This is mainly due to the fact that the conventional methods of determining astronomical co-ordinates, besides other reasons, are time consuming both in the field and in the office, insufficiently accurate, very costly and old fashioned and therefore not compatible with GPS. With the efficiency of GPS positioning and developments in computer aided astrometry based on high performance CCD cameras, a new, fast, fully digital and solid-state technology is now available to reconsider astronomical levelling.

This paper reports on research work going on in IfEN which is aimed at developing a system for full digital observation of the direction of the plumb line, a digital data transfer and numerical determination of the deflection of the vertical and local geoid by combination of CCD technology, differential GPS, GPS based time distribution and computer technology. Preliminary results indicate that an accuracy of the order of 0.3 arcsec or better in the astronomic values, required for "cm" level geoid determination, can be obtained

1 Astronomical Levelling

The relationship between natural (astronomical) co-ordinates and ellipsoidal (GPS) co-ordinates as shown in figures 1 and 2 is given by the following expressions (HEISKANEN and MORITZ 1967):

$$\xi = \Phi - \phi \quad (1)$$

$$\eta = (\Lambda - \lambda)\cos\phi \quad (2)$$

$$h = H + N \quad (3)$$

Also $\theta^2 = \xi^2 + \eta^2$ (4)

And $\varepsilon = \xi \cos\alpha + \eta \sin\alpha$ (5)

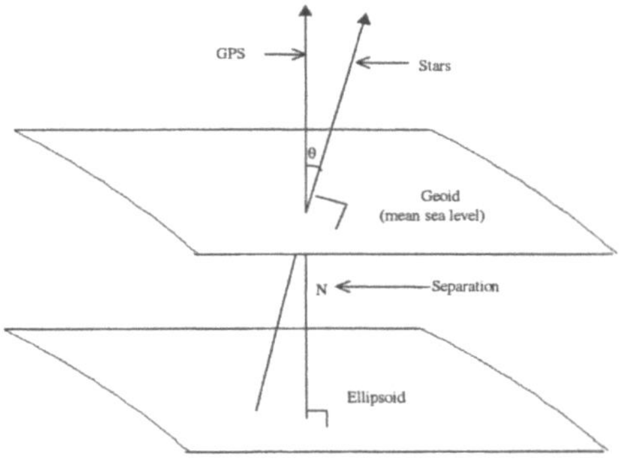

Fig. 1: Relation between geoid and ellipsoid (Helmert's definition for z, h (q))

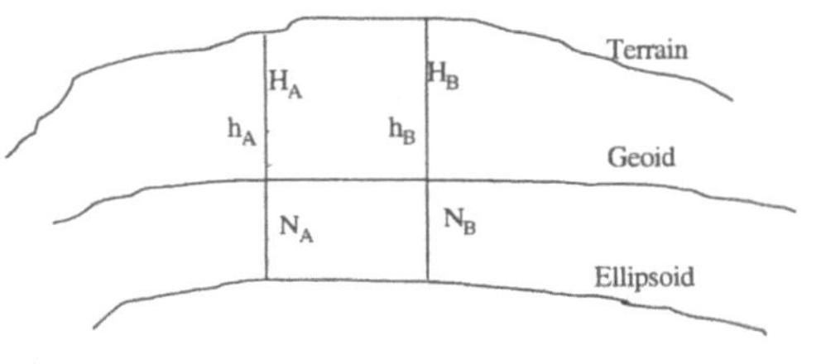

Fig. 2: Relation between orthometric, GPS and geoid height.

Since the deflection of the vertical (e) is the slope of the geoid (fig. 3), y integration over distance (i.e., astronomical levelling) we get the geoid height (N).

$$' = N_0 + \int_0^S \varepsilon ds + Corr. \quad (6)$$

where S is the distance and Corr. are corrections like curvature of the plumb line and dynamic path correction.

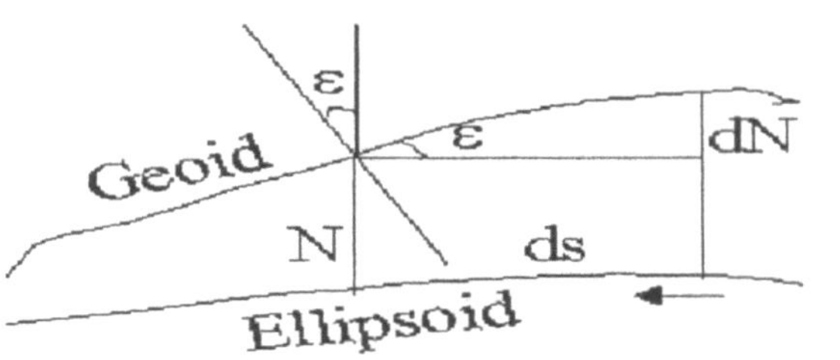

Fig.3: The slope of the geoid

Error analysis of equation (6) shows that if we can determine astronomical co-ordinates to an accuracy better than

[1] Collins Fosu, Bernd Eissfeller, Günter W. Hein: Institute of Geodesy and Navigation (IfEN), University FAF Munich, D-85577 Neubiberg, Germany, Tel.: +49 - 89 - 6004-3382,/-3017,/-3425, Fax: +49 - 89 - 6004-3019, email: fosu@ifen1.bauv.unibw-muenchen.de, bernd.eissfeller@unibw-muenchen.de, guenter.hein@unibw-muenchen.de

0.3 arcsec, we will be able to determine a precise geoid (s<1 cm) required by GPS users.

2. CCD and IST Operation

The charge coupled device (CCD) is a small photo-electronic imaging device, typical sizes range from a few mm to 20 mm, made from a crystal of silicon. It consists of individual sensitive picture elements (pixels) capable of storing charge created by absorption of light. The operation of a CCD involves photo electrons generated by incoming electrons. They are collected at each pixel and then after a user defined exposure, transferred to an output register and into an output amplifier where the electrons are counted and converted into an image suitable for storage or analysis by a computer. Electronic cameras using CCDs are also referred to as 'area photometers' because their performance and stability enable accurate measurements to be obtained over the entire area of view (MACLEAN 1989 p. 262). CCDs offer the following advantages over other detectors (FOSU et al 1998);

1 Desirable features such as: miniature size, stability and geometric regularity -Regularity in the sensor element spacing is in the order of 1/100thof the spacing which makes it an ideal measurement device.
2 Good quantum efficiency - This quantum efficiency(the ratio of the number of photo electrons released for each incident photon of light absorbed by a detector), means that CCDs make use of most of the light coming through a telescope at a given time (ZEILICK and SMITH 1987 p.177)
3 Linearity of response - It can measure light intensities very accurately (ibid.).
4 Natural transducer for optical input to a computer -The image emerges in a natural digital formready for image processing.
5 Unlimited use.

CCDs are the best current example of technology expanding the capabilities of smaller telescopes, in other words they enhance the power of small telescopes. These telescopes offer better accessibility of long term surveys and the ability to react to targets of opportunity.

For astral imaging the CCD is placed at the focus of an optical telescope (MACLEAN 1989, p. 42). The telescope has the purpose of gathering and combining the incoming light rays more rapidly to a focus while the CCD serves as a recording, measuring and storing device. Two main factors which define the imaging parameters of a telescope system are the diameter of the objective aperture (D) and the focal length (f). These largely determine the basic performance of the instrument, defined by factors such as limiting visual magnitude, theoretical resolution and image scale.

3 Instrumentation

Directions may be measured photographically by imaging against the background of stars. The idea here is to measure directly the direction of the plumb line (vertical) using CCD camera and GPS timing. The principle is to locate the position of the zenith (vertical) on an image which shows a number of stars surrounding the zenith. The directions to these stars are defined by the right ascensions and declinations which are known from spherical astronomy. By interpolation we find the right ascension and declination of the zenith. For a star at the zenith (a_Z, d_Z). The relations between a star and its sub-stellar point are given by the following equations (BOMFORD 1980 p. 316f.);

$$\Phi = \delta_Z \qquad LST = \alpha_Z \tag{7}$$

$$\Lambda = LST - GST \tag{8}$$

where LST is the sidereal time of observation and GST (or GMST) is the sidereal time which refers to Greenwich meridian. In practice, if the field of stars near the zenith (Z) in astronomical latitude (F) and longitude (L) is imaged for two positions of camera differing by 180° in azimuth and image co-ordinates are measured, the transformation of image co-ordinates to right ascension and declination system provides the co-ordinates (a_Z, d_Z) of the zenith, after corrections have been made for the level readings and averages taken, (GURTNER and BURKI 1986).

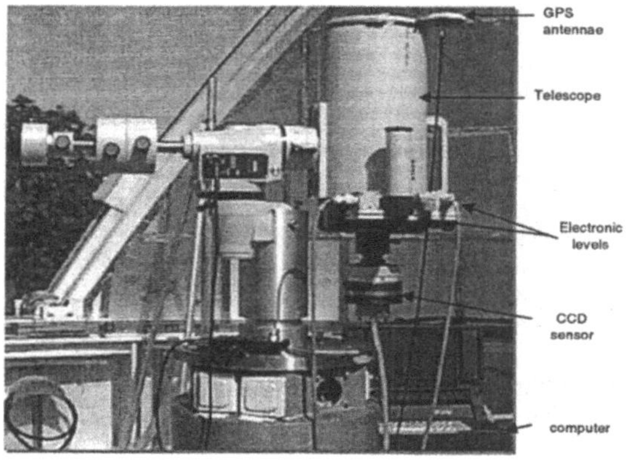

Fig. 4: The IfEN CCD integrated telescope

The overall system layout is as follows;

3 Telescope + mount
4 CCD
5 Host computer + peripherals
6 An electronic interface to computer
7 Levels
8 Timing system (GPS)
9 Image display system

The IfEN CCD integrated telescope shown in figure 4 is made up of;

- Zeiss Mak 180/18 Telescope + mount
- Photmetrics CCD camera system + software
- Kontron industrial computer + display system
- Novatel OEM GPS card and antennae.
- Focal reducers.

4 Error Budget

The error budget in astronomic positioning, using the zenith imaging method, from one set of observations is made up mainly of the following:

1. Errors in the stellar co-ordinates (Tab. 1)
2. Image co-ordinate measurement errors
3. Instrumental errors - These include: (A) both systematic and random errors of levelling and (B) timing errors.

Since all these errors sources are independent we can apply the law of error propagation to derive the error in the determined astronomical co-ordinates

Tab. 1: Star catalogues based on FK5 system

catalogue	Magnitude of listed stars	Accuracy (arcsec)	Remarks
Position and proper motion (PPM)	≤ 12	0.15	
Guide Star (GSC)	≤ 15	0.2 - 0.3	Accuracy can be worse than 0.5
Hipparcos and Tycho	≤ 12	0.002	

4.1 Timing

Relation between longitude and observed time, given in equation (8), can be expressed as

$$\Lambda = \alpha_Z - (UT1 + R) \tag{9}$$

and $$d\Lambda ='' -15dT \tag{10}$$

dt is in seconds of time while dL in arc-seconds.

From the above two equations the major problem is that errors in timing will translate fully into longitude error. To maintain precision less than 0.1 arcsec in longitude, the accuracy of our timing should be at the millisecond level. Furthermore timing should be in the UT1 system instead of UTC or GPS time scale. The timing device used for stellar observations in this project consisted of a Novatel OEM GPS card, GPS antennae and timing software on the host computer. The resolution of the timing was <1ms. The UTC time recorded was converted to UT1 using corrections obtained from the IERS bulletin. It is obvious here that there is no personal equation involved and the error due to timing is less than 0.1 arcsec.

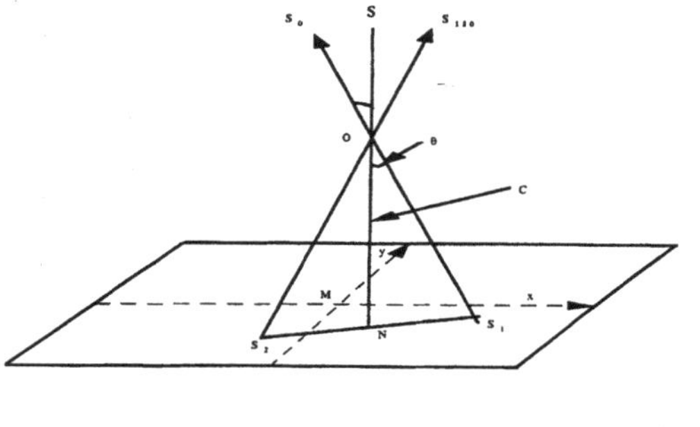

Fig. 5: Effect of small tilts of camera with respect to axis of rotation

4.2 Levelling Errors

There are two possible causes of mislevelling of camera axis:

1. Small tilt of the camera axis with respect to the axis of rotation. - Assuming the axis of rotation is vertical, the mislevelling will remain constant, as a matter of construction, but will change sign when the camera is rotated 180° in azimuth (figure 5)
2. Maladjustment of the level bubbles, so that the axis of rotation will be inclined from the centre when the bubbles are set at the centre of run. This error causes non-verticality of the camera axis. The effect of this cannot be removed or minimised by reversal (ROBINS 1976 p. 24f., BOMFORD 1980 p. 319).

In this work the approach to the problem of mislevelling was in two parts. First we used geodetic or precision level to level the instrument to within some specified accuracy, and applied the principle of reversal in the observations as mentioned in section [3]. Later we used rigorous mathematical modelling, based on the well known collinearity principle to determine inner orientation elements, i.e., shift (x_o, y_o), rotations or tilts (q_1, q_2, q_3) about OX, OY, OZ axes and correction to principal distance (dC) (see FOSU et. al. 1998)

5 Experimental Results

Results of experiments carried out using the IfEN CCD integrated telescope are shown in the tables and figures below. These results confirm that sub-pixel accuracy of image co-ordinate measurement can be achieved using centroiding algorithms. Furthermore the application of the principle of reversal and the mathematical model can eliminate or minimise levelling errors in zenith imaging.

Tab. 2: Precision of centroid determinations

Serial No.	Image name	Orientation (in deg.)	No. of common objects	Unit variance σ_0^2	Mean standard errors (in pixels)	
					σ_x	σ_y
1	Aug102	90	10	0.46	0.086	0.151
2	Aug10a2	90	14	0.44	0.054	0.196
3	Aug10a3	0	14	0.46	0.253	0.057
4	Aug10a4	270	10	0.32	0.060	0.186
5	Aug10b1	270	14	0.86	0.465	0.527
6	Aug10b2	0	13	0.39	0.207	0.071
7	Aug10b3	90	13	0.30	0.040	0.176
8	Aug10b4	180	13	0.25	0.159	0.061
9	Aug10c1	180	10	0.41	0.125	0.078
10	Aug10c2	90	9	0.50	0.134	0.247
11	Augbb2	0	14	0.41	0.245	0.063
12	Augbb4	180	10	0.25	0.115	0.080
13	Aug11a2	180	13	0.35	0.178	0.064
14	Aug11a3	180	11	0.23	0.091	0.040
15	Aug11a7	0	14	0.14	0.116	0.067
16	Aug11a8	0	7	0.11	0.070	0.066
17	Aug110	270	14	0.33	0.70	0.199
18	Aug111	270	14	0.56	0.054	0.157

Table 2 and figures 6 and 7 show that the measurement precision is of the order of 0.15 pixel. It is pertinent to note this accuracy is for one multiple image. Thus for four or more multiple images it is possible to obtain an accuracy better than 0.1 pixel corresponding to an accuracy better than 0.2 arcsec for portable imaging systems. (i.e., focal length between 80 and 150 cm., and pixel size <10 mm).

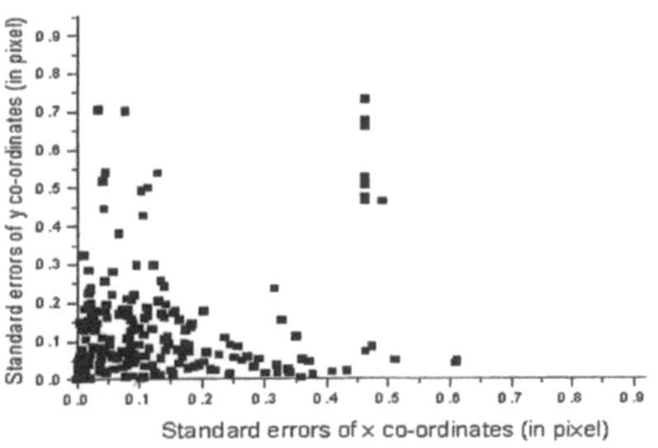

Fig. 6: Spread of image co-ordinate measurement errors

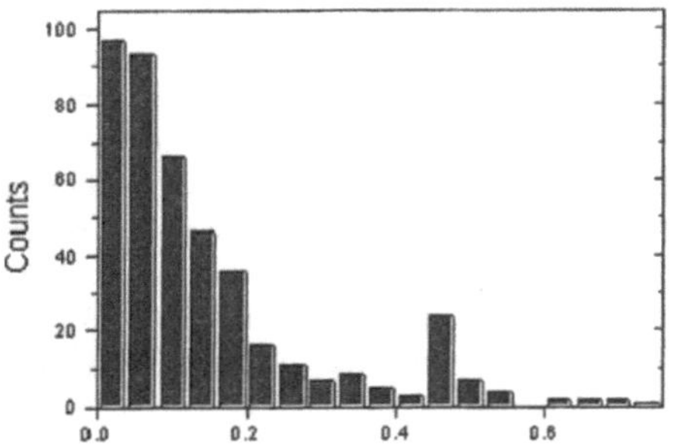

Fig. 7: Histogram of standard errors of co-ordinate measurements (mean = 0.44 pixel).

Fig. 8: One of the images used in the analysis. Exposure time 2 sec., Image size < 1cm x1cm, m<13.

From table 3 and 4 we see that the levels could not be brought to zero. This is due to the lack of fine levelling screws on the system. Secondly the tilt of the camera axis was far more than expected. Despite all these problems the results indicate very accurate astronomic positioning can be done by using a CCD integrated telescope. Also from figure 8 despite the small size of the CCD it could image enough stars to determine astronomical values.

Tab. 3: Astronomical latitude from zenith imaging

set	Image	β	Observed Φ (deg, min sec)	Mean Φ (deg min sec)	$d\Phi_m$ (secs)	Level reading (sec)	
						L0	L1
1	Aug10b2	0	48 19 41.11	48 04 42.02	-1.03	1.3	0.8
	Aug102	90	47 56 01.61			0.9	0.7
	Aug7s1	180	47 49 21.44			-4.0	11.0
	Aug10a4	270	48 11 43.94			0.5	0.7
2	Augbb2	0	48 19 40.81	48 04 43.27	0.22	1.4	0.7
	Aug10a2	90	47 58 03.64			1.0	1.4
	Augbb4	180	47 49 29.73			1.1	0.8
	Aug10b1	270	48 11 38.90			1.1	0.7
3	Aug11a7	0	48 19 18.86	48 04 43.09	0.04	-22.4	28.6
	Aug10c2	90	47 57 30.56			1.0	1.5
	Aug11a2	180	47 49 22.40			9.7	-12.4
	Aug111	270	48 12 40.54			16.0	20.0
4	Aug11A8	0	48 19 44.39	48 04 43.83	0.78	-22.0	29.5
	Aug10b3	90	47 57 42.76			1.4	-0.7
	Aug10c1	180	47 49 29.73			1.4	-0.9
	Aug7b2	270	48 11 58.45			5.0	3.0
			Mean of all the sets =	48 04 43.05			

Tab. 4: Astronomical longitude from zenith imaging

set	Image	β	Observed Λ (deg, min sec)	Mean Λ (deg min sec)	$d\Lambda_m$ (secs)	Level reading (sec)	
						L0	L1
1	Aug10b2	0	11 48 08.68	11 37 39.46	-0.69	1.3	0.8
	Aug102	90	12 00 03.41			0.9	0.7
	Aug10b4	180	11 27 11.46			-1.1	-1.0
	Aug10a4	270	11 15 14.27			0.5	0.7
2	Augbb2	0	11 48 07.33	11 37 40.96	0.81	1.4	0.7
	Aug10a2	90	11 59 57.77			1.0	1.4
	Augbb4	180	11 27 24.50			1.1	0.8
	Aug10b1	270	11 15 14.24			1.1	0.7
3	Aug10a3	0	11 48 01.51	11 37 40.03	-0.12	-0.9	1.5
	Aug10c2	90	11 59 51.69			1.0	1.5
	Aug10c1	180	11 27 11.78			1.4	-0.9
	Aug111	270	11 15 35.11			16.0	20.0
			Mean of all the sets =	11 37 40.15			

6 Conclusion

The above discussions, observations, analysis and results show that combining GPS and CCD will make available a new, fast, fully digital and solid state technology to modernise astronomical levelling which will enable us to overcome one of the main weakness of GPS positioning, the problem of height. Furthermore the combination will provide impetus for development of new technologies in the geo-science. Further developments are still going on in IfEN.

References

BOMFORD G. (1980): *Geodesy*. Clarendon Press,

FOSU C., EISFELLER B., HEIN G. W. (1998): *CCD to marry GPS*. Proceedings of ION-98, Nashville Tennessee.

GURTNER W., BURKI H. B.(1986): *Deviation of the vertical*. Proceedings of CERN accelerator School, Switzerland.

HEISKENEN W.A., MORITZ H.(1967): *Physical geodesy*. W. H. Freeman and Company, San Francisco

ROBINS A.R. (1976): *Field and geodetic astronomy*. Military engineering Vol. XIII Part IX, Hermitage U.K.

MACLEAN L.S. (1989): *Electronic and computer aided astronomy*. Ellis WoodLid, Chichester England.

SERVILLA M. J. (1980): *Determination of latitude and longitude from zenith observations*. Technica Topografica Vol.8 No.37

ZEILICK M., SMITH E. V. P.(1987): *Introductory astronomy and astrophysics*. CBS Publishing New York

Orbit Modelling Based on Laser and CCD-Observations

E. FRAGNER, R. WEBER[1], M. PLONER[2]

Abstract

The modelling of satellite orbits is often precondition for subsequent geodynamic investigations and belongs to the standard tasks of modern geodesy. Does the mission require a low altitude, e.g. in case of missions basically dedicated to the gravity field determination, the modelling proves to be extremely difficult. Considering a passive laser target like GFZ-1, the analyses suffers to a great extent from the small number of observations.

The paper discusses a Kalman filter approach to model the motion of the GFZ-1-satellite. This approach relies both on the integration of the Gauss equations of motion within the adopted force field as well as on the processing of current laser distance measurements, if available. In addition, data obtained from a second sensor, in particular a very sensitive CCD-camera, was taken into account. Analyzing CCD-images allows nowadays for the derivation of spherical satellite coordinates with an across-track accuracy of about +/-0.2"-0.3" (about +/-50cm in case of GFZ-1). In the along-track direction the accuracy of about +/-0.5" should be sufficient to improve the time-bias estimates to an order, which allows for the reacquisition to low-earth-orbiting laser targets. Simulation computations clearly demonstrate, that a network of about 7 stations equipped with CCD-cameras may bridge the geographical gaps of the laser network.

Orbit Model

In 1995 GFZ-1 was released from the MIR space station into the lowest orbit ever flown by geodetic satellites (initial altitude of 400 km, decay to 300 km within about 5 years). As already known, such low altitude satellite orbits are extremely sensitive to both, errors in the chosen gravity field expansion and the atmospheric model (surface forces). Even after a few revolutions they are causing major orbital errors (mainly in along track direction), which would become visible to an observer on the Earth as a shift in the rise time of the satellite (time bias). We decided, to describe the satellites orbit by a set of initial conditions ($x_s(t_o)$, position and velocity vector at t_o; respectively the corresponding set of Hill variables) and various conservative forces, enumerated in the sequence: Anomalous Earth's Potential (JGM-3, up to degree and order 40), gravitational attraction of the sun and the moon and solid Earth tides. We do not account for surface forces, although (in case of GFZ-1) their magnitude is comparable to forces imposed by sun and moon. Mean parameters of a drag function approximating the drag residuals, useful to improve the orbit predictions, can be obtained from *http://www.dgfi.badw-muenchen. de/ edc/edc_de.html*, but the overall uncertainties of the available atmospheric models are relatively large. Thus, the time bias increases very rapidly and can only be controlled in real-time, if at least a few laser range observations per day can be acquired and processed.

According to LANDAU (1988) we use the linearized form of an extended Kalman Filter, which takes advantage of a reference trajectory $\underline{x}'(t)$ for the satellite orbit (determined by numerical integration). The pure orbit-prediction within periods without observations relies on the continuous formulation (eqn.(1)).

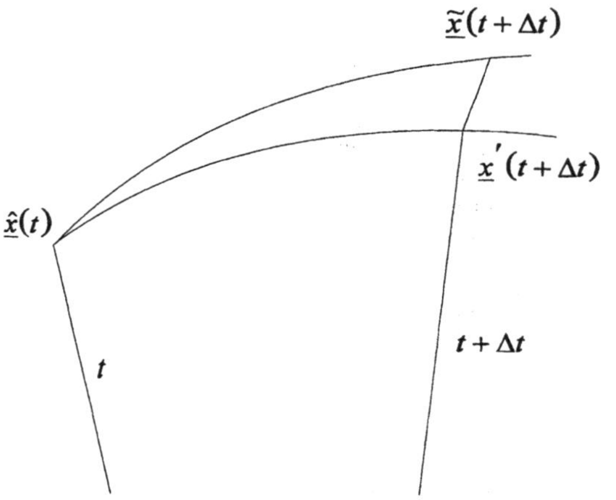

Fig.1: Reference trajectory and predicted orbit

State vector and covariance Prediction:

$$\tilde{\underline{x}}(t) = \int\limits_{t-1}^{t}\int \ddot{\underline{x}}[\hat{\underline{x}}(t-1),\, t]dt + B(t-1)) \cdot$$

$$\cdot(\hat{\underline{x}}(t-1) - \underline{x}'(t-1)) \qquad (1)$$

$$\tilde{Q}xx(t) = B(t-1)\cdot \hat{Q}xx(t-1)\cdot B^T(t-1) +$$

$$+Qnn(t-1) \qquad (2)$$

with

$\underline{x}(t)$	state vector
$B(t)$	transition matrix
Q_{xx}	covariance matrix of the state vector
Q_{nn}	system noise
~	prediction
^	update

[1] E. Fragner, R.Weber: Dept. for Theoretical Geodesy, University of Technology, Vienna, Austria; email: efragner@luna.tuwien.ac.at

[2] M. Ploner: Astronomical Institute, University Berne, Switzerland

For test purposes we computed the satellite orbit in the described force field from initial conditions over one day. Fig.2 - Fig.4 show the difference between the resulting orbit and precise ephemeris for GFZ-1, kindly placed to our disposal by the GFZ-1 Analysis Group (Potsdam). The calculation shows that above all in along track direction the predicted orbit is not satisfactory represented. The deviation is primarily produced by the atmospheric drag, as mentioned above.

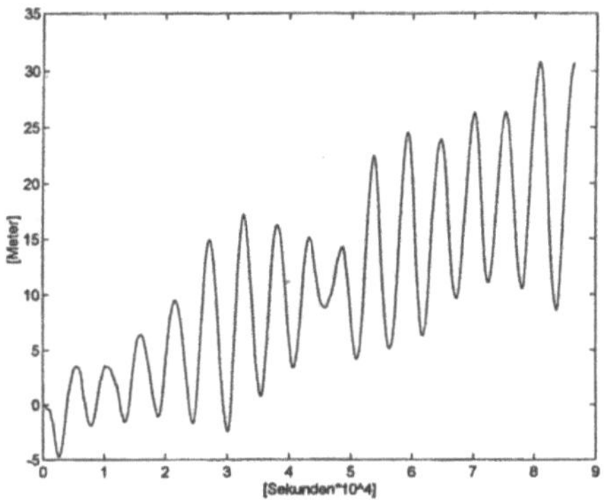

Fig.2: Residuals in radial direction

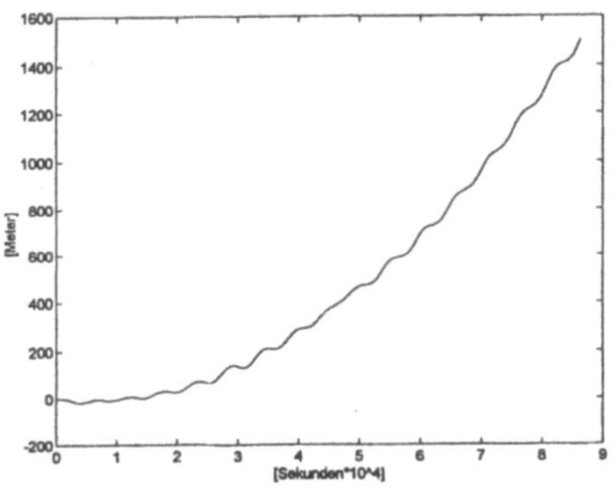

Fig.3: Residuals in along track direction

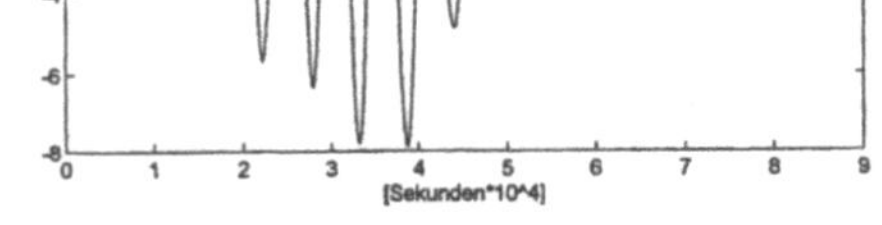

Fig.4: Residuals in out of plane direction

Subsequently actual laser distance measurements for supporting the position of the satellite were introduced. The laser ranges allow for the update of the state vector and the associated covariance matrix (eqns.(5 a-c)). The simplified observation equation reads:

$$d+v=\frac{1}{2}\Delta t\cdot c+\Delta d_0+\Delta d_s+\Delta d_b+\Delta d_r \qquad (3)$$

with

- d one-way range
- Δt the measured flight-time of the laser pulse between the start and stop signal
- c the speed of light
- Δd_0 the eccentricity correction on the ground
- Δd_s the eccentricity correction at the satellite
- Δd_b the signal delay in the ground system
- Δd_r the refraction correction
- v remaining systematic and random observation errors

Usually the position of the laser station $\underline{X}_{St}$ will be expressed in an Earth fixed reference frame ITRFxx (currently ITRF96, see BOUCHER et al., 1998). On the other hand the satellite position, obtained through an integration of the equations of motions, shows up in the inertial reference frame (ICRF). Therefore $\underline{X}_{St}$ must be transformed to the ICRF by means of the well known relation

$$\underline{X}_{st} = PN(t)U(t)W(t)\underline{X}_{st} \qquad (4)$$

where $PN(t), U(t)$ and $W(t)$ are the transformation matrices arising from the motion of the Celestial Ephemeris Pole (CEP) in the ICRF, from the rotation of the Earth around the axis of the CEP ($U(t)=R_3(-\theta_{T,Gr})$) and from polar motion ($W(t)=R_1(y_p)R_2(x_p)$).

We use form matrix A to linearize the distance measurement at the reference trajectory,

$$A=\left(\frac{\partial d}{\partial \Delta x_1}\;\frac{\partial d}{\partial \Delta x_2}\;\frac{\partial d}{\partial \Delta x_3}\;0\,0\,0\right)$$

with

$$\frac{\partial d}{\partial \Delta x_1}=\frac{\Delta x_1}{d'} \qquad \frac{\partial d}{\partial \Delta x_2}=\frac{\Delta x_2}{d'} \qquad \frac{\partial d}{\partial \Delta x_3}=\frac{\Delta x_3}{d'}$$

calculate the Kalman gain matrix and perform the update step by evaluating eqns. (5b) and (5c).

State vector and covariance Update (Observations):

$$G(t)=\tilde{Q}_{xx}(t)\cdot A^T(t)\cdot\left[A(t)\cdot\tilde{Q}_{xx}(t)\cdot A^T(t)+Q_{dd}(t)\right]^{-1}$$
$$\hat{\underline{x}}(t)=\tilde{\underline{x}}(t)+G(t)\cdot\left[d(t)-d'(t)-A(t)\cdot(\tilde{\underline{x}}(t)-\underline{x}'(t))\right]$$
$$\hat{Q}_{xx}(t)=\left[I-G(t)\cdot A(t)\right]\cdot\tilde{Q}_{xx}(t) \qquad (5\text{ a-c})$$

with

$d(t)$ range observation

$d'(t) = \left| \underline{x}'_s - \underline{x}_{st} \right|$

$A(t)$ Form matrix

Q_{dd} measurement noise

$G(t)$ Kalman gain matrix

$\Delta \underline{x}$ difference vector between satellite position and ground station

The small number of tracking stations involves, that quasi-simultaneous observations from two or more sites are very rare. On the other hand such multiple observations would be extremely helpful for stabilizing the satellite position in along-track as well as in radial and out-of-plane direction. In absence of quasi-simultaneous measurements and due to Fig.2 - Fig.4, we decided to relate each observation in eqn.(5b) to a sole correction of the along-track component. Furthermore pure distance measurements are not suited to update the velocity information (see form matrix A). Thus,

we relate position updates to a force acting solely in F_2^g -direction (see eqns.(6)) in the Gauss-system (equals along-track direction for eccentricity=0). Accounting for this additional disturbing force parameter within the integration steps provides us at every instant with a complete set of Hill variables, which can easily be related to the current position and velocity vector.

Equations of motion expressed in Hill variables

$$\frac{d\dot{r}}{dt} = F_1{}^g + \frac{G^2}{r^3}$$

$$\frac{dG}{dt} = r \cdot F_2{}^g$$

$$\frac{dH}{dt} = r\left(\cos i \cdot F_2{}^g - \sin i \cdot \cos u \cdot F_3{}^g\right)$$

$$\frac{dr}{dt} = \dot{r} \qquad (6\text{ a-f})$$

$$\frac{du}{dt} = \frac{G}{r^2} - \frac{r}{G} \cdot \cot i \cdot \sin u \cdot F_3{}^g$$

$$\frac{d\Omega}{dt} = \frac{r}{G}(\sin i)^{-1} \cdot \sin u \cdot F_3{}^g$$

with

i inclination

r the radius vector

$\dot{r}$ the radial velocity

G the angular momentum

H the z-component of the angular momentum

u the argument of latitude

Ω the longitude of the ascending node

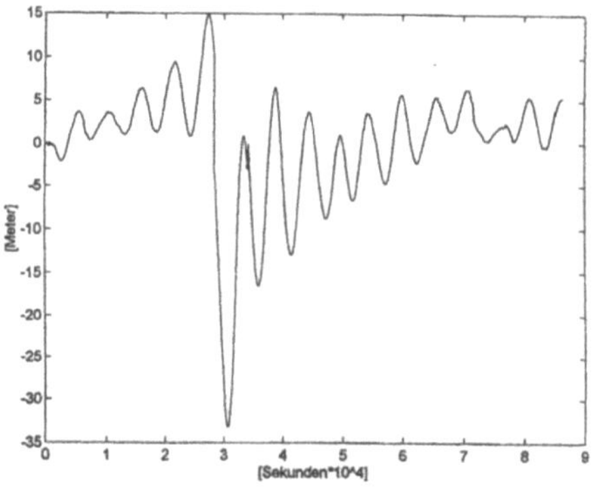

Fig.5: Residuals in along track direction

Eqns.(1,2) along with eqns.(5 a-c) are used to compute the estimate of the state vector, incorporating range measurements spread very inhomogeneous over the day. Fig.5 shows the difference between the orbit obtained from the Kalman filtering algorithm and the precise ephemeris for GFZ-1. Compared to the corresponding Fig.3, it is obvious, that the observations improve the situation dramatically.

In case of a period of bad weather conditions or if the satellite is in daytime for e.g. all stations in the Northern Hemisphere (problem of the gap in the distribution of the tracking stations in the Southern Hemisphere), the time-bias can neither be determined nor predicted with sufficient accuracy. This deficiency may prevent the reacquisition of the satellite and causes a lack of tracking data.

Subsequently the idea was born, to take additionally into account satellite positions obtained from a CCD-camera to calculate the timebias during and after such periods.

Spherical coordinates (α, δ) of the satellite can be calculated by means of eqns.(7)

$$\tan(\alpha - \alpha_0) = \frac{\xi}{\cos\delta_0 - \eta \sin\delta_0}$$

$$\tan\delta = \frac{\sin\delta_0 + \eta\cos\delta_0}{\sqrt{\xi^2 + (\cos\delta_0 - \eta\sin\delta_0)^2}} \qquad (7\text{ a, b})$$

with:

α_0, δ_0 Direction pointing towards the central point of the CCD-image

ξ, η Tangential coordinates of the satellite resulting from a 8-parameter transformation of the image coordinates (pixel) and subsequently transformed to the ICRF

To introduce this position in eqns.(5) we need the form matrix:

$$A = \begin{pmatrix} \dfrac{\partial\delta}{\partial x_1} & \dfrac{\partial\delta}{\partial x_2} & \dfrac{\partial\delta}{\partial x_3} & 0 & 0 & 0 \\ \dfrac{\partial\alpha}{\partial x_1} & \dfrac{\partial\alpha}{\partial x_2} & \dfrac{\partial\alpha}{\partial x_3} & 0 & 0 & 0 \end{pmatrix}$$

with

$$\frac{\partial\delta}{\partial x_1} = -\frac{x'_1 \cdot x'_3}{r'^2 \cdot \sqrt{x'^2_1 + x'^2_2}} \qquad \frac{\partial\delta}{\partial x_2} = -\frac{x'_2 \cdot x'_3}{r'^2 \cdot \sqrt{x'^2_1 + x'^2_2}} \qquad \frac{\partial\alpha}{\partial x_2} = \frac{x'_1}{x'^2_1 + x'^2_2} \qquad \frac{\partial\alpha}{\partial x_3} = 0$$

$$\frac{\partial\delta}{\partial x_3} = \frac{\sqrt{x'^2_1 + x'^2_2}}{r'^2} \qquad \frac{\partial\alpha}{\partial x_1} = -\frac{x'_2}{x'^2_1 + x'^2_2} \qquad r' \quad |\underline{x}'|$$

Fig.6: GFZ-1 06.10.1997 17.4730UTC Graz BMK 75 0.5s

Spherical satellite coordinates derived from a CCD-image achieve an across track accuracy level of ±0.2"-0.3" (about +/- 50cm in case of GFZ-1), while the along track accuracy is at the +/-0.5" level.

Conclusions

A Kalman Filter Model is very well suited for the combination of data, obtained from various techniques, at the observation level. Introducing present distance observations allows for overcoming deficiencies in the applied force field and thus for a reduction (respectively determination) of the time-bias. This procedure, usually performed by the laser community, reduces in our computations the along-track error of GFZ-1 form several hundred meters to less than 10 meter over the period of 24 hours. In addition, gaps within the laser-tracking data, caused by daylight periods or bad weather conditions may be bridged by high accuracy astrometric positions (CCD-images), gained from few stations situated especially in the southern hemisphere. In this case the presented results are of course preliminary. Detailed studies concerning both an improved parametrization of the orbit model and evaluating an optimal geographic distribution of CCD-sites are under way.

References

BOUCHER C., ALTAMIMI Z., SILLARD P.(1998): *Results and Analysis of the ITRF96* IERS Technical Note 24, Observatoire de Paris

BRAMMER K., SIFFLING G. (1985): *Kalman-Bucy-Filter - Deterministische Beobachtung und stochastische Filterung*. 2.Auflage, R.Oldenbourg Verlag, München, Wien

CUI C., MAREYEN M.(1992): *Gauss's equations of motion in terms of Hill variables and first application to numerical integration of satellite orbits Manuscripta Geodetica*, Volume 17, Nr. 3 /1992

KÖNIG R., BODE A., CHEN Z., GRUNWALDT L., NEUBERT R., REIGBER CH., SCHWINTZER P.(1996): *The GFZ-1 Mission: Design*, Operations and First Results Scientific Technical report STR96/09; GeoForschungsZentrum Potsdam

LANDAU H. (1988): *Zur Nutzung des Global Positioning Systems in Geodäsie und Geodynamik*, Heft 36, Schriftenreihe des Studienganges Vermessungs-wesen der Univ. der Bundeswehr, München

Unification of the GPS Work in Egypt

A. SHAKER[1], D. ALNAGGAR[2], A. A. SAAD[1]

Abstract

GPS proved itself as a powerful surveying tool. It is used in Egypt in many surveying and geodetic applications. Three main GPS networks are considered here. Each of these networks is made separately and treated in different way. The differences in the coordinates among the three networks at the common points in the three projects are investigated.

The aim of this work is to make use of all the GPS work in Egypt and unify it under one system of coordinates.

1. Introduction

GPS positioning techniques play an important role in establishing, extending or strengthening of geodetic networks. They also achieve a fundamental goal of geodesy through the determination of absolute position with uniform accuracy at all points on the earth's surface. It is well known that GPS coordinates are given with respect to a world geodetic coordinate system (WGS 84). Not all GPS work is the same and not all results (coordinates) can be said that they are related absolutely to WGS 84. To verify this, the following factors should be regarded during the observations:

- Observation session: the longer session at the station during the observations, the more accuracy of the obtained results.
- The used receivers; the receivers are single and dual frequencies. Dual frequency receivers give more accurate results than single frequency receivers do, especially in the long base lines and the way the tropospheric and the ionospheric delay are treated.
- The connection to IGS stations; IGS stations are absolutely related (defined) to WGS 84. These stations are worldwide distributed and they have receivers working full time.
- Using precise ephemeris; suitable surveying results can be obtained using broadcast ephemeris, but more accurate results can be obtained using the precise ephemeris to overcome the orbital errors.

2. Data under Consideration

The available data here are data from three projects, some information about them are the following:

a) HARN project; this project is made by ESA to cover the territory of Egypt by a zero order geodetic network. The network covers the area of Egypt in a very good geometry with spacing of about 200 km between the stations of class A [1]. Zero order networks consisting of 29 stations (class A) are observed and tied to the IGS stations and precise ephemeris were used in the processing, fig. (1). For the densification, another 113 stations were observed and tied to the zero order stations.

b) Project 1; the aim of this project was establishing a precise geodetic control network in different sites in Egypt. In the way of doing that, 21 primary stations of ESA were observed (to connect the new work to the national first order net) with dual frequency GPS receivers and tied to 4 IGS stations. The distribution of these stations is shown in fig. (1). The work in this project is nearly done in the specification of the HARN project [2]. Beside the 21 primary stations, more than 20 stations were observed in different places in Egypt.

c) Project 2; Finmap The objective of this project was to provide a base for 1:50 000 topographic mapping covering the Eastern Desert of Egypt. Primary net, which includes 31 stations, of which 6 are initials, is shown in fig. (1), and a densification net that includes 356 stations [3]. Station interval in the primary net is approximately 120 km. Stations for densification surveys are located along traverses between primary stations. Approximate interval between densification stations is about 20 km. The observations were accomplished using six single-frequency, five channel receivers. Point positioning was done at initial station E7 to position the network.

3. Transformation Model

Similarity transformation, using Bursa model, is used to compute the 7 transformation parameters relating two geodetic systems to each other. In Bursa model the relationship between the two systems is expressed by three translations (Xo, Yo, Zo), three rotations (Rx, Ry, Rz) and scale factor (Δ), [4], [5]. The transformation equation between the two systems is given in vector form as:

$$\bar{F} = \bar{T} + (1+\Delta) R \bar{G} - \bar{D} = 0 \qquad (1)$$

The final form of the transformation equation between the two systems after omitting terms containing the products of scale and rotation is given as:

$$F_K = \bar{T} + \Delta \bar{G} + R1 \bar{G} + \bar{G} - \bar{D} = 0$$

[1] Ahmed Shaker, Abdalla A. Saad: Shoubra Faculty of Eng. 108 Shoubra St. Cairo, Egypt

[2] Dalal Alnaggar: Survey Research Inst. 308 Alharam St. 12111-Giza, Egypt

Where:

T is translation (shift) vector between the origins of the two systems.

Δ is a scale factor.

R is the product of three rotation matrices about the second system axes

R1 = R - I

G is the vector of the second system coordinates

D is the vector of the first system coordinates

4. Estimation of the Transformation Parameters

Different least square solutions were carried out using Bursa model. The coordinates of the common points were used in two groups:

- Group 1 is the common points (4 points) between HARN and project 1
- Group 2 is the common points (5 points) between HARN and project 2

The following solutions were made for both groups:

- Solution 1: To solve for all 7 parameters
- Solution 2: To solve for 3 shift parameters
- Solution 3:To solve for 3 shifts and scale factor
- Solution4:To solve for 3 shifts and Rz parameters.
- Solution 5: To solve for 3 shift, Rz, and scale parameters.

For every solution and in both groups, the residuals at the common points are calculated as (V_φ, V_λ, V_h). For every solution, max, min, mean, RMS of these residuals are tabulated. The summation of the absolute residuals is also shown, tab. 1.

Tab. 1: Solution 1, all parameters are considered

	Group 1			Group 2		
	V_j	V_λ	V_h	V_φ	V_λ	V_h
max	0.017	0.019	0.036	0.306	1.245	0.594
	-0.016	-0.019	-.042	-0.297	-0.991	-0.931
	0.001	0.000	0.000	0.009	0.005	0.000
	0.014	0.016	0.037	0.244	0.909	0.612
	0.044	0.046	0.124	0.961	2.459	2.418

Tab. 2: Solution 2, 3 shift parameters are considered

	Group 1			Group 2		
	V_j	V_λ	V_h	V_φ	V_λ	V_h
max	0.029	0.027	0.038	0.783	1.003	0.527
min	-0.021	-0.028	-0.042	-0.344	-0.869	-0.990
mean	0.001	0.000	0.001	0.003	0.011	-0.017
RMS	0.025	0.024	0.037	0.455	0.769	0.697
Σ|V|	0.086	0.079	0.124	1.552	3.065	2.940

Tab. 3: Solution 3, 3 shift and scale parameters are considered

	Group 1			Group 2		
	V_j	V_λ	V_h	V_φ	V_λ	V_h
max	0.016	0.017	0.036	0.395	1.033	0.537
min	-0.016	-0.019	-0.043	-0.266	-0.822	-0.973
mean	0.001	0.000	0.000	0.002	0.009	0.000
RMS	0.014	0.015	0.037	0.287	0.785	0.693
Σ|V|	0.043	0.045	0.124	1.176	3.182	2.938

Tab. 4: Solution 4, 3 shift and Rz parameters are considered

	Group 1			Group2		
	V_φ	V_λ	V_h	V_φ	V_λ	V_h
Max	.011	0.012	0.019	0.548	0.806	0.372
Min	-.014	-0.022	-0.018	-0.336	-0.850	-0.235
Mean	.000	0.000	0.000	-0.003	0.003	0.000
RMS	.011	0.015	0.017	0.362	0.745	0.247
Σ|V|	.029	0.043	0.054	1.452	3.051	0.990

Tab. 5: Solution 5, 3 shift, Rz, and scale parameters are considered

	Group 1			Group 2		
	V_φ	V_λ	V_h	V_φ	V_λ	V_h
Max	0.029	0.028	0.038	0.694	1.214	0.580
min	-0.022	-0.027	-0.041	-0.245	-1.038	-0.948
mean	0.001	0.000	0.001	0.010	0.007	-0.017
RMS	0.025	0.025	0.037	0.389	0.914	0.615
Σ|V|	0.086	0.081	0.124	1.339	3.638	2.420

Looking at the above 5 tables, the solutions can be arranged according to their minimum residuals as follows:

- Solution 1 (all parameters) shows minimum residuals in both groups.
- Solution 3 (3 shift and scale) is the second one
- Solution 5 (3 shift, Rz, and scale) is the third one
- Solution 2 (3 shift) is the fourth one
- Solution 4 (3 shift and Rz) is the fifth one.

5. Results and Analysis

Transformation parameters from solution 1 are chosen to transform the coordinates in both projects to the coordinate system of HARN. In both cases, project 1 and 2, new values are computed and subtracted from their corresponding original values producing residuals. The statistics of these residuals are computed and tabulated as follows:

Tab. 6: Statistics of the differences between original and transformed values at all stations of both projects

	Group 1			Group 2		
	V_φ	V_λ	V_h	V_φ	V_λ	V_h
Max	-0.024	0.015	-0.285	-3.235	8.65	12.64
min	-0.066	-0.042	-0.386	-4.286	7.26	10.17
mean	-0.043	-0.022	-0.331	-3.830	8.14	11.31
RMS	0.011	0.015	0.035	0.332	0.24	0.7

Tab. 6 shows small residuals in the horizontal positioning (few Cm) and (0.331 m) in the heights as a mean shift in group 1. This is expected because the observations were taken nearly in the same specifications of HARN. These small differences may refer to the differences in the occupation time and the points of connection to IGS stations and the used way of adjustment.

Tab. 6 shows also big residuals in all coordinates of group2. This is also expected because of the different followed techniques between project 2 and HARN:

- Project 2 used single frequency receivers and HARN used dual frequency ones
- Project 2 used broadcast ephemeris and HARN used precise one
- Project 2 depended on point positioning at one station and HARN is connected to IGS stations.

The residuals are big but they relatively represent a shift between the two sets of coordinates in HARN and project 2. This is clear from the values of the mean and RMS of the differences in every direction (-3.830± 0.332, 8.149± 0.248, 11.316±0.697).

In both projects, the residuals are not correlated with the distance of the point from the common points. This is very good indicator that the GPS results are consistent.

Project 1 has 6 common points with project 2. To check the strength of the transformation unification procedure, differences between the transformed values in both projects are computed at these common points. Statistics of these differences are shown in tab. 7 as follows:

Tab. 7: Statistics of the differences of the transformed coordinates at the common points in both projects

	V_φ	V_λ	V_h
max	0.394	0.904	0.184
min	-0.162	-0.890	-0.561
mean	0.150	0.156	-0.205
RMS	0.214	0.673	0.303

Looking at tab. 7 the following can be concluded:

- The differences in the transformed latitudes in both projects are ranging from (-0.162 m) to (0.394 m) with (0.150 m) mean and bigger RMS
- The differences in the transformed longitudes are bigger than the differences in latitudes. This is expected because the common points in the solutions cover almost the whole territory of Egypt in the north-south direction and cover just a strip in the east-west direction (30 36 E: 34 8 E) where Egypt ranges from longitude 25 to 37 E, fig. (2), (3).

6. Conclusion

Three of the GPS projects in Egypt are HARN, project 1, and project 2. HARN has 142 stations cover the area of Egypt in good geometry and very good precision. HARN has 29 zero order stations tied to the IGS stations and the rest of 142 stations are tied to zero order stations. Project 1 has 21 stations from the primary net of Egypt and more than 20 stations in different places in Egypt. Project 1 is done nearly within the specifications of HARN. Project 2 was in the Eastern Desert of Egypt and has 387 stations among them are 31 first order stations.

When GPS projects are done in different specifications, so it can not be said that the results are related to WGS 84 with the same way (weight). In this research, project 1 and 2 are transformed to the High Accurate Reference Network (HARN) of ESA.

Bursa model of similarity transformation is used at the common points between the projects.

After unification, 570 GPS stations are covering Egypt within good geometry, densification, and precision and they are related to the coordinate system of HARN, which is WGS 84. So it is recommended here that all GPS projects done in Egypt should be unified in the same way with HARN to be consistent with the specifications of the national work in ESA. It is also recommended that in any new GPS work, users should use stations of the unified GPS network as base (reference) stations in their work.

7. References

[1] ESA (1997) *"Final Seminar Schedule Introducing The New Egyptian Datum-1995"*. ESA, Giza, Egypt.

[2] Personal Communication, Bolbol S., Shoubra faculty of engineering, Cairo, Egypt.

[3] Finmap (1988) *"Final Report of 1:50 000 Topographic Mapping in Eastern Desert"*, ESA, Giza, Egypt.

[4] ADAM J., HALMOS F., VARGA M. (1979) *"On the Concept of Combination of Doppler Satellite and Terrestrial Geodetic Networks"*. Has, Sopron.

[5] LEICK A., VAN GELDER B.H.W.(1975) *"On Similarity Transformations and Geodetic Network Distortions Based on Doppler Satellite Observations"*. OSU Report 235.

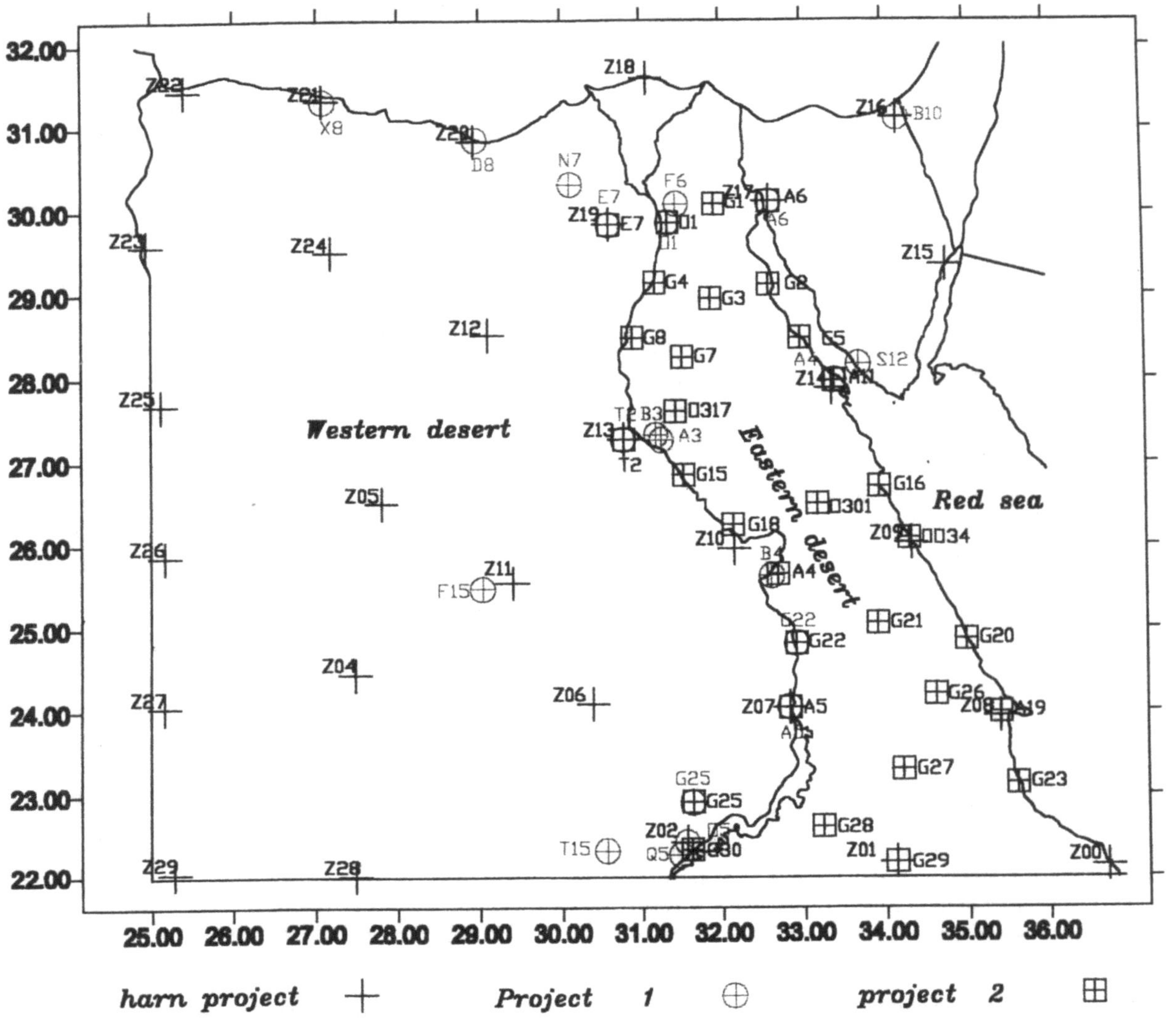

Figure 1

Main GPS networks in Egypt

Earth Orientation Parameters Measured by Space Geodesy Techniques

R. Devoti, M. Fermi, C. Ferraro, R. Lanotte, V. Luceri, A. Nardi, R. Pacione, P. Rutigliano, C. Sciarretta[1], G. Bianco, F. Vespe[2]

Introduction

Nowadays the space geodetic techniques are providing the geophysicists with a lot of highly accurate products, as the EOP series. Even if at an international level a huge effort to standardize some guidelines to be followed in the analyses has been done (e.g. IERS standards), the peculiar features of each technique permeate the results and make their comparison and/or integration a topic to be handled with care.

For instance, in the frame of the EOP determination problem, it is well known that the drifts derived from the differentiation of two different EOP time series must mirror, in a well defined way, the time derivatives of the rotation parameters derived from the differentiation of the two relevant velocity fields. Often, instead, also in the case of the solutions submitted to the IERS to have a combined unique solution for the TRF, some discrepancies are evident. Besides, also our SLR and VLBI solutions show this kind of discrepancy at the 0.1 mas/yr level in one of the component. That general problem makes the question arise, whether the space geodetic techniques recognize the EOP with equal sensitivities.

The above premise motivates this work, aiming to characterize the accuracy level obtainable by the different space techniques in estimating the EOP and the time span necessary to stabilize it.

Dilution of precision: the concept

Our analysis strategies of the various space technique data rely mainly upon the Bayesian Least Squares. In this approach the evaluation of the error budget is very straightforward; if, as usual, A is the design matrix of the problem to be solved and W the weight matrix of the observations, the quantity $(A^TWA)^{-1}$ is the covariance of the estimated parameters, namely the model propagation of the variances/covariances on the measurements to the variances/covariances on the estimated parameters.

Neglecting observations weights, we can define as Dilution of Precision factors the square root of the sum of selected diagonal elements of the $(A^TA)^{-1}$, spanning all the cases between the trivial ones (the trace of the matrix and the single diagonal entries). These factors say roughly how an average error on the observations maps on the formal error of a group of estimated parameters.

If we restrict our point of view just to the relationship between the measurements and the EOP, namely x and y of the pole and UT, the partial derivatives appearing in the design matrix are as follows:

$$A = \begin{pmatrix} \frac{\partial m_1}{\partial x} & \frac{\partial m_1}{\partial y} & \frac{\partial m_1}{\partial UT} \\ \cdot & \cdot & \cdot \\ \cdot & \cdot & \cdot \\ \cdot & \cdot & \cdot \\ \frac{\partial m_n}{\partial x} & \frac{\partial m_n}{\partial y} & \frac{\partial m_n}{\partial UT} \end{pmatrix}$$

where the dimension is nx3, being n the number of observations.

For EOP, the dependence of the design matrix on the acquisition geometry is very strong. For range observations, as in SLR and GPS cases, the entry of the design matrix relevant to the observation i and to the parameter j can be written as:

$$A_{ij} = \frac{\partial m_i}{\partial p_j} = \frac{\partial m_i}{\partial \mathbf{X}_{Sta}} \cdot \frac{\partial \mathbf{X}_{Sta}}{\partial p_j} \approx \mathbf{k}_i \cdot \frac{\partial \mathbf{X}_{Sta}}{\partial p_j}$$

where $\mathbf{k}_i$ is the direction Station/Satellite.

In the case of VLBI, where the observation is a Time Delay between two stations as receiving the same signal, we can write:

$$A_{ij} = \frac{\partial m_i}{\partial p_j} \approx \mathbf{k}_i \cdot \frac{\partial (\mathbf{X}^1{}_{Sta} - \mathbf{X}^2{}_{Sta})}{\partial p_j}$$

where $\mathbf{k}_i$ is the direction Stations baseline/Radiosource.

The difference between the characterization of GPS/SLR observations vs the VLBI ones implies that, while in the first case, the geometrical key factor in designing the network is the site distribution, in the second case the key factor is the length and the orientation of baselines between couples of stations.

We refer to the above approach, where the measurements are related just to the EOP, as 'partial' approach in the following. We define the relevant DOP factors as:

1 R. Devoti, M. Fermi, C. Ferraro, R. Lanotte, V. Luceri, A. Nardi, R. Pacione, P. Rutigliano, C. Sciarretta: Telespazio SpA - CGS Matera Italy

2 G. Bianco, F. Vespe: ASI - Agenzia Spaziale Italiana - CGS Matera Italy

$$DOP_x = \sqrt{\left[\left(A^T A\right)^{-1}\right]_{11}}$$

$$DOP_y = \sqrt{\left[\left(A^T A\right)^{-1}\right]_{22}}$$

$$DOP_{UT} = \sqrt{\left[\left(A^T A\right)^{-1}\right]_{33}}$$

In designing an experiment the DOP factors from the 'partial' approach should be minimized. For SLR/GPS, in the case of polar motion, the best observations seem to be those ones at low elevations. Specializing for polar stations, it means that inclined satellites are the best ones, while for mid-latitude stations the polar satellites are favourite. In general, stations placed on the Greenwich meridian are sensitive to rotations about the y-axis (x pole), while stations placed on the meridian orthogonal to the reference one are sensitive to rotations about the x-axis (y pole). For the rotation parameter, a general policy foresees as favourite the low elevation E-W observations, inclined satellites and non-polar stations. In the VLBI case, for what stated above, N-S baselines help the polar motion determination, while E-W baselines help in the case of rotation.

This 'partial' approach gives obviously unrealistic accuracy values: other parameters, 'nuisance parameters', are necessarily involved in the data reduction, even if the main aim is, as in this case, the estimation of EOP. To take into account also their influence, the design matrix has to be generalized:

$$A = \begin{pmatrix} \frac{\partial m_1}{\partial x} & \frac{\partial m_1}{\partial y} & \frac{\partial m_1}{\partial UT} & \frac{\partial m_1}{\partial p_1} & \cdots & \frac{\partial m_1}{\partial p_m} \\ \cdot & \cdot & \cdot & \cdot & \cdot & \cdot \\ \cdot & \cdot & \cdot & \cdot & \cdot & \cdot \\ \frac{\partial m_n}{\partial x} & \frac{\partial m_n}{\partial y} & \frac{\partial m_n}{\partial UT} & \frac{\partial m_n}{\partial p_1} & \cdots & \frac{\partial m_n}{\partial p_m} \end{pmatrix}$$

where the dimension is nx(3+m), being n the number of observations and m the number of nuisance parameters.

This approach will be referred to as 'complete' approach. The nuisance parameters included for the three techniques are respectively:

- VLBI: clock parameters, tropospheric zenith path delay, nutation corrections;
- GPS: tropospheric zenith path delay, station clock constant bias, initial ambiguities;
- SLR: satellite positions, satellite residual accelerations, station measurement bias.

If we assume to have measurement derivatives w.r.t. EOP parameters in the first three columns of the design matrix, the DOP factors of the 'complete' approach are defined as previously. Summarizing, they can tell, after the careful analysis of the geometry constraints by means of the 'partial' approach, how a more realistic accuracy level is degraded by the nuisance parameters.

DOP analysis: results

The DOP factors as described above are handy tools to characterize the accuracy level in time of EOP as derived from the three techniques in 'laboratory' conditions, more or less near to reality, as in 'complete' and in 'partial' approach. Besides the DOP factors time series, also the correlations time series among the three EOPs and with the other parameters can enlighten the strength and the weakness of each technique in deriving them.

We analyze three typical test cases of global networks, well balanced geographically, acquiring few days of observations for the three techniques.

The analyzed SLR network acquires five-day observations from the LAGEOS satellites; the VLBI one acquires one-day observations from six well distributed stations; the GPS one acquires one-day observations from fifteen well distributed stations.

All the data reductions have been carried out along the lines of the consolidated analyses performed at Matera Space Geodesy Center, following the IERS recommendation.

In Fig.1 are plotted the DOP factors vs number of observations from the three networks above.

To quantify the comparison among the results, we fit the data with an exponential function of the kind $y=ax^b$, being a and b the solve-for-parameters. To assess for the reliability of the model assumption, we estimate also the coefficient of multiple determination R^2: the more near to unity, the better the model fitting.

$$R^2 = 1 - \frac{\sum\left(y_i - \hat{y}_i\right)^2}{\sum\left(y_i - \bar{y}\right)^2}$$

being $\bar{y}$ the average of y_i data and $\hat{y}_i$ the modeled value corresponding to y_i.

In the case of the 'complete' approach the b coefficients are farther from -1/2 than those from the 'partial' approach; the exponent -1/2 is the one characterizing the problem of deriving the formal error for an average of data each one with its uncertainty. As that average problem has a very simple design matrix (a unit vector), it means that, if b is far from that value, the actual estimation problem is more complex than that and in particular the correlations with the nuisance parameters are not negligible.

In the case of 'partial' approach, the coefficient of multiple determination is near to unit for all the techniques; the 'complete' approach provides values close to the previous ones except for the VLBI technique in the case of UT determination: in this case the presence of nuisance parameters affects also the kind of model fitting the data.

Comparing globally the 'complete' and 'partial' approach, it seems evident that from a theoretic point of view the VLBI technique assures better accuracy at the same level of number of observations. Nevertheless, the VLBI technique is the one more degraded by the nuisance parameters in the case of pole coordinates, while it keeps its primacy in the case of UT.

In general, as expected, the 'complete' approach brings higher final DOP values under the same condition; the critical zone (high variation of the DOP factors at small variation of number of observations) in the 'complete' approach is shifted towards high values of number of observations; after a high number of observations (>100) the decay rate of the DOP factors approaches the rate observed in the 'partial' approach.

The translation of those results in terms of acquisition times says that after one day, GPS and VLBI have acquired a sufficient number of measurements to be in the 'stable' part of the DOP curve for all the EOP, while SLR technique is on the 'knee' of the curve. In terms of reached accuracy level, GPS seems to be the best one. In addition GPS and VLBI seem to be able to give information on the subdiurnal EOP determination.

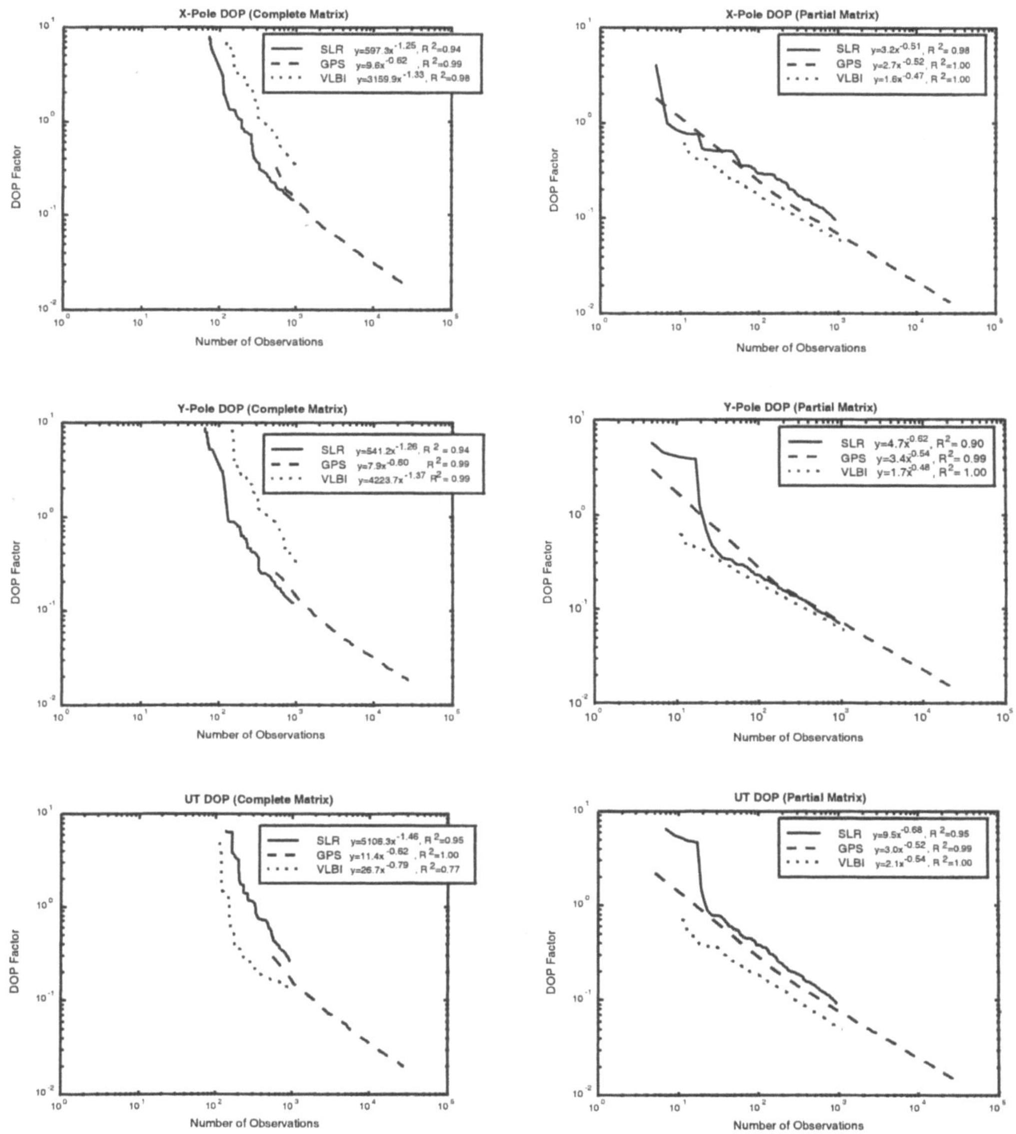

Fig. 1

Conclusions

The DOP approach is very general: if applied in a 'partial' sense, it allows to draw some strong considerations on the geographical design of the geodetic network and its aptness to optimize the accuracy of estimation for parameters relevant to some physical problem; even if purely geometrical, this approach allows also to derive a lower bound in the data time span needed to achieve the requested accuracy.

A more realistic treatment of the design matrix, by adding at least the nuisance parameters of the considered space geodetic technique, allows to quantify how that accuracy level is degraded in the real case.

Another aspect of the optimization of estimates vs time, not related in this paper, resides on the study of the correlation of parameters vs time. In fact, even if the continuous accumulation of observations makes the DOP lower, after some time, characteristic of the problem, some parameters can become correlated.

New Method for the Selection of Additional Sites for the Homogenisation of an Inhomogeneous Cospherical Point Distribution

H. HASE[1]

Abstract

Global reference frames are based on the coordinates of observatories on the Earth's surface employing geodetic space techniques. In a first order approximation the observatories represent a cospherical point field. The existing global reference frames are based on an inhomogeneous distribution of the observing sites. It is possible to improve the homogeneity of the cospherical point distribution in an optimized manner by adding new observatories at the center point of the largest empty circle on the sphere.

The sequential adding of new platforms will converge to a homogeneous point distribution. This method was applied to the distribution of fundamental stations in order to find geometrically the best sites on which possibly the Transportable Integrated Geodetic Observatory (TIGO) could be operated.

1. Method of Densification of Inhomogeneous Cospherical Point Fields

Observatories around the globe represent the backbone of global geodetic reference frames. Most of their sites have been selected due to national interests. For a *global* reference frame the distribution is adverse and a more homogeneous site distribution is required in order to:

- understand the system Earth and test global plate tectonic models,
- avoid systematics in the determinations of satellite orbits,
- provide globally the same level of accuracy of the reference frame.

Additional observatories are required. Where should they be located? Besides geological, climactical, social and political arguments it exist also a strict geometric approach to homogenise the unequal distribution.

In a first order approximation we can look at the existing platforms being located on a sphere. If one is able to locate the most remote point from the other platforms, that would be the best candidate to setup a new site for the homogenisation process. With infinite iterations by adding always the most remote point (or a site being closely located to the most remote point in case it is in the ocean) the inhomogeneous site distribution converges to more homogeneity.

In order to solve this problem mathematically, one has to find the largest empty circle, in which no existing observatory lies. A circle is defined by three points. The center of the largest empty circle is the most remote location from the existing point distribution. Therefore the densification process can also be called the **method of the largest empty circle**.

The Delaunay-triangulation provides the closest point neighbours of any given point. If the point distribution is cospherical, then the normals of each Delaunay triangle indicate the Voronoi-vertex. The geometrical dual to the Delaunay triangulation is the Voronoi-diagram, which contains the Voronoi-ridges showing the lines of maximum distance to the given point distribution (O'ROURKE: *Computational Geometry in C*, 1994). The Voronoi-vertices are the candidates for the center of the empty circles formed by three surrounding points. One of the Voronoi-vertices is the center of the largest empty circle.

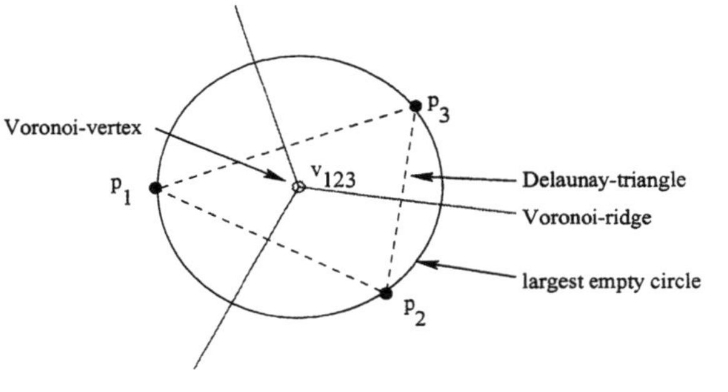

Fig. 1: Delaunay-triangle $\overline{p_1p_2p_3}$ and Voronoi-vertex v_{123} in a plane. The center of the empty circle is the Voronoi-vertex on a sphere. The Voronoi-ridges mark the lines of maximum distance from the given point distribution.

2. Distribution of Fundamental Stations

The key role in the generation of terrestrial reference frames like the ITRF is played by fundamental stations, which contain all important space geodetic techniques as such as VLBI, SLR, GPS and possibly DORIS at one site and which are operated on a long term basis emitting high quality data. The concept of fundamental stations allows the determination of the eccentricity vectors between the geodetic reference points by the means of a local terrestrial survey.

[1] Hayo Hase: Bundesamt für Kartographie und Geodäsie, Fundamentalstation Wettzell, D-93444 Kötzting, Germany; Email: hase@wettzell.ifag.de

In 1998 the situation shows that permanent observations with VLBI, SLR, GPS are done at the fundamental stations of

- **Wettzell**, Germany,
- **Matera**, Italy,
- **KeyStone Tokyo**, Japan,
- **Greenbelt**, U.S.A.

In addition there are quasi-fundamental stations, where the distances of the different instruments are larger than 1 km. They are making the local survey more difficult since it should be one magnitude more accurate than the (inter) continental baseline determinations provided by the geodetic space techniques. However the results from these stations represent the same region in a global reference frame. The list of these stations consists of

- **Shanghai**, P.R. China. It exists a distance of ≈ 2 km between SLR and VLBI station. The SLR-station will be moved to the VLBI-radio telescope site soon.
- **Canberra - Tidbinbilla**, Australia. It exists a distance of ≈ 27 km between the SLR-observatory and the DSN-station, which supports VLBI-observations in Australia.
- **Kokee Park - Maui**, Hawaii. The distance between the SLR-observatory and the VLBI-station is ≈ 387 km. Due to the importance of this site in the pacific ocean and the high quality of data these stations were also included in the list of fundamental stations.

The idea of collocating instruments was picked up by NASA-GSFC. Mobile SLR-stations can be used to form fundamental stations, if they are operated continuously and collocated to VLBI-radio telescopes. Hopefully in the near future a new fundamental station will be formed:

- **Hartebeesthoek**, South Africa. Negotiations are on the way to use MOBLAS-6 of NASA-GSFC permanently next to the VLBI-radio telescope.

Another fundamental station which will contribute to global reference frames in the future is the **Transportable Integrated Geodetic Observatory**. It is the first time, that a complete fundamental station can be positioned with respect to the maximisation of the gain for global reference frames.

3. TIGO - a Transportable Fundamental Station

The Transportable Integrated Geodetic Observatory (TIGO) was design to serve the terrestrial reference frames at remote locations in the southern hemisphere. TIGO consists of a

- radio telescope for geodetic VLBI,
- an optical telescope for SLR,
- permanent GPS receivers,
- super conducting gravity meter,
- broadspectrum seismometer,
- meteorological sensors.

For transportation all instruments will be housed in six standard containers. During the operation the containers are transformed to the operation rooms and workshops (see TIGO-Poster).

Fig. 2: TIGO at platform in Wettzell during 1998.

4. Proposed Sites for Fundamental Stations according to the Method of the Largest Empty Circle

The site selection according to the method of the largest empty circle considers geometrical aspects only. Other aspects regarding the use of TIGO needs to be considered:

- partner institute in the hosting country for a joint operation of TIGO,
- geophysical stability of the site,
- climatic conditions of the site,
- property of partner institute to build the platform for TIGO,
- provided infrastructure by the partner institute,
- educational level of operators.

For the proposal of possible locations for the fundamental station TIGO one has to consider the global distribution of existing fundamental stations. For the following computations the set of station coordinates given in Tab. 1 had been used.

Tab. 1: Averaged coordinates of fundamental stations based on the ITRF96. After reduction to a unit sphere they have been used for the largest empty circle computation in order to find the best candidates for TIGO sites.

Name	Long. [°]	Lat. [°]	Alt. [m]
Wettzell	12.878	49.145	662.996
Matera	16.704	40.649	534.371
Hartebeesthoek	27.697	-25.888	1486.177
Shanghai	121.255	31.122	15.656
Tokyo	139.489	35.710	122.447
Canberra/Tidb.	148.962	-35.517	934.191
Kokee Park	-159.665	22.126	1167.349
Greenbelt	-76.827	39.022	14.505

The repeated computation of the largest empty circle on a global sphere with adding one new site in each step is given in the following subsections. The fundamental stations listed in Tab. 1 had been used as the initial site distribution.

The Voronoi-diagrams below indicate the fundamental stations with stars. The center points of the three largest empty circles are marked with the ordering number and a small circle around the intersection of three Voronoi-ridges, which are the lines of maximum distance between two neighbouring sites. The table to each plot lists the geographical location of the Voronoi-vertices and the radius of the largest empty circle as baseline to its closest sites.

The results of this analysis give a first clue for sites in the southern hemisphere, which deserve a fundamental station for the improvement of the distribution in the global network.

The final decision about the abroad sites for TIGO has to consider the other above mentioned and not computable aspects as well.

4.1 Computation of the 1st Proposed Site

The largest empty circle has a baseline radius of 8648 km and covers almost a complete hemisphere of which most of the area is in the southern hemisphere (Tab. 2). This fact emphasizes the lack of a fundamental station in the southamerican-pacific area. Since the center point of the largest empty circle is in the ocean, the university town of Valdivia in the south of Chile was chosen as a suitable first proposed site for the subsequent computations.

Tab. 2: Coordinates of the Voronoi-vertices and baseline radii of their cocircle (Fig. 3).

vertex	long. [°]	lat. [°]	basel. rad. [km]
1	-84.295	-46.276	8648.410
2	-102.360	-38.654	8280.435
3	-29.389	-2..019	6349.454
4	88.310	-16.316	6076.100
5	5 67.505	11.028	5732.164
...	...	...	...
11	164.131	2.352	4391.572
12	139.744	-0.423	3936.786

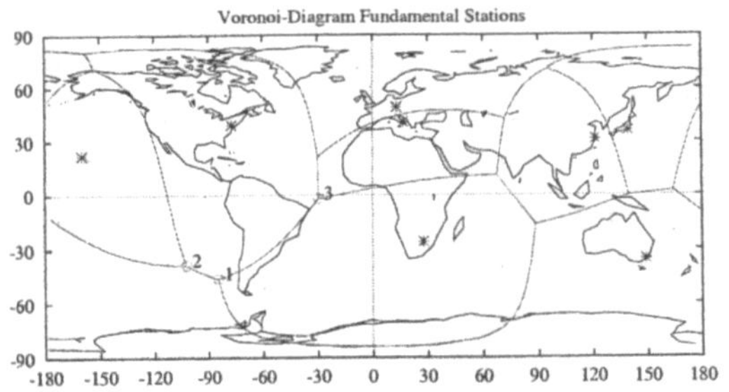

Fig. 3: Voronoi-diagram with fundamental stations. The numbered Voronoi-vertices label the order of the largest empty circle (1=largest, 2= second largest, 3 = third largest empty circle). There is a lack of fundamental stations in the southamerican pacific area. As first proposed site the close by town of Valdivia in the south of Chile was selected in the first step.

4.2 Computation of the 2nd and 3rd Proposed Site

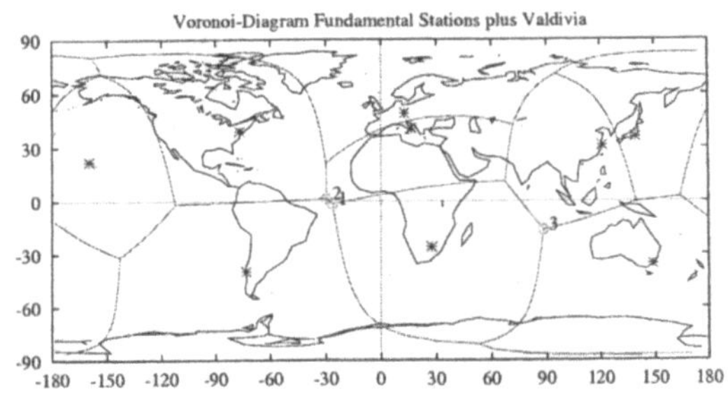

Fig. 4: Voronoi-diagram with fundamental stations with respect to an additional site in Valdivia. The largest empty circles of this configuration recommend sites in the Atlantic and in the Indian Ocean. The islands of Ascension and Cocos Island have been chose to be the next proposed sites.

Tab. 3: Coordinates of the Voronoi-vertices and baseline radii of their cocircle with respect to an additional site in Valdivia (Fig. 4).

vertex	long. [°]	lat. [°]	basel. rad. [km]
1	-26.231	-1.410	6111.460
2	-29.454	1.300	6106.756
3	88.311	-16.316	6076.101
4	52.964	-80.901	5999.382
5	-142.908	-31.504	5974.156
...	...	...	...
13	164.131	2.353	4391.573
14	139.744	-0.424	3936.790

The addition of Valdivia drops the baseline radius of the largest empty circle down by 30% to 6111 km. Consequently the largest empty circles distribute much better. Two sites in the Atlantic and in the Indian Ocean compete as sites to host a second fundamental station. As before the best points are in the water, therefore **Ascension** and **Cocos Island** as close by land surfaces have been chosen to be included in the next computation steps. The representing areas of these two sites are uncorrelated Voronoi-regions and their baseline radii of the first two largest empty circle are about the same. Therefore both stations can be added at once in order to compute the next optimal site.

4.3 Computation of the 4th Proposed Site

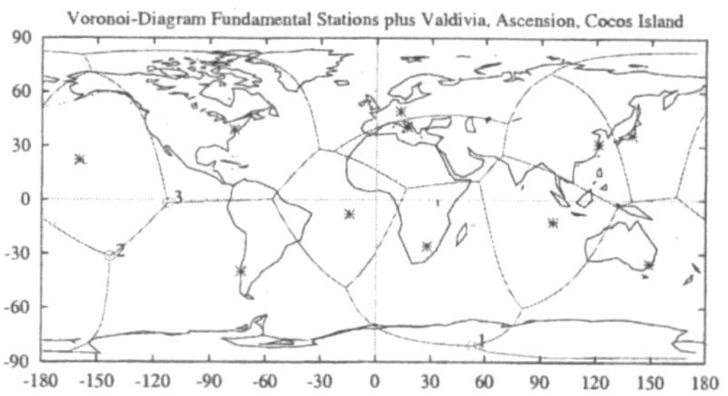

Fig.5: Voronoi-diagram with fundamental station with respect to three additional sites in Valdivia, Ascension and Cocos Island. With the excursion of Antarctica follows the 4th site on the pacific Pitcairn Island.

If Cocos Island is considered to be the third candidate then the Voronoi-diagram points in the neighbourhood of **Pitcairn Island**. The final computation can be done with Valdivia, Ascension, Cocos Island and Pitcairn Island as additional funentdamal stations.

Tab. 4: Coordinates of the Voronoi-vertices and baseline radii of their cocircle with respect to three additional sites in Valdivia, ns Asceion and Cocos Island (Fig. 5).

vertex	long. [°]	lat. [°]	basel. rad. [km]
1	52.964	-80.901	5999.380
2	-142.908	-31.504	5974.156
3	-112.255	-1.640	5585.564
4	-158.713	80.499	5437.318
5	80.292	-60.083	5341.655
...	...	...	...
17	-23.212	26.693	3901.791
18	16.522	6.471	3728.283

4.4 Computation of the 5th Proposed Site

With the four additional sites in the southern hemisphere a fifth site would be either in **Antarctica** or **Arctica**. From Tab. 5 it can be seen that the baseline radius of the largest empty circle is about 6000 km without and about 5000 km with occupations in the arctic regions. The map of Fig. 6 shows with four additional points already a much more homogeneous site distribution in comparison with that of Fig. 3.

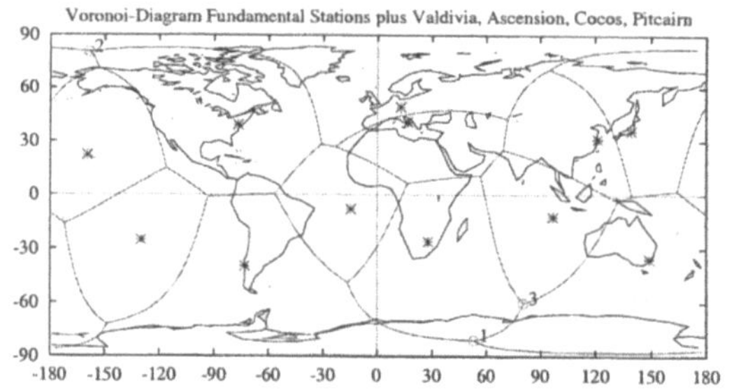

Fig. 6: Voronoi-diagram with fundamental station with respect to four additional sites in Valdivia, Ascension, Cocos Island and Pitcairn Island. In this configuration the recommended sites are located in arctic or antarctic regions. Those areas needs to be occupied with special equipment.

Tab. 5: Coordinates of the Voronoi-vertices and baseline radii of their cocircle with respect to four additional sites in Valdivia, Ascension, Cocos Island and Pitcairn Island (Fig. 6).

vertex	long. [°]	lat. [°]	basel. rad. [km]
1	52.964	-80.901	5999.380
2	-158.713	80.499	5437.318
3	80.292	-60.083	5341.655
4	-153.364	71.160	5320.694
5	-148.574	-71.997	5208.071
6	56.585	10.709	4987.213
7	69.151	25.092	4980.047
8	-56.485	0.332	4651.939
9	-93.404	-1.062	4644.672
10	94.938	70.517	4588.103
11	-115.973	14.714	4567.480
12	-16.083	-48.828	4464.338
13	71.847	42.463	4429.093
14	164.131	2.353	4391.573
15	-171.801	-16.080	4350.850
16	-30.907	28.080	4294.909
17	139.744	-0.424	3936.793
18	131.828	-3.324	3913.007
19	-23.212	26.693	3901.791
20	16.522	6.471	3728.283

5. Conclusion

The homogenisation of the existing distribution of sites in the global reference frame can be achieved using the method of the largest empty circle on a sphere to define new locations for additional sites. A transportable fundamental station is the ideal tool to optimise the global network configuration under this aspect. If the configuration changes, a new computation with the presented method will point to a new optimal location in the global network.

Westford VLBI to GPS Vertical Tie and Implications for the TRF

A. E. NIELL[1]

Abstract

The difference between the heights of the VLBI and GPS antennas at Westford as found in ITRF96 differs from the optical survey value by approximately 3 cm. The height of the GPS antenna is, however, a function of the minimum elevation angle cutoff with a sensitivity of approximately -3 mm/°. A sample of other Dorne Margolin choke ring antennas of the global GPS network shows similar sensitivity, while two Trimble antennas have more severe sensitivities which are of the opposite sign. Until the elevation dependent phase error can be measured *in situ* there is no way to determine the absolute height of the current GPS antennas. As a consequence, the scale of the Terrestrial Reference Frame cannot be determined by GPS to better than several parts per billion, and the heights of individual sites are uncertain at the several centimeter level.

1. Introduction

With the accuracy of the Terrestrial Reference Frame approaching 1 ppb, the requirements placed on eccentricity measurements among collocated techniques cannot be stressed enough. In order that these critical quantities be exempt from limiting the accuracy of the TRF, the intra-site ties must be known to better than ~1 mm in the horizontal and 3 mm in the vertical for the current TRF accuracies. Jim Ray has pointed out the serious discrepancy of ~3 cm between the ITRF96 height difference and the vertical tie between the GPS antenna, WES2, and the VLBI antenna at the Westford site. In order to investigate the source of this discrepancy ten days of GPS data were taken with a Dorne-Margolin choke ring antenna mounted over a surveyed reference mark much nearer the VLBI antenna than WES2. In this paper the ITRF96 VLBI to GPS height difference is compared to the optical geodetic surveys and the GPS measurements, and the serious limitation of the GPS antennas for use in defining the TRF is illustrated.

2. The Westford Site

The Dorne-Margolin choke ring antenna for WES2 is mounted on a steel tower, 11 meters tall, approximately 50 meters northeast of the Westford VLBI antenna. The TurboRogue receiver is located in the same control room as the Mark IV VLBI recorder and electronics rack. The GPS data and meteorological data are transferred to the NOAA CIGNET facilities once per day.

An optical geodetic survey was conducted by NGS in 1977 to connect the Westford and Haystack VLBI antennas for comparison with the VLBI measurement of the vector baseline between the two (CARTER et al 1980). Several other marks were observed in the survey, including the mark Westford1977, which is approximately 20 meters south of the Westford VLBI antenna at the level of the concrete base. (The relative locations of WES2, Westford1977, and the VLBI antenna are indicated in Figure 1.) At the time of the survey the radome was off of the VLBI antenna, allowing direct access to the reference points. The resulting tie between Haystack and Westford agreed with the VLBI determination, after correcting for gravitational deflection of the Haystack antenna, to 2mm, 4mm, and 6mm (in magnitude) in local East, North, and Up, thus validating the uncertainties of the geodetic tie at the estimated values of ~3mm. This is important because all eccentricities involving the VLBI antenna at Westford must rely on the tie from the VLBI reference point to another surveyed point, such as Westford1977. Because of the radome it is difficult to remeasure the tie to the VLBI reference point.

Two optical geodetic surveys have been made in order to relate the position of the WES2 antenna to the Westford1977 mark. The results of a survey in 1988 by Bendix Field Engineering Corporation have been reported by Carpenter and Nelson (1997), along with the results of their survey in 1996. The height from Westford1977 to WES2 differs by 5mm. In this paper only the height is discussed.

In ITRF96 the derived height difference between the VLBI and GPS reference marks for the Westford site differs by ~ -5 cm from the IERS eccentricity given in IERS#24 and by ~ -3 cm from the optically re-measured height difference (Carpenter and Nelson 1997). The IERS eccentricity is now known to be in error by 35mm (summarized in NGS note by W. E. STRANGE). In order to assess the magnitude of possible systematic errors in the GPS-determined position, the height difference between WES2 and the mark Westford1977, adjacent to the VLBI antenna, was measured using the GPS data from WES2 and data from a tripod-mounted Dorne-Margolin choke ring (DMCR) antenna over the reference mark.

The positions of WES2 and WF77 (the antenna reference point (ARP) for Westford1977) were determined for ten days using point positioning for minimum elevations of 5° to 30° in 1° intervals. The apparent height of WES2 was found to decrease by 3 cm as the minimum elevation was increased from 5° to 15° (Figure 2). For WF77 no data were received below 11° elevation due to surrounding trees. The apparent height of WF77 changed with elevation by less than 2 cm when averaged over the 10 days (Figure 3).

The measured height differences for the VLBI and WES2 GPS reference points are shown in Figure 4. The perplexing

[1] Arthur E. Niell; MIT Haystack Observatory, Westford, MA, USA 01886; Tel.: +1-978-692-4764; Fax: +1-781-981-0590; Email: aniell@mit.haystack.edu.

feature is that for an elevation cutoff of 15°, the GPS measured height difference agrees with the optical value, but the difference from ITRF96 disagrees by almost 3 cm. In view of the advertised uncertainties of ~1cm for each of the measurement types (VLBI, GPS, and optical), however, the difference is less than 2 combined standard deviations and, in spite of the seemingly large discrepancy, cannot be considered significant.

3. Global network

These differences in the sensitivity of the measured height to the minimum elevation angle cutoff for the *same type of antenna* are caused by the local electromagnetic environment, characterized by both scattering in the near-field and by multipath (ELÓSEGUI et al 1995; JALDEHAG et al. 1996; NIELL 1997). In order to investigate such differences among the sites of the global network, six other antennas (not intended to be globally distributed) were evaluated for sensitivity to minimum elevation for one of the days reported for WF77. The heights relative to the value for 5° minimum elevation are shown in Figure 5. Of the six, GOL2 and CIT1, also DMCR antennas, have characteristics similar to WES2, changing in apparent height by -7 cm for elevation cutoff increase from 5° to 30°. JPLM, a DMCR, is similar to WF77 for minimum elevations greater than 10°.

ZIMM and TAEJ, which are Trimble antennas, but of different types, have much larger sensitivity and of opposite sign, reaching +21 cm at 30° for ZIMM and +17 cm for TAEJ. Although no elevation dependent phase corrections were applied for the Trimble antennas, none are incorporated in the JPL analysis for the determination of orbits and clocks. (The Trimble antennas TAEJ, HOFN, and MDVO have been used in the JPL global solutions in 1998, but ZIMM has not (D. JEFFERSON, private communication).

A striking illustration of the possible effect of changing receivers is given by the systems at Goldstone, California. GOLD is a Rogue receiver connected to the same antenna as GOL2, a TurboRogue receiver. However, no data from GOLD are accepted below about 20°. As a consequence the estimated height of the GOL2 antenna is 40 mm higher than the GOLD antenna for a 5° cutoff and 18 mm higher for a 20° cutoff due to the elevation dependence of the antenna and the difference in data for the two receivers.

Thus it appears that the heights of the current GPS antennas cannot be determined with an *accuracy* of better than several centimeters, and perhaps much worse. (In addition to the observed dependence on minimum elevation, there may be a dependence of phase on sine(elevation) which cannot be separated from the height of the antenna.) With this sensitivity to the distribution of data included in the analysis, whether due to choice or to physical limitations of the antenna and receiver, the accuracy of the terrestrial reference frame as determined by GPS must be considered to be no better than ~3 ppb.

References

CARPENTER TROY V., VAUGHN E. NELSON: *"Westford-Haystack Check Survey"*, AlliedSignal Technical Services Corporation note, Feb 1997.

CARTER W. E., ROGERS A. E. E., COUNSELMAN C. C. III, SHAPIRO I. I.,: *"Comparison of Geodetic and Radio Interferometric Measurements of the Haystack-Westford Base Line Vector"*, JGR, Vol 85, No B5, 2685-2687, May 10, 1980.

ELÓSEGUI P., DAVIS J.L., JALDEHAG R.T.K., JOHANSSON J.M., NIELL A.E., SHAPIRO I.I.: *"Geodesy Using the Global Positioning System: The Effects of Signal Scattering on Estimates of Site Position"*, JGR, Vol. 100, No. B7, 9921-9934, June 10, 1995.

JALDEHAG R.T.K., JOHANSSON J.M., RÖNNÄNG, ELÓSEGUI P., DAVIS J.L., SHAPIRO I.I., NIELL A.E.: *"Geodesy Using the Swedish Permanent GPS Network: Effects of Signal Scattering on Estimates of Relative Site Positions"*, JGR Vol 101, No. B8, 17,841-17,860, August 1996.

NIELL A. E.: *"Near-field Problems for GPS Antennas"*, in The Global Positioning System for the Geosciences, National Academy Press, Washington, D.C. 1997.

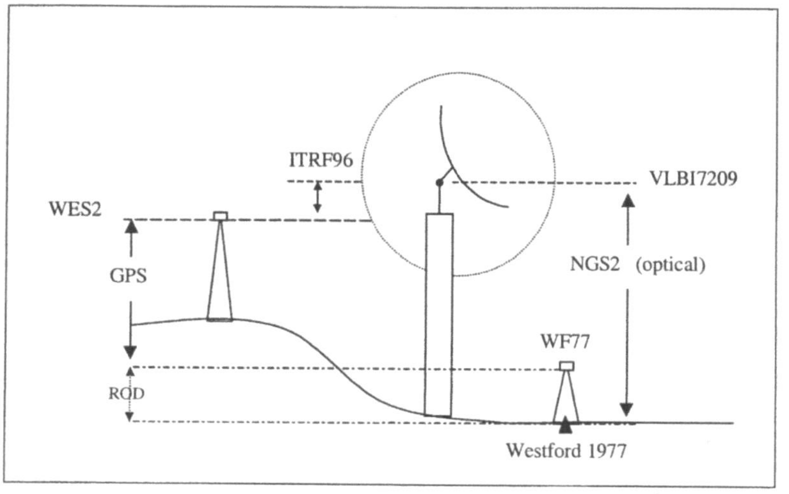

Fig. 1: The relative locations of the VLBI and GPS antennas at the Westford site

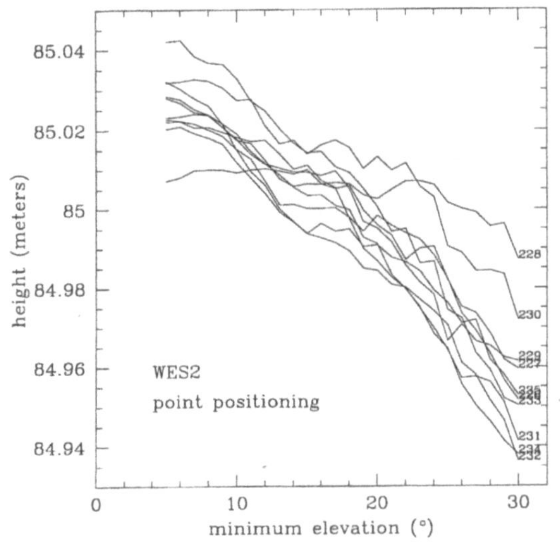

Fig. 2: The apparent height of WES2 as the minimum elevation of included data is changed.

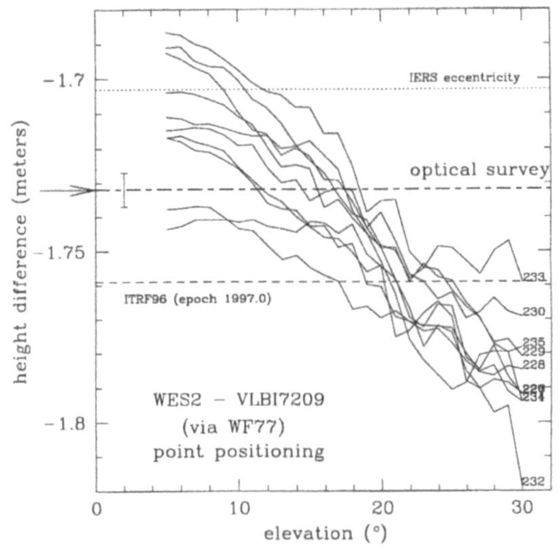

Fig. 4: The measured height difference from the VLBI reference point to WES2 as the minimum elevation of included data is changed. The values are also shown for the IERS eccentricity, the optically measured height difference, and the height difference from ITRF96.

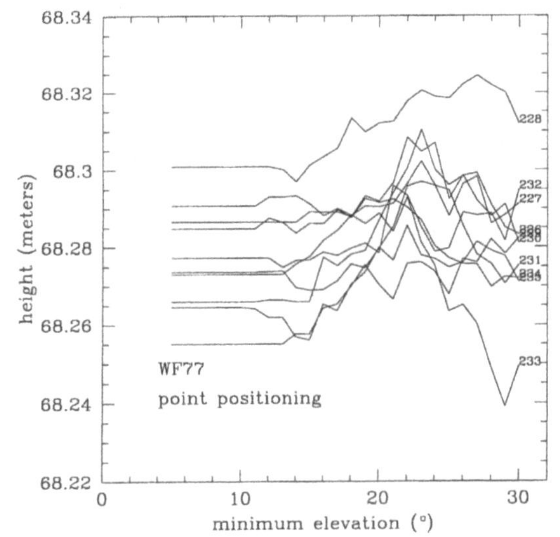

Fig. 3: The apparent height of WF77 as the minimum elevation of included data is changed.

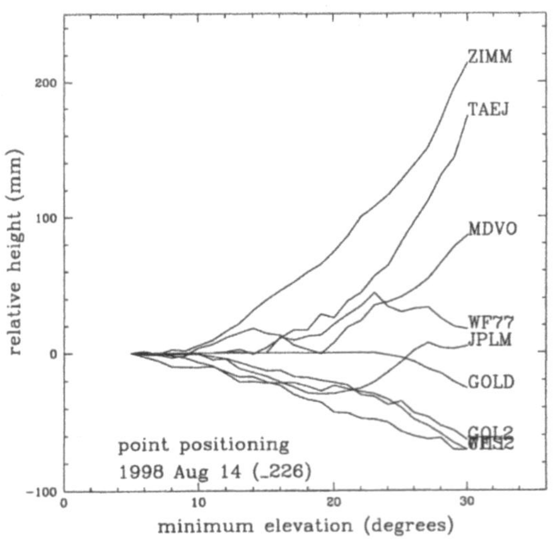

Fig. 5: The estimated height as the minimum elevation of included data is changed for six antennas of the global GPS network. ZIMM and TAEJ are Trimble antennas. The others are Dorne-Margolin choke ring antennas. No elevation dependent phase corrections were applied.

Posters – Session C

Upcoming Gravity Field Satellite Missions

Impact of terrestrial data on future satellite gravity field solutions

J. KUSCHE, K. H. ILK, S. RUDOLPH[1]

Abstract

This study investigates the benefit of terrestrial data in regional focusing strategies for recovery of the earth's gravity field from new satellite missions. The use of terrestrial data in a space–localizing approximation technique is discussed in a reproducing kernel Hilbert space. A simulation study reveals the characteristics of combined gravity field solutions as compared to satellite–only solutions.

1. Upcoming Missions

The task of the envisaged SST/SGG–missions CHAMP, GRACE, and GOCE is the computation of a global gravity field with high resolution and precision and – if possible – with repetition in time. Global recovery approaches are aimed at the computation of high resolution spherical harmonic coefficients. But open questions are related to a violation of an ideal data coverage or to not well–defined boundary surfaces. These problems are caused by non–polar orbits, by interruption of the data flow by orbit manoevres, or by varying satellite altitudes. Furthermore, the global support of the spherical harmonics does not allow to adapt the precision of the gravity field determination to areas of geodynamical or economical interest. The latter aspects can be properly solved by regional focusing solution techniques, where the true gravitational field is approximated by space–localizing kernel functions. Combination with terrestrial data could optimally supplement the satellite derived regional field information, which is not a trivial problem in case of global recovery strategies. Finally, if the "regional" solution covers the whole earth, spherical harmonic coefficients may be derived by a simple summation process.

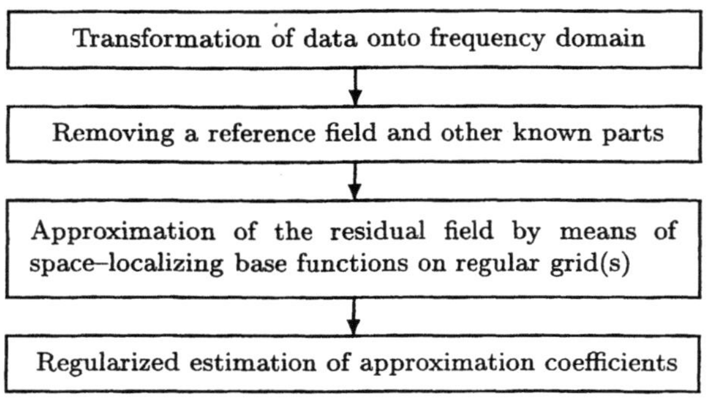

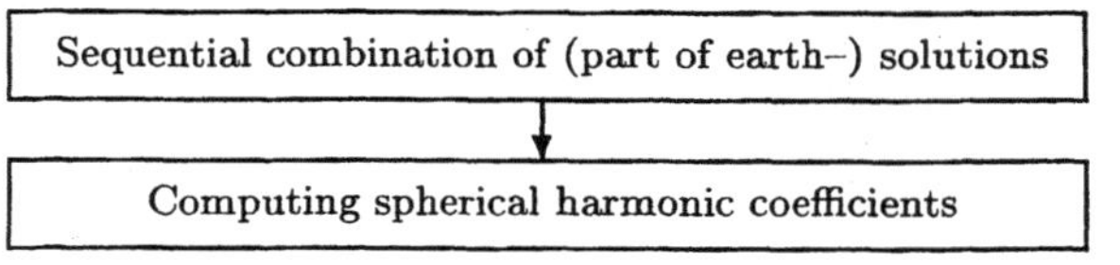

Fig. 1 - Scheme of two–step analysis.

2. Modelling

Data processing. The primary observation equation, based on eigenfunction expansion (Schneider 1984) of the intersatellite or gradiometry baseline $r(\tau) = \bar{r}(\tau) + \sum_\nu r_\nu \sin(\nu\pi\tau)$, reads

$$r_\nu = -2\left(\frac{\Delta t}{\nu\pi}\right)^2 \int_0^1 \sin(\nu\pi\tau')\Big(\frac{1}{r}(|\dot{\boldsymbol{r}}|^2 - \dot{r}^2) + \boldsymbol{e}\cdot\boldsymbol{D}(V_{ref}+T)\Big)d\tau' \quad (1)$$

with

$$\boldsymbol{D}_{SST} = \boldsymbol{\nabla}_2 - \boldsymbol{\nabla}_1 \qquad \boldsymbol{D}_{SGG} = \boldsymbol{r}\cdot\boldsymbol{\nabla\nabla}\,. \quad (2)$$

These expansion coefficients may be computed from SST range, range–rate, or SGG differential acceleration

$$\begin{aligned} r_\nu &= 2\int_0^1 \sin(\nu\pi\tau)\,\Big(r(\tau) - \bar{r}(\tau)\Big)\,d\tau \quad (3)\\ r_\nu &= 2\frac{\Delta t}{\nu\pi}\int_0^1 \cos(\nu\pi\tau)\,\dot{r}(\tau)\,d\tau \\ r_\nu &= -2\left(\frac{\Delta t}{\nu\pi}\right)^2\int_0^1 \sin(\nu\pi\tau)\,\ddot{r}(\tau)\,d\tau \end{aligned}$$

allowing a simple common processing of different data types as well as a certain degree of data compression.

Space–localizing base functions. The anomalous potential T is supposed to be a member of a reproducing kernel Hilbert space H. We approximate T using space–localizing base functions B_j (isotropic kernels) $\in H$ with $(B_j, B_k) = (\boldsymbol{P})_{jk}$

$$\tilde{T} = \sum_j \chi_j\, B_j\,. \quad (4)$$

Noisy observations are considered as bounded linear functionals: $l^j + \epsilon^j = \mathcal{A}^j T = (A_j, T)_H$. A meaningful solution should minimize a Tykhonov–functional weighting LS and H–norm of T

$$\mathcal{F}_{\gamma^2} = ||\boldsymbol{A}\chi - \boldsymbol{l}||^2_{\boldsymbol{P}_{ll}} + \gamma^2||\tilde{T}||^2_H \quad (5)$$

[1]Jürgen Kusche, Karl Heinz Ilk, Stephan Rudolph; University of Bonn, Institute of Theoretical Geodesy, Nußallee 17, 53115 Bonn, Germany, Tel.: +49-228-733578, Fax: +49-228-733029, Email: kusche@uni-bonn.de

leading to the well-known regularized (biased) estimate

$$\hat{\chi} = \left(A^T P_{ll} A + \gamma^2 P\right)^{-1} A^T P_{ll}\, l \qquad (6)$$

with covariance matrix of the $\hat{\chi}_i$

$$\hat{C}_{\hat{\chi}\hat{\chi}} = \hat{\sigma}^2 \left(A^T P_{ll} A + \gamma^2 P\right)^{-1} \qquad (7)$$
$$A^T P_{ll} A \left(A^T P_{ll} A + \gamma^2 P\right)^{-1} .$$

If, on the other hand, the regularized estimate is viewed as (bayesian) estimate with a-priori-covariance matrix $\gamma^{-2} P^{-1}$, a-posteriori we simply have

$$\hat{C}_{\hat{\chi}\hat{\chi}} = \hat{\sigma}^2 \left(A^T P_{ll} A + \gamma^2 P\right)^{-1} . \qquad (8)$$

Finally spherical harmonic coefficients for $\tilde{T}$ are given by

$$\tilde{T}_{nm} = (\tilde{T}, Y_{nm})_{L^2(\Omega)} = b_n \sum_j \chi_j \, Y_{nm}(Q_j) , \qquad (9)$$

where b_n are the Legendre coefficients of the base functions B_j.

Terrestrial data. Block mean anomalies after reductions read

$$\delta g(\Delta\Omega_j) = \frac{1}{\Delta\Omega_j} \int_\Omega \xi_{\Delta\Omega_j}(Q') \qquad (10)$$
$$\int_\Omega \mathcal{S}\, R(P', Q'')\, T(Q'')\, d\Omega(Q'')\, d\Omega(Q') ,$$

where $\xi_{\Delta\Omega_j}$ denotes the index function on a block area $\Delta\Omega_j$, $\mathcal{S}$ is the Stokes operator, and R is the Poisson kernel of upward continuation. We assume that the Stokes-kernel also serves for generating the base functions, $B_j(Q) = S(Q, Q_j)$, and the observation point approaches the reference sphere. A single direct observation $l^j = \delta g(\Delta\Omega_j)$ of χ_j weightet by p_j produces a correction

$$\hat{\chi}^{(+)} - \hat{\chi} = \frac{l^j - (\hat{\chi})_j}{(p_j)^{-1} + (c)_j}\, c \qquad (11)$$

where c abbreviates the jth column vector of $C_{\hat{\chi}\hat{\chi}}$.

3. Simulation Study

To investigate the characteristics of satellite-only solutions as well as combined solutions a simulation study was carried out. Fig. 2 provides the baselines of the study. More details can be found in (Kusche et al 1998).

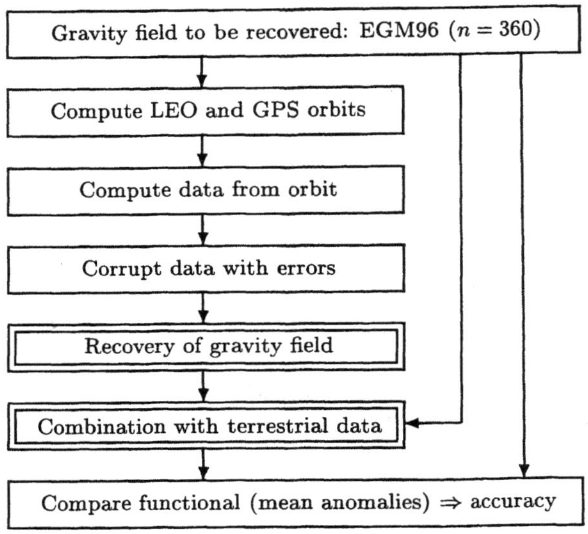

Fig. 2 - Scheme of simulation study.

The basic mission configurations were *i)* a GRACE scenario consisting of 2 LEO's with baseline $300km$ and 25 GPS satellites, *ii)* a CHAMP scenario: 1 LEO and 25 GPS satellites. A LEO orbit was chosen sun-synchronous by $a = 6732266.20m, e = 0.001, i = 97.29^o$, mean altitude $354km$. According to the "snapshot" objective of GRACE a mission duration of $31days$ was simulated. The number of observations are 20,000 in case of LEO-LEO and 140,000 for each of the both LEO-GPS links. A white noise level of $\sigma = 1\mu m/s$ for LEO-LEO range-rates was assumed and of $\sigma = 3cm$ for LEO-GPS ranges. All orbits were integrated numerically applying the EGM96 gravity field model. The area under consideration is defined to $(\lambda, \varphi) \in [0^o, 10^o] \times [0^o, 90^o]$ with base function grid spacing of $1^o \times 1^o$.

Finally three $1^o \times 1^o$ mean gravity anomaly blocks were computed from the EGM96 model, located at different geographical latitude. They served for error-free — and, of course highly idealized — terrestrial input data.

Fig. 3 illustrates the accuracy of recovery in a GRACE-satellite-only experiment. *Left side:* mean residual gravity anomalies computed from the EGM96 model with reference to the low-degree model, *middle:* GRACE recovery result, *right side:* absolute deviations. In fig. 4 results are displayed for the idealized CHAMP mission, *left side:* mean $5^o \times 5^o$ anomalies from the EGM96 input, *middle:* CHAMP recovery result, *right side:* absolute deviations.

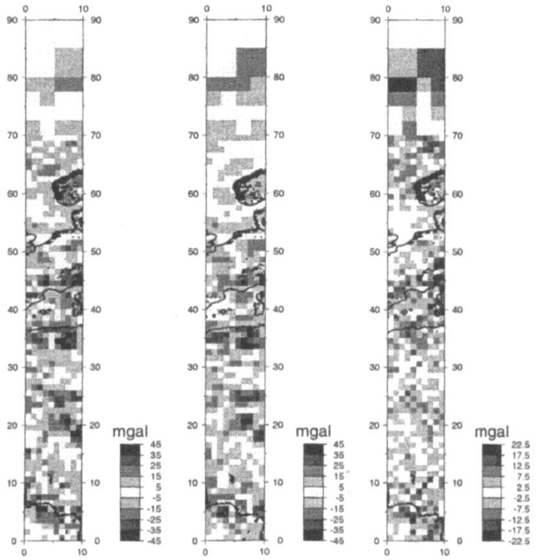

Fig. 3 - Recovery of mean anomalies from GRACE.

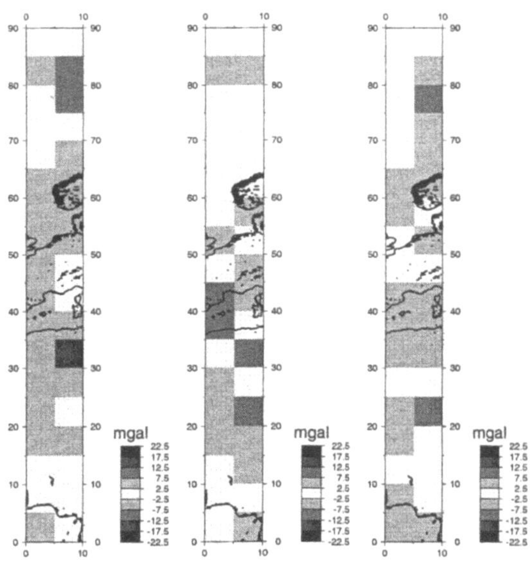

Fig. 4 - Recovery of mean anomalies from CHAMP.

In fig. 5 corrections of the satellite–only solution due to the surface gravity anomaly blocks are given, computed from (11) using either (7), *left side*, or (8) on the *right side*. All six images are plotted using the same greyscale but for each *row* - each data element - the correction was different but of positive sign. Note that the correction pattern is dependent of geographical latitude due to varying satellite coverage. More obvious is the fact that using the biased estimate's covariance matrix (7) produces much larger oscillations in the correction than working with (8).

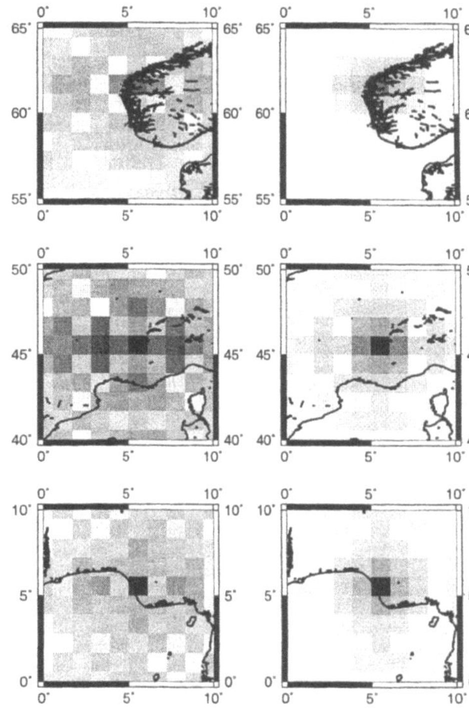

Fig. 5 - Correction from isolated surface data

4. Results and Conclusions

Aiming for a spatial resolution of $1^o \times 1^o$ $(100km)$ in a GRACE/$31days$ recovery solution, we found an overall rms value of $6 - 7mgal$ for the satellite–only solution. The accuracy degrades in areas of rough gravity field, as clearly indicated in fig. 3. When removing features of less than $300km$ from the EGM96 anomalies by truncating the expansion at $n = 120$, we found the deviations distributed randomly with an rms value of $3 - 4mgal$. For the CHAMP satellite–only scenario we had an rms value of $4mgal$ for 5° mean anomalies.

Isolated surface data of course cannot provide an overall improvement of the satellite–only solution: each observed block affects only few "surrounding" approximation parameters χ_i. The local correction pattern strongly depends on the regularization matrix and the expression chosen for the covariance matrix, where (7) seems to produce unrealistic oscillations. For the idealized GRACE case we had under consideration we could not identify a significant improvement *in the neighborhood* of the observed surface blocks.

References

ILK K. H., RUMMEL R., THALHAMMER M., (1995), *Refined method for the regional recovery from GPS/SST and SGG.* In: CIGAR III/2 Study of the gravity field determination using gradiometry and GPS, ESA contract No. 10713/93/F/FL

KUSCHE J., ILK K. H., RUDOLPH S., *Two-step data analysis for future satellite gravity field solutions: A simulation study.* Paper presented at the 2nd Joint Meeting of IGC/IGeC, September 1998, Trieste, Italy

SCHNEIDER M., 1984, *Observation equations based on expansions into eigenfunctions.* manuscripta geodaetica, 9:169–208

XU P., RUMMEL R., 1994, *Generalized ridge regression with applications in determination of potential fields.* manuscripta geodaetica 20:8–20

SAGE: An Italian Project of Satellite Accelerometry

F. SANSO[1], A. ALBERTELLA[1], G. BIANCO[2], A. DELLA TORRE[3], M. FERMI[4], V. IAFOLLA[5], A. LENTI[6], F. MIGLIACCIO[1], A. MILANI[7], A. ROSSI[8]

Abstract

SAGE is a project for a satellite accelerometric mission aimed at determining with high precision the low-medium coefficients of the earth gravity field and their time variations.

In the international panorama, SAGE would be a prosecution of CHAMP, both in time (in case of approval, SAGE launch date would be within the year 2001) and in space (SAGE would fly polar, filling in the polar gaps).

SAGE has gone through a Phase A study commissioned by the Italian Space Agency (ASI), which lasted from December 1997 to October 1998.

1. The Project of the Mission SAGE and its Relation to the Other Gravity Field Missions

SAGE is a fully Italian project aiming at determining the gravity field of the Earth by exploiting the concept of high-low satellite-to-satellite tracking (SST). This means that the satellite orbit is determined by GPS, while the non-gravitational perturbations are determined by a three-axes accelerometer which, integrated back two times, gives the "virtual" orbit followed under gravitational effects only.

The outline of the mission SAGE and the general requirements are as follows:

- Launch date	2001
- Lifetime	5 years (minimum)
- Altitude	between 450 and 350 km
- Inclination	87° (possibly polar)
- Nominal eccentricity	e = 0
- Bus	MITA (C. Gavazzi Space)
- Payload	Italian Spring Accelerometer (ISA) GPS receiver Lagrange Laser Retroreflector Array (by GFZ Potsdam)

Several Italian research institutions and industries participated to the SAGE Phase A study. Namely, the following research centers were involved:

- ASI/CGS Matera, for telemetry and GPS analysis;
- IFSI/CNR Roma, for the development of the ISA accelerometer;
- Politecnico di Milano – IGeS, for the study of the spacewise reduction strategy of the data;
- Università di Padova, for the GPS data analysis;
- Università di Pisa – CNUCE, for the study of the timewise reduction strategy of the data and the orbital analysis;
- Università di Trento, for complementary studies on the spacewise reduction strategy.

The industrial partners were:

- Carlo Gavazzi Space S.p.A., developing the study of the MITA bus and of the navigation and data communication segment;
- Laben S.p.A., involved in the accelerometer engineering and in the development of the Lagrange GPS receiver;
- Telespazio S.p.A., for the GPS data analysis, data downloading and telemetry.

[1] Fernando Sanso, A. Albertella, F. Migliaccio: Politecnico di Milano, DIIAR Sezione Rilevamento, Piazza L. da Vinci, 32, I-20133 Milano, Tel.: +39-2-23996-504,/-509,/-507, Fax: +39-2-23996-530, e-mail:fsanso@ipmtf4.topo.polimi.it; /alberta@ipmtf4.topo.polimi.it, /federica@ipmtf4.topo.polimi.it

[2] G. Bianco: ASI – Centro di Geodesia Spaziale di Matera, P.O. Box 11, Loc. Terlecchia, Italy Tel.:+39-835-377209, Fax:+39-835-339005, e-mail:bianco@asi.it

[3] A. Della Torre: Carlo Gavazzi Space S.p.A., via Gallarate, 139, Milano, Italy; Tel.: +39-2-38048255, Fax: +39-2-3086458, e-mail: cgssw@interbusiness.it

[4] M. Fermi: Telespazio S.p.A., Via Tiburtina 965, Roma, Italy, e-mail: marco@sidus.mt.asi.it, antonio@sidus.mt.asi.it

[5] V. Iafolla IFSI/CNR, Via del Fosso del Cavaliere, I-00133 Roma, Italy; Tel: +39-6-49934391, Fax: +39-6-40796202; e-mail:iafolla@ifis.rm.cnr.it

[6] A. Lenti: Laben S.p.A., S.S. Padana superiore 290; I-20090 Vimodrone (MI), Italy; Tel.: +39-2-25075370, Fax.: +39-2-2505515, e-mail: lenti.a@laben.it

[7] A. Milani: Dip. Matematica, Università di Pisa, Via Buonarroti 2, I-56100 Pisa, Italy; Tel.:+39-50-844254, Fax: +39-50-844224, e-mail: milani@dm.unipi.it

[8] A. Rossi: CNUCE/CNR, Via Santa Maria 36, I-56126 Pisa, Italy; Tel.: +39-50-593-220, Fax: +39-50-904-052, e-mail: a.rossi@cnuce.cnr.it

It is interesting to remark that SAGE would be logically and temporarily interwoven with CHAMP, as the two missions are based on the same scientific principle. Specifically, SAGE would aim at:

- improving the CHAMP goals aiming at the useful determination of coefficients up to degree 70;
- extending in time the CHAMP mission allowing for a high-low SST sampling of the gravity field at an altitude of 400 km with an overall time span of 7 to 8 years, which could give more reliable information on time variations of the field;
- cross-checking the two missions one with respect to the other, because they will fly together more or less at the same altitude for at least two years, allowing for a reciprocal (indirect) calibration of the two instruments, which are based on different technologies.

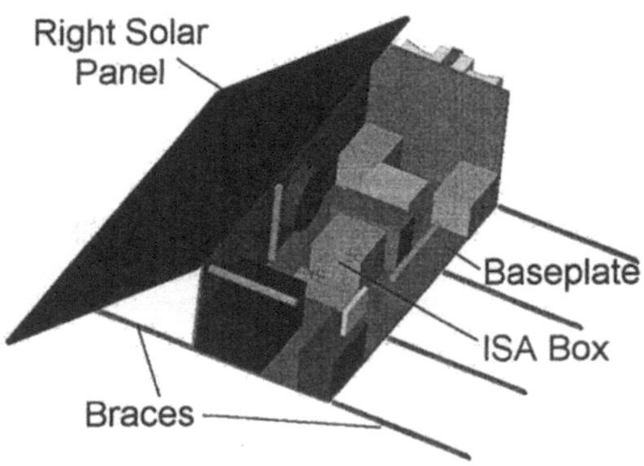

Fig. 1: SAGE structure and internal accommodation, showing the position of the ISA accelerometer next to the centre of mass of the spacecraft.

The relation to the other two gravity missions, GRACE and GOCE, is the same as that of CHAMP (namely, SAGE works on medium-high wavelengths, increasing the accuracy of the estimated coefficients), but strengthened by the fact that CHAMP and SAGE together can provide a highly reliable and dense data set at an average altitude of 400 km.

In case SAGE should fly polar, it could also reach the important aim of filling the "polar gaps", which are normally not avoided by similar missions.

2. Instruments

The main instruments designed to constitute the payload of SAGE are made after Italian technology.

The ISA accelerometer has been developed at the IFSI Institute (Istituto di Fisica dello Spazio Interplanetario) of the CNR (National Research Council). This accelerometer has been designed to be the fundamental element of a spaceborne, room temperature, gravity gradiometer having a sensitivity of 10^{-2} EU/ $\sqrt{Hz}$. In this case the accelerometers must have a sensitivity of 5 10^{-13} g/ $\sqrt{Hz}$. Even if this target requires further effort, the sensitivity of 10^{-9} g/ $\sqrt{Hz}$, required for the SAGE experiment, has been reached at the present state of the art. A prototype of an accelerometer having sensitivity better than 10^{-9} g/ $\sqrt{Hz}$ has been built and is operating in the underground laboratory of Gran Sasso (L'Aquila). The fundamental part of this equipment is a mechanical harmonic oscillator. Typical displacements of the oscillating proofmass due to accelerations of 10^{-9} g are of the order of 2 10^{-11} m. A capacitor transducer in a bridge configuration, followed by a low noise amplifier, provides the detection of the signal.

Fig. 2 Prototype of the Italian Spring Accelerometer (ISA).

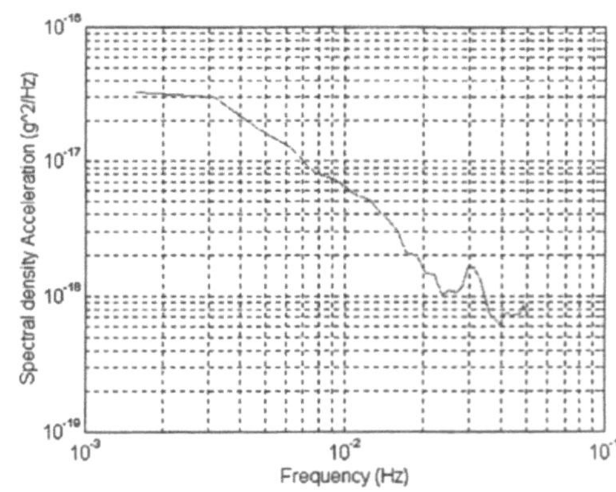

Fig. 3 Power spectral density of the seismic noise measured at the Gran Sasso laboratory by the spring accelerometer ISA, at the level of 10^{-9} g/ $\sqrt{Hz}$.

The Lagrange instrument is under development at Laben S.p.A.. This equipment has been designed as a GPS/GLONASS receiver to be used not only for navigation purposes but also for scientific objectives like Precise Orbit Determination (POD). The error of Lagrange is less than 1 mm at a sampling rate of 10 Hz.

3. Scientific Objectives of the Mission

To describe the overall scientific performance of the SAGE gravimetry mission we can use the maximum degree ℓ for which the error is larger than the signal; i.e., the RMS value of the error in the $(2\ell + 1)$ coefficients with the same degree (and different order m) is larger than the RMS value of the coefficients themselves. For the error we use the square root of the formal variance of the determined coefficients, that is the standard error according to the Gaussian theory. This formal error, however, must also be compared with the systematic errors, in particular with those resulting from dynamic model errors and those resulting from coloured noise. This check has been performed with two different methods for the two different approaches used to treat the SAGE simulated data (timewise and spacewise) and it turns out that the systematic errors are not dominant for the timewise approach, while they can be significant for some noise models in the spacewise approach.

The largest complete simulation we have performed has used six months of data with the timewise approach. The maximum degree ℓ_{MAX} for which the signal exceeds the error has been found to be 62. If this result is extrapolated to a data span ten times longer, by using the Gaussian statistics which is well known to be applicable in this kind of simulations, then ℓ_{MAX} can be estimated to be about 72. Taking also into account one year of extended mission, when SAGE should be orbiting at a lower altitude (about 360 km), the value of ℓ_{MAX} could be as high as 74.

The largest simulation with the spacewise approach has been performed with one year of data. The results of the spacewise treatment are similar to the ones of the timewise approach, if the noise is assumed to be white; a complete solution, however, has not yet been computed with an accurate enough algorithm. The influence of coloured noise could further degrade the results by a factor of about 2 in accuracy. The differences between the results with the two approaches are not yet fully understood, and are certainly worth a deeper theoretical and numerical investigation; nevertheless, they are not large enough to result in different evaluation of the scientific merit of the mission.

The possibility to obtain a solution up to a comparatively high degree with only six months of data implies the possibility of detecting time variations. The geopotential coefficients are not expected to change by 100 % of their value, however, and the significant result is the possibility to detect changes of 1 % with periods of one year and longer, up to and even beyond degree 25. This should be compared with the present knowledge of secular trends for the coefficients of degree 2 and 3 only.

The essential goal is to obtain a gravitational field complete to a given resolution, and therefore a reference geoid, which is reliable, with the possibility of an external and independent check of the absolute accuracy of the results. The simultaneous measurements from CHAMP and SAGE for at least two years should allow a direct comparison of the results, providing an objective assessment of the reality of the solution. In this respect the spacewise approach, with its transformation of the observations into geographically distributed quantities, can provide an interesting comparison tool, directly at the satellite altitude.

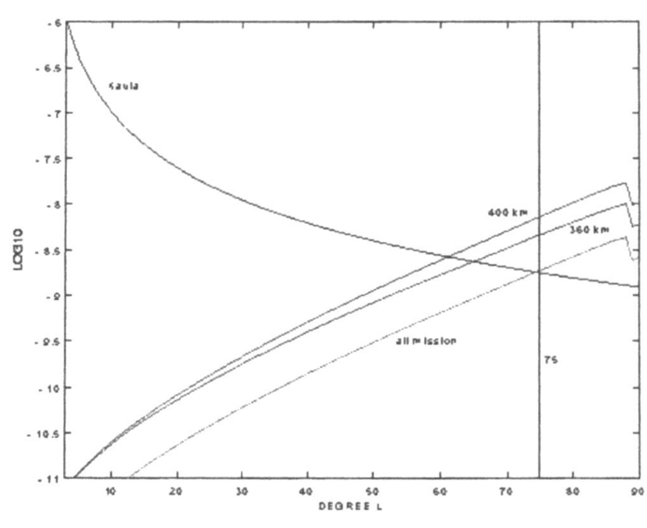

Fig. 4 The performance expected over the entire mission, assuming 5 years at an average altitude of 400 km plus one year at an average altitude of 360 km. The lower line is the extrapolated performance for such a data set, the upper lines are the formal standard deviation for 180 days of data at 400 and 360 km respectively.

4. Conclusions

The Phase A studies and simulations of the mission showed that SAGE can meet its targets.

Though SAGE is not developing a new scientific concept, its scientific targets are new: this mission would extend the gravity sampling of CHAMP for three more years and will improve the spatial resolution of high low SST; in case SAGE would fly polar, the polar caps will be surveyed and filled with data; moreover, the accuracy of the measurements and the data analysis in SAGE will allow for the determination of a much higher degree than CHAMP.

All the technological requirements of SAGE are based on existing instrumentation fully developed in Italy. In particular, the measurement principle of the spring accelerometer ISA is different from that of any other instrument employed in other projects, which makes the mission more valuable.

In case of approval, SAGE would open new perspectives for the scientific policy of the Italian Space Agency, while the development of a fully Italian technology could be the first step for future, more advanced targets: the Italian Spring Accelerometer, once qualified for space missions, could become an Italian Spring Gradiometer.

References

ALBERTELLA A., MIGLIACCIO F., SANSÒ F., SONA G. (1996). *The problem of polar caps in a gradiometric mission.* International Geoid Service Bulletin n. 5, DIIAR, Politecnico di Milano.

BALMINO G. (1974). In: *The Use of Artificial Satellites for Geodesy and Geodynamics*, ed. G. Veis, Academic Press, Orlando.

BALMINO G., PEROSANZ F., RUMMEL R., SNEEUW N., SUENKEL H., WOODWORTH P. (1998). *European views on dedicated gravity field missions: GRACE and GOCE.* European Space Agency.

DAVIES G.W. (1997). *Exploring the limits of GPS-based precise orbit determination.* Navigation, 44, n. 2, 181-193.

ESA (1998). *European Views on Dedicated Gravity Field Missions: GRACE and GOCE*, ESD-MAG-REP-CON-001.

FULIGNI F., IAFOLLA V., NOZZOLI S. (1992). *Gradiometro gravitazionale IFSI Frascati*, Nota IFSI, 92.1, Gennaio 1992

IAFOLLA V., MORBIDINI A., NOZZOLI S. (1992). *Tecniche realizzative di un gradiometro gravitazionale a tre assi e prime prove sperimentali.* Nota IFSI, 92.11, Settembre 1992.

IGeS (1997). *The earth gravity model EGM96: testing procedures at IGeS.* International Geoid Service Bulletin n. 6, DIIAR, Politecnico di Milano.

KAULA W.M. (1966). *Theory of Satellite Geodesy.* Blaisdell, Waltham, Massachusetts.

MIGLIACCIO F., SANSÒ F. (1989). *Data processing for the Aristoteles Mission.* Proc. Italian Workshop on the European Solid Earth Mission Aristoteles, Trevi, Italy.

MILANI A., NOBILI A.M., FARINELLA P. (1987). *Non-gravitational perturbations and satellite geodesy.* Adam Hilger, Bristol.

MORITZ H. (1989). *Advanced physical geodesy*, Part A. Wichmann, 50-56.

MORITZ H., HEISKANEN W.A. (1990). *Physical geodesy.* Reprint, Institute of Physical Geodesy, Technical University Graz, Austria.

ROSBOROUGH G.W., TAPLEY B.D. (1987). *Radial, transverse and normal satellite position perturbations due to the geopotential.* Celest. Mech., 40, 409-421.

SANSÒ F., RUMMEL R. (eds.) (1989). *Theory of satellite geodesy and gravity field determination.* Lecture notes in earth sciences. Springer-Verlag, Berlin.

TAPLEY B.D. et al. (1994). *Precise orbit determination for TOPEX/Posedion.* J. Geophys. Res., 99, n. C12, 24, 383--424, 404.

WAGNER C. (1987). *Geopotential orbit variations: applications to error analysis.* J. Geophys. Res., 92, 8136-8146.

WUNSCH C. (1993). *Satellite Altimetry in Geodesy and Oceanography.* Lecture notes in earth sciences, (F. Sansò and R. Rummel eds.), Springer-Verlag, Berlin.

Accelerometry Aboard CHAMP

P. SCHWINTZER, Z. KANG, F. PEROSANZ [1]

Abstract

One of the key instruments aboard the CHAMP geoscientific small satellite is the three-axes STAR accelerometer, which shall measure the resultant acceleration of the spacecraft due to non-gravitational forces. Then, the purely gravitational signal in CHAMP's orbit, measured with GPS satellite-to-satellite tracking, can be extracted for use in Earth gravity field modelling. Besides the noise in the accelerometer measurements, environmental systematic error sources may affect the accelerometer data: e.g. accommodation at CoG, orientation knowledge and spurious, linear and rotational accelerations from the attitude control system and due to boom oscillations. A crucial point in the exploitation of the measured accelerations is the determination of the instrument's calibration parameters, i.e. bias, and scale factor which shall be done with an in-flight data adjustment. Results of numerical simulations of CHAMP's tracking and accelerometer scenario, taking into account the GPS and accelerometer error sources and the estimation of calibration and nuisance parameters, are presented with regard to impacts on orbit and gravity field recovery accuracy. The objective is to develop an optimal data processing approach which can be applied to real mission data after the launch of CHAMP, end of 1999.

1. Introduction

CHAMP (Challenging Mini-Satellite Payload for Geoscience and Application) is a German small satellite mission for (REIGBER et al. 1998):

- global long- to medium-wavelength Earth gravity and magnetic field recovery and its temporal variations,
- atmosphere/ionosphere GPS limb sounding

The characteristic orbit parameters are near-polar (i=87 deg), near-circular, with an initial altitude of 450 km. The design lifetime is 5 years. The launch of the CHAMP satellite is planned for end 1999 with a Russian COSMOS launch vehicle.

For precise orbit determination and global gravity field recovery, CHAMP carries an accelerometer to measure directly the hard to model non-gravitational accelerations, mainly air drag, which then can be applied in CHAMP's dynamic orbit restitution with GPS satellite-to-satellite (SST) and ground-based laser tracking to extract the purely gravitational orbit perturbations. However, a central problem in using the measured accelerations is the determination of the instrument's calibration parameters, i.e. bias, and scale factor for each of the three axes. The calibration parameters shall be adjusted simultaneously with the orbit parameters during the orbit computation process. Therefore the goals of the simulation study presented here are:

- to investigate the solvability of accelerometer calibration parameters: bias, scale factor
- to develop an optimal data processing approach

The simulated data which include both, GPS-SST and accelerometer observations, will be used for the parameter estimation in dynamic orbit integration and adjustment.

2. The Accelerometer

The STAR accelerometer (Fig. 1), which is supplied by the French space agency CNES and manufactured by ONERA, France, is going to fly on the CHAMP satellite. The measurement principle of accelerometer is based on the electrostatic suspension of a parallel-epipedic proof-mass inside a cage. It measures only non-gravitational accelerations on the CHAMP satellite. These forces include air drag, solar radiation pressure, Earth albedo and attitude manoeuvres. The 3-axes accelerometer has a dynamic range of $\pm 10^{-4}\,\mathrm{ms}^{-2}$ with a resolution of $3 \cdot 10^{-9}\,\mathrm{ms}^{-2}$ in the normal and tangential axes, and $3 \cdot 10^{-8}\,\mathrm{ms}^{-2}$ in the radial axis, respectively. The measurement bandwidth ranges from $2 \cdot 10^{-4}$ (1/1rev) to 10^{-1} Hz.

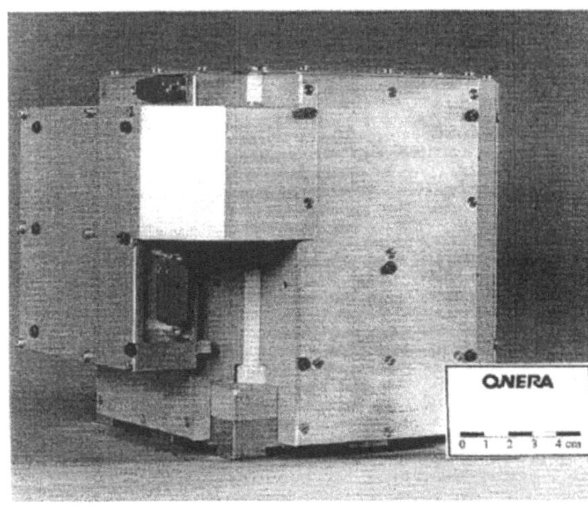

cnes ONERA

Fig. 1. STAR Accelerometer

Generally, the accelerometer output is not an absolute value. The measurement has to be calibrated applying a bias and a scale factor to each axis. Because these parameter can't be obtained on ground under the 1g-environment, they have to be determined from the flight data. This is the motivation for the investigation presented here: to determine the accelerometer parameters in CHAMP's orbit adjustment using simulated data generated with realistic assumptions.

[1] Peter Schwinzer, Z. Kang, F. Perosanz: GeoForschungsZentrum Potsdam AB 1, Telegrafenberg, D-14473 Potsdam, Germany;
Tel.: +49-331-288-1130, Fax: +49-331-288-1169, e-mail:psch@gfz-potsdam

3. Simulation Assumptions

The baseline tracking system for CHAMP is GPS-SST, which generally consists of the GPS satellite constellation, the GPS ground receiver network and the CHAMP flight GPS receiver. Measurements are taken between the high flying GPS satellite and the low flying CHAMP spacecraft, and between the GPS satellites and the ground receivers. In addition, the non-gravitational accelerations are measured with an on-board accelerometer. The processing is carried out using the two-step approach (KANG et al. 1995): first, GPS orbits and clock parameters are determined using only the data from the GPS ground receivers. In a second step, the low-Earth satellite orbit and other relevant parameters then are estimated using GPS-SST observations from the on-board GPS receiver and accelerometer measurements with GPS orbits and clock parameters fixed. In practice, the first step is dispensable. One can obtain the GPS ephemerides and clock parameters directly from IGS (BEUTLER et al. 1994). Therefore, the determination of the GPS orbits and clocks parameters is not part of the simulation. One important error source arises from the attitude thrusters operating on CHAMP. The systematic acceleration errors generated by thruster pair mismatches during one pulse can be up to 1.2 E-6 m/s^2 and cannot be extracted from the accelerometer readings. The thrust duration varies between 0.1 - 1 seconds. The average number of pulses per orbit revolution is 9, 2 and 4 in roll, pitch and yaw direction, respectively. The total error model is shown in Table 1 and the simulation scenario is given in Table 2. The simulated GPS-SST observations are carrier phases and pseudo-ranges. In addition, the accelerometer observations are simulated. The assumed observation equations are as follows:

Code pseudo-range observation equation:

$$P_k^i = \rho_k^i + \Delta\rho_{ion,k}^i + \Delta\rho_{tro,k}^i + c\Delta t_k - c\Delta t^i$$

where

P_k^i ... code pseudo-range for a receiver k and a GPS satellite i

ρ_k^i ... geometrical distance between satellite i and receiver k

$\Delta\rho_{ion,k}^i$ ... ionospheric refraction

$\Delta\rho_{tro,k}^i$ ... tropospheric refraction

c ... velocity of light

Δt_k, Δt^i ... receiver, satellite clock error, resp.

Carrier phase observation equation:

$$L_k^i = \rho_k^i - \Delta\rho_{ion,k}^i + \Delta\rho_{tro,k}^i + c\Delta t_k - c\Delta t^i - \lambda N_k^i$$

where

L_k^i ... carrier phase between a receiver k and a GPS satellite i

λ ... wavelength;

N_k^i ... unknown number of integer cycles ambiguity)

Accelerometer observation equation:

$$a_{ci} = a_i + k_i a_{mi} \quad i = 1, 2, 3 \text{ (x-, y-, z-axis)}$$

where

a_{ci} ... calibrated acceleration in i-direction

a_i ... accelerometer bias in i-direction

k_i ... scale factors in i-direction

a_{mi} ... measured acceleration in i-direction

Tab. 1: Error Models

GPS position error	5 - 8 cm each component
GPS clock error	2 cm uncertainty in GPS observ.
Thruster firing	disturbing linear acceleration generated by thruster pair mismatch: $1{,}2\cdot10^{-6} ms^{-2}$, at about every 10 min over 0.1 - 1 sec.
Measurement Noise	on-board GPS receiver: 30 cm Pseudo-range, 5 mm Carrier Phase accelerometer: 3.E-9 m/s^2 in along track and cross track 3.E-8 m/s^2 in radial direction

Tab. 2: Scenario

Orbit (circular):	altitude: 450 km; inclination: 87 deg
Number of GPS Satellites:	24
Flight Antenna Field of View:	hemispherical
Receiver Tracking Capacity:	12 Channels (L1 and L2)

Simulation test

The orbit computation software package EPOS developed at GeoForschungsZentrum Potsdam (GFZ), was used for the simulation study, which uses a purely deterministic approach (SCHWINTZER et al. 1995):

- Generation of observations from real-world orbits and parameters (pseudo-range, carrier phase and accelerations)
- Application of error models to observations and underlying models and dynamic orbit integration
- Least squares orbit adjustment
- Comparison between real-world and estimated parameters

A summary of the assumptions used in the simulation is given in Table 3, and the estimated parameters are shown in Table 4.

Tab. 3: Assumptions used in the simulation

Data types	GPS-SST pseudo-range, carrier phase and accelerations (3D)
Arc length	1 day
Number of arcs	5
Data sampling rate	30 sec for GPS-SST observation 10 sec for accelerometer data

Tab. 4: Estimated parameters

Parameter	Treatment
CHAMP state vector at initial epoch	once per arc
Accelerometer bias	three axes
Accelerometer scale factor	three axes
Thruster parameters	6 constant thruster pair disturbing acceleration vectors
GPS ambiguities	constant over a continuous pass
GPS receiver clock bias	one bias per 30s-epoch
Geopotential coefficients complete to degree and order 20 (except GM and the degree1 terms)	constant in 1 and 5 day accumulated normal equation system

4. Results and Discussion

The simulation for the parameter estimation is carried out based on the method and assumptions described above. The geoid (l, m=20) recovery accuracy is about 10 cm for a one day and 6 cm for a 5 day solution, whether acceleration bias, scale factor, and thruster parameters are adjusted or fixed to default in solution. The correlations between accelerometer calibration parameters or thruster parameters exceeding the ± 0.7 level for a one day and five day solution, respectively, are given in Table 5. For the one day solution, there exist large correlations between the scale factor and bias in along track and cross track, respectively. But for a five day solution, the correlations between the accelerometer calibration parameters are fairly small. For the solved-for thruster accelerations, higher correlations can be observed, also when estimated over a longer time duration. But taking into account these perturbations is absolutely necessary in precise orbit determination. It is important that there are no large correlations between the gravity field parameters and the estimated accelerometer/ thruster parameters.

Tab. 5: Correlations within accelerometer/thruster parameters (if exceeding 0.7) and w.r.t. gravity coefficients

Parameter pair	r_{ij}	Parameter pair	r_{ij}
one-day solution			
AST - ABT	0.95	ASN - ABN	0.74
THR 3 - THR 1	-0.76	THN 3 - THN 1	-0.81
THT 6 - THT 2	-0.81	A/T - Grav. C/S	< 0.5
five-days solution			
THR 3 - THR 1	-0.73	THN 3 - THN 1	-0.74
THT 5 - THN 4	0.74	THR 6 - THR 2	-0.76
THT 6 - THT 2	-0.87	A/T -Grav. C/S	< 0.5

Note: A: accelerometer; S: scale factor; B: bias; R: radial; T: along track; N: cross track; TH: thruster: 1, 2, ... 6: thruster; pair number; r_{ij} : correlation

Table 6 shows that the calibration parameters in the five-day solution are improved not only in its formal error, but also in approximating the 'true value' compared to the one-day solution. In the frame of this simulation, the calibration parameters can be determined on the specified resolution level by using only five days of data. They have of course to be stable over at least this time period.

Tab. 6: True and Estimated Accelerometer Calibration Parameters

Parameters		True value	Estimated value		Formal error	
			1 day	5 days	1 day	5 days
Bias (m/s^2)	Radial	0.0	1.4E-8	-8.1E-10	3.9E-9	1.5E-9
	Along track	0.0	-1.3E-9	2.0E-10	1.0E-9	1.0E-10
	Cross track	0.0	-1.0E-8	1.2E-8	4.4E-9	1.1E-9
Factor	Radial	1.0	0.699	0.999	0.084	0.002
	Along track	1.0	1.003	1.000	0.013	0.00001
	Cross track	1.0	0.766	1.008	0.340	0.003

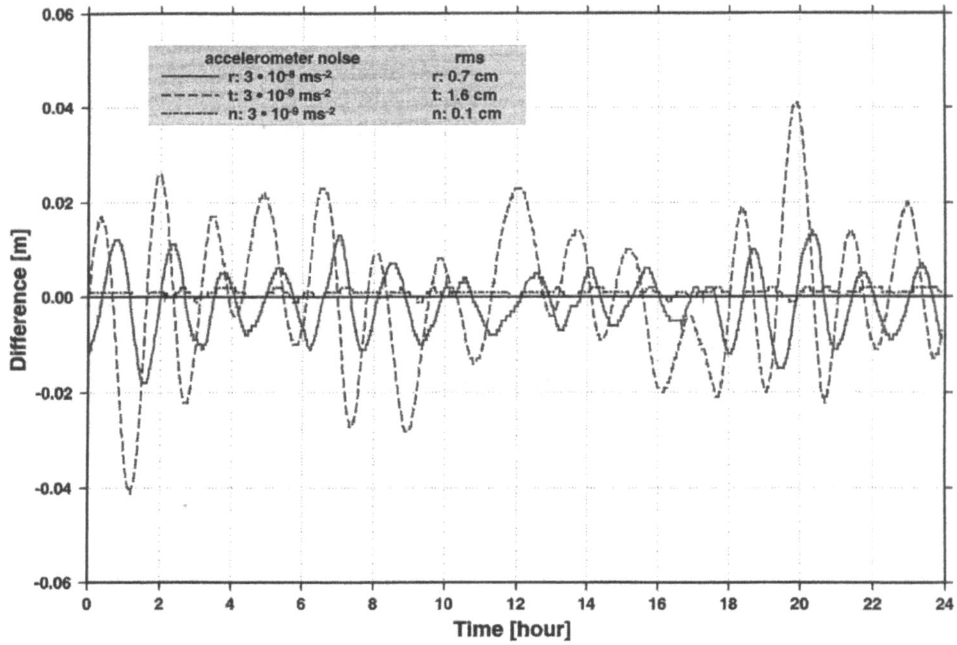

Fig. 2: CHAMP numerical orbit recovery simulation (impact of accelerometer noise)

Fig. 2 shows that the effect of the accelerometer noise on CHAMP orbit recovery accuracy is fairly small: 0.7 cm, 1.6 cm and 0.1 cm root mean square in the radial, transverse and normal component, respectively. It is about the same as the impact of GPS orbit and clock errors on CHAMP orbit restitution. Thus, the measurement accuracy of the accelerometer, as specified above, meets the CHAMP mission requirements.

5. Conclusions

The CHAMP accelerometer calibration parameters and disturbing attitude thruster accelerations were estimated in dynamic orbit adjustment using simulated GPS and accelerometer observations. The following conclusions can be drawn:

- The accelerometer calibration parameters (bias and linear scale factor) are solvable with the specified accuracy within the assumed mission and tracking scenario from a 5 days' worth of data. The thruster induced linear perturbing accelerations can be absorbed by estimating a systematic error vector per thruster pair.
- There are no large correlations between spherical harmonic gravitational coefficients (except GM, which has an one-to-one correlation with the radial bias) and the accelerometer calibration and thruster parameters. Therefore, the coefficients can be separated from those parameters.
- The impact of the accelerometer noise on CHAMP orbit restitution is only 0.7 cm, 1.6 cm and 0.1 cm root mean square in the radial, transverse and normal component, respectively, which is in balance with the orbit restitution accuracy to be expected from GPS-SST.

References

BEUTLER G., MUELLER I.I., NEILAN R.E. (1994). *The international GPS service for geodynamics (IGS): development and start of official service on 1 January 1994*, Bull. Geod. 68, 43-51.

KANG Z., SCHWINTZER P., REIGBER CH., ZHU S.Y. (1995). *Precise Orbit Determination for TOPEX/Poseidon Using GPS-SST Data*, Adv. Space Res. Vol. 16, No. 12, pp. (12)59-(12)62 .

REIGBER CH., SCHWINTZER P., LUEHR H. (1998). *The CHAMP Geopotential Mission*, Bollettino Di Geofisica, submitted

SCHWINTZER P., KANG Z., REIGBER CH. (1995). *Global Field Improvement from SST/GPS and Other Data*. Technical Report, GIGAR III, Phase 2, Final Report.

Simulation of the GOCE Gravity Field Mission

H. OBERNDORFER[1], J. MÜLLER[1], R. DOROBANTU[1], C. GERLACH[1], R. RUMMEL[1], N. SNEEUW[1], R. KOOP[2], P. VISSER[2], P. HOYNG[3], A. SELIG[3], M. SMIT[3]

Abstract

GOCE (Gravity Field and Steady-State Ocean Circulation Explorer) is one of the four selected ESA Earth Explorer Missions. The main objective of GOCE is the determination of the Earth's gravity field with high spatial resolution and with high homogeneous accuracy. For this purpose, two observation concepts will be realised. Satellite-to-Satellite Tracking (SST) in high-low mode will be used for the orbit determination and for the retrieval of the long-wavelength part of the gravity field. Satellite Gravity Gradiometry (SGG) will be employed for the derivation of the medium/short-wavelength parts of the gravity field..

For the realisation, a GPS receiver, a 3-axis gradiometer and auxiliary instruments are needed; e.g. star trackers to control the orientation of the spacecraft or thrusters for attitude and drag-free control. Each instrument exhibits, its own error behaviour which affects the measurements and the final products in some typical way.

The SID consortium consists of three organisations: the Institut für Astronomische und Physikalische Geodäsie (IAPG), Delft Institute for Earth-Oriented Space Research (DEOS) and the Space Research Organization Netherlands (SRON). They have the goal to provide a reliable error budget for the GOCE system, identify it's weaknesses and limitations and make a complete error analysis of the whole mission.

A complete simulator of the gradiometer has been developed. The simulator describes the test mass motion as a mass-spring system. The output are gravity gradients (V_{ij}^{out}) as if they were measured, based on input gravity accelerations, S/C position, orientation and disturbing forces. The corresponding error PSD's (Power Spectral Densities) representing the effect of different error sources are computed. The PSD's are propagated to spherical harmonic error spectra, geoid heights and gravity anomaly accuracies to assess effects of instrument and measurement errors on gravity field determination. The different tasks within the consortium are described by means of a flow chart.

1. Introduction

The launch of the ESA mission GOCE is envisaged for the year 2003 (ESA 1996). The project is one of four selected missions in ESA's Earth Explorer Programme (GOCE Phase A started July 1998). Opposite to the previous two other dedicated gravity missions CHAMP and GRACE (see corresponding literature), the basic observable for GOCE is satellite-gravity-gradiometry (SGG), i.e. the measurement of differential accelerations over short baselines. For GOCE the baseline is either 40 or 70 cm, and observations are performed in three independent directions. This concept allows high resolution determination of the gravity field with homogeneous accuracy. A centimeter-geoid will be feasible for wavelengths down to 160 km. To support the determination of the long wavelength part of the gravity field, the orbit perturbations are measured by GPS antennae (high-low SST). The GOCE orbit, is designed to be circular and sun-synchronous (≈ 97°) at a nominal height of 270 km. The spacecraft will be partially kept drag-free. Mission duration is limited to 9 months.

2. END-TO-END Simulation by the SID Consortium

Investigations and simulations concerning the GOCE mission are taking place in a consortium consisting of the following institutes:

- **S**pace Research **O**rganization **N**etherlands (SRON),
- **I**nstitut für **A**stronomische und **P**hysikalische **G**eodäsie (IAPG), and
- **D**elft Institute for **E**arth **O**riented **S**pace Research (DEOS).

From the initials of these institutes, the consortium is coined SID. The consortium is aided by the space division of Alenia Aerospazio, Torino, which is also the prime contractor of the GOCE Phase A Study.

Goal of the consortium is a full-fledged simulation of the mission, beyond the usual degree variance analysis or error covariance propagation. "Full-fledged" refers to the fact that the mission as a whole is simulated. The gradiometer is modelled by differential equations, basically coupled mass-spring systems, excited by gravitational and non-gravitational forces. In a backward step, a gravity gradient signal is derived from the integrated equations of motion. Purpose of the end-to-end simulations is to be able to identify weaknesses and limitations in an early stage, and to provide a reliable error budget for the system. In particular the effects of misalignment, instrument errors, calibration strategy, non-perfect drag-free control, etc. can be investigated. Moreover, the control loops for attitude and orbit determination (AOCS) and the drag-free orbit (DFC) are incorporated. If, for example, the star sensors observe an unwanted attitude motion, the thrusters have to control this motion.

[1] H. Oberndorfer, J. Müller, R. Dorobantu, C. Gerlach, R.. Rummel, N. Sneeuw, Institut für Astronomische und Physikalische Geodäsie (IAPG), Technische Universität München, 80290 München, Arcistr. 21 (e-mail: helmut@alpha.fesg.tu-muenchen.de, Tel: 0049-89-2892-3185)

[2] R. Koop, P. Visser: Delft Institute for Earth-Oriented Space Research (DEOS), Delft University of Technology, NL

[3] P. Hoyng, A. Selig, M. Smit: Space Research Organization Netherlands (SRON), Utrecht, NL

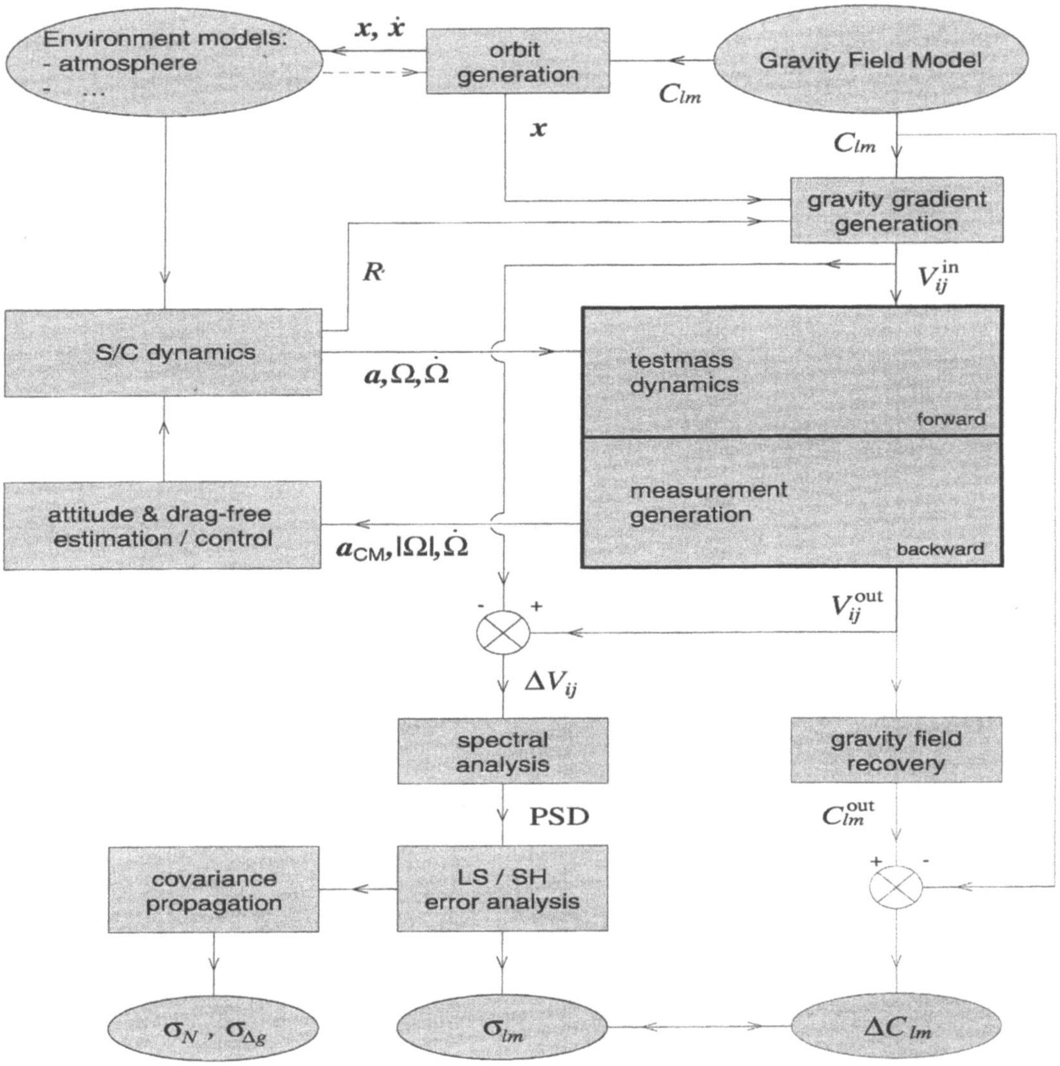

Figure1: flow chart about the SID activities

Or if the accelerometers sense non-gravitational orbit perturbations, the orbit has to be reconstituted within prescribed margins. These controls, on the other hand, affect the orbit, the observations, and finally the end-products. . Thus the simulator has to take care of the control loops, that is the simulation has to take place in *closed-loop*. The difference between input and output gradients results in error Power Spectral Densities (PSD's) that serve as input for further error analysis of end-products (spherical harmonic coefficients, geoid heights, gravity anomalies). Besides error analysis, the coefficients themselves are recovered. For this line of simulation the name *end-to-end* is used. Tasks have been divided within the SID consortium such that specific knowledge of the institutes is optimally used. The central component, the gradiometer model, is developed and implemented by SRON (Hoyng, 1997). It simulates gravity gradients *as measured*, based on position (perturbed orbit), input gravity gradients, orientation (attitude errors) and disturbing forces. These quantities are provided by DEOS, supported by Alenia. From the output gradients, DEOS recovers the potential coefficients. IAPG determines error PSD's from the differences between input and output gravity gradients for many case studies. These serve as basis for further covariance propagation onto end-products like potential coefficients (spectral) or geoid heights (spatial). Moreover, IAPG validates the numerical results, as supplied by the gradiometer model, by means of analytical modeling.

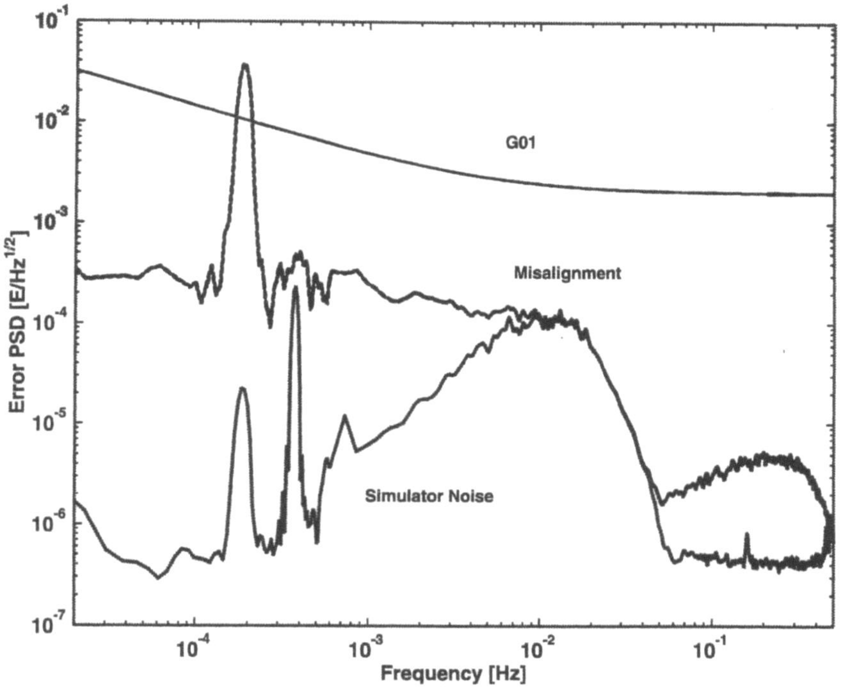

Figure 2: PSD's of the Vyy observable compared to the mission requirement.

3. Calculation of Power Spectral Densities

A power spectral density represents the spectral distribution of measurement errors or instrument accuracy. Integrated over the measurement bandwidth it yields the error for a single observation. Both calculated (input) gradients and observed (simulated output) gradients are available during the simulation process as time-series. The difference signal, i.e. gradient errors, represents all errors that have been built into the simulator for investigation purposes. The only error source that cannot be helped is numerical integration error. It can be reduced at will, though, by a proper time-step. The different time-series can be transformed now into PSD's (Figure 2). Case 'Simulator Noise' does not contain error sources. It solely reflects the numerical integration errors which are considered as simulator noise. Curve GO1 represents the mission requirement of 5 mE/√Hz and 1/f-behaviour below 1 mHz. Since this curve is well below GO1, a further reduction of the simulator noise by reducing the time step or complicating the integrator is not necessary. Clearly visible are the main peaks at orbit frequency and twice the orbit frequency. The noise level increases with frequency until it drops. The drop is caused by the lack of gravity field excitation in the high frequencies, since a gravity field up to maximum degree L=360 has been used in this simulation. In reality, the expected resolution of a GOCE gravity field, near degree 250, limits the required measurement bandwidth to, say, 0.1 Hz. Case study 'Misalignment' incorporates the effect of cross-coupling of other tensor components onto V_{yy} through misalignment. Similarly, other error sources are investigated. This procedure allows the determination of a full error budget, based on all types of realistic error sources, that have been and are identified during the whole project. Such an error budget is described in the spectral domain. During the mission the error sources are either measured with the same accuracy or compensated numerically or controlled.

4. Propagation onto End-Products

Primary goals of the GOCE mission are the determination of a gravity field model of spherical harmonic coefficients and the corresponding global geoid heights and gravity anomalies. These quantities are denoted here as *end-products*. The gradient accuracies for several mission scenarios and error case studies, as expressed by their PSD's, are propagated into a spherical harmonic error spectrum (SNEEUW et al., 1996) by least squares error analysis. The so-called lumped coefficients approach is followed, which is a frequency domain approach. This is the reason why PSD's are optimally suited as input for stochastic modeling. A further error propagation transforms the results into spatial variances and covariances of the geoid and gravity field. Several options for graphical representation of the results exist. The error degree RMS represents the average error size per degree. The GO1 curve (mission requirement) shows at least 3 orders of magnitude improvement in the low degrees. Extrapolating to the right, one can expect to obtain a resolution, loosely defined as the crossing point of signal (KAULA) and noise (GO1), near degree 250, equivalent to 160 km wavelength. Accumulation over the degree and multiplication by the Earth radius returns the geoid commission error (fig.3, right), showing the geoid error up to a specific degree. For instance the GOCE geoid up to degree 100 would be better than 1 mm, cf. the curve labeled GO1. The curves 302 and 303 correspond to the above mentioned test cases 'simulator noise' and 'misalignment'.

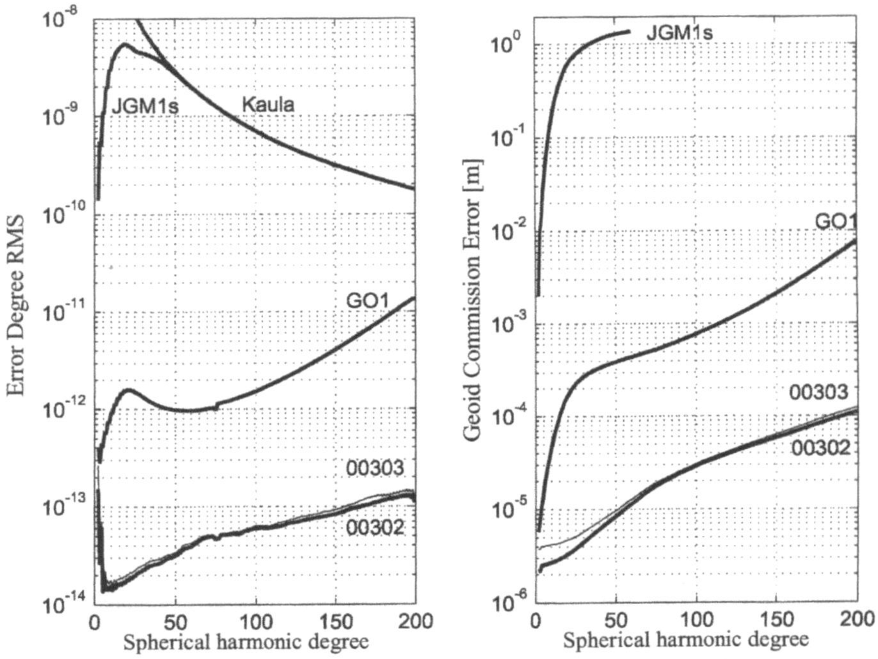

Figure 3: Error degree RMS (left) and geoid commission error of GOCE

5. Conclusion

In preparation of the GOCE mission a detailed and extensive simulation study takes place by the SID consortium. The simulation study corroborates mission requirements, identifies potential weaknesses and supplies a full spectral error budget. The simulation procedure is end-to-end, i.e. goes through the full cycle from test-mass motion through forcing and error terms to gradient retrieval, and finally to end-product assessment. The simulation procedure is in closed-loop in order to incorporate the interactions between the several system components (sensors, control loops, actuators).

The investigations have shown that GOCE is the first satellite mission to recover the static gravitational field with high spatial resolution and accuracy. The centimeter-geoid can be attained at a resolution of about degree 250 (160 km wavelength).

References

ESA (1996): *Gravity Field and Steady-State Ocean Circulation Mission*, Reports for Assessment: The Nine Candidate Earth Explorer Missions, ESA SP-1196(1), ESA Publications Division, ESTEC, Noordwijk

HOYNG P.(1997): *High tides for GOCE*, SRON Document I-G-001SR/97

REIGBER C., LUEHR H., SCHWINTZER P. (1998): *Status of CHAMP Geopotential Mission,* this issue

RUMMEL R.: (1986) Satellite Gradiometry, in: H. SÜNKEL (ed.), *Mathematical and Numerical Techniques Physical Geodesy*, Lecture Notes in Earth Sciences, Vol. 7, Springer Verlag, Berlin

SNEEUW N., KOOP R., SCHRAMA E.: (1996): *Global Gravity Field Analysis for the STEP Geodesy Co-Experiment Using GPS and Gradient Observations* Proceedings of the STEP Symposium, 6-8 April 1993, Pisa, Italy, ESA WPP-115

TAPLEY B.D. (1998): *The Gravity recovery and Climate Experiment (GRACE)*, this issue

Fundamentals and Applications of the Gravity Field Mission GOCE

N. SNEEUW*, C. GERLACH, J. MÜLLER, H. OBERNDORFER, R. RUMMEL

Abstract
This paper discusses the requirements that geosciences pose on global gravity field determination in terms of accuracy and spatial resolution. To be precise, the requirements from geodesy, oceanography and geophysics are presented. Particularly the gravity field recovery that can be attained by ESA's GOCE mission (Gravity Field and Steady-State Ocean Circulation Explorer) is emphasized. This implies that the underlying discussion is limited to the stationary Earth gravity field. The discussion is preceded by a brief introduction to the GOCE satellite and to gradiometry in general.

1 Introduction

GOCE (Gravity Field and Steady-State Ocean Circulation Explorer) is one in a series of complementary space missions devoted to map the Earth's gravity field, cf. (ESA, 1996). The other missions are CHAMP (Challenging Micro-Satellite Payload for Geophysical Research and Application), cf. (REIGBER , LUEHR AND SCHWINTZER, this issue), and GRACE (Gravity Recovery and Climate Experiment), cf. (TAPLEY, this issue). CHAMP will improve the present knowledge of the gravity field by one order, whereas GRACE not only will improve the low-degree part of the gravity field further, but will also monitor its time variation. In (BALMINO ET AL., 1998) these three missions and their underlying concepts are compared in terms of recovery capability of the static gravity field.

The main objective of GOCE is the determination of the Earth's gravity field with high spatial resolution and with high homogeneous accuracy. For this purpose, two observation concepts will be realised. Satellite-to-Satellite Tracking (SST) in high-low mode will be used for the orbit determination and for the retrieval of the long-wavelength part of the gravity field. Satellite Gravity Gradiometry (SGG) will be employed for the derivation of the medium/short-wavelength parts of the gravity field.

Many geoscientific disciplines will benefit from the GOCE gravity field: oceanography (ocean circulation from combination with altimetry; further input for climatology), geophysics (gravity field related to mechanisms in the Earth's interior) and geodesy (unification of height systems worldwide, levelling with GPS). Each of these disciplines requires knowledge of the geoid, gravity anomalies or geoid slopes at different spatial scales and at different accuracy levels. This will be addressed in more detail furtheron.

*Nico Sneeuw; Institut für Astronomische and Physikalische Geodäsie (IAPG); Technische Universität München; D-80290 Germany; Tel: +49-89-289 23193; Fax: +49-89-289 23178; Email: sneeuw@step.iapg.verm.tu-muenchen.de

2 Gradiometry by GOCE

The GOCE orbit will be near circular and sun-synchronous ($I \approx 97°$) at an altitude of 270 km. The mission duration is planned to be about 8 months, during which the orbit will be kept drag-free. The orbit will be determined by GPS with an accuracy of about 1 cm/$\sqrt{\text{Hz}}$. The measurement precision of the 3-axis gradiometer can be assumed at the 10^{-3} Eötvös/$\sqrt{\text{Hz}}$ level. The GOCE baseline mission parameters are summarized in figure 1. For more technical details it is referred to (REBHAN, this issue) and to (OBERNDORFER ET AL., this issue), in which the interplay between AOCS, drag-free control, error sources and the gradiometer is investigated.

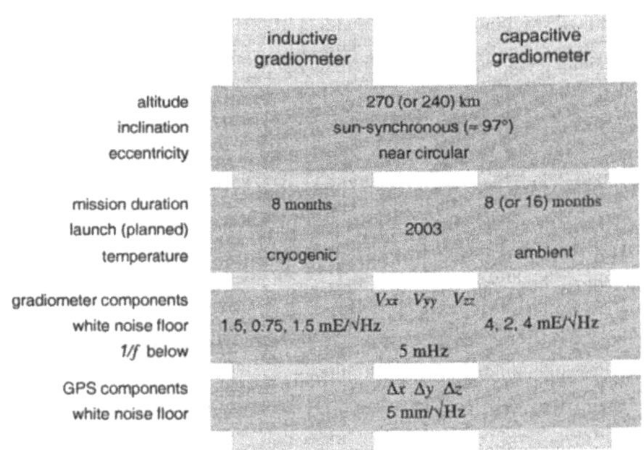

Figure 1: GOCE mission parameters

The low orders of the gravity field are to be recovered by space-borne GPS tracking (high-low SST). The main gravity field information, though, will come from the gradiometer. It measures the tensor—at least the diagonal terms—of the second spatial derivatives of the gravitational potential. Let the geopotential be written in complex-valued notation as:

$$V(r,\theta,\lambda) = \frac{GM}{r}\sum_{l=0}^{L}\left(\frac{R}{r}\right)^{l}\sum_{m=-l}^{l} K_{lm}Y_{lm}(\theta,\lambda) \ , \tag{1}$$

in which $[r, \theta, \lambda]$ are the spherical coordinates [radius, co-latitude, longitude] and R the Earth's equatorial radius. The base functions Y_{lm} are the fully normalized surface spherical harmonics and K_{lm} their corresponding coefficients. The index l is the spherical harmonic *degree* and m the *order*. The above series has been cut off at maximum degree $l = L$. The spherical harmonic representation is global by nature, which is suitable for satellite applications. A further advantage of the Y_{lm}, apart from their orthogonality, is that they represent a complete set of eigenfunctions for a large set of observable gravity functionals (RUMMEL AND VAN GELDEREN, 1995), commonly used in physical geodesy.

The 2nd radial derivative—for example—now reads:

$$\begin{aligned} \frac{\partial^2 V}{\partial r^2}(r,\theta,\lambda) &= V_{rr}(r,\theta,\lambda) \\ &= \frac{GM}{r^3}\sum_{l=0}^{L}(l+1)(l+2)\left(\frac{R}{r}\right)^{l} \cdot \\ &\quad \sum_{m=-l}^{l} K_{lm}Y_{lm}(\theta,\lambda) \ . \end{aligned} \tag{2}$$

The potential V, whose deviation from a normal potential is proportional to the geoid through Bruns's equation, is required at the Earth's surface. The measurements, however, are made at satellite altitude h, i.e. $r = R + h$ km. From equation (2) it is seen that $(R/r)^l$, the spectral symbol of the upward continuation operator, attenuates the signal for increasing degree l. This term explains why the satellite orbit should be as low as possible.

The expression $(l + 1)(l + 2)$, which is characteristic for the 2nd radial derivative, amplifies the signal with increasing degree. It thus counteracts the upward continuation. This is—in a nutshell—the rationale for satellite gravity gradiometry. It is demonstrated graphically in figures 2-4. Starting from the EGM96 geoid, cf. (LEMOINE ET AL., 1998), the 2nd radial derivative at the Earth surface, cf. figure 3, is seen to be a high-frequency signal. Continuing this signal upward to satellite height, a tremendous smoothing of the short wavelengths takes place, as becomes clear from figure 4. Note that the other tensor components are characterized by $\mathcal{O}(l^2)$-terms as well, cf. (RUMMEL AND VAN GELDEREN, 1995).

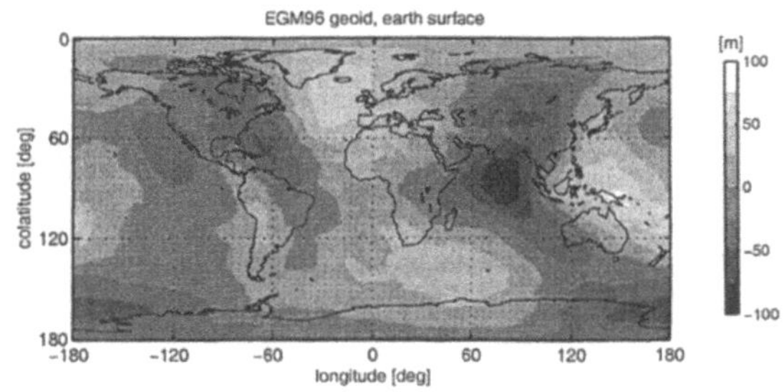

Figure 2: EGM96 geoid

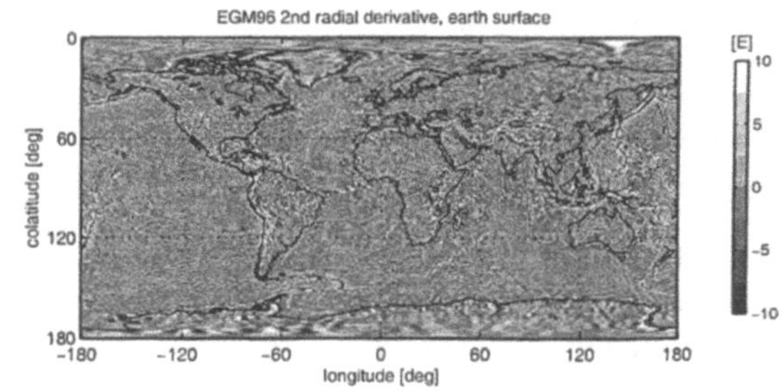

Figure 3: EGM96 2nd radial derivative at Earth surface

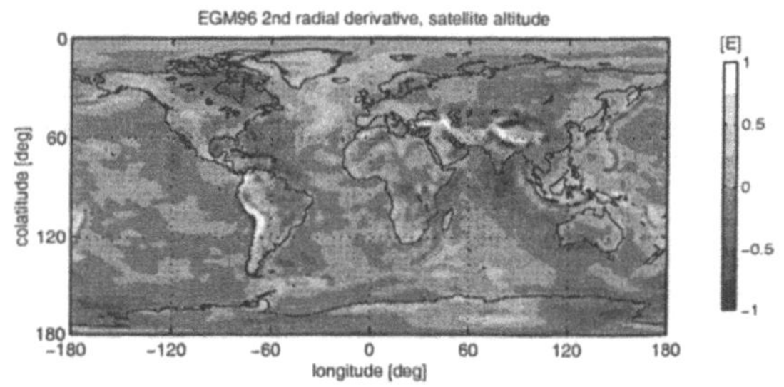

Figure 4: EGM96 2nd radial derivative at satellite altitude, $h = 270$ km

3 Requirements From Geoscience

The geoid and its corresponding gravity field are highly valuable quantities in several geoscientific disciplines. In this work we focus on the benefit of high quality geopotential modeling to geodesy, oceanography and geophysics.

1. In geodesy precise geopotential knowledge is required for the following purposes: inertial navigation, orbit determination, global unification of height systems, and levelling by GPS, where geometric GPS-derived heights are transformed into physical heights through the geoid.

2. For oceanography a precise geoid serves as equipotential surface. Altimetry provides the geometric sea-surface. From differences between

both surfaces the ocean circulation can be determined at several scales: from the Antarctic circumpolar current (ACC) and western boundary currents (WBC), like the Gulfstream and the Kuroshio, to small-scale coastal currents. In particular the slope of the geoid is the quantity of interest in oceanography.

3. Geophysics requires an accurate gravity field for several purposes: for modeling phenomena of the oceanic lithosphere (ridges, trenches, seamounts), the continental lithosphere (mountain ranges, roots, ...) to large-scale mantle effects (convection, ...). Also the determination of the global ice balance requires gravity field knowledge.

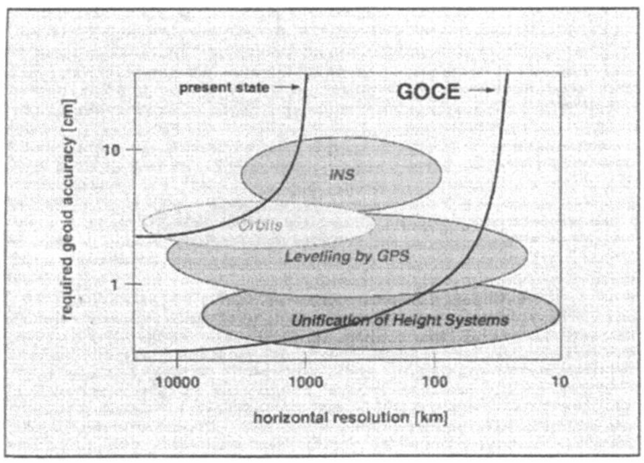

Figure 5: Requirements for geodetic purposes

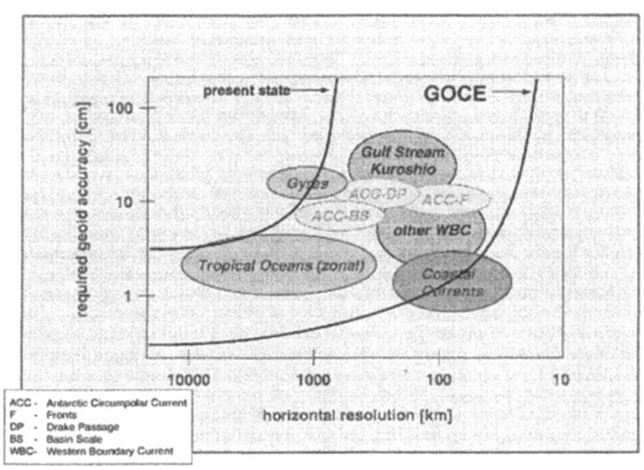

Figure 6: Requirements for ocean circulation determination

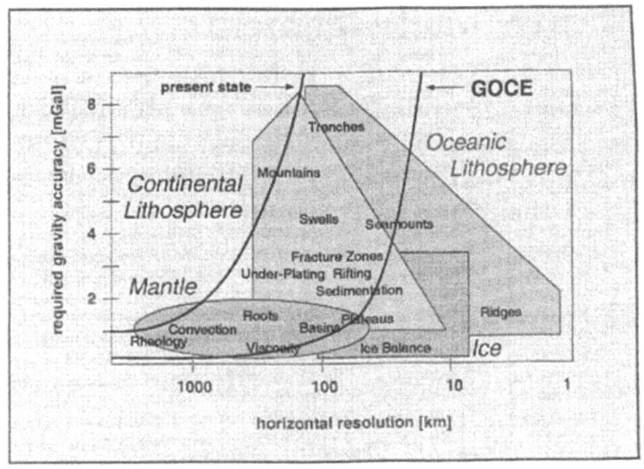

Figure 7: Geophysical requirements

Figures 5-7 summarize the requirements for the three mentioned geosciences. These figures are based on similar graphics in (BALMINO ET AL., 1998). Requirements are defined as 10% of the signal to be expected. Although this level is quite arbitrary, it serves its purpose of demonstrating the lack of present knowledge and need for an improved gravity field well enough. Every quantity or phenomenon has been pictured in two dimensions: length-scale and required accuracy. For precise orbit determination, eg. figure 5, geoid knowledge is required at the level of a few centimeters over scales from global down to a few hundred kilometers. Levelling by GPS requires better accuracy and a better resolution, i.e. shorter wavelengths.

In all three graphs a curve **present state** denotes the state of art of gravity field knowledge. Every phenomenon below and/or right of this curve cannot be detected at the 10% level with present knowledge. The **GOCE** curve indicates the improvement of the gravity field after a GOCE mission. For each of the three geosciences a wealth of new phenomena will become visible.

4 Summary

The satellite mission GOCE (Gravity Field and Steady-State Ocean Circulation Explorer) has the potential to recover the high frequencies of the Earth's gravitational potential. Measuring the 2nd spatial derivatives of the potential partially makes up for the smoothing effect due to measuring at satellite altitude. Thus satellite gravity gradiometry counteracts the upward continuation.

Figures 5-7 show the tremendous benefit of a mission like GOCE to geodesy, oceanography and to various geophysical disciplines. The phenomena below the curve denoted with **present state** will be revealed by GOCE, up to the level implied by the **GOCE** curve. These figures, and correspondingly the above conclusions, are limited, though, to the static gravity field.

5 Acknowledgements

We would like to thank PHIL WOODWORTH (Proudman Oceanographic Laboratory) and GEORGES BALMINO (CNES/GRGS) for providing us with input for the requirement figures.

References

BALMINO G., PEROSANZ F., RUMMEL R., SNEEUW N., SÜNKEL H., WOODWORTH P. (1998). *European Views on Dedicated Gravity Field Missions: GRACE and GOCE.* ESA Earth Sciences Division Consultation Document ESD-MAG-REP-CON-001.

ESA (1998). *Gravity Field and Steady-State Ocean Circulation Mission.* ESA SP-1196(1).

LEMOINE F.G. ET AL. (1998). *The Development of the Joint NASA GSFC and the National Imagery and Mapping Agency (NIMA) Geopotential Model EGM96.* NASA/TP—1998–206861.

OBERNDORFER H., GERLACH C., MÜLLER J., RUMMEL R., SNEEUW N., KOOP R., VISSER P., HOYNG P., SMIT M. (1998). *Simulation of the GOCE Gravity Field Mission.* this issue.

REBHAN H., JOHANNESSEN J.A. (1998). *The ESA Gravity Field and Steady-State Ocean Circulation Explorer Mission.* this issue.

REIGBER CH., LUEHR H., SCHWINTZER P. (1998). *Status of the CHAMP Geopotential Mission.* this issue.

RUMMEL R., VAN GELDEREN M. (1995). *Meissl scheme – spectral characteristics of physical geodesy.* manuscripta geodaetica 20:379-385.

TAPLEY B.D. (1998). *The Gravity Recovery and Climate Experiment (GRACE).* this issue.

Effects of Inhomogeneous Data Coverage on Spectral Analysis

R. Pail, W.-D. Schuh

Mathematical Geodesy and Geoinformatics (MGGI)
Technical University Graz, Steyrergasse 30, A-8010 Graz, AUSTRIA

ABSTRACT: The present study investigates the effects of irregular data distribution on the corresponding frequency representation. For a basic understanding it starts with a simple 1D-model and provides a description of effects emerging from transition of the infinite continuous to the finite discrete case. Considering regular (gridded) and irregular data distributions and even gap regions without data coverage, the corresponding results in frequency domain and the numerical stability are analyzed using trigonometric functions as base functions.

1 Introduction

In most cases the spectral analysis is closely linked with continuous or discrete equispaced sampled functions. Each arbitrary continuous function $f(x)$ (signal) can be represented by an infinite series of *sin*– and *cos*–functions (*Fourier series*),

$$f(x) = {\sum_{k=0}^{\infty}}' a_k \cos kx + \sum_{k=1}^{\infty} b_k \sin kx \tag{1}$$

with an infinite set of parameters a_k and b_k [1]. The determination of these parameters a_k and b_k

$$a_k = \frac{1}{\pi}\int_{-\pi}^{\pi} f(x) \cos kx dx \tag{2}$$

$$b_k = \frac{1}{\pi}\int_{-\pi}^{\pi} f(x) \sin kx dx \tag{3}$$

results from the orthogonality of the base functions *sin* and *cos* with respect to the integration in the interval $[-\pi,\pi]$ (*Fourier integrals*).

Similar relations also hold if N equispaced discrete samples of the function $f(x_i)$ are known. In connection with a finite and discrete set of data locations N, the reconstruction of the maximum involved frequency is bounded to the Nyquist frequency $(N-1)/2$. The usual equispaced (complete) discretization performs an orthogonal and thus numerically stable reconstruction of the signal. The unknown coefficients a_k and b_k can be determined parameterwise by

$$a_k = \frac{2}{N} {\sum_{i=0}^{N-1}}' f(x_i) \cos ki \, \frac{2\pi}{N} \tag{4}$$

$$b_k = \frac{2}{N} \sum_{i=1}^{N-1} f(x_i) \sin ki \, \frac{2\pi}{N} \, . \tag{5}$$

Considering that the relations $\frac{2}{N} = \frac{\Delta x}{\pi}$ and $x_i = i\frac{2\pi}{N}$ hold, this formula can be interpreted as the discrete solution of the Fourier integrals (2), (3), resp. However, on the other side this computation can also be seen as inverse problem due to equation (1). Each data point contributes a linear equation (reproduction condition) and yields an orthogonal design matrix (cf. fig. 1a).

Most of the discussions on spectral analysis deal with these two cases, which are closely linked to the continuous and discrete *Fourier transform*, and some effects emerge from the transition from the continuous to the equispaced discrete case (aliasing, leakage) and the restriction to a finite or special interval (Gibbs-phenomena, edge-effects[2]). In practice only a finite number of generally not equispaced data is available. Therefore, we have to distinguish between a variety of different cases (cf. tab. 1).

Table 1 illustrates the large variety of spectral analysis techniques.

max. involved frequency L	number of data locations N	max. resolved frequency P	number of parameter $M=2P+1$	
∞	∞	∞	∞	cont. Fourier analysis
L	$N=2L+1$	L	$M=N$	classical sampling (disc. Fourier analysis)
	$N<2L+1$	**incomplete discretization - aliased problems**		
		$P=\frac{N-1}{2}$	$M=N$	definite aliased
		$P>\frac{N-1}{2}$	$M>N$	underdetermined aliased (Backus&Gilbert)
		$P<\frac{N-1}{2}$	$M<N$	overdetermined aliased (least squares adjust.)
	$N>2L+1$	**complete discretization - overdetermined problems**		
		$P=L$	$N>M$	well-balanced
		$\frac{N-1}{2}>P>L$	$N>M$	overparameterized
		$P<L$	$N>M$	underparameterized (spectral leakage)

Table 1: Spectral Analysis Techniques.

The most important item within the analysis represents the maximum frequency L inherent in the signal $f(x)$. A strict reconstruction of the signal requires a sufficient sampling (cf. tab. 1, complete discretization). Only a higher sampling rate gives access to higher frequencies, and as an extreme case an infinite number of measurements is necessary to

[1]The prime in the first sum refers to the convention that the term $k = 0$ should be taken with *half weight*.

[2]A special form of leakage caused by an under-representation of long wavelengths.

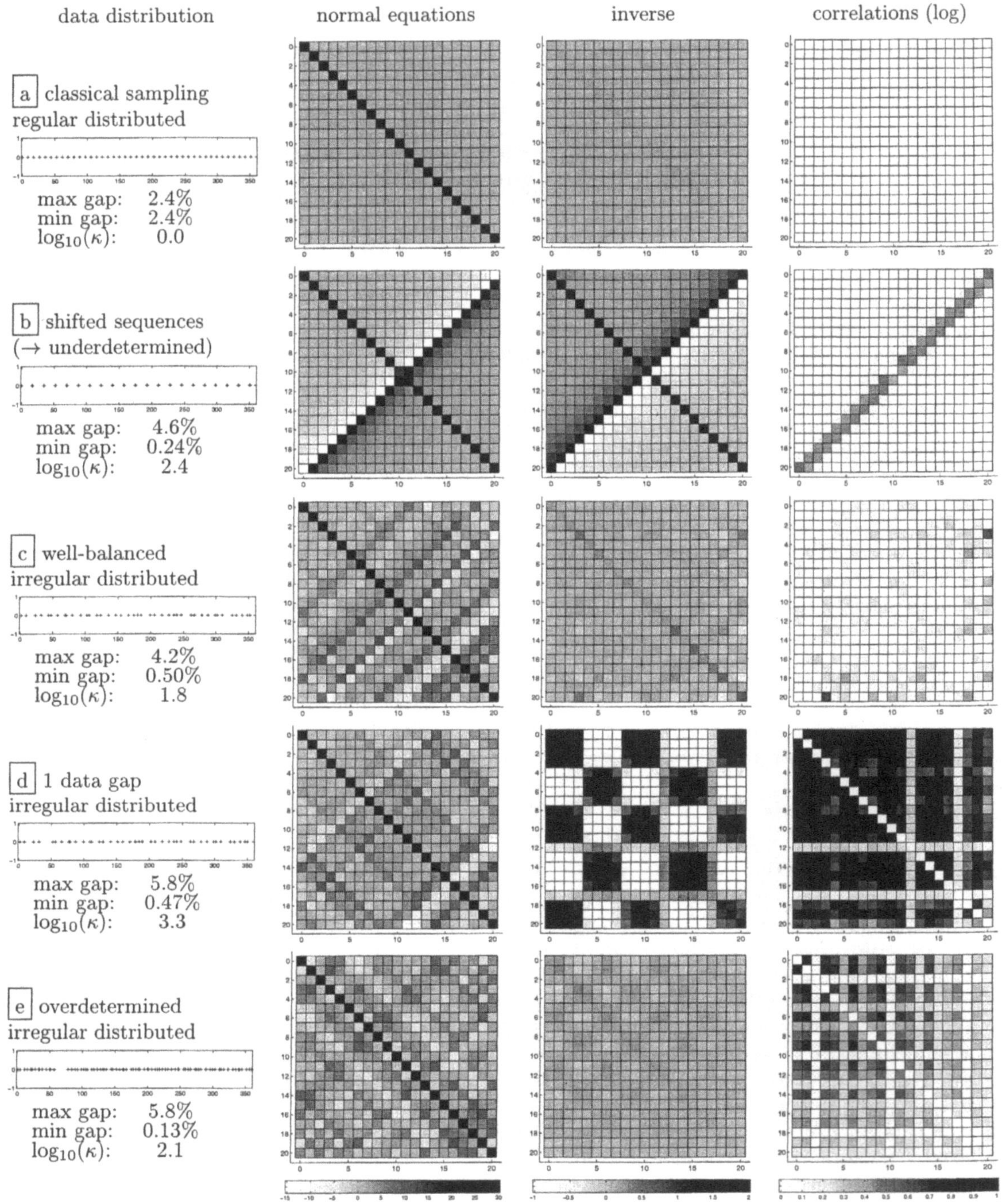

Figure 1: This figure shows the numerical behaviour of spectral analysis due to typical data distributions. In addition to some alphanumeric information on the left, the normal equations, the inverse of the normal equations and a rescaled correlation matrix ($\log_{10}(1 - \mathrm{abs}(\rho_{ij}))$) are plotted. Besides the classical equispaced sampling (fig. 1a), different types of data distributions are studied. Sampling by shifting equispaced sequences generally allows a stable reconstruction of the signal. In the normal equations this relative shift is reflected by non-zero entries in the second main diagonal, whose amplitude grows with increasing shift rate (fig. 1b). Simultaneously the condition number worsens. In the extreme case the two equispaced sequences coincide, resulting in an underdetermined and thus singular system.

Many test computations with irregular data coverage (fig. 1c) yield that the numerical behaviour is dominated by the size of the maximum data gap. This fact is underpinned by simulation fig. 1d, where the maximum data gap is enlarged compared with simulation fig. 1c, leading to an increase in the condition number representing the stability of the system. In general, the extension of the gap size influences the numerical behaviour much more extensively than a distance reduction between adjacent data locations. The stability can be improved by additional measurements, resulting in an overdetermined system demonstrated in simulation fig. 1e and consequently in a reduction of the condition number.

resolve a signal with infinite frequency content. Otherwise an underdetermined problem occurs (Backus & Gilbert problem, cf. [1] or [5]) and additional information (e.g. degree-variances, cf. [3]) is required to determine an unique set of parameters.

As mentioned above, the number of equispaced data locations steers the maximum resolvable frequency (Nyquist frequency). The same fact holds also for irregular distributed data locations. Each location x_i supplies an equation of type (1). The reconstruction of the parameters a_k and b_k can again be performed by a discrete version of the Fourier integrals or by the solution of the $N \times N$ equation system. However, if some disturbances due to higher frequencies inherent in the signal or some disturbances in the discrete function values occur, these two approaches pursue different principles. The integral approach tries to reconstruct an orthogonal base by a weighting strategy due to the data location. Therefore, the recreation of an individual frequency forms the main target. On the other side, the inversion (adjustment) procedure tries to find a best reconstruction of the signal, regardless of the recreation of each individual frequency.

Irregular data distributions lead to a loss of orthogonality and thus to non-diagonal systems of normal equations. The stability (condition number) of the design matrix on the one side and the frequency of the residuals on the other side form the main criteria. Fig. 1 reflects the behaviour of typical situations.

2 Data Gaps

Detailed investigations have been performed to explore the connection between data gaps and the numerical stability of the system of normal equations. Many test computations with irregular data coverage yield that its numerical behaviour is dominated by the size of the maximum data gap. Therefore we investigate in this section the simplified configuration of one growing data gap with equispaced sampling intervals beyond the gap region. The examples of section 3 will confirm that similar results can also be obtained from irregular data coverage.

Fig. 2 shows the stability of the corresponding system of normal equations, expressed by the decadic logarithm of the condition number κ, as a function of the relative gap size r, which is defined by

$$r = \frac{\text{absolute gap}}{\text{analysis interval}} \ . \tag{6}$$

P and μ denote the maximum resolved frequency (cf. tab. 1) and a factor ($1 \leq \mu < \infty$) describing the redundancy of the system, respectively. Consequently, for the number of data follows

$$N = 2\mu P + 1 \ . \tag{7}$$

Fig. 2 yields an approximately linear relation between the relative gap size r and the logarithm of the condition number κ up to the numerically computable limit of about 15 digits and thus provides a first possibility to predict the numerical stability.

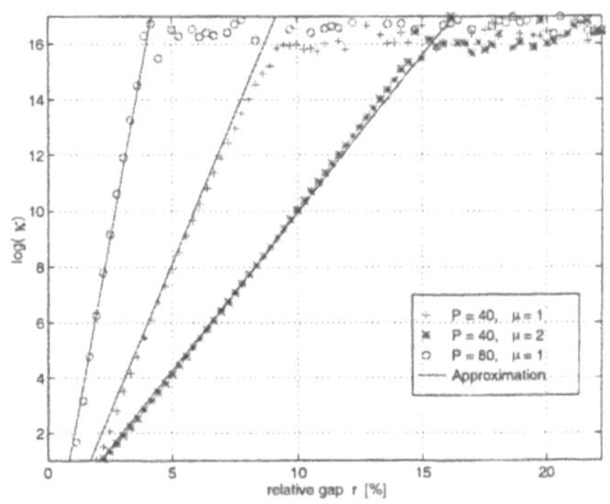

Figure 2: condition κ vs. relative gap size r, based on a given maximum resolved frequency P and a redundancy μ.

This nearly linear behaviour is used to normalize the logarithmic condition number $log_{10}(\kappa)$ by the relative gap size r. The resulting normalized quantity characterizes the slope of the curves shown in fig. 2. Plotting these slopes for various redundancies μ versus the maximum resolved frequency P, again an approximately linear behaviour becomes obvious.

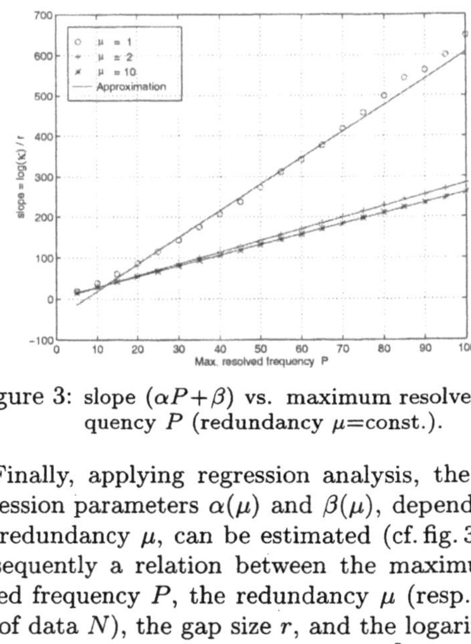

Figure 3: slope ($\alpha P + \beta$) vs. maximum resolved frequency P (redundancy μ=const.).

Finally, applying regression analysis, the linear regression parameters $\alpha(\mu)$ and $\beta(\mu)$, depending on the redundancy μ, can be estimated (cf. fig. 3), and consequently a relation between the maximum resolved frequency P, the redundancy μ (resp. number of data N), the gap size r, and the logarithm of condition number κ can be established[3]:

$$\log_{10}(\kappa) = (\alpha P + \beta)\left(r - \frac{1}{2P+1}\right) \tag{8}$$

with

$$\alpha = \frac{0.32}{\mu - 0.92} + 2.55 \ ; \quad \beta = \frac{-4.00}{\mu - 0.92} + 3.30 \ . \tag{9}$$

[3]The term $1/(2P + 1)$ in (8) describes the fact that the curves in fig. 2 do not cross the origin of the coordinate system.

The solid lines in figs. 2 and 3, respectively, yield that this rule of thumb fits the strict solutions very well. Inversely to fig. 3, fig. 4 demonstrates that

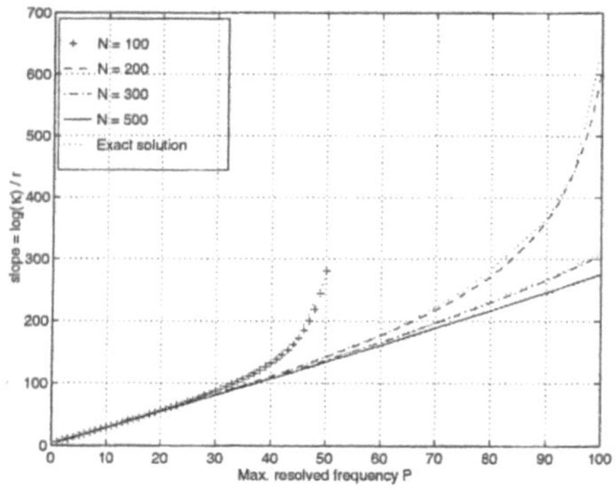

Figure 4: slope ($\alpha P+\beta$) vs. maximum resolved frequency P (number of data N=const). The approximation results are compared with the exact solutions (dotted curves).

for a given number of data points N within a fixed analysis interval, a reduction of the maximum resolved frequency P (corresponding to an increase of redundancy μ) leads to a marked stabilization of the system and consequently to a decrease of condition number κ. For example, concerning the dashed line ($N = 200$) in fig. 4, we can estimate a reduction of the slope from 600 for the max. resolved frequency $P = 100$ to about 140 for $P = 50$, corresponding to the redundancy of $\mu = 2$. Both relations can also be extracted from fig. 3. For the well-balanced case ($\mu = 1$) using the curve indicated by circles ($P = 100 \rightarrow$ *slope* ~ 600) and the two-fold overdetermined case ($\mu = 2$) taking the crossed curve into account ($P = 50 \rightarrow$ *slope* ~ 140). Both figures illustrate very impressively the marked improve of the numerical behaviour already for small redundancies up to two. A further redundancy increase improves the stability only slightly.

Consequently, the rule of thumb (8) as a function of κ, r, P, μ, allows the estimation of one parameter, if the other three parameters are fixed. This is demonstrated on the basis of a few examples in the following section.

3 Examples

Example 1 demonstrates the straightforward prediction of the condition number based on the given parameters P, r and μ. The last line provides the results for a strict computation with respect to regular and irregular data coverage beyond the gap.

IN	max. degree	P	40
	relative gap	r	5%
	redundancy	μ	1
OUT	condition	$\log_{10}(\kappa)$	8.1
STRICT	$\log_{10}(\kappa)$	regular	7.9
		irregular	8.4

Example 1.

This situation is also treated graphically by the crossed curve in fig. 2.

Example 2: Which additional number of data is required to remain under a given condition number $\kappa = 10^5$? The rule of thumb estimates a redundancy factor of 1.24. A recalculation using this configuration yields condition numbers of $10^{4.8}$ (regular case) and $10^{5.4}$ (irregular case), respectively.

IN	max. degree	P	40
	condition	$\log_{10}(\kappa)$	5
	relative gap	r	5%
OUT	redundancy	μ	1.24
STRICT	$\log_{10}(\kappa)$	regular	4.8
		irregular	5.4

Example 2.

Example 3: How many parameters (max. frequency) can be resolved if we postulate a condition $\kappa = 10^5$, gap size $r = 5\%$ and redundancy $\mu = 2$? Applying the resulting maximum frequency estimate $P=45$, the recalculation reflects the postulated stability.

IN	redundancy	μ	2
	condition	$\log_{10}(\kappa)$	5
	relative gap	r	5%
OUT	max. degree	P	45
STRICT	$\log_{10}(\kappa)$	regular	4.9
		irregular	5.0

Example 3.

4 Two Data Gaps

In section 2 we mentioned that in the case of irregular data coverage the numerical behaviour is dominated by the size $r = r_1$ of the maximum data gap. In most cases the numerical stability worsens with a growing second gap r_2 and markedly depends on the relative position of the two gaps. As an example fig. 5

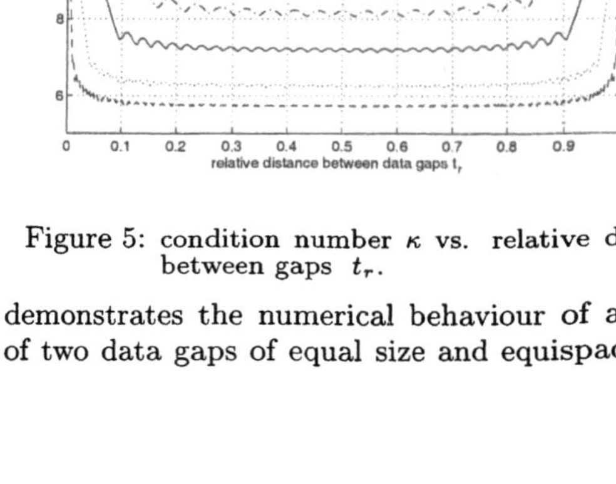

Figure 5: condition number κ vs. relative distance between gaps t_r.

demonstrates the numerical behaviour of a system of two data gaps of equal size and equispaced data distribution in between as a function of the distance between the two gaps and for different redundancy factors.

In the *well-balanced case* ($\mu = 1$), the condition number grows with decreasing distance and in this example even reaches the numerical limit in the range of $\kappa = 10^{15}$. The minimum occurs for a periodic configuration. In this special case of two periodic gaps of equal size the condition number turns out to be lower by about 10 to 15 % compared with only one gap of the same size. As the contrary extreme case the two gaps unify, $r = 2\,r_1$, and thus the rule of thumb (8) provides the worst case estimate for two data gaps[4].

Please note the marked drop of the condition number in the case of *overdetermination* ($\mu > 1$). In this case a whole interval of distances yields approximately equal condition numbers, where this interval extends with increasing degree of redundancy. As a consequence, already a small number of additional measurements leads to an extensive improvement of the system's numerical stability.

5 Conclusions

Given a continuous function, its frequency properties - represented by the Fourier coefficients - can be uniquely determined by solving the Fourier integrals. In the case of the transition to a discrete, equidistantly sampled function, the reconstruction of coefficients can be performed either by a discrete analogon of the Fourier integrals or by the solution of an orthogonal system of equations. These two approaches differ only if certain disturbances are inherent in the signal.

Concerning irregular data coverage, numerical problems emerge due to the loss of orthogonality properties of the base functions, leading to non-diagonal normal equation systems, whose numerical behaviour turns out to be dominated by the size of the maximum data gap. Consequently, we propose a simple rule of thumb based on this one gap configuration, which gives a good estimate (±10%) for the numerical stability. Since this rule of thumb can also be applied in the presence of large data gaps, it may help to facilitate feasibility studies by estimating whether a certain data configuration is numerically stable, and how to modify the parameter model (max. resolved frequency, redundancy) in order to stabilize the system, respectively. The rule of thumb is applied in a selection of examples in order to demonstrate its practicability.

[4] An improved estimate for arbitrary ranges results from a linear approximation between the periodic case (rule of thumb value for data gap size r_1, diminished by about 10 %) and the case of gap unification (rule of thumb value for gap size $2\,r_1$), which finally reads

$$\log_{10}(\kappa) = (6.55P - 46.7)\left[2r - \frac{1}{2P+1} - \left(2.2r - \frac{0.2}{2P+1}\right)t_r\right] \quad (10)$$

with

$$t_r = \frac{\text{absolute gap distance}}{\text{analysis interval}}\;; \quad t_r \in [0,\,1/2]\,. \quad (11)$$

References

[1] BACKUS G., F. GILBERT (1970): Uniqueness in the Inversion of Inaccurate Gross Earth Data. Philos. Trans. R. Soc. London, 266, pp. 123-192.

[2] BRIGHAM E.Oran (1974): The Fast Fourier Transform. Prentice-Hall, Inc., Englewood Cliffs, New Jersey.

[3] SCHWARZ K.P. (1976): Least Squares Collocation for Large Systems. Bollettino di Geodesia e Science Affini, Anno XXXV, N.3, pp. 309-324.

[4] SNIEDER R. (1990): Global inversions using normal modes and long-period surface waves. In: Seismic tomography, edited by H. M. Iyer and K. Hirahara, pp. 23-63, Prentice-Hall,London.

[5] TARANTOLA A. (1987): Inverse Problem Theory. Elsevier, Amsterdam.

Geodetic Applications of the ROCSAT-3/COSMIC Mission

E. C. PAVLIS[1], B. F. CHAO[2], C. W. HWANG[3], C. C. LIU[4], C. K. SHUM[5], C. L. TSENG[6], M. YANG[6]

Abstract

The jointly planned Taiwan-US space mission ROCSAT-3/COSMIC (Constellation Observing System for Meteorology, Ionosphere and Climate) is scheduled for launch in 2001. A 6-8 low Earth orbiting satellite constellation will conduct atmospheric limb-sounding using GPS radio occultation. Although not the primary scientific objective, it is envisioned that the initial low altitude orbits (~400 km) of the satellites (some in tandem) will provide unique opportunities for gravity mapping and precision orbit determination experiments. Simulations show that, depending on the orbit evolution scenario, the use of these data can yield at least an order of magnitude improvement over the state-of-the-art global gravity model EGM96 out to degree and order 20-40 (spatial resolution of 1000-500 km). This improvement would be very useful in anticipation of missions that focus on the temporal variations of the field, such as GRACE, and for current and future altimeter missions. Additionally, useful signals of low-degree temporal variations can be obtained if non-conservative forces remain "modelable" at the final altitude of above 700 km. Some of the geodetic benefits that can be expected from ROCSAT-3/COSMIC include:

(a) simultaneous improvement of the GPS satellites' orbits within a few hours from data collection, essential to all GPS applications (a factor of two improvement can be expected over the current accuracy at 5-10 cm),

(b) improved estimation of the global surface pressure fields, useful for geodynamic and altimetric studies such as Earth rotation, geocenter motion, time-varying gravity, and ocean circulation, and

(c) inclusion of satellite laser ranging reflectors, proposed as a secondary tracking system, will help establish local datum ties to the conventional terrestrial reference frame.

1. Introduction

COSMIC is jointly planned by Taiwan and US primarily as a meteorology space mission. The tracking data that will be collected for the precise orbit determination of the COSMIC spacecraft however will contain valuable information about the terrestrial gravity field. Mapping of the terrestrial gravitational field has seen great advances in data acquisition and modeling techniques over the past four decades ever since the first days of the space age. The driving impetus of these advances is the quest for higher precision, higher spatial resolution, as well as the prospect of reasonable temporal resolution for its time-varying components. International efforts to address these areas led to a series of new models, most recently for example the *Earth Gravity Model 1996* -- EGM96, [LEMOINE et al., 1998], a spherical harmonic expansion complete to degree and order 360 corresponding to a spatial resolution of about 50 km. Such high resolution models incorporate terrestrial surface gravity data and ocean altimetry data for short and medium wavelength components, but the "back-bone" long-wavelength signals are derived from the analysis of the tracking data from near-Earth orbiting satellites. These data carry the imprint of the perturbations experienced by the spacecraft due to the non-spherical nature of the terrestrial field.

2. The COSMIC mission

The eight COSMIC satellites will be launched together in early 2001 into an initial orbit at a 400-km altitude and an inclination of about 78 degrees. On-board thrusters will subsequently boost different satellites to different altitudes ranging from 400 to 800 km. A nominal scenario is that 3 pairs will be raised to 500 km, 600 km, and 700 km altitudes, respectively, while one is boosted up to 800 km and one remains at 400 km. The corresponding different nodal precession rate at different altitudes will then gradually drift the orbit planes apart. After 2-3 months, the paired satellites will in turn be separated by raising one of the two to about half way to the next higher orbit. The differential precession continues until a more-or-less even distribution of the eight orbit planes around the globe is achieved. This whole process will take about a year, during which time the satellites will already start collecting atmospheric data. At the end of this orbit-adjusting phase all eight satellites will be raised to the final altitude of 800 km, at which point the normal operation phase commences.

As far as obtaining gravity information is concerned, it is the first 2-3 months, when some of the satellites orbit in tandem, that are most valuable for the following reason. COSMIC satellites do not carry drag compensating mechanisms and accelerometers will not be carried either. For COSMIC, we use the GPS double-difference scheme to take advantage of the "common-mode cancellation" as a third way of dealing with residual non-gravitational forces.

1 Erricos C. Pavlis: Joint Center for Earth Systemtechnology, JCET-UMBC and NASA Goddard SFC, Space geodesy Branch, Code 926, Greenbelt, MD, USA 20771-0001 Tel.: +301-286-4880, Fax: +301-286-1760, e-mail:epavlis@Lighthouse.JCET.umbc.edu

2 B. F. Chao: NASA Goddard SFC, Greenbelt, MD, USA

3 C. W. Hwang: National Chiao Tung University, Taiwan, R.O.C.

4 C. C. Liu: Academia Sinica, Taiwan, R.O.C.

5 C. K. Shum: Ohio State University, Columbus, Ohio, USA

6 C. L. Tseng, M. Yang: National Cheng Kung University, Taiwan, R.O.C.

The double-difference scheme, depicted in Figure 1, is a standard technique used in reducing GPS tracking data [PAVLIS and OLSON, 1995]. The original purpose was to eliminate any difference, and hence resultant errors, of the receiver clocks relative to the GPS satellite clocks and minimize the effect of any force model errors in the GPS satellites' orbits. We, however, here take the additional advantage of the fact that when flying in close tandem a satellite pair is subject to similar non-gravitational forces, which largely cancel as common mode errors in a GPS double-difference scheme. Note further that during this period the relatively low altitude orbits (compared with the final orbit) are advantageous in obtaining higher spatial resolution.

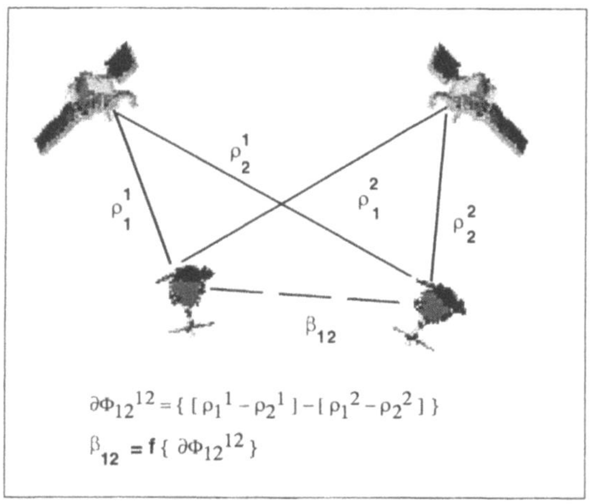

Fig. 1: The double-difference concept for GPS data.

Obviously a longer lifetime, say up to 5-10 years, would be highly desirable for any temporal variation experiment. Data collected during this phase will not require the common-mode cancellation as much due to the higher orbit altitude. At the 800-km altitude atmospheric drag, which at low altitude is the largest non-gravitational error source, is much weaker so that its empirical modeling alone can yield satisfactory results.

3. Simulation studies

The simulation results that we describe herein fall in two categories: results from simplified *analytical* evaluations of various configurations; and more elaborate, *numerical* solutions using the same software and models that are used for real data analysis. The first, analytical approach, being fast but still accurate enough, is used for the quick, relative evaluation of mission designs with regard to the type and accuracy of the observed quantity, the sampling rate, the altitude of the orbit(s), and the duration of the mission [PAVLIS et al., 1996].

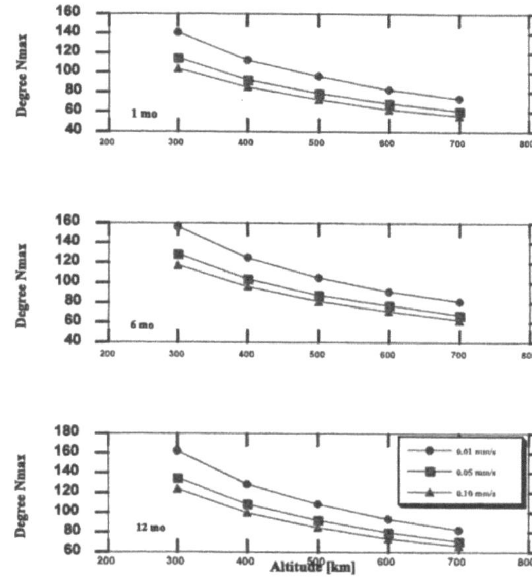

Fig. 2: Dependence of maximum resolvable degree on the mission duration and orbital altitude of a COSMIC pair.

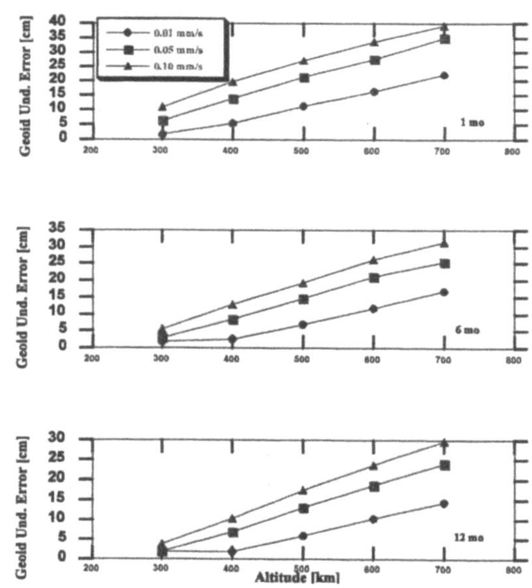

Fig. 3: The mean (2°x2°) geoid undulation error as a function of mission duration and orbital altitude of a COSMIC pair.

Analytical simulations

We used the analytical approach to examine the inter-relationship between measurement precision, mission altitude and duration, highest resolution reached and the corresponding cumulative commission error mapped in geoid space. For the measurement precision we have used three values, 0.10, 0.05 and 0.01 mm/s, that we expect to be the most conservative estimates (i.e. worst case scenario) for SST range-rate measurements with the GPS technology likely to be used on a 2001-launch mission. The range of altitudes we consider is from the lowest feasible at 300 km up to 700 km. In terms of mission duration we chose three values, one month, six months and one year, to be representative of various stages in the obit adjusting phase described above.

The results are summarized in Figures 2 and 3. Note that these results are for a single pair of COSMIC spacecraft. Three independent pairs should improve the situation by a factor of 3. A separate graph shows the results for the different mission duration. Within each graph, three different curves correspond to the three measurement noise levels. The graphs in Fig. 2 show the maximum resolution as a function of the mission altitude. Those in Fig. 3 differentiate the various scenarios on the basis of the cumulative geoid commission error; they are mapped into more easily visualized geoid space. The utility of these graphs is to aid the mission design team in evaluating scenarios and trade-off alternatives.

Numerical simulations

For the numerical simulation, we studied a case based on an early description of the constellation. This simulation was done at NASA's Goddard Space Flight Center (GSFC) using the GEODYN software [Pavlis et al., 1996]. We assumed a near circular, near polar orbit at an altitude of 420 km, with 8 COSMIC spacecraft in tandem at spacing of 400 km. In addition to white noise we included some atmospheric error in the form of atmospheric density model differences: we simulated the data using the French DTM87 model, while for the recovery we used GSFC's MSIS86 model.

The simulated data were doubly differenced ranges for reasons we explained above. We assumed that they were collected with GPS receivers similar to the GPS/MET-class design but with the capability to track up to 12 GPS spacecraft simultaneously on both frequencies. A minimum of 8 GPS spacecraft were in sight at every event and we used a 1 cm noise (to include some systematic errors, which otherwise would be too difficult to do) per observation collected at the rate of 1 Hz. The actual data were grouped into 5-min normal points, with the appropriate accuracy. Data were generated between 7 independent spacecraft pairs using a 0°-elevation cut-off angle at the spacecraft horizon. In this simulation there are no data that involve ground tracking. In the reduction of the data we assumed that the GPS spacecraft orbits will be provided by the IGS service and we adopted a very conservative 50 cm error in each component. Calibration of these products indicate that their accuracy is closer 5 cm, [PAVLIS, 1995].

Four, 8-day arcs were simulated, spanning 32 consecutive days. They resulted in some 300,000 doubly differenced ranges. During the reduction process, in addition to the usual spacecraft state-vector and ambiguity parameters, we adjusted the static gravity harmonic coefficients up to (20,20) and time-varying terms at seasonal periods up to (4,4).

The primary result of this simulation is shown in Figures 4 and 5 in the form of an error spectrum for the recovered coefficients and the associated geoid error. In these figures we included spectra of several geophysical signals that are useful in the evaluation of the results. In Figure 4 we show the signal spectra of the oceanic and atmospheric circulation, as well as the post-glacial rebound. In Figure 5 the associated geoid error spectra are shown in terms of geoid height. Here we have included the signal spectra of the dynamic ocean topography and that of the expected effect of vertical datum errors on the low degree and order harmonics. Both quantities are prime examples of areas where geodesy, oceanography and other scientific disciplines would like to see significant improvement in the very near future.

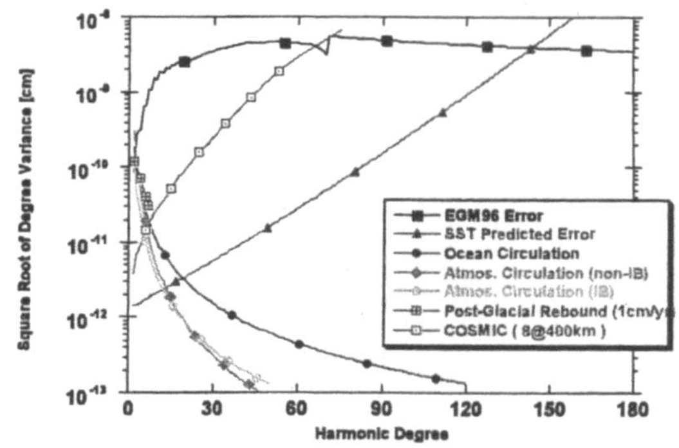

Fig. 4: Harmonic coefficient error spectra for EGM96, COSMIC and a GRACE-class mission, along with some geophysical signal spectra.

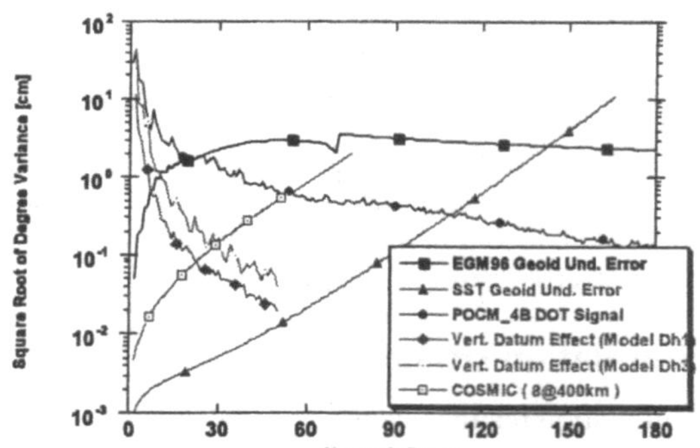

Fig. 5: Error spectra for EGM96, COSMIC and a GRACE-class mission, along with some geophysical signal spectra.

The error spectrum of the latest, state-of-the-art gravity model, EGM96 [LEMOINE et al., 1998], and the expected error spectrum of a model that could be derived from data of the future gravity mapping mission GRACE [TAPLEY, 1997] are also shown. From these results we can summarize that the static part of the model will be improved significantly (by about an order of magnitude) for the low degree and order terms up to degree 40 or so, with lesser but still significant improvement extending as high as out to degree 60.

Ideally, using a single month's data set we can deduce a (low-degree) gravity solution every month. Based on the results for the precision of such a monthly solution we can forecast the expected decrease in the standard deviation of these terms as a function of the mission duration. Table 1 shows that in the case of signals with seasonal frequency a mission that produced monthly solutions for five consecutive years would give us estimates with a standard deviation as small as 4% of that of the monthly estimate, with less improvement for lower frequency terms. It seems then possible that signals as low as ~$5x10^{-13}$ for the secular rates could be reliably resolved within the first five years.

Obtaining information about the (small) temporal variation of the gravity field will address very important meteorological and geophysical problems, such as those associated with climate change and the derivation of constraints on the climate-driven mass redistribution on Earth at intra-seasonal to secular scales. In particular, our results show that after a 5-year mission, a majority of large-scale, temporal geophysical signals [CHAO, 1994; NAS, 1997] could be encompassed in the signals observed by COSMIC.

Tab. 1: The gain in the precision of the seasonal variation terms as a function of the duration of the COSMIC mission.

Component	% Gain in precision*				
Lifetime (y)	1	2	3	4	5
Secular	89	33	18	12	9
Annual	63	23	13	8	6
Semi-annual	51	19	10	7	5
Seasonal	40	15	8	5	4

* With respect to the monthly error estimate

4. Summary

The geodetic utility of COSMIC is to provide new sets of SST data that will be used in the improved modeling of the static terrestrial gravity field and some of its temporal variations. The present scenario takes advantage of the very precise double differenced phase observations between a pair of COSMIC s/c and a pair of GPS satellites to mitigate the adverse effects of non-conservative force mis-modeling.

Acknowledgments.

The COSMIC Geodesy team wishes to thank Drs. Lou Lee, Bill Kuo, and Chris Röcken for encouragement and discussions. This work is supported by NASA Solid Earth and Natural Hazards Program on the U.S.A. side and by the National Science Council on the Taiwan, R.O.C., side.

References

CHAO B. F. (1994). *The Geoid and Earth Rotation*, in Geophysical Interpretations of Geoid, ed. P. VANICEK and N. CHRISTOU, CRC Press, Boca Raton.

LEMOINE F. G., PAVLIS N. K., KENYON S. C., RAPP R. H., PAVLIS E. C., CHAO B. F.(1998). *New high-resolution model developed for Earth's gravitational field*, EOS, Trans. Amer. Geophys. Union 79, 113-118.

NAS (National Science Council) (1997). *Satellite Gravity and the Geosphere: Contributions to the study of the solid Earth and its fluid envelope*, J. O. DICKEY (ed.), Washington, D.C.

PAVLIS D. E., et al. (1996). *GEODYN Operations Manual, 5 Volumes*, Raytheon ITSS Corp., Greenbelt, Maryland.

PAVLIS E. C. (1995). *Comparison of GPS s/c Orbits Determined from GPS and SLR Tracking Data*, Adv. Space Res., 16, (12)55-(12)58.

PAVLIS E. C., OLSON R. T. (1995). *Geopotential Improvement From Explorer Platform Single-Frequency GPS Tracking*, Eos Trans. AGU, 76, Fall Meeting Suppl., 147.

PAVLIS E. C., CLARK T. A., BILLS B.(1996). *Gravity modeling with the Hummingbird Constellation*, Eos Trans. AGU, 77, Spring Meeting Suppl., S40.

TAPLEY B. (1997). *The Gravity Recovery and Climate Experiment (GRACE)*, Eos Trans. AGU, 78, Fall Meeting Suppl., 163.

Estimation of Ocean Mass Redistribution by Means of Altimetry and Circulation Models and its Impact on the Gravity Field

TH. GRUBER, CH. REIGBER, J. WÜNSCH[1]

Abstract

With the upcoming new satellite missions CHAMP and GRACE, it will be possible for the first time to observe the time variable gravity field. Besides other sources, ocean mass redistribution is one of the major factors contributing to the overall time varying gravity signal. For estimating this time variable gravity signal, simulation studies with altimeter, sea surface temperature and ocean circulation data have been performed.

Monthly mean sea surface height models were computed over a period of three years combining altimetry and the corresponding mean sea surface temperature fields, which were used to remove the thermal water expansion. Based on these models monthly mass redistributions were computed in terms of water heights. The attraction of these water masses was computed and transformed into gravity field coefficients by spherical harmonic analysis with a resolution of up to degree 100 (corresponding to 400 km wavelength). By analyzing the complete monthly time series, amplitudes and phase lags for each of the spherical harmonic coefficients can be detected. Another method is the analysis of spherical harmonic coefficients derived from bottom pressure fields from the Parallel Ocean Circulation Model (POCM) up to degree and order 6 (provided by T. JOHNSON, CSR) for an 8 years period. Annual, semiannual and third annual amplitudes of variation of the Stokes coefficients are determined by Fourier analysis. The paper shows results of both simulation methods, their intercomparison, as well as comparisons with the expected error budget for both new gravity field missions.

1. Introduction

Apart from the Earth's magnetic field and climate research, a main objective of the upcoming small-satellite missions CHAMP (Challenging Mini-Satellite Payload for Geophysical Research and Application) (REIGBER et al., 1997) and GRACE (Gravity Recovery and Climate Experiment) (TAPLEY, 1997) is the high precision global gravity field determination. In both missions the satellites shall be used in low orbits as a proof mass with inter-satellite measurements: in the case of CHAMP to the GPS satellites and in the case of GRACE between two LEO-satellites. With the long mission duration of 5 years each and due to the continuous tracking information together with the accelerometers on-board of all satellites to measure the non-gravitational forces, a homogeneous sequence of gravity field models will be available with accuracy improvements of one order of magnitude and more with respect to current models. By analyzing the gravity field sequence (e.g. with monthly solutions), an overall signal of mass variations from a variety of sources can be determined. The separation of this signal into its individual source components will be one of the challenges of future gravity field research. This investigation quantifies gravity variations induced by oceanic mass redistributions by calculating the gravity variation signal using altimeter data and ocean bottom pressure data from an ocean circulation model. Comparing the signal with the expected performances of CHAMP and GRACE its sensitivity to these satellites can be determined.

2. Altimetry Approach

From reprocessed ERS-1 altimeter data monthly sea surface height models for the period of April 1992 to March 1995 were computed (ANZENHOFER et al., 1998). Residual sea surface heights with respect to the three years mean, together with residual sea surface temperatures, which were calculated from the monthly US. National Meteorological Center (NMC) temperature fields, were used to estimate the thermal expansion component in the residual sea heights. For computing the thermal expansion, monthly correlations between both data sets were performed after an additional data editing and smoothing. The mean of these monthly gradient fields was in the end multiplied with the residual temperature fields to remove the thermal expansion component from each monthly residual sea surface height field. Assuming that the remaining signal is exclusively from oceanic water mass redistributions, the residual attraction in terms of gravity is calculated from the water volumes. For this a standard ocean water density of 1028 kg/m^3 was used. From each monthly residual gravity signal a spherical harmonic series of up to degree and order 100 was computed by a block-diagonal least squares approach COLOMBO, 1981).

Figure 1 summarizes the complete processing chain from the altimeter data to the residual gravity field solutions. Each resulting spherical harmonic series represents a residual gravity field with respect to the three years mean. By analyzing the sequence of these residuals gravity fields in terms of monthly coefficient differences, the impact of oceanic mass redistribution from month to month on the gravity coefficients can be determined. Finally, by analyzing the signal degree variances of the monthly coefficient differences together with the error spectra of the new gravity missions and by performing spectral analysis of individual coefficients time series, the sensitivity of the new missions and the major frequencies of the signal can be determined.

[1] Th. Gruber, Ch. Reigber, J. Wünsch: GeoForschungsZentrum Potsdam (GFZ), Division 1, Telegrafenberg A6, 14473 Potsdam, Germany; Contact: Reigber@gfz-potsdam.de, Tel. +49 (0) 331 2881100, Fax +49 (0) 331 2881111

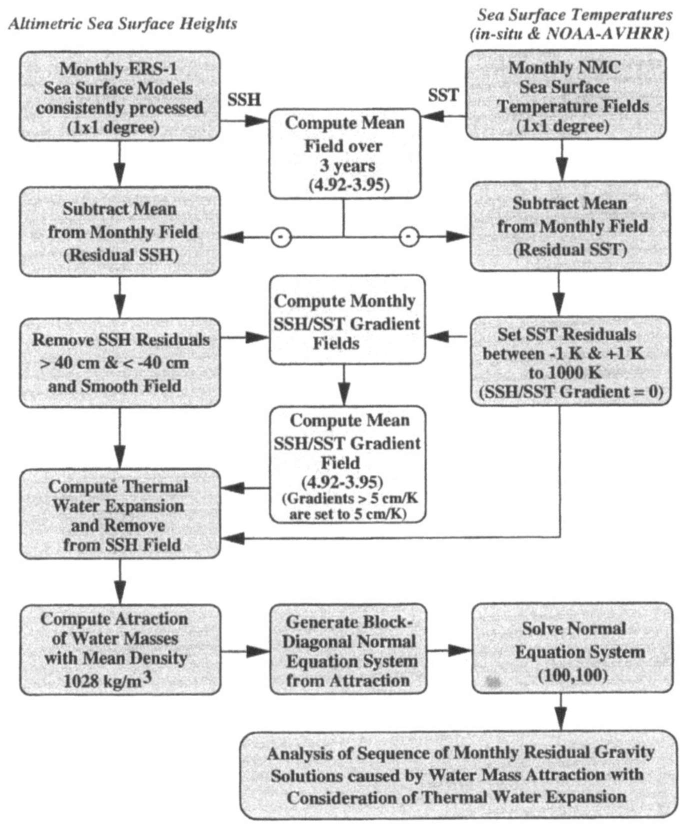

Fig. 1: Schematic Overview of Residual Gravity Field Computation from Altimeter Data

Figure 2 shows the square root of monthly coefficient differences degree variances of the oceanic signal and the error spectra for CHAMP and GRACE in terms of geoid heights. Apparently we can expect a signal of up to degree 12 from oceanic water mass redistributions in the CHAMP gravity field solutions and of up to degree 50 to 60 (depending on the satellite altitude) in the GRACE gravity field models.

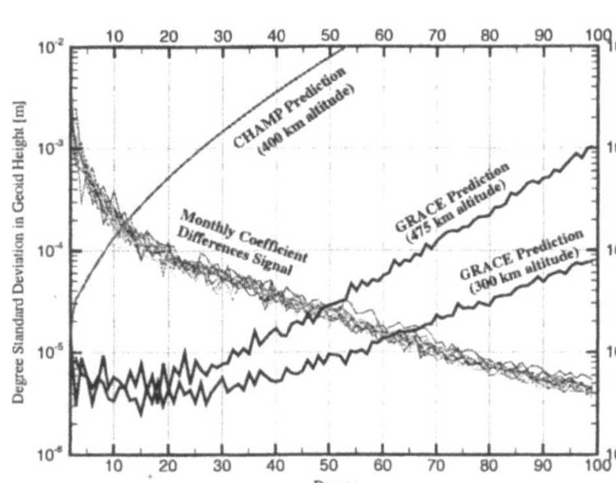

Fig. 2: Monthly Altimetric Ocean Signal versus Mission Sensitivity

The gravity variation signal in terms of residual geoid heights is computed by spherical harmonic synthesis of the coefficient differences. Monthly large-scale geoid variations of up to 1 cm caused by ocean water mass redistributions can be observed with the altimetry approach. Performing spectral analysis of the coefficients time series, the amplitude spectrum for each coefficient was computed. Figure 3 shows as a sample the amplitude spectra for all C-coefficients of order 2. The figure illustrates the rapid decrease of amplitudes with increasing degree. Beyond degree 40 no signal in the amplitude spectra can be seen anymore for the given z-axis resolution of 10^{-12}.

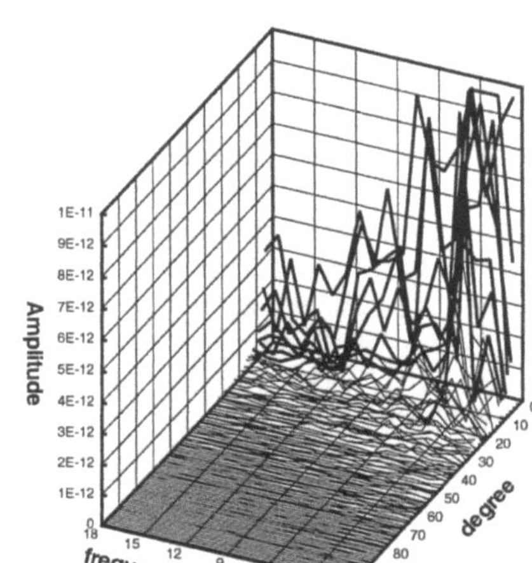

Fig. 3: Amplitude Spectra for Order 2 C-Coefficients

For a more detailed analysis some amplitude spectra for specific C-coefficients of order 2 are plotted separately. Figure 4 shows the amplitude spectra for all coefficients from degree 3 to degree 10 of order 2.rrr

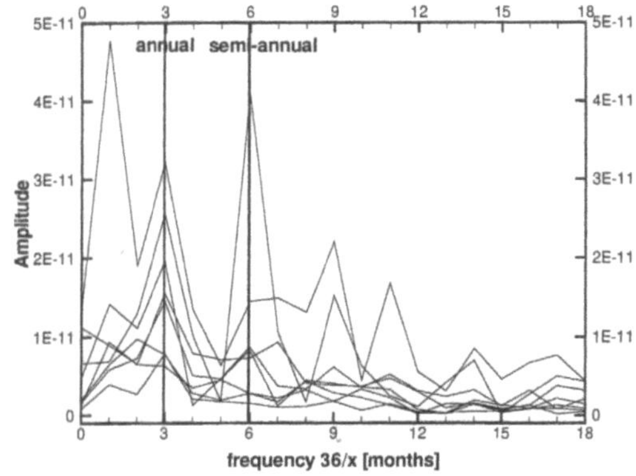

Fig. 4: Amplitude Spectra for C-Coefficients of Order 2 and Degree 3-10

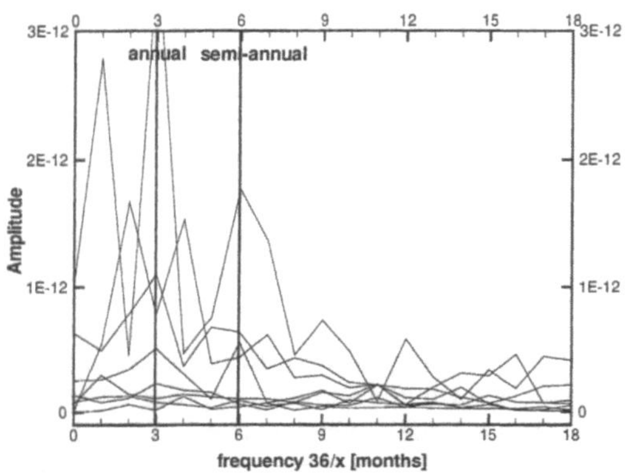

Fig. 5: Amplitude Spectra for C-Coefficients of Order 2 and Degrees 15,20,25,30,35,40,45 and 50

In addition, Figure 5 shows some selected amplitude spectra for C-coefficients of order 2 and degrees 15, 20, 25, 30, 35, 40, 45 and 50.

Figures 4 and 5 show clear peaks for annual and semi-annual frequencies with respect to the noise level for each coefficient. For some lower-degree terms a small peak for the 4 months frequency (x-axis=9) can be identified, which is probably caused by some frequency folding during the FFT analysis. No physical reason can be found for this wavelength. From Figure 5 it can be seen as well, that for higher- degree coefficients (degrees 40, 45 and 50), which are represented by the bottom lines, no annual or semi-annual signal is detected anymore.

3. Ocean Model Approach

As a second source for estimating the gravity variation signal from ocean mass redistributions, ocean bottom pressure fields calculated from the Parallel Ocean and Climate Model (POCM) (SEMTNER et al., 1992) were used. The 1x1 degree gridded bottom pressure fields were derived from all 20 levels of ocean temperature and salinity for an 8 years period (1988-1995) with 6 days resolution. Gravity coefficients from these fields up to degree and order 6 were inverted by Tom Johnson from the Center for Space Research at the University of Texas in Austin and further analyzed in this paper. Figure 6 schematically shows the processing chain.

As for the altimetry-derived coefficients the 8 years time series of gravity coefficients was also analyzed in the spectral domain. Figure 7 shows as an example the amplitude spectra for all coefficients of order 2 up to degree 6.

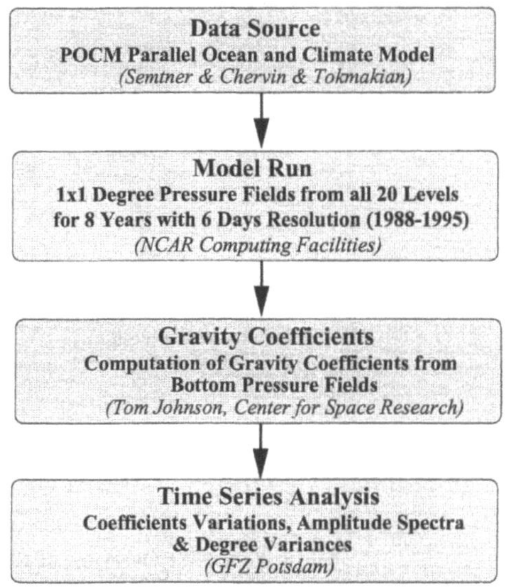

Fig. 6: Schematic Overview of Ocean Model Approach

As for the altimetry approach, the annual and semi-annual peaks can be clearly identified. In addition slightly increased amplitudes with respect to noise level can also be seen for some higher frequencies, that cannot be addressed to specific seasonal phenomena. Comparing this result with Figure 4, where the amplitude spectra for some identical coefficients from the altimetry approach are plotted, much smaller values are determined from the ocean model approach.

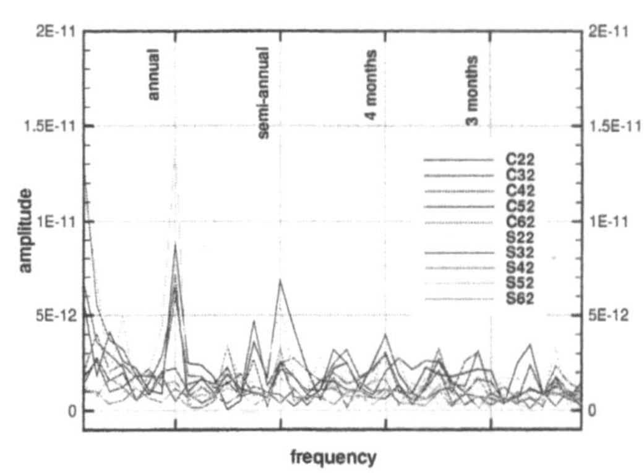

Fig. 7: Amplitude Spectra for C- and S-Coefficients of Order 2 estimated from POCM Model

4. Comparison of Results

Spectral analysis results in Chapters 3 and 4 show, that the annual and semi-annual frequencies are detected in the gravity signal in both approaches. But as already mentioned, the amplitudes for these frequencies differ significantly between both data sets. For further analysis of this difference, the square root of degree variances for two sample gravity coefficients data sets is compared in Figure 8.

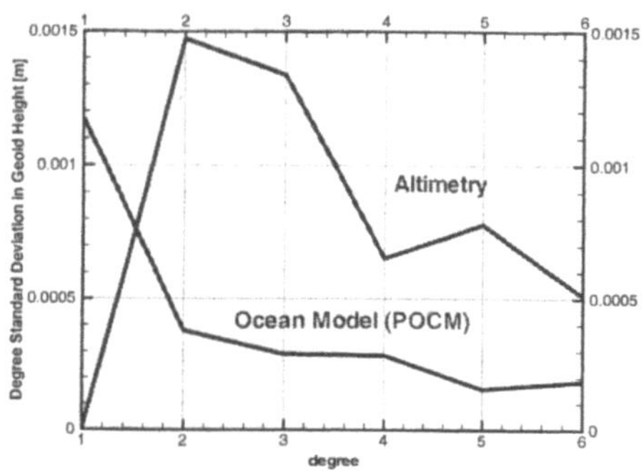

Fig. 8: Square Root of Degree Variances in Geoid Heights from both Approaches

This comparison also illustrates, that for the low- degree harmonics the ocean model approach provides a smaller signal than the altimetry-derived coefficients. The reason for this difference probably has to be attributed to the particularities of both approaches. On the one hand there are indications that the POCM is underestimating the amplitudes of ocean variability by a factor of 2 to 4 (STAMMER et al., 1996) and on the other hand the block-diagonal approach for gravity field estimation could cause an overestimation especially in the low-degree harmonics due to the land-ocean distribution. In addition, for the altimetry approach it is not absolutely sure how reliable the thermal expansion component is estimated and whether other signals which cannot be addressed to ocean mass redistributions are still in the residual sea surface fields. To answer these questions further analysis with different altimeter and ocean temperature data sets and other ocean circulation models have to be performed.

5. Conclusions

Concluding this investigation the following statements can be made:

- From altimeter data analysis together with sea surface temperatures for removal of the thermal expansion component, residual gravity coefficients representing ocean mass redistributions can be estimated up to high degrees by using the block-diagonal technique.
- From ocean bottom pressure fields derived from an ocean circulation model low-degree gravity coefficients for ocean mass redistributions can be calculated.
- Both approaches enable the identification of the major amplitudes for annual and semi-annual signals.
- The amplitudes and the degree variances differ significantly in their size. The altimetry approach provides by a factor of 3 larger signals than the POCM model approach.
- The reason for these differences probably has to be attributed to the approaches themselves. There are indications that the POCM model is underestimating the ocean variability, while the altimetry approach possibly overestimates the coefficients due to the block-diagonal technique and/or an incomplete thermal expansion reduction.
- From the altimetry approach it is supposed that the CHAMP mission will be sensitive up to degree 12 and the GRACE mission up to degree 50-60 to ocean mass redistributions. Even if an overestimation of the signal by a factor of two is assumed, with the CHAMP mission such features will be detectable up to degree 8 and with the GRACE mission up to degree 40-50.

6. References

ANZENHOFER M., GRUBER TH.: *Fully reprocessed ERS-1 altimeter data from 1992 to 1995: Feasibility of the detection of long term sea level change;* Journal of Geophysical Research, Vol. 103, p. 8089-8112, 1998

COLOMBO O.: *Numerical methods for harmonic analysis on the sphere*; The Ohio State University, Department of Geodetic Science, Report No. 310, Columbus/Ohio, 1981

REIGBER CH., LÜHR H., KANG Z., SCHWINTZER P.: *The CHAMP mission and its role in observing temporal variations of the geopotential fields;* Supplement to EOS Transactions of the American Geophysical Union, 78 (46), F163, 1997

SEMTNER A.J., CHERVIN R.M.: *Ocean circulation from a global eddy-resolving model;* Journal of Geophysical Research, Vol. 97, No. C4, p. 5493-5550, 1992

STAMMER D., TOKMAKIAN R., SEMTNER A., WUNSCH C.: *How well does a 1/4 degree global circulation model simulate large-scale oceanic observations?;* Journal of Geophysical Research, Vol. 101, No. C10, 25779 -25811, 1996

TAPLEY B.D.: *The gravity recovery and climate experiment (GRACE)*; Supplement to EOS Transactions of the American Geophysical Union, 78 (46), F163, 1997

GFZ and DLR Contribution to a GPS Ground Network to Support the CHAMP Mission

CH. REIGBER[1], R. GALAS[1], R. KÖNIG[1], N. JAKOWSKI[2],
J. WICKERT[2], A. WEHRENPFENNIG[2]

Abstract

The paper discusses various scientific/technical issues as network distribution, requirements for autonomous and permanently operating GPS stations, near real time data transfer and operational requirements and data flow for a GPS ground station network being currently installed in preparation of the CHAMP mission.

1. Introduction

CHAMP is a geoscientific mission for the determination of gravity field, magnetic field and for GPS-based atmospheric sounding (REIGBER et al., 1996). The satellite is scheduled for launch for end of 1999.

Both the gravity field experiment as well as the GPS based radio occultation measurements require high precision orbit estimation. The highest precision of orbit parameters can be achieved by dynamic orbit determination procedures supported by a fiducial network of GPS ground stations. Additionally, the so obtained GPS observations shall enable double differencing to compensate satellite clock errors in radio occultation data analysis.

GPS ground data for the GPS occultation experiment of the CHAMP mission will be delivered by a dedicated global network.

The network stations will be established by NASA/JPL and GFZ Potsdam. GFZ jointly with its partner DLR/DFD Neustrelitz suggests a sub-network, composed of 8 stations. This ground tracking network shall support also other forthcoming low Earth orbiting missions, e.g. GRACE (TAPLEY and REIGBER, 1998).

2. GPS Ground Network Requirements

Besides the precise orbit determination the ground station network serves primarily to compensate clock errors of the TRSR receiver onboard CHAMP for a precise analysis of radio occultation measurements, the network has to be high operational and shall ensure global coverage of radio occultation measurements. To achieve an even data coverage of radio occultation measurements, some geometrical relationships between CHAMP satellite, ground stations, occulting and reference GPS satellites have to be taken into account. Furthermore, the infrastructure of selected sites has to fulfil some criteria such as secure housing, continuous power supply, stable communication links via internet. So one has to find a compromise to ensure a satisfying data coverage on one hand, and to guarantee on the other hand a reliable operation of the GPS receivers and a fast data transfer via internet to the GFZ Potsdam. Taking into account these various aspects, GFZ Potsdam plans to realize a network topology illustrated in Fig. 1. To meet operational requirements with a data latency of less than 3 hours, the ground station data should be available at the processing data centers in GFZ and DLR within 1 hour after data taking. The only way to meet this requirement is the construction of an autonomous tracking station described in the following section. Because the radio occultation measurements need a rather high sample rate of 50 Hz, the sample rate of GPS phase measurements at ground stations is required to be at least ^{3} 1 Hz, i.e. the ground network has to be equipped with an appropriate GPS receiver. Further requirements for a ground tracking station are: quick set-up, self control, high reliability, easy operating and maintenance. Moreover a remote control by an operator located in a network operation center must be possible.

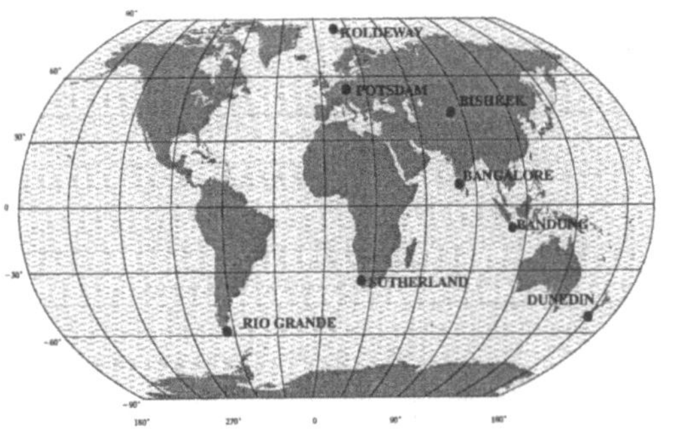

Fig. 1: Planned GFZ High Rate GPS Tracking Network. Station distribution according to demand for global coverage of the occultation events

3. Ground Tracking Station

To fulfil the above requirements a CHAMP ground tracking station (GALAS and BURGHARDT, 1999) is composed of:

- GPS Bench Mark receiver from AOA, collecting data with 1 second sampling rate.
- meteorological station of type TM200,

[1] Christoph Reigber, Roman Galas, Rolf König: GeoForschungsZentrum Potsdam, Kinematik u. Dynamik der Erde, Telegraphenberg A 17, 14473 Potsdam, Germany; Tel.:+49 - 331-288 -1100,-1101,/-1146, Fax: +49 - 331-288-1111,/-1169; e-mail: reigber@gfz-potsdam.de, wuen@gfz-potsdam.de

[2] N. Jakowski, J. Wickert, A. Wehrenpfennig: Deutsches Zentrum für Luft- und Raumfahrt e.V. (DLR), Deutsches Fernerkundungsdatenzentrum (DFD), Kalkhorstweg 53, 17235 Neustrelitz; Contact: Norbert.Jakowski@dlr.de, Tel. +49 (0) 3981 480151, Fax +49 (0) 3981 480123

- industry standard personal computer running under Linux operating system,
- power controller with a relays and watch dog,
- a battery with re-charger.

The monitoring computer has a true Internet address and can be accessed any time using a dedicated line.

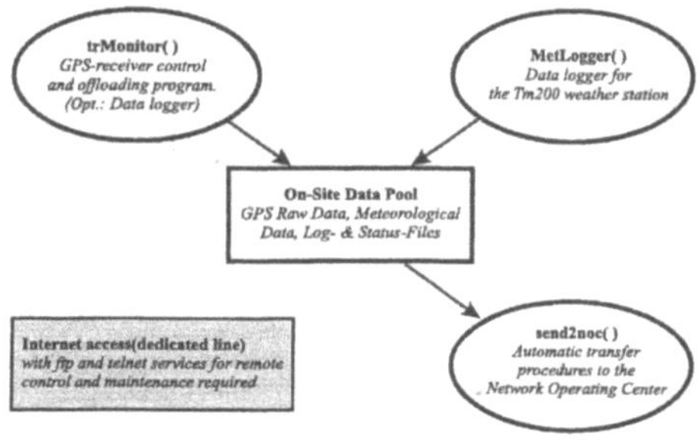

Fig. 2: Data Flow at an Autonomous Tracking Station

The flow diagram in Fig. 2 shows the relationships between the software modules running on a field computer. Each of the software modules is implemented as a separate program and works pretty much independently of the other ones. The main module tasks are:

- Monitoring the GPS receiver, trouble shooting, GPS data downloading
- Data logging from a meteorological station
- GPS data compression
- Automatic data delivery to a network operating center (including all log-files)

This software architecture allows development, tests, upgrade or adding of new modules in a very comfortable way and guarantees high system reliability.

The GPS data are downloaded onto PC hard disk every 15 minutes, then compressed by a very efficient procedure (KÖHLER, 1999) and sent via Internet to GFZ.

4. Network Data Flow and Processing

A number of programs, taking care for the data handling, will continuously run at the CHAMP Network Operating Center. The data handling software (see Fig.3) is now implemented and under test. All GPS- and log-files are handled. The GPS files are converted to a receiver independent binary format and merged into hourly data files.

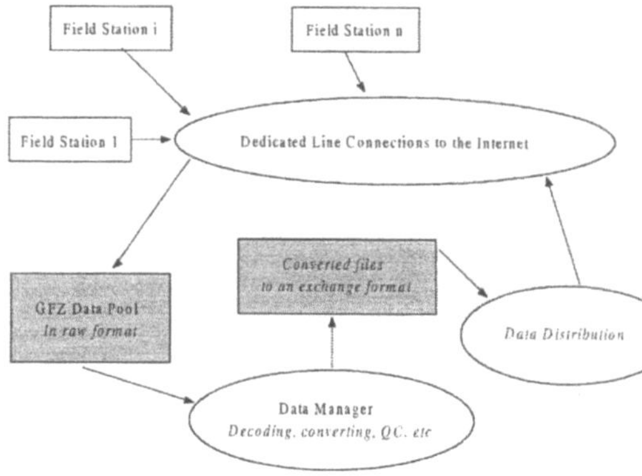

Fig. 3: Data Flow in the GFZ Tracking Network

Next the pre-processing and quality control procedures are applied. As a result for each data file a log-file, containing information about cycle slips, bad data etc., is created. Additionally, the hourly GPS files are sampled at 30 second rate and merged into daily files. All received log-files from the tracking stations are automatically analyzed and some network status reports are created.

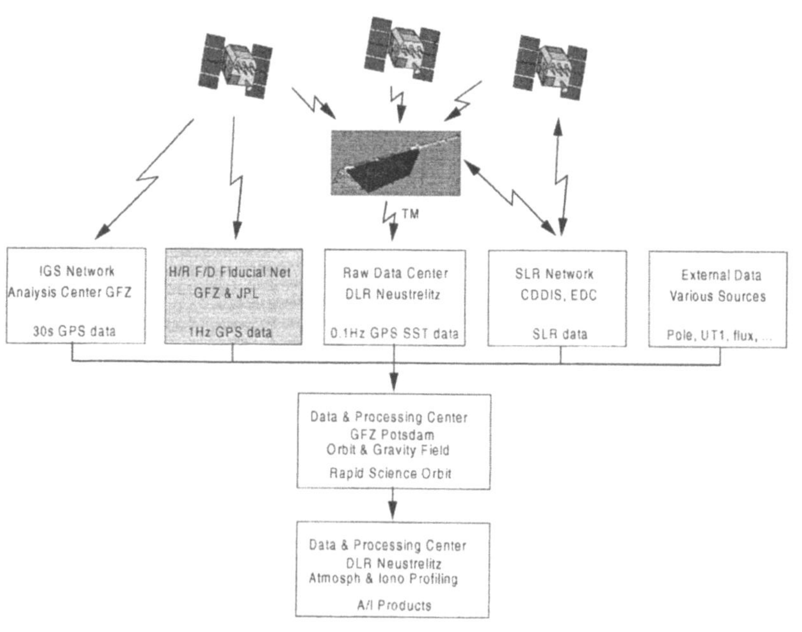

Fig. 4: Orbit Processing Scheme

Sampled high-rate fiducial network data provide the fast and reliably available data base for precise orbit determination, complemented by the IGS 30 s GPS data, on-board SST POD data, and precision SLR range data. The orbit processing chain (see Fig. 4) comprises acquisition and preprocessing of all these observational and auxiliary data, determination of precise orbit products for the occultation processing chain, and generation of special orbit predictions for the SLR network and for mission planning. Before the launch of CHAMP, a validated and ready to operate S/W system will be available to generate a dynamic orbit solution. The dynamic solution evaluates the satellites equation of motion by accounting for gravitational and non-gravitational forces. The GFZ approach realized in the EPOS-OC S/W package is characterized by using undifferenced pseudo-range and carrier phase measurements from GPS ground station and on-board receivers for the estimation of GPS orbits and clock parameters and the LEO ephemeries. GFZ experience in the SST technique is a result of gravity field modeling activities from TOPEX/POSEIDON and GPS-MET SST data (e.g. KANG et al., 1997). The benefits from synergies of GPS and SLR data have been established by ZHU et al., 1997.

The dynamic solution allows to estimate physical parameters (e.g. gravity field parameters as one of the main science objectives of the CHAMP mission) in addition to the parameters primarily looked at in occultation processing, by adjusting the trajectory solution from physical and empirical force modeling to the tracking observations in the least squares sense. The dynamic solution also provides an easy means to predict the orbit by forward integration based on the parameters derived before in the adjustment step.

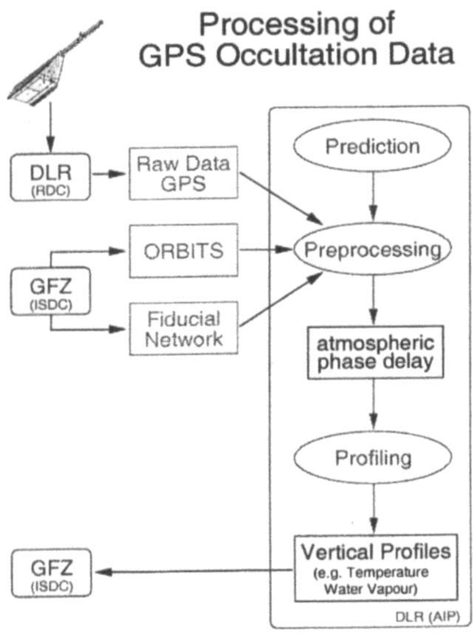

Fig. 5: Processing of GPS Occultation Data

Atmospheric and ionospheric data products will be generated by an automatically working processor system, which is under development and will be installed at GFZ Potsdam and DLR/DFD Neustrelitz. For retrieving accurate data products high precision orbits are needed.

5. Conclusions

The coverage of occultation events was simulated by applying the Radio Occultation Simulation Tool (ROST) developed at DLR/DFD Neustrelitz (JAKOWSKI et al., 1998).

Using the station distribution as displayed in Fig.1, ROST computations provide globally distributed radio occultation events satisfying the mission requirements to a high degree (see Fig. 6). Further improvement of global coverage will be reached by combining the GFZ station configuration with the JPL network. As it has been shown, with a data latency less than 1 hour, the above described network fulfils all operational requirements. So it can be also used for future operational multi-satellite radio occultation missions without any significant modifications.

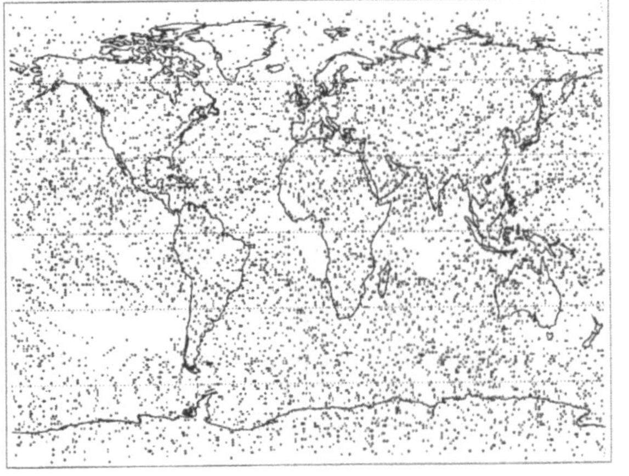

Fig. 6: Global distribution of occultation events (30 days)

References

GALAS R, BURGHARDT W., (1999). *Autonomous CHAMP Ground Tracking Station: System Documentation and Description*, GFZ Scientific Technical Report (in preparation), GFZ Potsdam

JAKOWSKI N., WICKERT J., HOCKE K., REIGBER CH., FÖRSTE CH., KÖNIG R. (1998). *Atmosphere/Ionosphere Sounding Onboard CHAMP*, 1998 Western Pacific Geophysics Meeting, EOS Suppl., Transactions, AGU, 79, No. 24, W20

KANG Z., SCHWINTZER P., REIGBER R., ZHU S.Y., (EOS). *Precise Orbit Determination of Low-Earth Satellites Using SST Data*, Adv. Space Res., 19, 1667-1670, 1997

KÖHLER W., (1999). *Compression algorithms for GPS binary format*, GFZ Scientific Technical Report, (in preparation)

REIGBER CH, BOCK R., FÖRSTE CH., GRUNWALDT L., JAKOWSKI N., LÜHR H., SCHWINTZER P., TILGNER C. (1996). *CHAMP Phase B*, Executive summary, GFZ Scientific Technical Report STR96/13, GFZ Potsdam

TAPLEY B. D., REIGBER CH. (1998). GRACE: *Satellite-to-satellite tracking geopotential mapping mission*, 1998 AGU Fall Meeting, Suppl. to EOS, Transactions, AGU Vol. 79, No. 45, 1998.

Zhao C. (1998). *The Sensitivity of Atmospheric Temperature Retrieved from Radio Occultation Technique to Orbit Errors of GPS and LEO Satellites*, GFZ Scientific Technical Report STR98/23, GFZ Potsdam.

ZHU S.Y., REIGBER CH., KANG Z. (1997). *Apropos Laser Tracking to GPS Satellites*, J. Geodesy, 71, 423-431, 1997.

The ESA Gravity Field and Steady-State Ocean Circulation Explorer Mission: Impact on Solid Earth Physics

H. REBHAN, J. JOHANNESSEN, M. AGUIRRE, A. TOBIAS[1]

Introduction

The Gravity Field and Steady-State Ocean Circulation Explorer (GOCE) is a candidate Earth Explorer Mission which ESA is considering for the post 2000 Era. The Mission aims to provide a precise and high resolution measurement of the gravity field. In so doing it will open a completely new range of spatial scales of the Earth's gravitational field spectrum for research in solid Earth physics, geodesy and oceanography.

We summarise the essential scientific objectives and mission definition and compare the GOCE mission concept and simulated performance with other planned gravity missions. Further on the benefits of an improved gravity field for advancing the quantitative estimates of geodynamic and tectonic processes are discussed. Finally we give a brief indication of the current status and future developments.

Scientific objectives

The scientific objectives as defined for the GOCE mission are:

- to advance the understanding of the physics of the Earth's interior and develop better global and regional models of the Earth's gravity field.
- to implement a better global height reference system for datum connection.
- to provide, for the first time, the precise reference equipotential surface (geoid) at high spatial resolution needed for the quantitative determination of absolute ocean currents and their transport of heat and other properties from satellite altimetry.

Performance requirements

The mission performance requirement ranges for the different scientific aims and it is expressed in terms of geoid heights and gravity anomalies versus spatial resolution expressed in half wavelength (Table 1). The values shall be understood as 'most likely' values i.e. as equal to the standard deviation assuming a normal distribution (after GOCE, 1996).

Tab. 1: Performance requirements as defined for the objectives of GOCE.

	Required Accuracy		Spatial Resolution (half wavelength in km)
	Geoid [cm]	Gravity [mgal]	
Geodynamics			
- continental lithosphere (thermal structure, post-glacial rebound)		1-2	50-400
- Mantle composition, rheology		1-2	100-5000
- Ocean lithosphere and interaction with asthenosphere		5-10	100-200
Geodesy			
- Ice and land vertical movements	2		100-200
- Rock basement under polar ice sheets		1-5	50-100
- World-wide height system	< 5		50-100
Ocean Circulation			
- Small scale	2		100
	0,2		200
- Basin scale	< 0,1		1000

Comparison of GOCE and other Gravity Field Missions

The time may now have come for a series of complementary space missions devoted to map the gravity field of the Earth. Beside the planned GOCE two other gravity mission are approved:

NASA announced the selection of a Gravity Recovery and Climate Experiment (GRACE) as one of the first of their Earth System Science Pathfinders. The mission scenario is planned as a lo-lo satellite-to-satellite tracking (SST) concept combined with a hi-lo SST by utilization of GPS satellites. GRACE aims at monitoring monthly to annual or even inter-annual time variations of the long wavelength part of the

[1] Helge Rebhan, J. Johannessen, Miguel Aguirre, Alberto Tobias: ESTEC Earth Sciences Division, Postbus 299; NL-2200AG Noordwijk, The Netherlands; Tel: +31-71565-5846,/5846,/-3828,/-3992, Fax:+31-71565-5675, e-mail:hrebhan@jw.estec.esa.ni, jjohanne@jw.estec.esa.nl, maguire@estec.esa.nl, atobias@estec.esa.nl

gravity field. A launch in 2001 is envisaged with a mission duration of five years.

The German Challenging Minisatellite Payload for Geophysical Research and Application (CHAMP) is intended to improve our knowledge of the global gravity field for the long- to medium wavelength spectra. The CHAMP concept is based on hi-lo SST. The planned launch is in 1999.

Figure 2 (next page) summarises schematically the impact of all three missions on the study of features with high geophysical interest in terms of gravity signals and simulated model errors at different wavelengths.

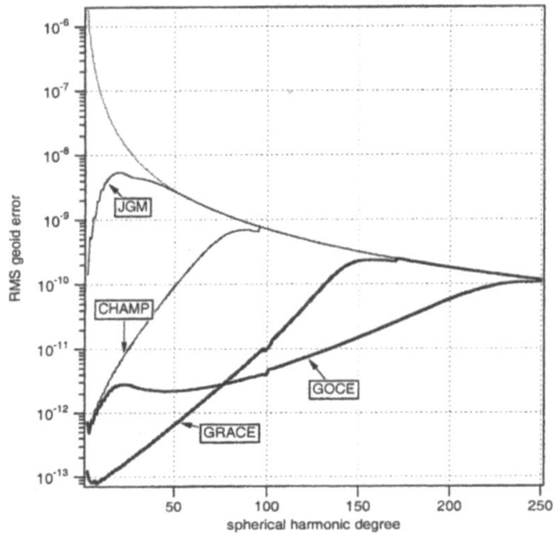

Fig. 1: Performance comparisons for the different mission concepts.

A recent comparative performance study (BALMINO et al., 1998) showed that GRACE is excellent on the mid resolution range, whereas GOCE is excellent in the high resolution range (see Figure 2).

The expected marine geoid error at different wavelengths between 1000 km and 160 km are listed in Table 2. From this comparison it is concluded that GOCE and GRACE are complementary with a cross-over point at about degree- and order 80, equivalent to ≈ 450 km wavelength.

Tab. 2: Performance comparison at specific wave-lengths (BALIMNO et al., 1998)

Wavelength [km]	Harmonic Expansion [deg]	GRACE cumulated geoid error	GOCE cumulated geoid error
1000	40	0,01 cm	0,07 cm
400	100	0,1 cm	0,1 cm
270	150	10 cm	0,2 - 0,6 cm
160	250	Not covered	3 -6 cm

Preliminary results from the Phase A instrument studies are showing very promising performance values which are clearly below the assumed 5 mEotvos for the GOCE versus GRACE simulation study (Table 2).

Tab. 3: Total performance errors in mEotvos for both instrument concepts.

	XX	YY	ZZ
Capacitive Instrument	3,378	1,668	3,04
Inductive Instrument	1,057	0,516	0,904

Solid Earth Physics

Precise determination of the Earth's gravity field is of fundamental importance for understanding the structure and dynamics of the solid Earth. Gravity data is used in combination with topography to resolve the compositional, thermal and mechanical structure of the deep crust and upper mantle and to evaluate fundamental aspects of the plate tectonic process. The current precision of the gravity models such as EGM96 is insufficient for many geophysical applications. A few examples shall illustrate the benefits of improved gravity field knowledge:

Understanding of *mantle processes*, in particular convection patterns, and of post-glacial mass readjustment, will greatly benefit from improved and more detailed knowledge of the Earth's gravity field. Improvements can also be expected for the knowledge of the rheology of the mantle that controls the dynamics of the interior and of the whole Earth. Mantle rheology impacts not only the long-time scale convective pattern of our planet but its comprehension also enables the interpretation of post-glacial rebound data and of present-day sea-level changes.

For the understanding of the dynamics of the *continental lithosphere* the accurate and detailed determination of the anomalous gravity field plays a key role. One important process that occurs on the continental lithosphere is rifting. Rifts are common tectonic features because the strength of the lithosphere is least under extension, and can be found in a variety of tectonic settings. The GOCE mission with an accuracy of 1 – 2 mgal at 50 – 400 km spatial resolution will enhance our knowledge on the major geodynamical causes of rift formation and evolution.

An important example for the dynamic of the *oceanic lithosphere* are subduction zones. They are characterized by two wavelengths, associated with the broad back-arc basin (several hundreds of kilometers) and the narrower uplifting arc and trench (about 50 – 100 km). Of special interest is the great variability within the subducting plate between the state of expansion that dominates at shallow depths and the state of compression at great depths, as inferred from deep seismicity. The GOCE measurements will provide new insight on the dynamic interaction between the subducting plate, the mantle and overriding plate and the gravity anomalies in the trench region with an accuracy of 5 – 10 mgals at a spatial resolution of 100 – 200 km.

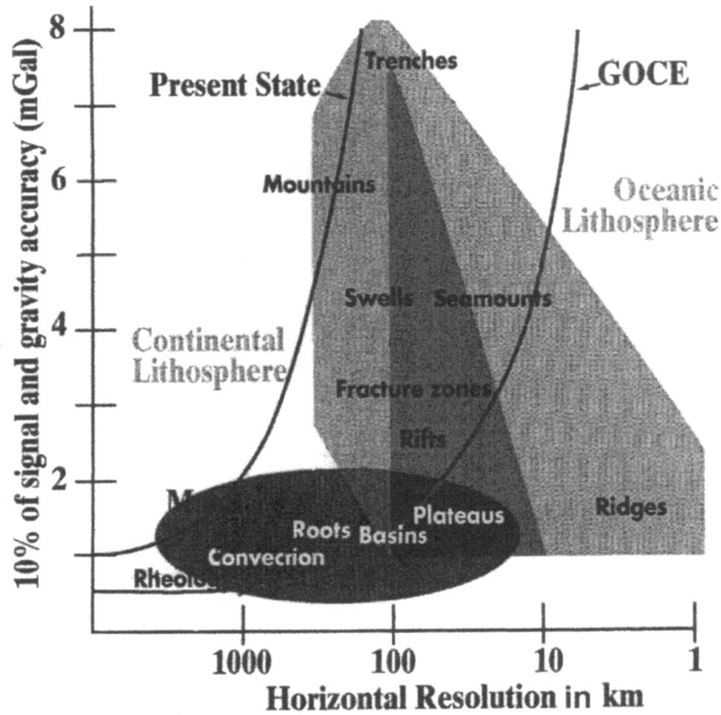

Fig. 2: Schematic description of the required accuracy (estimated at approximately 10% of the gravity signal) as a function of horizontal resolution necessary to resolve the quoted geodynamical and tectonic features. The curves represent the current and potential accuracy obtained from planned gravity missions.

Geodesy

An improved geoid is significant for two important scientific goals in geodesy: The unification of regional height systems and the transformation between GPS derived ellipsoidal heights and orthometric heights.

The *unification of height systems* is an essential premise for the assessment of a global sea-level rise and for uniform hydrographic mapping and tidal modelling. Geodetic height systems have currently their zero-values fixed through a mean sea level as calculated at or near tide-gauges. Due to changes in the sea surface topography this means that the height systems at present differ with values of the order of decimeters between two islands and between an island and a continent having distances up to a few hundred km. Between continents the differences may be up to a meter. These height systems (national datum) should be unified by determining the differences relative to an accurate geoid based on data from a gravity field mission enhanced with local data.

Modern space borne methods for positioning and mapping like GPS or SAR provide geometric heights above the rotation ellipsoid. For most purposes (i.e. link to historical measurements) these heights need to be *converted to orthometric heights* for which a precise knowledge of the geoid is necessary. The practical determination of the geoid is nowadays conducted using the long wavelength information combined with local gravity and topographic information. This determination is currently disturbed by errors in the longer wavelengths, often of the order of 0.5 - 2.0 m. The goal is to enable the computation of geoid height differences to better than about 0.02 m for distances of 100 to 200 km. This can be achieved if local gravity data is combined with an improved spherical harmonic expansion.

Such an improved geoid surface would in turn serve as a stationary reference for the study of all topographic processes, including dynamic ocean topography (and therefore ocean circulation), the evolution of the ice sheets and land surface topography.

Status and future Developments

Based on the outcome of the User Consultation Meeting in Granada in May 1996 the Earth Science Advisory Committee (ESAC) endorsed in the GOCE mission to proceed to Phase A. During the initial period of the Phase A study which started in July 98 a selection of the gradiometer instrument design will be taken. As can be seen in Table 3 the inductive (cryogenic) and the capacitive (ambient temperature) instrument design are both below 5 mEotvos currently used as upper performance limit. A schematic illustration of the satellite concept with both instrument solutions is shown in Figure 4.

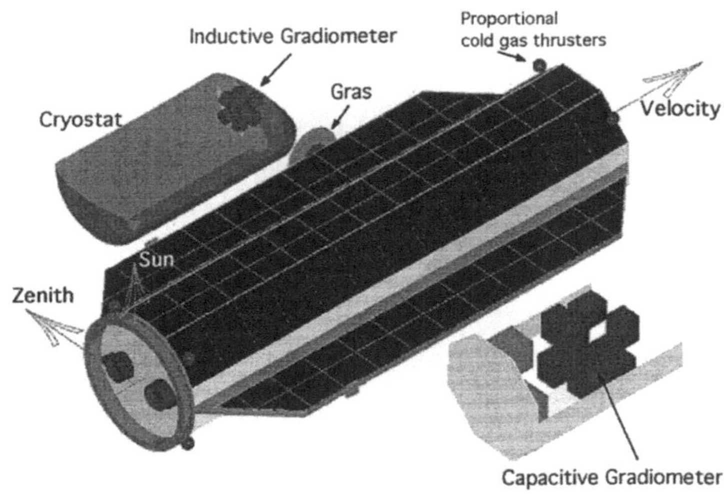

Fig. 3: Schematic illustration of the GOCE satellite concept

In parallel with the Phase A several scientific studies will be carried out to obtain clearer quantitative insights into the implications of the GOCE mission for the different scientific applications. Beside studies related to oceanography a refinement study of the observation requirements for geodynamics and geodesy will also be carried out. In particular this includes:

- How improved gravity field models might provide new and better estimates of dynamic processes of the Earth interior such as the thermal structure, isostasy, mass balance and the post glacial rebound.
- How precise gravity models might lead to advanced applications for world-wide height systems, levelling by GPS, inertial navigation, separation of steric from non steric fluctuations of sea level and ice-mass balance.

During the phase A study it will also be urgent to establish the boundary condition between the GOCE mission and other gravity and magnetometry missions, i.e. CHAMP and GRACE, planned to be flown prior to GOCE. In view of a "minimum competition and maximum complementary" scenario this should also consider and identify the optimum operation-time of GOCE.

References

BALMINO G., PEROSANZ F., RUMMEL R., SNEEUW N., SÜNKEL H., WOODWORTH P. (1998). *European views on dedicated gravity field missions: GRACE and GOCE*, ESA Earth Sciences Division publication, ESA-ESTEC.

GOCE (1996). *Gravity Field and Steady-State Ocean Circulation Mission*, Reports for Assessment, The nine Candidate Earth Explorer Missions, ESA SP-1196/1.

Posters – Session D

Implication for Earth Sciences

The Regular Character of the Plate Motion: Implication for Earth Sciences

YU. V. BARKIN[1]

Abstract

In the present geological epoch, the plate motion has a regular and ordered character. In particular, the elegant order in positions and motions of the plates is clearly observed in the inclined lithosphere reference system (LRS) [BARKIN, 1994]. The main plane of the LRS (the lithosphere equator plane) is inclined at an angle of 26°.7 to the Earth's equator plane and has a longitude of the ascending node of 3°.4W. The geometrical, kinematical, and dynamic regularities of the plate motion were found and described. As a basis, we used a plate model with averaged powers of the oceanic and continental areas of the plates [BARKIN, 1995] and current geological and satellite theories of their motion HS2-NUVEL1 [GRIPP and GORDON, 1990], GPS-PM [LARSON et al., 1977] and others. The geodynamic meaning and nature of the regularities discovered are very important for many Earth's studies. They reflect dynamic asymmetry, gravitational and other interactions of the Earth's envelopes, their inclined relative rotations, and displacements of their centers of mass. Similar regularities and peculiarities are observed in the structure and evolution of the Earth's physical fields and all the Earth's envelopes, including Vernadskij's biosphere and noosphere. For example, the centers of ancient civilizations are situated in the vicinity of the lithosphere equator.

Certainly, the regularities of the plate motion must be taken into account in the realization of new satellite geodesy projects, in particular for siting of GPS stations.

1 The Regular Character of the Plate Motion

A detailed investigation of geometrical and dynamical characteristics of the plates has shown that the plate arrangement and motion are quite ordered, not chaotic, as many researchers believed previously. This new phenomenon in the global tectonics is characterized by certain geometrical, kinematical, and dynamical regularities [BARKIN, 1995, 1996a, b]. In fact, these regularities describe some properties of symmetry in the distribution of the plates and their motion with respect to the inclined lithosphere reference system (LRS).

1.1 LRS

This reference system is introduced in different ways, for example, on the basis of locations of triple junctions of the plates, epicenters of the plates' centers of mass (for brevity, we will call them plate centers), other geological structures, including the Pacific islands and other formations. All the definitions are in a good agreement with each other and allow to quite reliably introduce inclined reference system $Ox_Ly_Lz_L$.

This reference system is geocentric and immovable with respect to the mantle. Intersections of coordinate planes Ox_Ly_L, Ox_Lz_L, and Oy_Lz_L with the spherical surface of the Earth are great circles: E_L is the lithosphere equator, O_L is the zero meridian of LRS, and M_L is meridian 90° of LRS. The plane of the lithosphere equator (circle E_L) is inclined to the Earth's equator plane by 26°.7, and the longitude of its ascending node is 3°.4W. Axis Ox_L lies in the Earth's equatorial plane and is directed along the line of intersection of the equator planes of the Earth and lithosphere toward the west coast of Africa. Great circle M_L is meridian 86°.6E in ERS and meridian 90°E in LRS. Polar axis Oz_L is directed toward Hudson Bay, and its pole has geographic coordinates 93°.4W, 63°.3N. The opposite end of axis Oz_L has coordinates 86°.6E, 63°.3S.

The main method of definition of LRS was based on an analysis of the positions of the plate centers: their coordinates were obtained from an approximate calculation of corresponding volume integrals for a certain plate model [BARKIN, 1996b, c]. The model proposed is rather simple, it takes into account only essential features and peculiarities of the dynamic structure of the plates. The plate boundaries are defined by a map of positions of seismic belts, rift zones, and transform trenches in the present geological epoch; this map was used in our interpretation of the data of GPS observations in GFZ-Potsdam [BARKIN, 1996b, c]. The model pays attention to different dynamical structure of the oceanic and continental parts of the plates [BARKIN, 1996a, b]; for their powers, we took averaged effective values. With this, the lithosphere mass and polar moment of inertia are in good agreement with their values for the well-known PREM model of the Earth [DZIEWONSKI and ANDERSON, 1987], constructed on the basis of multiple seismic observations.

1.2 Positions of the Plate Centers in the LRS

The centers of nearly all the major, medium, and some minor and micro- plates are located near three great circles on the Earth's spherical surface, E_L, O_L, and M_L. Only the center of the Australian plate occupies an isolated position. This may reflect the peculiar evolution and fast motion of the Australian plate during the last 43 millions of years. The location of lithosphere equator E_L is clearly delineated by the centers of the plates: African, Arabian, East-China, Philippines, Pacific, Nasca, south microplate of the Easter Island. A detailed description of geometric regularities, in

[1] Yuri V. Barkin; Sternberg Astronomical Institute, Moscow State University, 13 Universitetskij prospekt, Moscow, 1119899 Russia; Email: barkin@sai.msu.su

the location of the plate boundaries and centers together with positions of various geological structures is given in BARKIN, 1998.

1.3 Ordering in the Locations of Triple Junctions of the Plates

The points of triple junctions of the plates lie on selected parallels and meridians of LRS, forming a peculiar grid structure on the Earth's surface. For instance, all of the 40 junctions considered are located near seven LRS meridians. The origin of the structure discovered has a dynamic nature and is due primarily to the global inclined rotation of the lithosphere, with interaction and relative motion of the plates.

1.4 Order in the Plate Motion

Certainly, the above-mentioned geometrical peculiarities of the plates could have persisted at geological time intervals only under the condition that the motion of the plates themselves is also ordered. This phenomenon was discovered as a result of an analysis of the well-known geological kinematic theories of the plate motion NNR-NUVEL1 [ARGUS and GORDON, 1991], HS2-NUVEL1 [GRIPP and GORDON, 1990] and modern satellite measurements of the GPS stations' motion [LARSON et al., 1997]. In particular, it was shown that the centers of the lithospheric-belt plates (enumerated in section 2) have a prevailing trend in their motions: along the lithosphere equator E_L.

1.5 Analogs of the Kepler and Titius-Bode Laws in the Plate Tectonics

Interesting regularities were established as a result of an analysis of the kinetic momentum vectors of the plates and the plate momentums in the hotspot reference system. In particular, the ratio of kinetic momentums of an arbitrary pair of plates is equal to the ratio of the modules of their momentum vectors (an analog of the 2nd Kepler law in celestial mechanics). The plates' kinetic momentums (as well as vectors of their momentums) are rationally commensurable. The formula describing this empirical relationship reminds by its form the well-known Titius-Bode law for planetary distances in the Solar System.

1.6 Asymmetry of the Lithosphere

In the first turn, it is due to an asymmetric distribution of oceanic and continental parts of the plates. The center of mass of the lithosphere is shifted by 772.2 km toward the geographical point 41°N, 36°E (in the Black Sea region). The poles of the minor, medium, and major main moments of inertia of the lithosphere are: 49°.1N, 85°.2E, 26°.8, 139°.6 and 28°.4N, 213°.8E). Thus, the axis corresponding to the greater of the main moments of inertia forms an angle of 61°.6 with the Earth's rotation axis.

1.7 Inclined Rotation of the Lithosphere

It was interpreted as rotation of the Tisserand's system of reference, in which the overall kinetic moment of all the particles of the lithosphere plates is zero. The global inclined rotation of the lithosphere is a spherical motion with an angular velocity of 0.268° Myr^{-1} around a geocentric axis, directed toward the geographic point 46°.3S, 58°8E.

1.8 The Structure of the Earth's Gravity Field

The structure (e.g., given by the GRIM4 model) has a property of symmetry in the distribution of main anomalies and isolines with respect to the equator plane E_L; naturally, this expresses the global asymmetry of this field in the geographic Earth's system of reference.

1.9 The Structure and Variations of the Magnetic Field

They are tightly connected with LRS. For example, the plane of the lithosphere equator, established from geometrical and kinematical characteristics of the plates, coincides with the plane of orbit of an eccentric dipole modeling the main component of the Earth's magnetic field and its slow westward drift.

2 Implications for the Earth Science

2.1 Geodynamics

Above we have mentioned only the main phenomena and regularities of those found. They clearly and unambiguously indicate an asymmetric structure of the lithosphere and other Earth's covers, an important role of their mutual gravitational interactions and, as a consequence, their slow relative inclined motions. It is known that these motions manifest themselves in the plate motion, global rotation of the lithosphere, rotation of the fluid core and solid core with respect to the mantle. The phenomenon discovered is apparently caused by interaction of individual plates and the lithosphere as a whole (as dynamically-nonspherical celestial bodies) with each other, with other covers of the Earth as well as with external celestial bodies. Dynamical studies of this phenomenon may make a base of a new geodynamic concept, in which gravitational interactions of the Earth's covers occupy the central place.

2.2 Satellite Geodesy

Certainly, the peculiarities of plate location and motion (their regular character) must be taken into account in interpretation and reduction of the data of modern VLBI, SLR, and GPS observations as well as in development of new space projects. Apparently, the GPS station siting must be oriented to a uniform distribution with respect to LRS. Of special interest are the stations located in the vicinity of the lithosphere equator. Already now a number of the operating GPS stations are located virtually on the lithosphere equator. If the GPS stations are appropriately placed along the great circles and meridians of LRS, the problem of secular and periodic variations of the Earth's mean radius may be solved.

2.3 Geophysics

All physical fields possess the property of asymmetry in the geographic system of coordinates. Perhaps, this is determined by asymmetry of the dynamic structure of the Earth's covers, inclined character of their slow relative motions, displacements of their centers of mass, and the general trend of the global tectonic process. On the contrary, in LRS itself many global structures of the Earth's covers (in particular, those of the plates and lithosphere) and physical fields possess symmetry, for example, often with respect to the lithosphere equator plane. In addition to the gravitational and magnetic field, these properties of symmetry-asymmetry have been established for the field of heat flux from the Earth's interiors, for the field of amplitudes of the lunar-solar deformations of the mantle, for the distribution of elastic parameters over the Earth's surface, for the field of the boundary-surface motions of the fluid core, for the velocity field of the ocean level variations, etc.

2.4 Biosphere and Noosphere Sciences: On the Planetary Distribution of the Centers of Ancient Civilizations

Apparently, the above-mentioned peculiarities and regularities of the structure and evolution of the Earth's covers and physical fields essentially manifested themselves in formation and development of the Earth's `biological cover', in formation and development of its civilizations.

A simple analysis of geographical locations of ancient mythologies and cultures quite clearly indicates their association, in the first turn, to the geodynamic equator and possibly to certain meridian circles of LRS.

It is sufficient to indicate the well-known regions of ancient civilizations and cultures, lying along the geodynamic equator: ancient Upper Egypt (the cities of Thebes, Luxor, Karnak, Abu-Simbel), the south-west of the Rub-el-Hali desert (the country of the Adita people and the mythological land of Wabar), Mesopotamia (the Kingdom of Shumer and Akkad, Babylonia), the oldest Indian civilization in the Indus river basin (the cities of Mohenjo-Daro, Harappa), Central and Northern Arabia (the city of Mecca), Polynesia, Micronesia, the Easter Island, the Nasca Desert region, southern and central regions of Benin (the Dahomey culture), etc.

Thus, an investigation of the regularities of the structure and evolution of the biosphere in its unity and interaction with other envelopes and physical fields of the Earth is a task of primary importance.

Acknowledgments

This work was supported by the Russian Foundation for Basic Research (project code 96-05-65015) and the Russian Universities Program (topic 9-5240).

References

ARGUS D.F., GORDON R.G. (1991). *No-net-rotation Model of Current Plate Velocities Incorporating Plate Motion Model NUVEL-1.* Geophysical Research Letters, Vol. 18, No. 11, pp. 2039-2042.

BARKIN YU.V. (1995). *Particularities in the Plates' Motion and Global Structure of the Earth's Physical Fields.* International EUG8 Conference. European Union of Geosciences (April 9-13, 1995, Strasbourg, France). Abstracts.

BARKIN YU.V. (1996a). *Global Properties of the Structure, Evolution, and Interconnection of the Lithosphere and other Earth's covers.* The 1st Russian Conference on Interaction in the Lithosphere-Hydrosphere-Atmosphere System (Moscow, November 28-29, 1996), Abstracts of Papers, Physics Department, Moscow State University, p. 22.

BARKIN YU.V. (1996b). *Regularities of the Plate Motion and the Global Asymmetry of the Earth's Covers.* XXI General Assembly of EGS (The Hague, May 6-10, 1996). Annales Geophysicae, Supplement I to Volume 14, Part I, p. C36.

BARKIN YU.V. (1996c). *Phenomenon of Ordering in the Plate Positions and Motions.* Russian Conference on Modern Problems and Methods in Astrometry and Geodynamics (October 23-27, 1996, St. Petersburg). Abstracts of Papers. St. Petersburg, Institute of Applied Astronomy, Russian Academy of Sciences, pp. 130-131.

BARKIN YU.V. (1998). *New Regularities in the Lithosphere Structure.* Geotectonics (in press).

DZIEWONSKI A.M., ANDERSON D.L. (1981). *Preliminary Reference Earth Model.* Physics of the Earth and Planetary Interiors, Vol. 25, pp. 297-356.

GRIPP A.E., GORDON R.G. (1990). *Current Plate Velocities Relative to the Hotspots Incorporating the NUVEL-1 Global Plate Motion Model.* Geophysical Research Letters, Vol. 17, pp. 1109-1112.

LARSON K.M., FREYMUELLER J.T., PHILIPSEN S. (1997). *Global Plate Velocities from the Global Positioning System.* Journal of Geophysical Research, Vol. 102, No. B5, pp. 9961-9981.

About Global Rotation of the Lithosphere

YU. V. BARKIN[1]

Abstract

Global rotation of the lithosphere (GRL) is interpreted as rotation of the Tisserand's reference system, connected with lithosphere in such a way, that w.r.t. this system (TRS) the full angular moment of the plate motion is equal to zero. The GRL parameters w.r.t. the reference system of the hot spots, given by kinematical theory HS2-NUVEL1 (GREEP, GORDON, 1990), have been obtained. This rotation takes place with an angular velocity of 0.268° 1/m.a. about a geocentric axis with the pole 46.3° S, 58.8° E. Also, the parameters of the lithosphere rotation w.r.t. the reference system given by the NNR-NUVEL1 kinematical theory (ARGUS, GORDON, 1991) were obtained: 0.067° 1/m.a., 55.2° N, 84.0° W. The GRL parameters in the Earth's reference system (ERS 94), in which the modern satellite theory of the plate motion on the basis of GPS data was constructed (LARSON, FREYMUELLER, PHILIPSEN, 1997), are determined: 0.033° 1/m.a., 47.6° N., 120.1° E. With respect to the hot spots reference system the rotation of the ERS 94 is given by parameters: 0.301° 1/m.a., 46.5° S, 58.9° E. In the calculations, the values of the components of the tensor of inertia of the plates in their model with average powers of the oceanic and continental areas are used (BARKIN, 1996).

1 Introduction: Tisserand's reference system

Lithosphere is an ensemble of the 10-12 major and medium plates and of a set of small and micro plates. To a good approximation the plates can be considered as thin rigid envelopes with different power, executing slow relative spherical motions. Rotation of the plate w.r.t. to a definite Earth's reference system is given by the angular velocity and pole coordinates of the corresponding axis of rotation. Naturally, that these parameters depend on the reference system.

The purpose of the paper is to define the global component in the spherical motion of all the plates, characterizing the global rotation of the lithosphere as a united body w.r.t. a definite reference system (for example, in the hot spots reference system). Here the classical presentation about the character of rotation of the deformable body will be used. The rotation of the Tisserand's geocentric reference system

$Ox_T\, y_T\, z_T$ is compared to the GLR (TISSERAND, 1990; LIOUVILLE, 1858). In this reference system, the angular moment of the relative motion of the plates is equal to zero.

For determination of the GLR I use the values of the inertia tensor components of the plates, defined for their simplest models in the form of the thin rigid spherical envelopes with average powers of their oceanic and continental areas. The values of the axial and centrifugal moments of inertia of the plates in the Earth's reference system (Greenwich reference system, ERS) were defined as a result of the approximate calculation of the corresponding volume integrals by the trapezium method (BARKIN, 1996). These values are presented in the Tab. 1.

The plate motion is given by the current geological theories HS2-NUVEL1 (GREEP and GORDON, 1990) and NNR-NUVEL1 (ARGUS, GORDON, 1991). For a comparison and for a study of the relative rotations of the main reference systems, I also will consider the plate motion according to the current satellite GPS-PM theory (LARSON et al., 1997).

The plate motion on the HS2-NUVEL1 theory is given in the reference system connected with the average positions of the hot spots , that is relative of the astenosphere layers of the mantle (w.r.t. mantle reference system , MRS). NNR-NUVEL1 theory defines some mean lithosphere reference system (LRS). It is defined from a condition that the total moment of the forces of the viscous friction, acting on the all plates from the lithosphere, is equal to zero (CHASE, 1979; HARPER, 1986):

$$\oiint \bar{v} \times \bar{r} ds = 0 \qquad (1)$$

In (1) the integration is spread over all the spherical surface of the Earth, $\bar{r}$ and $\bar{v}$ are radius-vector and velocity of some elementary area of the foot of the plates ds. Condition (1), of course, is approximate. It does not take into account, for example, variability of the plate thickness and lateral variations of the astenosphere viscosity.

In the GPS-PM theory of the plate motion, the international Earth's reference system 1994 is used as basic (ERS 94). This reference system is defined by the most exact positions of the 240 geodesical stations. The coordinates of these stations were found by methods of the VLBI, SLR and GPS data.

NUVEL1 theory of the plate motion (DE METS et al., 1990) lying in the basis of the HS2 and NNR-NVEL1 theories is a geological theory and was constructed on the basis of data about spreading velocities in the middle-oceanic ridges, on the azimuths of the transform breakings and on the slipping vectors on the plate boundaries at earthquakes. In this, the average values of the velocities and directions in a «short» geological period of time 3.16 m.a. are used.

Satellite GPS-PM theory (LARSON et al., 1997) is based on the GPS observations in 1991 - 1996 years and is in good agreement with above mentioned geological theories.

[1] Yuri V. Barkin; Sternberg Astronomical Institute, Moscow State University, 13 Universitetskij prospekt, Moscow, 1119899 Russia; Email: barkin@sai.msu.su

Tab. 1: Axial and centrifugal moments of inertia of the plates (1 unit $=10^{-2}$ C, C is polar moment of inertia of the Earth).

N	Plate	A_σ	B_σ	C_σ	F_σ	E_σ	D_σ
1.	EA	1.072	0.805	0.683	-0.071	0.095	0.348
2.	AF	0.304	0.869	0.902	0.159	0.020	-0.036
3.	PA	0.402	0.678	0.704	0.134	-0.024	0.019
4.	NoA	0.653	0.483	0.299	-0.004	-0.043	-0.220
5.	SoA	0.461	0.318	0.534	-0.214	-0.093	0.102
6.	AN	0.387	0.341	0.456	-0.150	0.123	-0.153
7.	AU	0.583	0.535	0.128	0.018	-0.015	-0.041
8.	IN	0.163	0.025	0.162	0.034	0.008	0.035
9.	NA	0.133	0.024	0.117	0.005	0.001	0.039
10.	AR	0.065	0.058	0.087	0.042	0.026	0.027
11.	PH	0.027	0.026	0.042	-0.021	-0.010	0.010
12.	CO	0.028	0.004	0.029	-0.007	0.002	-0.007
13.	CA	0.025	0.001	0.025	0.002	-0.001	-0.004
14.	JF	0.002	0.002	0.001	0.001	-0.001	-0.001

With the lithosphere, we also connect the other reference system – Tisserand's reference system (TRS). The method of introduction of this system is analogous to the method of the NNR RS introduction (1) and is defined by the following condition:

$$\overline{G}_L^{(r)} = \sum_{\sigma=1}^{N} \overline{G}_\sigma^{(r)} = \sum_{\sigma=1}^{N} \iiint_{\tau_\sigma} \rho\, \overline{v} \times \overline{r}\, d\tau = 0 \qquad (2)$$

where $\overline{G}_\sigma^{(r)}$ is the angular moment of all plates in the TRS, $\overline{G}_\sigma^{(r)}$ is the angular moment of the separate plate, N is the number of the plates, $\overline{r}$ and $\overline{v}$ are radius-vector and velocity of the elementary volume of the plate $d\tau$ in the TRS, ρ is density in this volume. Integration in (2) is spread over all the volume of the plate τ_σ.

Integrals in (2) can be calculated if some model of the density distribution, of the boundaries and contours of the plates are given. In the paper, the model of the plates with average values of the powers of the continental and oceanic areas are used. For conditional average density 3.3g/cm^3 corresponding thicknesses of these areas are 117 km and 47 km respectively (BARKIN, 1996).

Contours of the plates are defined by positions of the seismic belts, rifting zones, transform zones and both. in the current geological epoch. The model is characterized by definite values of the masses, of the coordinates of the mass centers, of the axial and centrifugal moments of inertia, of the coefficients of different harmonics of their gravitational potentials, both for separate plates and for the lithosphere in full. In this paper, we will use the values of the axial $A_\sigma, B_\sigma, C_\sigma$ and of the centrifugal $F_\sigma, E_\sigma, D_\sigma$ moments of inertia of the plates, which were obtained for the above mentioned model (BARKIN, 1996). Their values are presented in Tab. 1 for the main and medium plates, considered in the paper.

In Tab. 1 abbreviations of the full titles of the main and middle plates are given: Eurasian (EA), African (AF), Pacific (PA), North-American (NoA), South-American (SoA), Antarctic (AN), Australian (AU), Indian (IN), Nasca (NA), Arabian (AR), Philippines (PH) , Cocos (CO), Carribean (CA), Juan de FCA (JF)

Let us point out, that conditions (1) and (2) are equivalent only in the particular case where all plates are homogeneous and have identical power for all the areas of the lithosphere.

2 Rotation of the lithosphere in the hot spots reference system

Let the rotation of the every plate in the mantle reference system (MRS), connected with the system of the hot spots, be given by the projections of the angular velocity $\overline{\Omega}_\sigma$ on the Greenwich's axes O_x, O_y and O_z:

$$\Omega_{\sigma x}, \Omega_{\sigma y}, \Omega_{\sigma z} \qquad (3)$$

(their numerical values in accordance with the HS2-NUVEL1 theory are presented in Tab. 2).

We denote the projections of the angular velocity $\overline{\Omega}_L$ of the GLR as

$$\Omega_{Lx}, \Omega_{Ly}, \Omega_{lz} \qquad (4)$$

As stated earlier, $\overline{\Omega}_L$ is angular velocity of the TRS w.r.t. the main reference system $Oxyz$. Since linear velocity of some particle of the plate σ relative the TRS is

$\overline{v} = (\overline{\Omega}_\sigma - \overline{\Omega}_L) \times \overline{r}$.

Tab. 2 Components of the angular velocities of the spherical motion of the plates in accordance to the HS2-NUVEL1 theory (1 unit= 1^0/m.a.)

N	Plate	$\Omega_{\sigma x}$	$\Omega_{\sigma y}$	$\Omega_{\sigma z}$
1.	EA	0.034	0.055	-0.064
2.	AF	0.144	0.013	-0.017
3.	PA	0.001	0.487	-0.846
4.	NoA	0.107	-0.023	-0.259
5.	SoA	0.030	0.107	-0.301
6.	AN	0.042	0.094	-0.026
7.	AU	0.564	0.506	0.129
8.	IN	0.491	0.199	0.158
9.	NA	-0.001	-0.320	0.329
10.	AR	0.492	0.165	0.155
11.	PH	0.683	-0.245	-0.847
12.	CO	-0.534	-1.098	0.406
13.	CA	0.081	-0.003	-0.156
14.	JF	0.392	0.679	-0.544

Then, after some algebra, vectorial equation (2) can be presented in the form of the following algebraic equations:

$$G_x = \sum_{\sigma=1}^{N} \{A_\sigma \Omega_{\sigma x} - F_\sigma \Omega_{\sigma y} - E_\sigma \Omega_{\sigma z}\} = A_L \Omega_{Lx} - F_L \Omega_{Ly} - E_L \Omega_{Lz}$$
$$G_y = \sum_{\sigma=1}^{N} \{-F_\sigma \Omega_{\sigma x} + B_\sigma \Omega_{\sigma y} - D_\sigma \Omega_{\sigma z}\} = -F_L \Omega_{Lx} + B_L \Omega_{Ly} - D_L \Omega_{Lz} \quad (5)$$
$$G_z = \sum_{\sigma=1}^{N} \{-E_\sigma \Omega_{\sigma x} - D_\sigma \Omega_{\sigma y} + C_\sigma \Omega_{\sigma z}\} = -E_L \Omega_{Lx} - D_L \Omega_{Ly} + C_L \Omega_{Lz}$$

From equations (5) we find the components of the angular velocity of the lithosphere (4):

$$\Omega_{Lx} = aG_x - fG_y - eG_z$$
$$\Omega_{Ly} = -fG_x + bG_y - dG_z \quad (6)$$
$$\Omega_{Lz} = -eG_x - dG_y + cG_z$$

where the projections of the angular moment of the absolute motion of the plates (in the MRS) G_x, G_y, G_z are defined by formulae (5), and $a, b, ..., d$ coefficients are expressed in terms of axial A_L, B_L, C_L and centrifugal F_L, E_L, D_L moments of inertia:

$$a = \frac{B_L C_L - D_L^2}{\Delta}, \qquad b = \frac{A_L C_L - E_L^2}{\Delta},$$
$$c = \frac{B_L A_L - F_L^2}{\Delta}, \qquad f = -\frac{E_L D_L + F_L C_L}{\Delta},$$
$$e = -\frac{F_L D_L + B_L E_L}{\Delta}, \qquad d = -\frac{F_L E_L + A_L D_L}{\Delta} \quad (7)$$

$$\Delta = A_L B_L C_L - A_L D_L^2 - B_L E_L^2 - C_L F_L^2 - 2F_L E_L D_L$$

Moments of the lithosphere are determined as sums of the corresponding moments of inertia of the plates, which are presented in Tab. 1.

In the paper, we analyze the relative rotation of the introduced reference systems: MRS, LRS, TRS and ERS94. We will denote vector of the angular velocity (and their components in the Cartesian reference system) of some reference system A w.r.t. reference system B by superscripts as $\Omega^{(A/B)}$.

For the values of the plate and lithosphere parameters presented in Tab. 1 and for components of the angular velocities of the plates in accordance with the HS2-NUVEL1 theory (see Tab. 2) we find:

$$\Omega_{Lx}^{(T/HS2)} = 0.096^0 / m.a.$$
$$\Omega_{Ly}^{(T/HS2)} = 0.158^0 / m.a. \quad (8)$$
$$\Omega_{Lz}^{(T/HS2)} = -0.194^0 / m.a.$$

Final conclusion: Mean Tisserand's axes of coordinates, in which the full angular moment of the plates is equal to zero, define the global rotation of the lithosphere w.r.t. the MRS with an angular velocity $\Omega_L = 0.268^0 / m.a.$ relative to a geocentric axis with the pole $46.3^0 S, 58.8^0 E$.

3 Rotation of the Tisserand's axes in the NNR-NUVEL1 lithosphere reference system

The LRS is given by the plate motion according to the NNR-NUVEL1 theory (ARGUS, GORDON, 1991). It rotates with an angular velocity $0.33^0 / m.a.$ relative to a geocentric axis with the pole $49.8^0 S, 65.0^0 E$. The components of this rotation vector $\overline{\Omega}^{(NNR/HS2)}$ have values:

$$\Omega_x^{(NNR/HS2)} = 0.092^0 / m.a.,$$
$$\Omega_y^{(NNR/HS2)} = 0.196^0 / m.a.,$$
$$\Omega_z^{(NNR/HS2)} = -0.249^0 / m.a.$$

From formula

$$\overline{\Omega}^{(T/NNR)} = \overline{\Omega}^{(NNR/H2)} - \overline{\Omega}^{(T/HS2)}$$

we find the following projections:

$$\Omega_x^{(T/NNR)} = -0.004^0 / m.a.,$$
$$\Omega_y^{(T/NNR)} = 0.038^0 / m.a., \quad (9)$$
$$\Omega_z^{(T/NNR)} = -0.055^0 / m.a.$$

In the LRS the lithosphere executes a slow rotation with an angular velocity of $0.067^0 / m.a.$ relatively a geocentric axis with the pole $55.2^0 S, 96.0^0 E$.

This slow rotation is executed about an axis close to the polar axis of the special inclined lithosphere reference system introduced by the author (BARKIN, 1996).

4 Lithosphere rotation in the Earth reference system 94

The plate motion theory was also constructed on the basis of the GPS data of the 1991 - 1996 years (LARSON et al., 1997). Authors of this paper note that their results are in good agreement with analogous data of the geological theories of the plate motion. In the satellite theory (GPS-PM) the motion of the plates are referred to the international Earth reference system 1994. This reference system (ERS94) is defined by more exact positions of the 240 geological stations , the coordinates of which were defined by the VLBI, SLR and GPS methods.

Using this reference system as the basis we define the components of the angular velocity $\overline{\Omega}^{(T/GPS)}$ of the lithosphere rotation in the ERS94 by means of formulae (5)-(7). In this case in formulae (5)-(7) we will use the projections of the angular velocities of the plates in the ERS94, which are given in Tab. 3.

Tab. 3 Projections of the angular velocities of the plate spherical motion according to the GPS-PM theory (1 unit= 1 °/m.a.).

N	Plate	Ω_{ox}	Ω_{oy}	Ω_{oz}
1.	EA	-0,032	-0,141	0,216
2.	AF	0.009	-0,167	0,199
3.	PA	-0,033	0.315	-0,624
4.	NoA	0.021	-0.219	-0,002
5.	SoA	-0,149	0.009	-0,057
6.	AN	-0.069	-0.096	0,209
7.	AU	0.395	0.34	0,318
8.	IN	-	-	-
9.	NA	-0.099	-0.522	0,456
10.	AR			
11.	PH	-	-	-
12.	CO	-	-	-
13.	CA	-	-	-
14.	JF	-	-	-

As a result, we obtain:

$$\Omega_x^{(T/GPS)} = -0.011^0 / m.a.$$

$$\Omega_y^{(T/GPS)} = -0.019^0 / m.a. \quad (10)$$

$$\Omega_z^{(T/GPS)} = 0.024^0 / m.a.$$

It means that in the ERS94, the lithosphere rotates with an angular velocity 0.032^0 / *m.a.* relative to the geocentric axis with the pole $48.1^0 N, 120.4^0 W$.

Using the values (8), (10) we can define the values of the corresponding projections of the angular velocity $\overline{\Omega}^{(GPS/HS2)} = \overline{\Omega}^{(T/HS2)} - \overline{\Omega}^{(T/GPS)}$:

$$\Omega_x^{(GPS/HS2)} = 0.107^0 / m.a.$$

$$\Omega_y^{(GPS/HS2)} = 0.177^0 / m.a.$$

$$\Omega_z^{(GPS/HS2)} = -0.218^0 / m.a.$$

Thus, ERS94 , in which the satellite kinematical theory of the plate motion was constructed, rotates w.r.t. the hot spots reference system HS2 at an angular velocity 0.301^0 / *m.a.* relative to a geocentric axis directed to the geographic point $46.5^0 S, 58.9^0 E$. This result points out that the NNR RS and ERS94 are close, although they execute a slow relative motion, as stated in the previous section.

Parameters of the lithosphere rotation as a single body are defined comparatively exactly. Consequently, this phenomenon is real. The study of the mechanical causes of this motion is an important scientific problem.

Acknowledgments

The work was supported by RFBR (project 96-05-65015), by the program of the SCST «Astronomy» (theme 1.8.1.3.), by the program «Universities of Russia» (theme 9-5240).

References

ARGUS D.F., GORDON R.G. (1991): *No-net-rotation Model of Current Plate Velocities Incorporating Plate Motion Model NUVEL-1.* Geophysical Research Letters, Vol. 18, No. 11, 2039-2042.

BARKIN YU.V. (1996): *Phenomenon of Ordering in the Plate Positions and Motions.* Russian Conference on Modern Problems and Methods in Astrometry and Geodynamics (Oct. 23-27, 1996, St. Petersburg). Abstracts of Papers. St. Petersburg, Institute of Applied Astronomy, Russian Academy of Sciences, 130-131.

BARKIN YU.V. (1998): *New Regularities in the Lithosphere Structure.* Geotectonics (in press).

CHASE C.G. (1979): *Asthenosphere Counterflow: a Kinematic Model,* Geophys. J.R. astron. Soc., 56, 1-18.

DE METS, C., GORDON, R.G. , ARGUS, D.F., STEIN S.(1990): *Current Plate Motions,* Geophys. J. Int., 101, 425-478.

GRIPP A.E., GORDON R.G. (1990): *Current Plate Velocities Relative to the Hotspots Incorporating the NUVEL-1 Global Plate Motion Model.* Geophysical Research Letters, Vol. 17, 1109-1112.

HARPER J.F. (1986): *Mantle Flow and Plate Motions,* Geophys. J. R. astron. Soc., 87, 155-171.

LARSON K.M., FREYMUELLER J.T., PHILIPSEN S. (1997): *Global Plate Velocities from the Global Positioning System.* Journal of Geophysical Research, Vol. 102, No. B5, 9961-9981.

LIOUVILLE J. (1858): *Dévelopements sur un chapitre de la Mecanique de Poisson.* Journal de Math. Pures et appliquées, 2 serie, tome III.

TISSERAND F.F. (1990): *Traité de Mecanique Celeste.* Tome II., Editions Jacques Gabay.

Least Squares Collocation Method for Space Geodetic Data Analysis

O. TITOV[1]

1 Introduction

Collocation technique has been developed by MORITZ [1980] for analysis of the Earth's gravitational field. He showed that the stochastic part of the field can be considered using special type of covariance functions for sphere.

Conventional least squares method (LSM) considers all parameters under analysis as a constant during 24-hour observational session. Practically, some parameters are not stable. Therefore other methods have been developed for space geodetic data adjustment to take into account stochastic character of variables.

Least squares collocation method (LSCM) will consider the unstable variable as a stochastic process if autocorrelation function is available. It allows to obtain the estimates with high temporal resolution corresponding to rate of experiment performance (3 – 5 minutes for VLBI technique) and study the spectrum of Earth rotation parameters (ERP) variations in range between 10 minutes and 2 – 3 days. If subdiurnal ERP variations are not removed during reduction calculations the ocean forced tidal terms will be present in resulting time series.

The same approach can be applied for space geodetic data analysis to improve the time resolution of results.

The oceanic tides cause variations of the Earth rotation parameters. The most of the variations appears to be at diurnal and semidiurnal frequencies. Some authors have been estimated the amplitudes of main terms both theoretically as well as from observations (VLBI, SLR, etc.). The agreement between results by different approach is about 1 - 3 ms for UT1 and about 7 – 12 mas for polar motion.

Usually all authors take the estimates from global adjustment of large set of experiments. But nobody considers an hypothesis that tidal term amplitudes can vary from year to year. Therefore making this paper I tried to check the hypothesis because large amounts of points after LSCM solution allows to estimate single-year estimates with appropriate accuracy.

Moreover post-fit residuals after the estimation contain very important information about ERP subdiurnal variations beyond ocean forced tidal terms. In time range from 10 to 3 days there are lot of natural phenomena which are able to cause any variations in ERP (for example, free oscillations of the Earth after large earthquakes; oceanic normal modes, etc.). The effects are small. Nevertheless the modern space geodetic technique have reached the necessary accuracy in determination of the ERP. The problem is to improve a time resolution of ERP estimates because almost all effects have short periods, from ~57 minutes (the strongest period of the free Earth oscillations) to 30 hours. Therefore improvement of the ERP resolution using LSCM allows to recognize the small short-period effects using regular VLBI observational data. After realization of CORE observational program (continuous VLBI observations) it would be possible to organize a monitoring of the ERP [CLARK et al., 1997].

2 Analytical foundation

Analytical expressions developed by Moritz can be used for adjustment of VLBI data because wet troposphere delay and clock offset are not constant during 24 hours. It is reasonable to divide all parameters under study into two groups: 'constant' and 'stochastic'. Stochastic parameters show unpredictable variations. Usually troposphere delay and clock offset only to be considered as a latter ones. But this paper the EOP will be added to the group of stochastic parameters because subdiurnal EOP variations is a subject of interest.

Let's consider the following two-parametric model of observational data;

$$Ax + By + w = l\ ,$$

A – the matrix of partial derivatives for 'constant' parameters;

x – the vector of 'constant' parameters;

B – the matrix of partial derivatives for 'stochastic' parameters;

y – the vector of 'stochastic' parameters;

w – vector of random errors;

l – vector of O – C values.

Following the conventional approach the function S should be minimized

$$S = w^T Q_W^{-1} w + y^T Q_y^{-1} y$$

A priori covariance matrixes Q_W and Q_y can be calculated from on the basis of corresponding stochastic process properties. The physically valid covariance functions for all stochastic parameters have to be known before adjustment procedure running.

Mathematician expectation for every stochastic parameters should be considered as a component of constant parameters vector and estimated as a part of vector $\hat{x}$. It provides the

[1] Oleg Titov: Astronomical Institute of Saint-Petersburg State University, 198904, Bibliotechnaya sq.,2, Petrodvorets, Saint-Petersburg, Russia; Fax: +7-812-428-71-29, e-mail: oleg_titov@usa.net

zero expectation of vector $\hat{y}$ every time. Equivalence of some parts of A and B does not prevent for solution making.

There are two formulae for both vectors estimates [MORITZ, 1980]

$$\hat{x} = [A^T (BQ_y B^T + Q_W)^{-1} A]^{-1} A^T (BQ_y B^T + Q_W)^{-1} l$$

$$\hat{y} = Q_y B^T (BQ_y B^T + Q_W)^{-1} (l - A\hat{x})$$

Obviously, to calculate $\hat{x}$ no necessary to know vector y itself. Only statistical information (containing in matrix Q_y) must be available. But $\hat{y}$ can be found after parameter x estimation only. It is worth of saying that the matrix $(BQ_y B^T + Q_W)$ is a non-diagonal one therefore the inversion process usually takes much time and RAM memory storage. After that one have to calculate post-fit residuals vector [TITOV, 1996]

$$\varepsilon = l - A\hat{x}$$

and so-called normalized chi-squared parameter

$$\hat{\sigma}_0^2 = \frac{\varepsilon^T (Q_W + BQ_y B^T)^{-1} \varepsilon}{N - n}$$

where N – the number of observations; n – the number of constant parameters. The value is applied for calculation of root mean square errors.

LSCM allows to trace all variations of stochastic para meters. Practically, the estimation of vector $\hat{y}$ is a sort of filtration where term $(l - A\hat{x})$ is an intermediate vector and the matrix in front of it – transient matrix of filtration. All smoothing properties of the matrix are depend on a priori covariance functions for stochastic parameters.

In comparison with alternative approach (like Kalman filter) the LSCM looks simpler because it does not use some additional information about type of accepted dynamic system. Therefore all analytical expressions as well as general perception of the LSCM is more visual and suitable for practical realization.

As to shortcomings of the LSCM, it takes too much RAM memory storage and computing time, especially for inversion of large non-diagonal matrix. But due to remarkable progress of hardware technology I believe that the memory storage as well as processor rate will not be an actual limits for the method application. Therefore the LSCM can become a strong tool for study of high frequency variations for the EOP in future.

3 Data analysis

EOP with high temporal resolution have been calculated using software OCCAM [TITOV, ZARRAOA, 1997]. All reductions corresponds to IERS 1996 Conventions [MC CARTHY, 1996] exception the part for subdiurnal EOP variations which must be resulted explicitly. For demonstration of LSCM performance I have chosen the CONT'96 VLBI experiment.

Five fundamental VLBI stations regularly have been working during the CONT'96 experiment: WETTZELL (the reference one), KOKEE, NY-ALESUND, NRAO20 and FORTALEZA. Analysis has been made in single-session mode. Nutation offset and station coordinates were considered as a constant parameters; troposphere delays, clock offsets and EOP were considered as a stochastic ones. Celestial reference frame has been fixed by radiosources coordinates in (ICRF 94). One-day a priori EOP values are in system by IERS C04. Station coordinates are from ITRF 94 solution. Direction WETTZELL – NRAO20 as well as vertical drift for KOKEE have been constraining to avoid the ambiguity in matrix of partial derivatives.

The results of UT1–UTC estimation for four 24-hour experiments from 16.09.96 till 20.09.96 are presented on fig. 1. Hourly estimates made by CALC/SOLVE software in USNO are presented here as well. Variations from LSCM have been decimated for seeking of clearness.

Earth orientation parameters (EOP) show a significant subdiurnal variations due to forced effects of ocean tides. The tidally driven variations in EOP for every time epoch T are given by

$$\Delta UT1_T = \sum_{i=1}^{N} [u_i^c \cos \varphi_i(T) + u_i^s \sin \varphi_i(T)]$$

where $\varphi_i(T)$ – so-called "fundamental" arguments.

Following the expression it is possible to remove the theoretical variations from our results and calculate post-fit residuals. IERS Conventions 96 recommend eight main tidal terms to be included in consideration. Fig. 2 demonstrates the post-fit residuals after removing of the tidal terms for the same time epoch. Obviously, there are signals in the time series which can be studied and analyzed by special way.

At first the conventional spectral analysis has been applied for the post-fit residuals to recognize the possible signals. Pictures of spectral density calculated by two alternative ways are shown on fig. 3 and fig. 4. There are significant signals with periods 11.5 and ~50 hours. Also some more small signals can be recognized in diurnal and subdiurnal time bands. But the spectral density is not stable and differ from day-to-day. For comparison I have calculated another subset of CONT'96 experiment (namely, 10 sessions from 04.11.96.

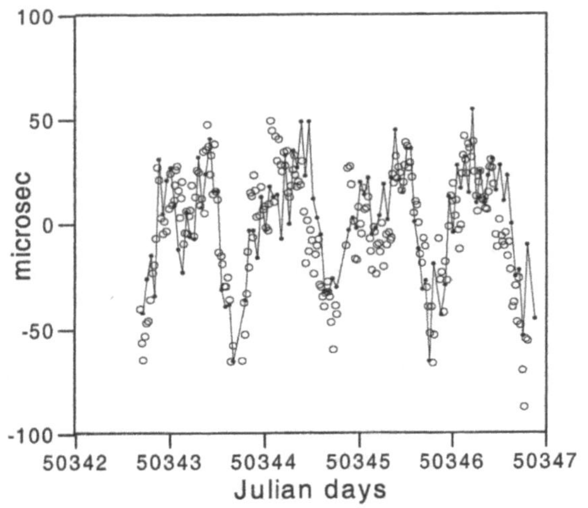

Fig. 1: UT1–UTC variations from CONT'96 (16.09.96 – 20.09.96)
Dash circles – OCCAM / LSCM
Dots and solid lines – CALC / SOLVE

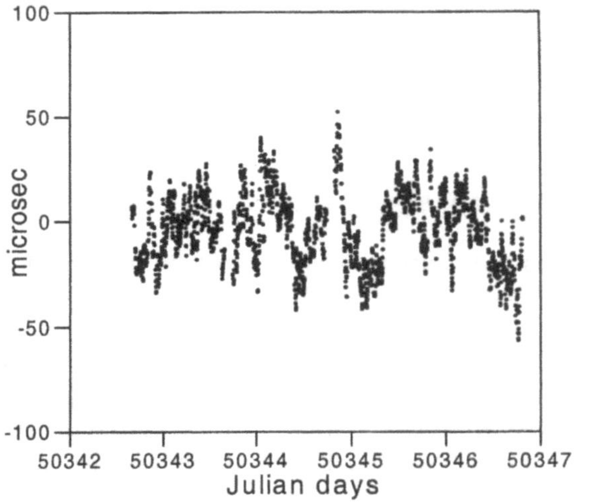

Fig. 2: UT1–UTC variations from CONT'96 (16.09.96 – 20.09.96). Post-fit residuals from UT1–UTC variations after removing of 8 main tidal terms

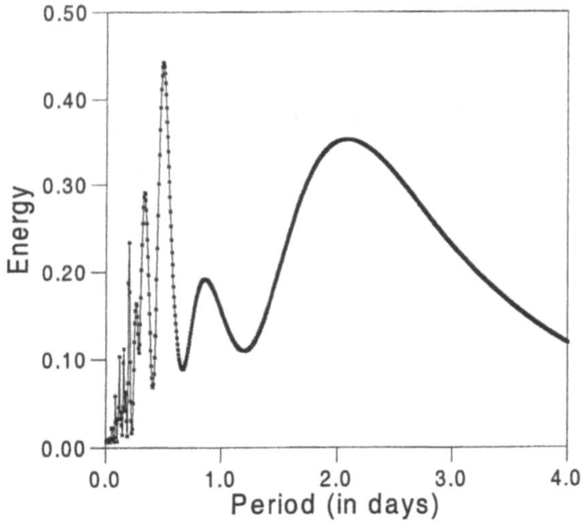

Fig. 3: Spectral density of post-fit residuals from UT1 – UTC after removing of 8 main tidal terms using correlogramm method. Experiment CONT'96 (16.09.96 – 20.09.96)

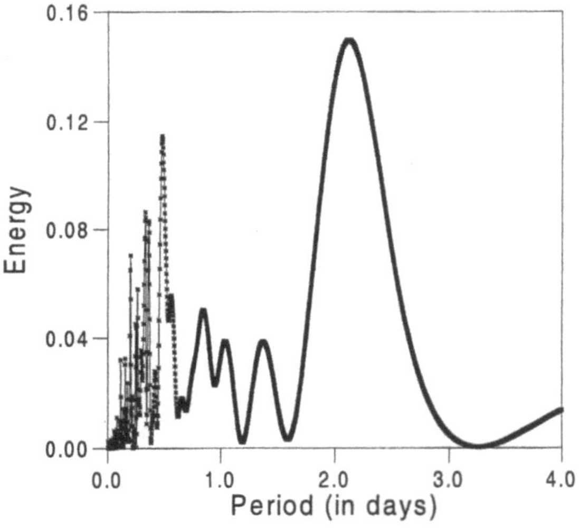

Fig. 4: Spectral density of post-fit residuals from UT1 – UTC after removing of 8 main tidal terms using method by Diming. Experiment CONT'96 (16.09.96 – 20.09.96)

BIFROST Project: Studies of Variations of Absolute Sea Level in Conjunction With the Postglacial Rebound of Fennoscandia

H.-G. SCHERNECK[1], J. M. JOHANSSON, R. HAAS

Abstract

Determination of changes in absolute sea level like in the BIFROST project (Baseline Inference for Fennoscandian Rebound Observations, Sea level, and Tectonics) is shown as an example where a number of space geodetic techniques might gain from achieving a higher level of integration. According to the BIFROST project proposition it is feasable to determine changes of absolute sea level from changes of relative sea level by adding the movement of the crust as obtained with space geodetic techniques to the variations determined by tide gauges. Measurements with the Global Positioning System (GPS) offer the advantage of continuous, inexpensive observations in dense networks, but may suffer from relatively weak constraints on the translation and orientation of the site reference frame. Satellite Laser Ranging (SLR) and Very Long Baseline Interferometry (VLBI) augment GPS by providing high accuracy ties to the frames. Particularly the link of SLR to the gravity centre as the physical orbit centre appears important. The integration of GPS and VLBI promises to stabilize the orientation of the site reference frame.

We give a general overview of BIFROST project and show preliminary results of vertical motion. We point out the problems and sketch where integrated space techniques could provide crucial input to resolving the sea level problem. First attempts to combine VLBI and GPS data using files in the Solution Independent Exchange (SINEX) format are presented.

1. Introduction

The goals of the BIFROST project are to determine three-dimensional crustal velocities associated with the glacial isostatic rebound in Fenoscandia, to infer the viscosity of the Earth's mantle below Fennoscandia, to constrain the space-time history of the Late Pleistiocene ice load in the region, to correct the extensive Fennoscandian tide gauge record for influences of vertical motion and to determine changes in absolute sea level.

Until now the BIFROST analysis in the space geodetic segment has primarily been using GPS observations. GPS data continuously observed with the permanent GPS arrays in Sweden, SWEPOS, and Finland, FinnRef, have been analyzed and daily solutions been derived since 1993. In order to strengthen the robustness of the methods employed to achieve the BIFROST goals, an integration of other space geodetic techniques into the analysis seems to be fruitful. Primarily the determination of absolute sea level changes will benefit from this integration. Here, the individual advantages of the techniques can be used to get a high level of integration. Satellite based techniques have their advantages concerning the Earth's center of mass while geodetic VLBI gives precise Earth orientation. A combination of different space geodetic techniques is possible using SINEX files with the GLOBK (HERRING et al., 1990, 1991) software package.

Figure 1 shows some space geodetic sites in Europe with the techniques that are used. This figure concentrates on GPS and geodetic VLBI sites. The sites of the dense permanent GPS arrays in Fennoscandia are marked with stars, European geodetic VLBI sites with triangles and sites of the International GPS Service for Geodynamics (IGS) network with circles. Some of the stations are equipped with several techniques and belong to several observing networks, e.g. Onsala is part of SWEPOS, IGS and Euro-VLBI.

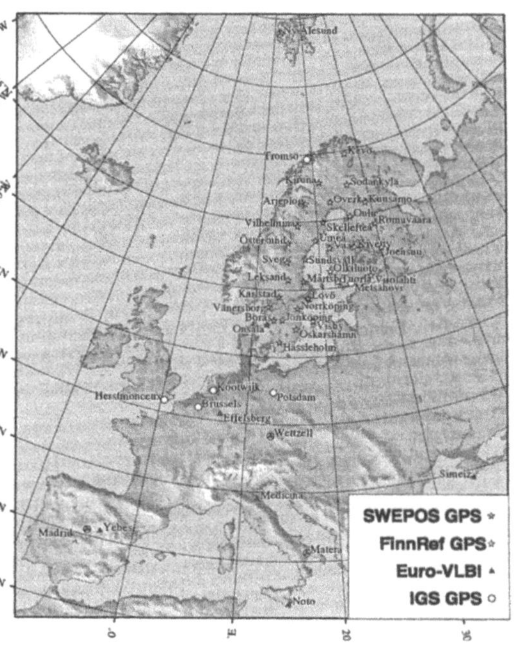

Figure 1: Space geodetic sites in Europe

[1] all authors at Onsala Space Observatory (OSO), Chalmers Tekniska Högskola (CTH), S-429 92 Onsala, Sweden; corresponding author Hans-Georg Scherneck; Tel. +46-31-772-5556; Fax +46-31-772-5590; e-mail hgs@oso.chalmers.se

2. BIFROST results using GPS

Until now, 4 years of GPS data from more than 40 permanent GPS stations in Europe, 34 of them in Fennoscandia, have been analyzed in the BIFROST project, and site motion has been inferred. Figure 2 shows as an example the topocentric motion of the SWEPOS site Umeå near the centre of the postglacial uplift region. Single-site determinations were obtained using the GIPSY-OASIS-II software (WEBB and ZUMBERGE, 1993) in a frame determined by the satellite orbits produced by the Jet Propulsion Laboratory (JPL). In a post-processing stage corrections to the evolution of the International Terrestrial Reference Frames ITRF92/93/94 have been applied in order to rectify the IGS orbits that had been used prior to 1996.

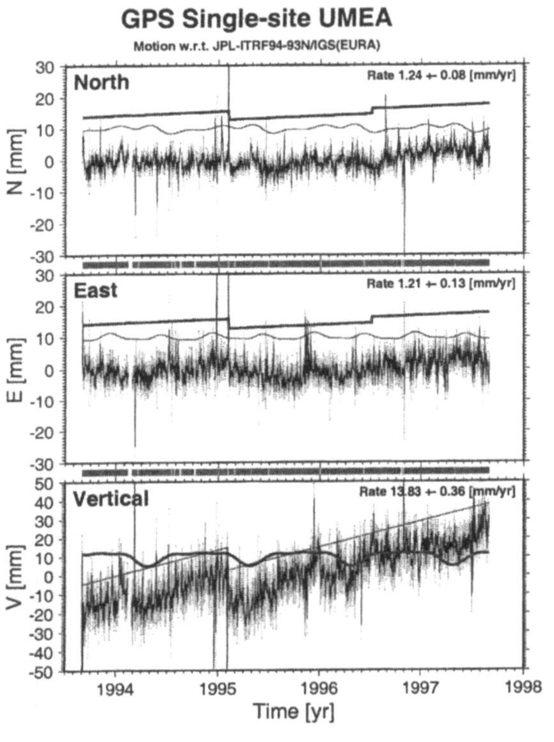

Figure 2: Topocentric motion of the SWEPOS site Umeå near the centre of the Fennoscandian uplift area. Results from singel-site solution obtained with GIPSY-OASIS-II software.

The GPS daily results for north, east and up motion in figure 2 are shown with their 95% confidence limits as thin black line on grey background. A periodic line shows a best fit of seasonal variations (annual, semi-annual and ter-annual harmonics) in each topocentric component. A straight line was estimated to infer the rate of topocentric motion. Jumps at events where the antenna radome at the station was changed were introduced and their height was simultaneously estimated. Umeå shows a large vertical rate of $+13.83 \pm 0.36$ mm/year, which is significantly larger than predicted model rates from a postglacial rebound model (MITROVICA et al., 1994). This trend appears to hold true for all northern stations, also the stations in Finland, and hence diverts arguments charging the simultaneous change of radomes in northern Sweden for the anomalous rates to other still unmodelled systematic effects.

Vertical uplift rates obtained from the BIFROST GPS analysis are compared to vertical uplift rates from tide gauge and levelling observations. As an example figure 3 shows the Empirical Orthogonal Functions (EOF) analysis of the tide gauge record at the station Ratan close to Umeå. The second curve from the top shows monthly mean tide gauge values. It is decomposed into the top curve representing a tidal signal dominated by a climatic annual oscillation, the third curve from the top showing the EOF-determined best combination of the Baltic sea tide gauge plus a constant rate determined at -8.35 ± 0.06 mm/year, and the residuals shown as bottom curve. The vertical rate obtained from tide gauge observations is significantly smaller than the rate obtained from GPS.

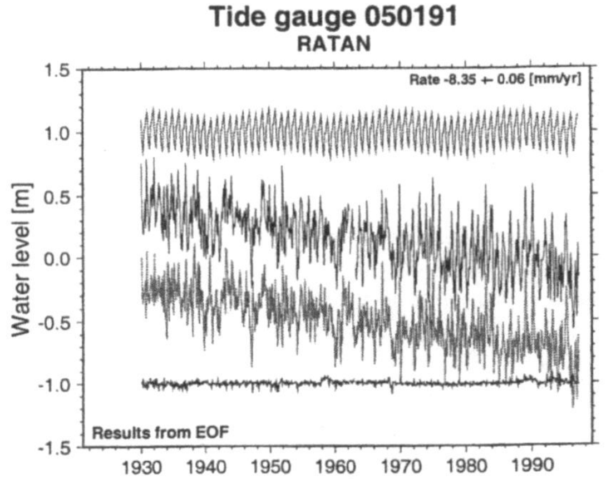

Figure 3: Tide gauge record at Ratan near Umeå

In figure 4 vertical rates from the regional Fennoscandian GPS analysis are shown versus tide gauge and levelling rates (EKMAN, 1996). Geoid rebound from a model (MITROVICA et al., 1994) was applied to the tide gauge and levelling data in order to infer the motion of the crust at the tide gauges. The GPS results are from EOF analysis employed for the purpose to reject common noise that appears coherently in the northern stations. The straight-line fit between these two data types determines a slope and an intercept. The slope is indicative for an uplift-dependent relation between the motion determined from GPS and the motion of the land surface inferred from the tide gauges. The intercept determines a regionally uniform, uplift-independent relation between the data types; a negative intercept would be indicative of a uniform sea level rise. The large slope that is found should caution the interpretation of the intercept, as it might still be severely biased.

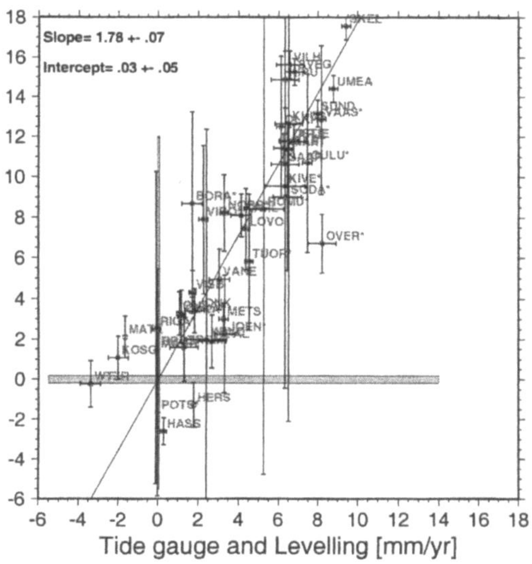

Figure 4: Vertical uplift rates form GPS versus vertical uplift rates from tide gauges and levelling. Geoid rebound according to the model (MITROVICA et al., 1994) is applied to the tide gauge and levelling data.

Figure 4 displays the remaining weaknesses in the GPS-only solution. The slope must be understood as a comparison of a short-term feature (4 years of GPS analysis) to a long-term feature (more than 50 years of tide gauges observations). It is significantly larger than unity. We suspect a number of systematics, ranging from technical issues like effects due to changes of the antenna radomes, time-dependent signal scattering in the neighbourhood of the GPS antennas, to climatic ones, e.g. hydrosphere loading.

The intercept is suspectable for offsets and rates in the centre of the GPS frame. For the projection of the GPS results, a limited number of IGS stations in Europe were used, which may introduce a regional bias into the vertical rates, a bias that cannot be distinguished from a global signal.

3. Combining VLBI and GPS

The idea is to combine several space geodetic techniques to benefit from their individual advantages. In a first attempt we combined VLBI data with GPS data using the GLOBK (HERRING et al., 1990, 1991) software package. The combination included 43 European VLBI sessions from January 1990 to April 1998 (see e.g. HAAS and NOTHNAGEL, 1998) on bi-monthly intervals and 140 days of GPS data from Scripps Orbit and Permanent Array Center (SOPAC) from two global (IGS1, IGS2) and a European network between January 1993 and February 1998, mostly at 14 days intervals. The combination based on SINEX files worked in four steps:

1. Step one were one-day GPS combinations of the three sets of SOPAC files.
2. Step two was the temporal GPS combination of the one-day combinations resulting from (1).
3. Step three was the temporal VLBI combination.
4. Step four was the final combination using the results from (2) and (3).

At all stages the networks were loosly constrained. After the final combination of the multi-year time span, the fix of the frames was carried out and the rates solution produced. The two techniques were tied to each other with a motion constraint of 3 mm/year imposed on the five colocation sites Ny Ålesund, Onsala, Wettzell, Madrid and Matera. Noto was excluded as colocation site, since it had a too short GPS data record to improve the solution. Earth rotation parameters were constrained to the results of the most recent global VLBI solution by National Aeronautics and Space Administration (NASA) Goddard Space Flight Center (GSFC) (MA and RYAN, 1998). Figure 5 shows the results for vertical rates in Europe obtained by the combination of GPS and VLBI.

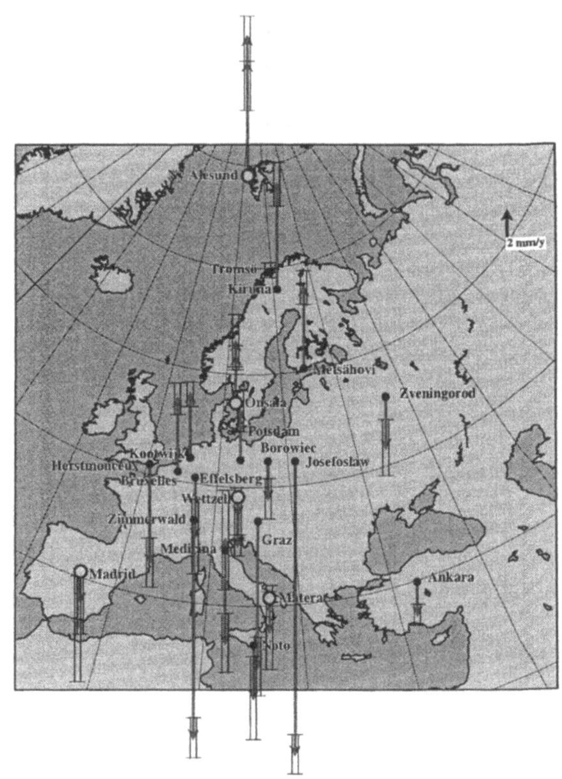

Figure 5: Combination of GPS and VLBI in Europe.

The results have to be treated as highly preliminary. They appear to suffer from a number of weaknesses. Some GPS sites included in the solution show insufficient stability. The recent extensions of the European VLBI network (Yebes, Simeiz) are still not reliable enough to increase the stability of the network. The number of colocated sites is too small and they form a network of weak geometry, mainly elongated north-south. The final combination step appeared to be very sensitive to small changes of input parameters which lead to widely different results.

4. Conclusions

We presented results form the BIFROST project with respect to vertical motion and sea level changes. The weaknesses of the present GPS-only analysis were pointed out. The GPS-only results for vertical rates show significantly larger values than predicted rates from a postglacial ulift model (MITROVICA et al. , 1994) and than rates from tide gauge and levelling observations (EKMAN, 1996). This may be due to unmodelled systematics in the GPS analysis concerned with technical issues, signal scattering and climatic influences.

In order to stabilize and improve the BIFROST analysis the idea is to include more observations from other space geodetic techniques. Especially the determination of abolute sea level changes is expected to benefit from the integration.

We presented a first attempt for the combination of data of different space geodetic techniques in order to serve the objectives of the BIFROST project. For this case study we used SOPAC GPS data from 1993 to 1998 and European geodetic VLBI data from 1990 to 1998. The results presented are highly preliminary and suffer from a number of weaknesses. On the one hand our experience using network combination techniques has to broardened and deepened. On the other hand some more general weeknesses have to be overcome. The number of colocated sites should be enlarged and more space geodetic techniques (e.g. SLR, Doppler Orbitography and Radiolocation Integrated by Satellite (DORIS), Global Navigation Satellite System (GLONASS)) should be included in the combination. Our experience shows, that SINEX files from some data centres do not include all necessary information for a useful combination, i.e. the estimated frame origin, orientation and satellite frame parameters.

Acknowledgements

We wish to thank Richard A. Bennett of the Center for Astrophysics for his tutorial visit to Onsala and his guiding through the software of GLOBK. Figures in this paper were made using the GMT software package (WESSELS and SMITH, 1991). This research has made use of NASA Goddard Space Flight Center's VLBI terrestrial reference frame solution number 1102g, 1998 August. Rüdiger Haas is supported by the European Union within the TMR programme under the contract FMRX-CT960071. This project is supported by grants from the Natural Science Research Council under G-AA/GU 03590.

References

BIFROST PROJECT (1996). *GPS Measurements to Constrain Geodynamical Processes in Fennoscandia.* EOS Transactions, AGU, Vol. 77, No. 34, 337–341

EKMAN M. (1996). *A consistent map of the postglacial uplift of Fennoscandia.* Terra Nova 8, 158–165

HAAS R., NOTHNAGEL A. (1998). *Crustal motion in Europe determined with geodetic Very Long Baseline Interferometry.* Proceedings of the 13th General Assembly of the Nordic Geodetic Commission, in press

HERRING T.A., DAVIS J.L., SHAPIRO I.I. (1990). *Geodesy by radio astronomy: the application of Kalman filtering to very long baseline interferometry.* Journal of Geophysical Research, Vol. 95, 12561–12581

HERRING T.A., DONG D., KING R.W. (1991). *Sub-millimeter determination of pole position using Global Positioning System data.* Geophysical Research Letters, Vol. 18, 1893–1896

MA C. RYAN J.W. (1998). *NASA Space Geodesy Program – GSFC DATA Analysis – 1998, VLBI Geodetic Results 1979-1998, August, 1998.* http://lupus.gsfc.nasa.gov/global/glb.html

MITROVICA J.X., DAVIS J.L., SHAPIRO I.I. (1994). *A spectral formalism for computing three-dimensional deformations due to surface loads, 2. Present-day glacial isostatic adjustment.* Journal of Geophysical Research, Vol. 99, 7075–7101

SOPAC (1998). *Scripps Orbit and Permanent Array Center* http://www.toba.ucsd.edu/

WEBB F.H., ZUMBERGE J.F. (1993). *An Introduction to GIPSY-OASIS-II Precise Software for the Analysis of Data from the Global Positioning System.* JPL, Pasadena, CA.

The Indonesian Gravity Field and the Geoid Model

A. KASENDA, A.M. KOMARA, S. SUTISNA[1]

Abstract

The Indonesian geoid model (INDGED96) has been constructed from the Indonesian gravity database of 5' grid interval. The accuracy of the geoid height is considered poor in some areas due to the lack of high resolution gravity data in the areas. In order to improve the geoid model, additional gravity data must be performed. Since the data gaps occur mostly in mountainous remote areas, performing terrestrial observation will be very time consuming. In this case, implementing airborne gravimetry is an alternative to accelerate the data acquisition. However, the accuracy of the airborne derived gravity data used for geophysical exploration purposes might not be suitable for the precise geoid determination. A case study of the airborne gravity data aimed at hydrocarbon allocation in Irian Jaya used to calculate the geoid height is undertaken. The preliminary result is presented.

1. Introduction

The Indonesian archipelagic country consists of 17 546 islands cover the area of more than 7 200 000 square kilometers. The determination of the geoid with high accuracy and resolution is required towards the unification of the national height system.

The Indonesian geoid model ever constructed was based on the existing land and marine gravity data combined with the earth potential model of OSU91A.

The computation was performed by using the remove-restore technique. The accuracy of the resulted geoid is uneven across the area. By comparison to the geometric geoid computed in several control points available in Jawa, Sumatera, Sulawesi and Irian Jaya, the agreements in Jawa and Sumatera are better than in the other islands. This is due to the fact that the existing gravity data in Jawa and Sumatera are denser.

In order to improve the geoid, it is required to fill the gaps especially in Kalimantan, Sulawesi and Irian Jaya.

At present, the data gaps are mostly occurred in mountainous terrain areas which covers by dense forest and jungle. Therefore, filling the gaps by means of terrestrial measurements would be very expensive and time consuming. However, the airborne technique can offer the opportunity to cover the remote and rugged areas in relatively shorter time.

2. The Indonesian Gravity Field

The Indonesian gravity surveys was started back in 1930s by Vening Meinesz which performed the pendulum survey in the Indonesian waters. Since then many gravity surveys were made aiming at locating structures related to hydrocarbon and mineral resources as well as for scientific purposes.

An attempt to compile all the existing data is to produce a regional Indonesian gravity database. An integration of marine and land was the first priority in order to produce a comprehensive database. The land data, however are very scattered and based on the various references.

In 1994 the primary Indonesian gravity database was constructed. Around 250 600 points of land surveys and 150 000 points of marine measurements are gridded in to 5' interval produced a data set of 136 000 points.

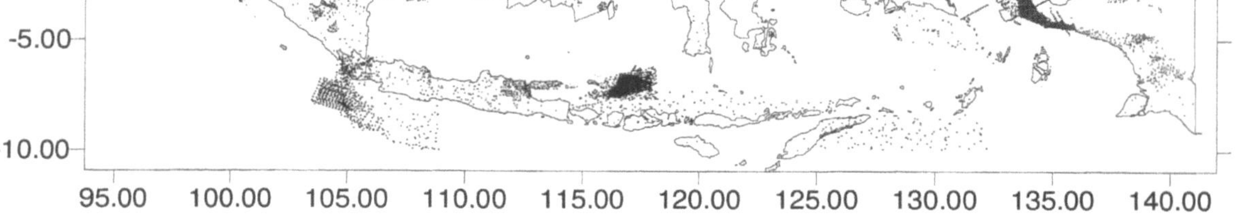

Fig.1 : The Indonesian Gravity Data

[1] A. Kasenda, A. M. Komara, S. Sutisna: National Coordination Agency for Surveys and mapping (BAKOSURTANAL), Jl. Raya Bogor, KM-46 Cibinong, Indonesia; Tel.: +62-21-8754654, Fax:+62-21-8752064, e-mailsobar@cbn.net.id

3. The Indonesian Geoid Model

The Indonesian regional geoid model known as the INDGED96 was constructed based on the gravity database of 5' grid interval. The undulation was computed by using the earth coefficients potential model of OSU91A as a reference to compute the medium to long wavelength component of the geoid, whereas the short wavelength component was calculated from the residual gravity anomalies. The result varies from 42.5 to 77.5 meters over the area with various accuracy across the country. By comparison with the geometric geoid in several GPS/Leveling control points available in major islands such as Jawa, Sumatera and Sulawesi it is showed that the relative comparison in Jawa and Suamtera are better than in Sulawesi. This is due to the fact that the original gravity data distribution in Jawa and Sumatera islands are better as compare to the data over Sulawesi island.

In order to fulfill the need of better geoid across the Indonesian area, densification of the gravity data is urgent.

In line with the exploration program of aimed at inventing the natural resources in Indonesia, the National Oil Company has applied airborne gravimetry to undertake a systematic geophysical mapping in eastern part of Indonesia. Upon successfully completion of these works, it is planned to extend the measurement to the area of less dense of the data.

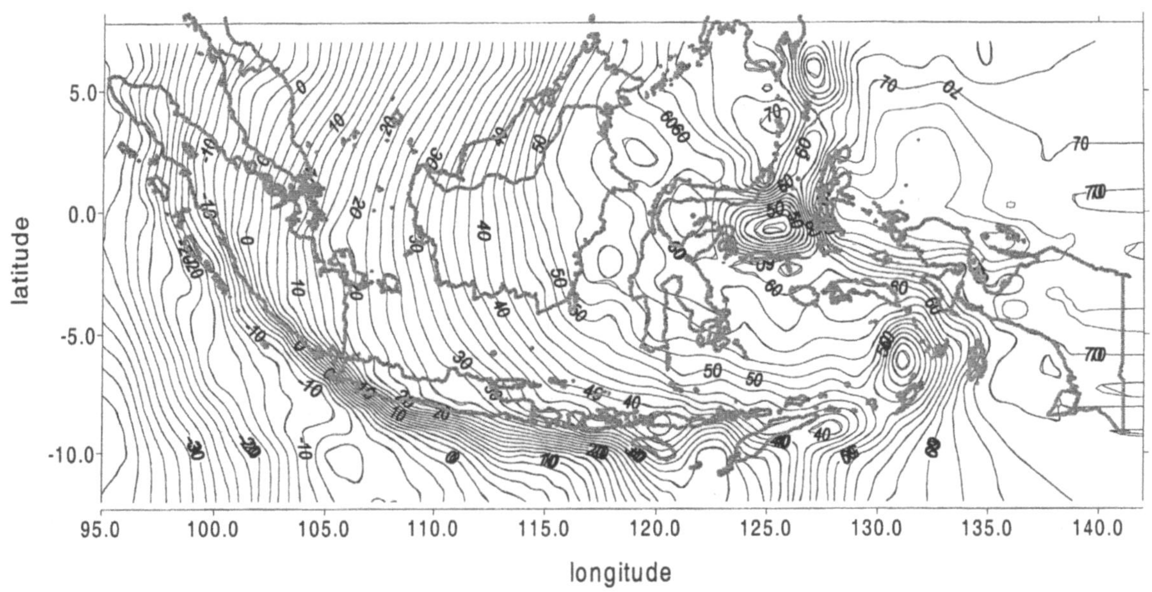

Fig.2 : The Indonesian Geoid Model

In conjunction with that, the in coming airborne gravity data can be used to update the geoid model with better coverage of gravity data. Prior to this plan, an initial test of using airborne gravity data was conducted to construct the local geoid model of the Lengguru area in Irian Jaya. The mean deviation from the INDGED96 model over the respective area is 0.45 meter with the rms deviation of 2.2. The relative accuracy is determined by comparing the airborne geoid to the geometric geoid which is planned to re-measure the Doppler/APR points in this area by GPS observations.

This initial test result is expected to allow the gravity anomalies derived from airborne technique aimed at geophysical exploration to be used for updating the geoid model of Indonesia.

References :

HEISKANEN W., MORITZ H., (1967): *Physical Geodesy*, W.H. Freeman and CO., San Francisco, Text Book.

KASENDA A., KEARSLEY A.H.W., (1993): *The geoid determination of Irian Jaya*, Unisurv. S-43, UNSW, Australia, 1-88.

KAHAR J., (1991), *Geoid dan pengaruhnya pada reduksi jarak*, Bakosurtanal, int. report, Cibinong.

KEARSLEY A.W.H, (1985), *Towards the optimum evaluation of the inner zone contribution to geoidal heights from gravimetry*, Reports No.20, university Uppsala

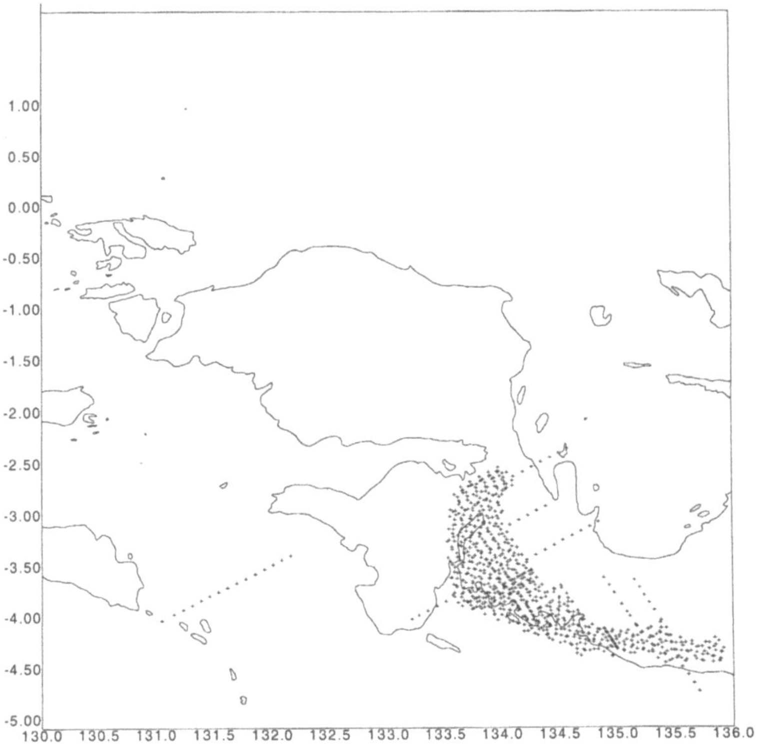

Fig. 3: The Airborne Gravimetry of Lengguru

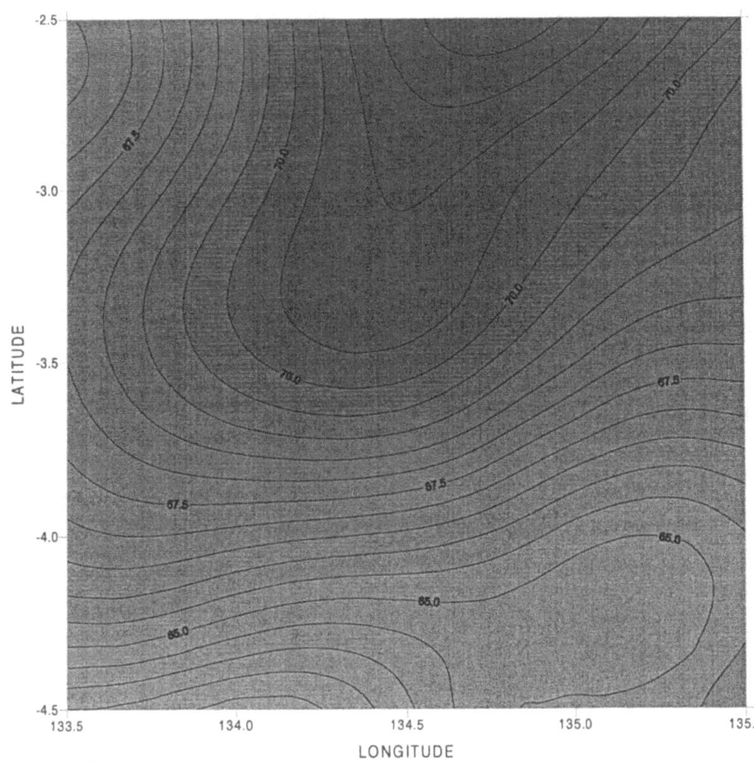

Fig. 4 : The Lengguru Airborne Geoid Model

ISRO Initiatives for Space Geodesy and Geodynamics in India

K. ELANGO, P. SOMA, K. RAMESHCHANDRA, S. RANGARAJAN[1]

Abstract

Indian Space Research Organisation (ISRO) is reviving the Space Geodesy by opening a Space Geodesy Division in ISRO Telemetry, Tracking and Command Network (ISTRAC). ISRO is already operating a PRARE and GPS station as a part of GeoForschungsZentrum (GFZ) global network. There is a proposal of establishing a third generation Satellite Laser Ranging (SLR) station at the same site thereby elevating Bangalore as a fundamental reference station for Geodesy and Geodynamic studies. With the long experience ISRO has gained earlier, the data from the above techniques can be effectively used for geodetic purpose. As geodesy is a nascent field in India, ISRO proposes to conduct an international workshop on Space Geodesy, which will provide more exposure and bring the national and international experts on a common platform to evolve utilisation and research using these data for the next decade. ISRO will collaborate with academic and research institutes to evolve a strong and vibrant science group in India carrying out studies on different aspects of Space Geodesy such as orbital dynamics, geophysics, earth dynamics, ocean dynamics, plate tectonics etc. ISRO has round the clock operation at SCC which is more suited for this programme and a regional data centre can be established. The data from these systems can also be used for ISRO's future satellites such as OCEANSAT, CARTOSAT and CLIMATSAT which requires stringent ground imagery accuracy of the order of 2.5 m.

Introduction

Indian Space Research Organisation currently providing TTC support to multiple remote sensing satellites through one of its unit namely ISRO Telemetry, Tracking and Command Network (ISTRAC) ISRO is also operating a Precise Range and Rangerate Equipment (PRARE) and P-code GPS receiver station at Spacecraft Control Centre (SCC) campus, Bangalore under a cooperative programme with GeoForschungsZentrum (GFZ), Germany. There is a proposal to install a third generation Satellite Laser Ranging (SLR) station at the same site which will elevate Bangalore as one of the fundamental reference station for Geodesy and Geodynamic studies. With the long experience gained earlier, ISRO plans to foster Space Geodesy in this country by setting up a Space Geodesy Centre. A space geodesy division with dedicated man power (system hardware and software expertise) has been formed recently at ISTRAC as first step in this direction.

Relevance of Geodesy in ISRO

- ISRO has long experience in operating the first generation SLR at Kavalur. ISRO can revive the SLR program in India and gain further experience and also will become part of International SLR network and will have access to the data from other SLR stations and the data products. ISRO's future satellite IRS-P5 (Cartosat) carries a GPS and laser reflectors onboard which is scheduled for launch in early 2000. Global network stations (SLR and GPS) will be used for precise orbit determination
- Geodynamics is an important field in India. The data from GPS, PRARE and SLR will form valuable input for carrying out geo-dynamics applications (plate tectonics).
- A Space geodesy division has been established at ISTRAC recently.

ISRO's Experience and Facilities

- ISRO has relevant and adequate experience in operating SLR, GPS and PRARE and the infrastructure is designed for such operation
- ISTRAC has been involved in operational Orbit Determination for LEO satellites for more than two decades and has extensive experience in this area including near real time POD, immediately after the injection of the satellite into orbit
- ISRO has developed in-house software for tracking, data processing, orbit determination and prediction software, ephemeris generation, antenna pointing angle generation.
- ISRO has extensive experience in developing/ maintaining/operating tracking systems including SLR.

Future Plans

- ISRO will provide support for elevating Bangalore as one of the fundamental reference/core station for geodesy and geodynamic studies. Open an operational Data centre for these studies
- Create a regional data centre and collect formatted tracking data from several operational data centres, maintain local archives of the data and transmit this data to the global data centre.
- Open an analysis centre for data products such as precise orbits, earth rotation, summary files etc.

[1] K. Elango, P. Soma, K. Rameshchandra, S. Rangarajan: ISTRAC - Indian Space Research Organisation, Department of Space, Bangalore – 560 058, India; Tel.: +91-80-8395129, Fax: +91-80-8394249 & e-mail : elango@istrac.gov.in

- ISRO will collaborate with academic and research institutions to evolve a strong and vibrant science group in India carrying out studies in different aspects of satellite based geodesy such as orbital dynamics, geophysics, earth dynamics, ocean dynamics, dynamics of tectonic plates, gravity field modeling.
- SLR stations are established in north of India (Helwan, Maidan and Wuhan) and all are north of Himalayas where Indian plate collides with Eurasian plate. SLR data from Bangalore will be very useful for studies in Inter-continental movement and crustal studies.
- ISRO has proposal to conduct an International workshop on Space Geodesy at Bangalore during the third quarter of 1999 focussing on the following aspects:
  - blue print for specific initiatives in space geodesy in India during next five years
  - enough science materials to take up the work for the next 5 to 10 years on space geodesy
  - identification of resource persons who can take up studies and work on different aspects of geodesy applications.
  - This work will also provide a platform for India to evolve suitable international collaborations for future studies.

Conclusion

Reviving Space Geodesy at ISRO will be useful for geodesy and geodynamic studies. Elevating Bangalore as one of the fundamental reference station with PRARE, GPS and SLR stations will put into the international network loop for space geodesy. Opening a regional data centre will provide chances for scientists from various universities and research institutions to carry our research studies using these data.

Acknowledgement

Authors are very much thankful to Dr K. Karthurirangan, Chairman, ISRO/Secretary DOS for his keen interest shown in preparation of this paper. Authors are also thankful to Shri K.R. Sridharamurthy Scientific Secretary ISRO HQ, for the encouragement shown in preparing this paper. The Support provided by Shri S. Kalyanaraman, Programme Director, IRS Mission , ISAC is also acknowledged.

References

1. Presentation materials during ISRO - NASA Interface meeting held at ISRO HQ, Bangalore 25-27 Feb. 1998
2. Feasibility of locating SLR system in ISTRAC, Bangalore - a project proposal Report No: ISTRAC/TR/02/98

A Sketch towards an Integrated Global Geodetic Observing System (IGGOS)

Global Integrated Geodetic and Geodynamic Observing System (GIGGOS)

R. RUMMEL[1]

1. Introduction

It is meanwhile generally accepted that progress in earth sciences fundamentally depends on understanding the earth as a system, ((KANDEL, 1980), (Earth System Science-A Closer View, 1988), (A Unified Theory of Planet Earth, 1988), (Earth Observation User Consultation Meeting, 1991)). As a consequence deeper insight has to be established into the individual role of each system component - atmosphere, oceans, cryosphere, solid earth and biosphere - the interaction between the components and the effect of all exogenic processes. Major decisions facing human societies will depend on our understanding of the system earth, (Solid Earth Sciences and Society, 1993). To this end UNESCO in cooperation with the International Council of Scientific Unions (ICSU) and other international bodies has established large international programs such as the World Climate Research Program (WCRP), the International Geosphere Biosphere Program (IGBP), or recently the Global Ocean Observing System (GOOS). The general focus of these programs is on climate change and environment. Although it is undeniable that on all spatial and temporal scales solid earth and ice processes are in strong interaction with our climate, the earth environment and our living sphere, these programs leave to a large extent open how their effect is to be quantified and modeled.

Geodesy is in a position to fill this gap by offering a very concrete and central element, a Global Integrated Geodetic & Geodynamic Observing System (GIGGOS). At the same time such a system would serve as an important contribution to the understanding of the physics of the earth's interior. The objectives of GIGGOS are:

1. To provide in a well-defined and reproducible global terrestrial frame,
2. the integral effect on earth rotation of all angular momentum exchange inside the earth, between land, ice, hydrosphere and atmosphere, and between the earth and sun, moon and planets,
3. the geometric shape of the earth's surface (solid earth, ice, and oceans), globally or regionally, and its temporal variations, whether they are horizontal or vertical, secular, periodical or sudden, and
4. by adding the earth's gravity field – stationary and time variable – mass balance, fluxes and circulation.

In the sequel the four components of the observing system are referred to as frame, geo-kinematics, earth rotation, and gravity field. The observing system, in order to meet its objectives, has to combine greatest measurement precision (10^{-8} – 10^{-9}, relatively) with spatial and temporal consistency and stability over decades. The basic structure of GIGGOS is shown in Figure 1.

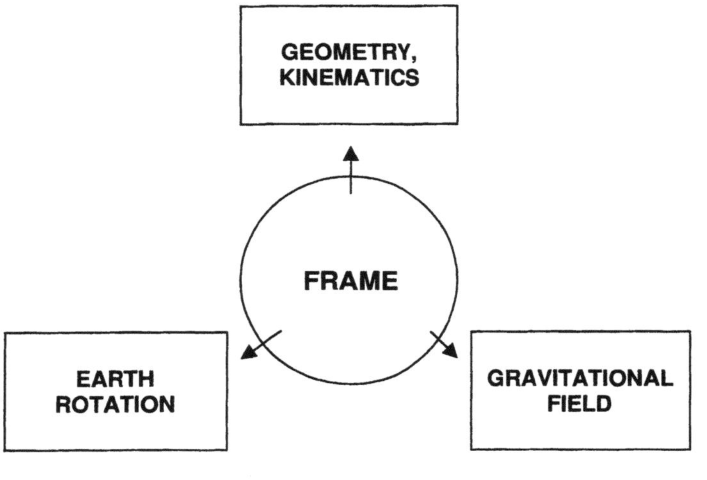

Figure 1: Global Integrated Geodetic & Geodynamic Observing System (GIGGOS)

[1] Reinhard Rummel: Institut für Astronomische und Physikalische Geodäsie, Technische Universität München, Arcisstraße 21, D-80290 München, Germany; Tel.:+49 - 89 - 289 - 2 3189/ Secr. 2 31 90, Fax: +49 - 89 - 289 - 2 31 78, e-mail: rummel@step.iapg.verm.tu-muenchen.de

In chapter 2 a discussion follows on the strategic role of such a system for the International Association of Geodesy (I.A.G.), chapter 3 deals with the relationship between the observing system and the technology development, in chapter 4 the geophysical value of the observing system is described. In chapter 5 four cases are sketched that discuss the role of the gravity field component of the observing system.

2. The Strategic Role of the Observing System for I.A.G.

The major international programs mentioned above attract currently the full attention in and outside earth sciences. The scientific issues addressed by these programs are highly relevant but at the same time they are very complicated. Progress is therefore rather slow. With the proposed observing system, geodesy could provide a very immediate and concrete contribution to the study of the earth system.

GIGGOS should be seen as geodetic contribution to earth system sciences. The proposal is timely because only now the potential arises in geodesy to investigate processes of such complicated nature. For example only now one can think of monitoring sea level rise and its regional variations in space and time from a combination of terrestrial, airborne and space techniques, and to separate sea level changes from vertical motion of the crust. It should be clear that the proposed GIGGOS, with its capability to integrate frame, earth rotation, geo-kinematics and gravity to one system, and to secure integrity and stability over decades on a 10^{-8} to 10^{-9} level, represents an enormous challenge for geodesy. Its success will depend on a very concentrated effort in terms of definition, logistics, coordination, sensor development and maintenance, theory and modeling, data collection, transfer and processing and interpretation.

On the other hand much of the proposed system exists already in one way or the other. The reference system as defined by the IERS conventions, (McCarthy, 1996), see also ((SEIDELMANN, 1992), the connection between terrestrial and celestial frame as recently formulated in (MA & FEISSEL, 1997), as well as the concept of the International GPS Service for Geodynamics (IGS), (ZUMBERGE et al, 1996), should be seen as the central elements of the proposed system. In GIGGOS these elements are to form one entity together with the parts provided by the other services (BGI, IGeS, ICET, PSMSL and BIPM), commissions, special commissions and special study groups.

The International Association of Geodesy can derive the following benefits from GIGGOS:

- **In its Relationship to Earth System Sciences:** If I.A.G. succeeds to establish GIGGOS as a recognized international program, geodesy's role in earth sciences will be substantially strengthened.

  (I am convinced GIGGOS is a realistic concept. Despite very valuable contributions of geodesy to geophysics in the past, such as the first in situ measurement of plate motion, the establishment of the link between the atmospheric angular momentum and earth rotation, the altimetric data base for the study of the ocean lithosphere, or the earth gravity field models, geodesy's role seems generally underestimated. With the observing system the contribution of geodesy to earth system sciences as a whole, in particular to the study of the dynamics of the earth's interior, the earth's state of departure from equilibrium, the interaction of land, ice and ocean processes, the evolution of sea level (a subject of high social relevance, too), and, last not least, to ocean circulation, and consequently to the study of energy transport in the oceans and related climate questions, would be focussed and strengthened.

  At the same time geodesy could demonstrate that its contribution to earth sciences reaches beyond data provision alone.)

- **For Geodesy itself:** GIGGOS could serve as a common and very challenging focus point for practically all current research activities inside I.A.G.

  Observatories and technology programs would become a justification from medium and long term activities of substantial relevance. At the same time their standard could be judged by their value to GIGGOS.

  The work of the services, very successful and probably that part of I.A.G. activities that receives most attention and appreciation from outside of geodesy, would be channelled towards one common goal. Also the theoretical work, the development of new analysis methods and of new measurement techniques, such as airborne gravimetry, could be placed under this joint umbrella. Developments in the diverse parts of geodesy would become visible as being complementary rather than competitive. Section IV would be faced with the very challenging task to derive a theoretical framework for the observing system as a whole and for its four elements individually.

- **For Young Geodesists:** The observing system presents a logical framework of future research themes and their priorities, whether applied or theoretical. Also geodesy's role in earth's sciences becomes clearer.

  Historically, in its fundamental mission of the determination of the figure of the earth, geodesy has available three basic elements, (1) techniques that provide the geometric shape of the earth e.g. in the form of a polyhedron, (2) astronomical techniques that relate celestial objects to terrestrial points (this part provides the orientation of the polyhedron on the globe) and (3) techniques that measure gravity and therefore allow to derive gravity potential (or height) differences between the points of the polyhedron. In 1878 BRUNS (1878), gave a beautiful account of this concept and of its shortcomings at that time. GIGGOS, as proposed here, is nothing but an elaboration of this concept on a 10^{-8} to 10^{-9} level, that, in addition, can make use of the most advanced space techniques. At that precision level the earth starts so-to-say to pulsate and a large variety of geophysical phenomena, relevant to the understanding of system earth, become visible. But the three underlying basic elements, that together make geodesy, are still the same. The future will require that geodesists loose their traditional hesitance to actively enter in their research into the various sub-disciplines of earth sciences such as geodynamics, geology, glaciology or oceanography.

(After one century of marriage, geodesy and surveying engineering will have to reflect on their relationship, their common fields as well as their individual goals. For surveying engineering priority number one seems to be the development of more and more versatile and intelligent geographic information systems (GIS), followed by geodetic navigation and control applications and real time processing techniques. For geodesy improvement in precision is still an issue as well as integration of techniques and models to increasingly complex systems. May be this system aspect as well as the need for global and globally consistent coordinate and height systems is their common basis. It would be worthwhile to reflect on this aspect, for only then one can convincingly and economically further develop our common curricula and research.)

3. GIGGOS and Geodetic Measurement Techniques

The four components of the observing system, frame, geokinematics, earth rotation, and gravity, are realized employing modern terrestrial, airborne, and space techniques. For three of the components, namely frame, kinematics and earth rotation, a

relative precision of 10^{-8} - 10^{-9}

is feasible already today. This implies, for example, the determination of intercontinental distances with millimeter to centimeter precision or of variations in length of day (l.o.d.) to milliseconds. For the gravity component 10^{-8} - 10^{-9} is achievable but only point-wise, i.e. at individual points. For a consistent system, however, gravity will be required as a field quantity at this precision level. There is good hope that this deficiency can be overcome with the planned gravity satellite missions in conjunction with airborne gravimetry and the available terrestrial gravity material. The main challenge will be, at this precision level to integrate the four components to one system and to realize consistent time series over decades. Seemingly minor aspects such as point identification, monumentation, calibration, datum definition and transformation or sensor stability will have to receive greatest attention.

The following main measurement techniques can be attributed to the four components of GIGGOS, see also Figure 2.

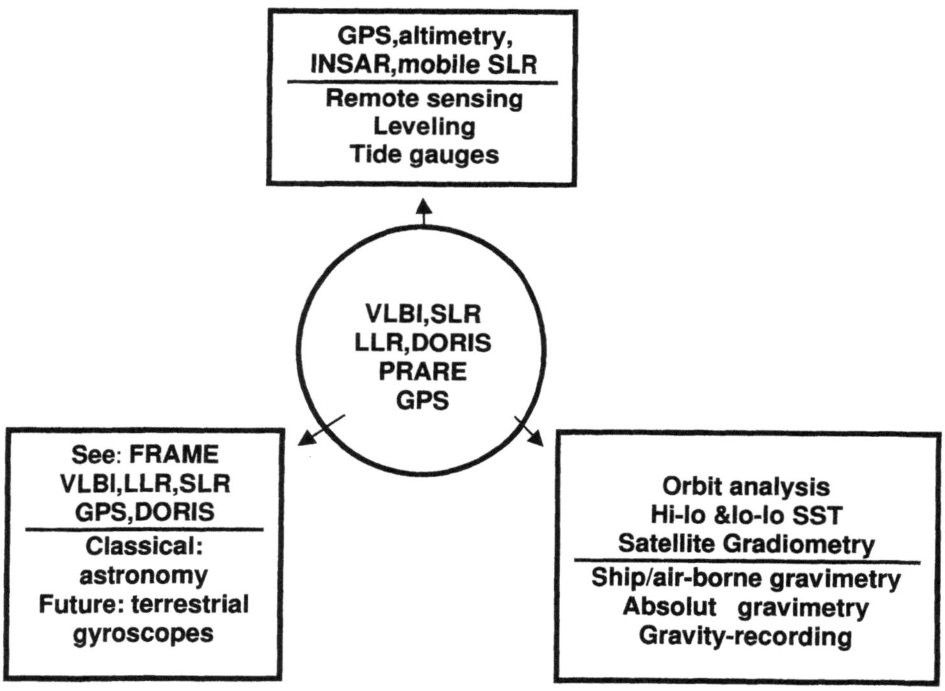

Figure 2: Measurement Techniques of the Integrated Observing System

a) Reference Frame

A selected network of geodetic fundamental stations determines the terrestrial reference frame, (MUELLER et al, 1995), into which the three other components are to be integrated. The frame secures continuity and consistency of all measurement series in time and space. The coordination and definition is done by the IERS in collaboration with the IGS. The fundamental stations are equipped with space techniques such as very long baseline interferometry (VLBI), satellite laser ranging (SLR), lunar laser ranging (LLR), global positioning system (GPS), DORIS, and PRARE, with terrestrial techniques such as superconducting gravimetry, and with equipment for precise time keeping and time transfer. A prototype transportable fundamental station is close to its completion.

b) Earth Rotation

The fundamental stations of component I have a second function. Apart from defining the frame fixed to our planet, they establish the time variable connection to the celestial frame. Earth rotation is divided into precession/nutation, polar motion and variations in l.o.d. Whereas in the old days the connection between terrestrial and celestial system was maintained by classical astronomical methods, it is now primarily based upon VLBI and, with the orbit of the moon or of selected artificial satellites as intermediary, on LLR, SLR, GPS and DORIS.

Purely terrestrial earth rotation sensors - without link to extra-terrestrial objects - are currently under development.

c) Geo-Kinematics

Connected to the terrestrial frame are a series of surface monitoring techniques, such as (land/ice/ocean-) altimetry, (differential) interferometric SAR, terrestrial GPS networks, satellite and airborne imaging, tide gauges, geodetic leveling etc. A large variety of phenomena are monitored in this manner: plate tectonics, intra-plate deformation, pre-seismic strain accumulation, volcanic deformation, vertical crustal motion of anthropogenetic or natural origin, ice and glacier flux, tides, sea level rise, and ocean variability.

Elements I and III together provide a comprehensive picture of the geometric figure of the earth - as expressed by a global polyhedron of selected points (sparsely spaced in some parts of the earth, densely spaced in areas of geophysical interest and practically continuously covered when applying altimetric and imaging techniques) - and of all temporal surface variations whether related to land, ice or ocean. The temporal variations of surface geometry may occur in horizontal or vertical direction and may be sudden, periodic or secular. The global monitoring system combines space-borne, airborne and ground based techniques. In this sense the Global Sea Level Observing System (GLOSS) must be seen as part of GIGGOS and will then permit separation of crustal motion from sea level rise/fall. Complementary weather, atmospheric, ocean, seismic etc. data sets are required for the modeling of signal delays, atmospheric loading, tides, micro-seismicity etc.

d) Gravity Field

The gravity field component comprises the determination of the global gravity field of the earth and of the geoid and of their temporal variation by satellite techniques, the regional high resolution gravity field determination by airborne/ship-borne gravimetry or gradiometry, and the high precision terrestrial gravimetry and gravity recording. Whereas the state-of-art of terrestrial absolute and relative gravimetry is rather high, major advances are expected in the near future for airborne and satellite techniques. Satellite missions employing the concepts of high-low satellite-to-satellite tracking (hi-lo SST), low-low satellite-to-satellite tracking (lo-lo SST), and gradiometry are close to their realization. Only then gravity field models will be of the precision, resolution and homogeneity required to be consistent with the three other components of the observing system.

Table 1 summarizes the elements of the Global Integrated Geodetic & Geodynamic Observing System (GIGGOS).

Table 1: Global Integrated Geodetic & Geodynamic Observing System (GIGGOS)

component	objective	technique	responsible
a) Terrestrial Frame	global cluster of fiducial points, determined at mm to cm level	VLBI, GPS, SLR, LLR, DORIS, PRARE, time keeping/transf., absolute gravimetry, gravity recording	international science community (I.A.G.: IERS, IGS)
b) Earth Rotation (nutation, polar motion, variations in l.o.d.)	integrated effect of changes in angular momentum (mass changes in atmosphere, cryosphere, oceans, solid earth, core/mantle; momentum exchange between earth system components)	classical astronomy, VLBI, LLR, SLR, GPS, DORIS, under development: terrestrial gyroscopes	international cooperation (I.A.U. & I.A.G.)
c) Geo-Kinematics	temporal variations of land/ice/ ocean surface (plates, intra-plates, volcanos, earthquakes, glaciers, ocean variability, sea level)	altimetry, INSAR, GPS-clusters, imaging techniques, leveling, tide gauges	international and national (e.g. GLOSS, WEGENER); space missions
d) Gravity Field potential (geoid) gravity anomalies and their temporal variations	adds mass balance to the observing system; (isostasy models, ice flux), quantification of ocean circulation, unification of height systems: for geodesy, sea level monitoring etc.	terrestrial gravimetry, ship- & airborne gravimetry and gradiometry, dedicated satellite missions: hi-lo SST, lo-lo SST, gradiometry	I.A.G., NASA, ESA, national space programs

4. The Observing System in its Context of Earth Sciences

The shape of the surface of the earth, defined as the boundary surface between solid earth, ice and oceans and the atmosphere, is an impressive manifestation of the dynamics of the earth system. Shape and change of shape result from the perpetual interaction of the driving forces: the convective heat exchange in the earth's interior (with plate motion, crustal deformation, earthquakes, volcano eruptions); the joint action of all exogenic forces: tides, solar radiation, winds, weather and storms; and gravitation (the driving force behind erosion in the broadest sense of its meaning and of flux). Depending on the viscosity of the considered system component: land surface, ice or oceans, this interaction of the driving forces results in almost perfect isostatic equilibrium for some parts of the lithosphere, in moving plates and internal stress for other parts, in slow or sudden motion of material, in collision, underplating, sedimentation, erosion etc. It results in ice flux accompanied by deglaciation or ice growth, or for the oceans in tides and stationary or variable circulation.

In the past only very sophisticated terrestrial control networks could be used to determine point configurations or detect their temporal variations. Meanwhile, with airborne and space-borne sensor systems, shape and changes in shape of land, ice and ocean surface can be determined accurately, very efficiently and with a high repetition rate. Even tiniest variations in e.g. the strain pattern around a volcano are detectable by (differential) INSAR nowadays. All geodetic methods concerned with the determination of the shape of the earth and its variation in time, regionally or globally are summarized under the heading geo-kinematics.

Modern space positioning techniques allow to tie regional geo-kinematic monitoring systems into one geocentric terrestrial reference frame. This way regional processes become comparable and accessible to global analysis.

Temporal variations in pole position and length of day reflect the integral effect of all angular momentum changes (mass, motion and forces) on the gyroscopic motion of the earth in space. They comprise the effect of all time variable deformations and mass rearrangements in the earth's interior and on its surface as well as torques due to external forcing. By now some excitation functions, like the atmospheric angular momentum, are well understood in their effect on earth rotation and define important constraints on the modeling of the energy exchange between the components of the earth system. Other potential excitation functions, like the electro-magnetic coupling at the core-mantle boundary, are still not well understood or speculative.

A connection of earth rotation with the geo-kinematic component of the observing system, i.e. with the monitoring of all surface changes, offers the unique possibility to distinguish "mass" from "motion" in earth rotation and to narrow down the potential excitation mechanisms. A higher temporal resolution of the earth rotation spectrum, e.g. with terrestrial gyroscopes, will prove very advantageous in this respect.

In earth system studies the gravity field plays a peculiar dual role. On the one hand it reflects the anomalous density structure in the earth's interior, anomalous being defined as the deviation from an ideal mass distribution, e.g. one of a slowly rotating fluid in hydrostatic equilibrium. In this role the gravity field together with magnetometry and seismic tomography form a triad of complementary data sources for the study of structure and dynamics of core, mantle and lithosphere. On the other hand, one equipotential surface of the earth's gravity field, the geoid, defines the hypothetical ocean surface at rest. In this role the gravity field defines, with the geoid as one of its level surfaces, a reference for all topographic processes, those of the solid earth, of the ice masses and of the ocean. In other words, topography and its time variations result from the combination of geometry (= shape of the earth's surface and its temporal variations) with gravity (= the geoid [or potential differences between surface points]). For example, combination of ocean altimetry, measuring the actual sea surface geometry, with the geoid yields directly dynamic ocean topography, and consequently ocean surface circulation. Back to the dual role: in one role the gravity field serves as an indicator of mass anomalies, i.e. of a state of imbalance, in its other role, as level surface, it defines "a water surface at rest" with the topography of land, ice, and ocean (together with their material properties) being a measure of the tendency of the actual surface to move or flow and with no fluxes at all if actual and level surface coincide (no erosion, ice flux, ocean circulation etc.).

There exists a link between all temporal mass changes, whether internal, on the surface, or as loading on the surface, and the gravity field signal, because any change in the mass distribution also results in an, albeit attenuated, change of the level surfaces, and an even more attenuated shift in the geo-center. In the near future, with the results of the up-coming gravity field satellite missions and with the further advancement if airborne gravimetry, if will become feasible to attain a global view of the temporal variations of the gravity field, in spatial resolution far beyond the already measurable changes in the terms J_2 and J_3. Then sea level changes, variations of the ground water tables, ice dynamics, postglacial readjustment, or atmospheric loading, monitored in one geocentric frame with the geo-kinematic component of the observing system, can be related to their gravitational signature with the ultimate goal to establish mass balances and to describe motion individually for each system component and in their interaction. Integral constraints, as derived from the earth rotation component, complete the picture.

Earth Tides, i.e. the gravitational attraction of sun, moon and planets on the masses of the earth relative to its geocenter, serve as perfect breadboard for GIGGOS. An extremely well defined external force, with a wide spectral range, deforms the yielding earth, its solid, ice, ocean, and atmospheric component, changes rhythmically its gravitational field (and its level surfaces) and affects the rotation of the earth. Here the interaction of the components of the observing system can be studied in well defined experiments, locally, regionally or globally.

In Figure 3 the role of the Global Integrated Geodetic & Geodynamic Observing System (GIGGOS) for earth system sciences is shown. Between the arrows of the system blocks items of added value of operating GIGGOS as an integrated system are listed.

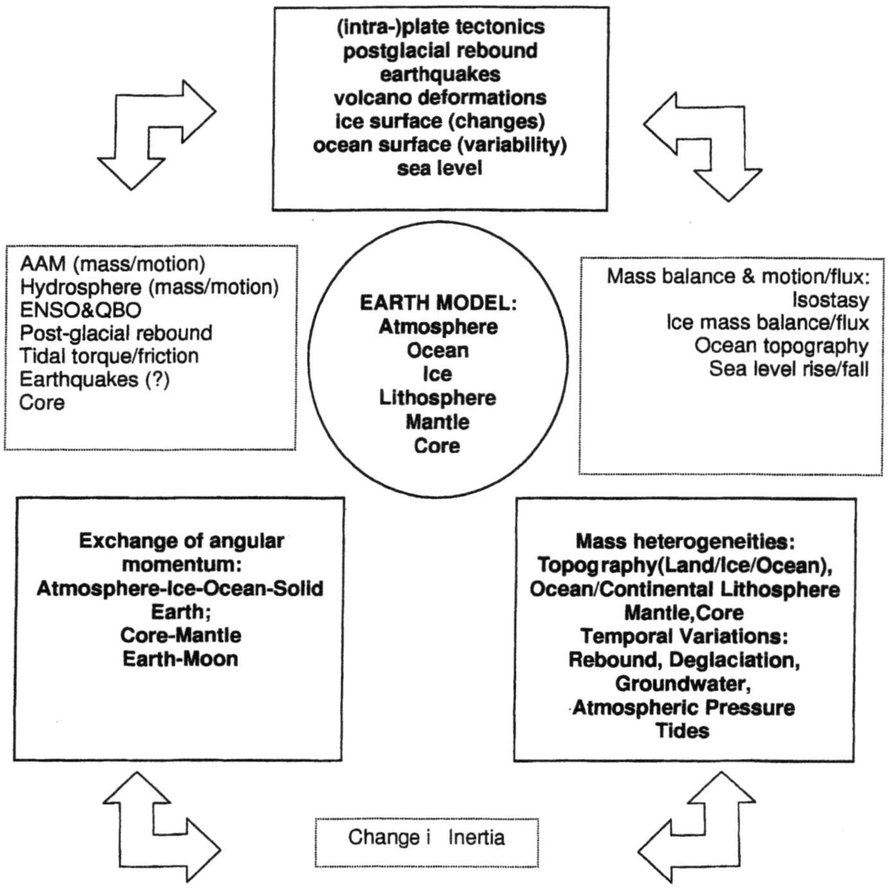

Figure 3: GIGGOS and Earth Modeling

5. The Role of the Gravity Field in GIGGOS: Four Cases

The interconnection of the components frame, earth rotation, and geo-kinematics is rather established already inside geodesy. Less established is what the component gravity field adds to the observing system. This will be discussed in the sequel. For this purpose four cases of increasing complexity will be presented. The fourth case, the most comprehensive one, discusses GIGGOS in its full capability.

Case A: Mantle, Oceanic and Continental Lithosphere

Mantle (ANDERSON, 1989), (HAGER & RICHARDS, 1989)

Observations of the geoid in conjunction with observations of seismic velocity heterogeneities place constraints upon the variation of mantle viscosity and the depth of mantle convection.

Required: Gravity field with low spatial resolution but high accuracy.

Ocean Lithosphere (CAZENAVE, 1994)

The gravity field structure as deduced from satellite radar altimetry yields information on compensation mechanisms, flexure, thermal evolution, hotspot swells, lineations.

Required: Gravity field (ocean geoid) and bathymetry with high spatial resolution but moderate accuracy.

Continental Lithosphere (Geophysical & Geodetic Requirements, 1987)

Information on deviation from hydrostatic equilibrium (models of isostasy); constraints on chemical and thermal composition, orogeny, underplating and delamination, thinning below sedimentary basins, hot spots.

Required: Gravity field and topographic heights with high resolution and medium to high accuracy.

Case B: Mass Balance Over Ice Sheets / Postglacial Re-adjustment / Temporal Variations of Gravity (TRUPIN, 1993), (GEGOUT & CAZENAVE, 1993), (HAN & WAHR, 1995), (CHAO, LEMOIN, KLOSKO & PAVLIS, 1996)

GIGGOS will have to establish the connection between the fundamental mass moments of degrees zero (GM), one (I_1, I_2, and I_3), defining the position of the geo-center and two (the components I_{ij} of the inertia tensor) and their temporal variations with (1) the coefficients of degree zero (usually one by definition), one (usually zero by definition) and two (partly zero by definition) of the earth's gravity field, and (2) with polar motion (in particular the mass term) and changes in l.o.d.(in particular motion term).

This topic is intimately related to the datum definition of the terrestrial frame (BAARDA, 1995) and its temporal evolution. It represents the - very challenging - stationary and secular component of the observing system. It will be very difficult to keep this part free of any non-geodynamic systematic distortions.

Required: Frame + Geokinematics + Earth Rotation + Gravity Field with ultimate accuracy.

Case C: Dynamic Ocean Topography (e.g. (NEREM et al., 1994))

The determination of "absolute" ocean topography from a precise, high resolution geoid in conjunction with satellite altimetry. This way the fundamental assumption of a level of known motion in ocean circulation computations could be removed, to yield in conjunction with ocean data absolute circulation at any depth. This would have significant implications on the understanding of ocean transport processes (bio-mass, pollutants, CO_2, heat) and on the understanding of the heat exchange between atmosphere and oceans.

The success of this approach depends on the right spatial resolution and accuracy, but also on that of satellite altimetry, hydrographic data, and ocean models. Recent discussions on these latter aspects from the oceanographic point of view can be found in (GANACHAUD et el., 1997), (HASSELMANN, 1996) and (MINSTER et al., 1996).

Required: Geoid with high spatial resolution and very high accuracy, altimetry with high temporal/spatial resolution, high accuracy and time stability.

Case D: Sea Level Change And Its Interaction With Ocean Dynamics, Ice and Solid Earth Processes (e.g. (CARTER et al., 1989), (TAPLEY, 1990), (NEREM, 1995a), (RAPP, 1996), (RAPP & ZHANG, 1996), (ZERBINI et al., 1996), (Gravity Field and Steady-State Ocean Circulation Mission, 1996))

The global monitoring of sea level change, its separation from vertical crustal movement, its relation to postglacial movement, deglaciation and thermal effects, its impact on the dynamics of the oceans.

Required: All components of the observing system with highest accuracy and time stability, gravity with highest spatial resolution.

Literature

ANDERSON, D.L.: *Theory of the Earth*, Blackwell, Boston, 1989.

A Unified Theory of Planet Earth, A Strategic Overview and Long Range Plan, Advisory Committee for Earth Sciences, NSF, Washington D.C., 1988.

BAARDA, W.: *Linking Up Spatial Models in Geodesy - Extended S-Transformations*, Netherlands Geodetic Commission, New Series, 41, 1995.

BAKER T., F.D.J. CURTIS, A.H. DODSON: *A new test of Earth tide models in central Europe*, Geophys. Res. Letters, vol. 23, no. 24, 3559-3562, 1996.

BRUNS, H.: *Die Figur der Erde*, Publ. des Königl. Preußischen Geodätischen Institutes, Berlin, 1878.

CARTER, W.E. et al.: *Geodetic Fixing of Tide Gauge Bench Marks*, Technical Report, WHOI- 89-31, Coastal Research Center, 1989.

CAZENAVE, A: *The Geoid and Ocean Lithosphere*, in: Geoid And Its Geophysical Interpretations (eds. P. Vanicek & N.T.Christou), 255-284, CRC Press, Boca Raton, 1994.

CHAO B. FONG: *The Geoid and Earth Rotation*, in: Geoid And Its Geophysical Interpretations (eds. P.Vanicek & N.T. Christou), 285-298, CRC Press, Boca Raton, 1994.

CHAO B. FONG, R. S. GROSS: *Changes in the Earth's rotation and low-degree gravitational field induced by earthquakes*, Geophys. J. R. astr. Soc. 91, 569-596, 1987.

CHAO B. FONG, F. LEMOINE, S. KLOSKO, N. PAVLIS: *Secular variations in the zonal harmonics and polar motion as geophysical constraints*, paper pres. Fall meeting AGU, 1996.

Report of the Earth Observation User Consultation Meeting, ESA SP-1143, ESTEC, Noordwijk, 1991.

Earth System Science - A Closer View, Report of the Earth System Sciences Committee, NASA Advisory Council, NASA Washington D.C., 1988.

GANACHAUD, A., C. WUNSCH, MYUNG-CHAU KIM, B. TAPLEY: *Combination of Topex/Poseidon Data with a Hydrographic Inversion for Determination of the Ocean General Circulation and its Relation of Geoid Accuracy*, Geophys.J.Int., 128, 708-722, 1997.

GEGOUT P., A. CAZENAVE: *Temporal variations of the Earth gravity field for 1985-1989 derived from Lageos*, Geophys. J. Int. 114, 347-359, 1993.

Geophysical and Geodetic Requirements for Gravity Field Measurements 1987 - 2000, Report of a Gravity Workshop, Geodynamics Branch, NASA, 1987.

Gravity Field and Steady-State Ocean Circulation Mission, reports for assessment, the nine candidate earth explorer missions, ESA SP-1196(1), Noordwijk, 1996.

HAGER, B.H., M.A.RICHARDS: *Long-Wavelength Variations in Earth's Geoid: Physical Models and Dynamical Implications.* In: Seismic Tomography and Mantle Circulation (eds.: R.K.O'Nions and B.Parsons) pp.309-327.The Royal Society, London, 1989.

HAN D., J. WAHR: *The viscoelastic relaxation of a realistic stratified earth*, and a further analysis of postglacial rebound, Geophys.J.Int. 120, 287-311, 1995.

HASSELMANN, K.: *A Review and Analysis of the Impact of an Improved Geoid on our Understanding of the Ocean Circulation and Climate*, Interim Report, ESTEC Contract No. 11528/95/NL/CN, 1996.

HECK, B.: *An Evaluation of Some Systematic Error Sources Affecting Terrestrial Gravity Anomalies*, Bulletin Géodésique, 64, 1, 88-108, 1990.

JACOBY W.R.: *Isostasy and density distribution in the crust and upper mantle*, Zeitschrift für Geophysik, Band 39, 79-96, 1971.

KANDEL, R.S.: *Earth and Cosmos*, Pergamon Press, Oxford, 1980.

KLEUSBERG, A.: *The Similarity Transformation of the Gravitational Potential Close to the Identity*, manuscripta geodaetica, 5, 241-256, 1980.

KOCH K.R., H.H. SCHMID: *Error study for the determination of the center of mass of the earth from Pageos observations*, Bulletin Géodésique 97, 1970.

MA, C., M. FEISSEL (eds.): *Definition and Realization of the International Celestial Reference Frame by VLBI Astrometry of Extragalactic Objects*, IERS Technical Note 23, 1997.

MARSHALL, J.A., S.M. KLOSKO, J.C. RIES: *Dynamics of SRL tracked satellites*, Reviews of Geophysics, Supplement, 353-360, 1995.

MATHER R.S.: *Four dimensional studies in earth space*, Bulletin Géodésique 108, 1973.

MATHER R.S.: *A world geodetic system from gravimetry*, Geophys. J. Int. 23, 75-100, 1970.

MCCARTHY, D.D.: *IERS Conventions (1996)*, IERS Technical Note 21, 1996.

MINSTER, J.-F., P. LEGRAND: *Impact of Geoid on Ocean Circulation Retrieval*, Interim Report, ESTEC, Contract no. 11528/95/NL, Noordwijk, 1996.

MUELLER, I.I., H. MONTAG, CH. REIGBER, P. WILSON: *An IUGG Network of Fundamental Geodynamic Reference Stations*, Potsdam, 1995.

NEREM, R.S. et al.: *Gravity model development for TOPEX/POSEIDON: Joint gravity models 1 and 2*, Journal of Geophysical Research, vol. 99, no. C12, 24, 421-24, 447, 1994.

NEREM, R.S.: *Global mean sea level variations from TOPEX/POSEIDON altimeter data*, Science, vol. 268, pp. 708-710, 1995.

NEREM, R.S.: *Terrestrial and planetary gravity fields*, Reviews of Geophysics, Supplement, 469-476, 1995a.

NEREM, R.S., C. JEKELI, W.M. KAULA: *Gravity field determination and characteristics: Retrospective and prospective*, Journal of Geophysical Research, vol. 100, no. B8, 15, 053-15, 074, 1995b.

RAPP R.H., C. ZHANG: *Sea level changes estimate from the analysis of TOPEX cycles 9 to 142*, submitted to Marine Geodesy, 1996.

RAPP R.H., C. ZHANG, Y. YI: *Analysis of dynamic ocean topography using TOPEX data and orthonormal functions*, JGR, vol. 101, 22, 583-22, 598, 1996.

SCHWINTZER, P. et al.: *Long-wavelength global gravity field models: GRIM 4-S4*, 189, 71/4, 1997.

SEIDELMANN, P.K. (ed.): *Explanatory supplement to the astronomical almanac*, University Science Books, 1992.

Solid Earth Sciences and Society, National Research Council, National Academy Press, Washington D.C., 1993.

TAPLEY, B.D.: *The Monitoring of Changes in Global Mean Sea Level Using Satellite Altimetry*, TOPEX/POSEIDON Science Working Team Meeting, Washington D.C., 1990.

TRUPIN, A.S.: *Effects of polar ice on the Earth's rotation and gravitational potential*, Geophys.J.Int. 113, 273-283, 1993.

ZERBINI, S. ET AL.: *Sea level in the Mediterranean: a first step towards separating crustal movements and absolute sea-level variations*, Global and Planetary Change 14, 1-48, 1996.

ZUMBERGE, J.F., M.P. URBAN, R. LUI, R.E. NEILAN: *1995 Annual Report*, International GPS Service for Geodynamics (IGS), IGS Central Bureau, JPL Pasadena, 1996.

Authors' Index

MIX
Papier aus verantwortungsvollen Quellen
Paper from responsible sources
FSC® C105338

If you have any concerns about our products,
you can contact us on
ProductSafety@springernature.com

In case Publisher is established outside the EU,
the EU authorized representative is:
Springer Nature Customer Service Center GmbH
Europaplatz 3, 69115 Heidelberg, Germany

Printed by Libri Plureos GmbH
in Hamburg, Germany